# Routledge Handbook of Global Sustainability Governance

The *Routledge Handbook of Global Sustainability Governance* provides a state-of-the-art review of core debates and contributions that offer a more normative, critical, and transformatively aspirational view on global sustainability governance.

In this landmark text, an international group of acclaimed scholars provides an overview of key analytical and normative perspectives, material and ideational structural barriers to sustainability transformation, and transformative strategies. Drawing on pivotal new and contemporary research, the volume highlights aspects to be considered and blind spots to be avoided when trying to understand and implement global sustainability governance. In this context, the authors of this book debunk many myths about all-too optimistic accounts of progress towards a sustainability transition. Simultaneously, they suggest approaches that have the potential for real sustainability transformation and systemic change, while acknowledging existing hurdles. The wide-ranging chapters in the collection are organised into four key parts:

- Part 1: Conceptual lenses
- Part 2: Ethics, principles, and debates
- Part 3: Key challenges
- Part 4: Transformative approaches

This handbook will serve as an important resource for academics and practitioners working in the fields of sustainability governance and environmental politics.

**Agni Kalfagianni** is Associate Professor at the Copernicus Institute of Sustainable Development, University of Utrecht, The Netherlands.

**Doris Fuchs** is Professor of International Relations and Sustainable Development and speaker of the Center for Interdisciplinary Sustainability Research at the University of Muenster, Germany.

**Anders Hayden** is Associate Professor of Environmental Politics in the Department of Political Science at Dalhousie University, Canada.

# Routledge Handbook of Global Sustainability Governance

*Edited by Agni Kalfagianni, Doris Fuchs,*
*and Anders Hayden*

Routledge
Taylor & Francis Group

LONDON AND NEW YORK

First published 2020
by Routledge
2 Park Square, Milton Park, Abingdon, Oxon OX14 4RN

and by Routledge
605 Third Avenue, New York, NY 10017

First issued in paperback 2021

*Routledge is an imprint of the Taylor & Francis Group, an informa business*

**Notices**
Practitioners and researchers must always rely on their own experience and knowledge in evaluating and using any information, methods, compounds, or experiments described herein. In using such information or methods they should be mindful of their own safety and the safety of others, including parties for whom they have a professional responsibility.

Product or corporate names may be trademarks or registered trademarks, and are used only for identification and explanation without intent to infringe.

Publisher's Note
The publisher has gone to great lengths to ensure the quality of this reprint but points out that some imperfections in the original copies may be apparent.

*British Library Cataloguing-in-Publication Data*
A catalogue record for this book is available from the British Library

*Library of Congress Cataloging-in-Publication Data*
Names: Kalfagianni, Agni, editor.
Title: Routledge handbook of global sustainability governance / edited by Agni Kalfagianni, Doris Fuchs, and Anders Hayden.
Other titles: Global sustainability governance
Description: New York : Routledge, 2020. | Includes bibliographical references and index.
Identifiers: LCCN 2019028224 (print) | LCCN 2019028225 (ebook) | ISBN 9781138048287 (hardback) | ISBN 9781315170237 (ebook)
Subjects: LCSH: Sustainability—Government policy. | Sustainable development—Government policy. | Environmental policy.
Classification: LCC HC79.E5 R6748 2020 (print) | LCC HC79.E5 (ebook) | DDC 338.9/27—dc23
LC record available at https://lccn.loc.gov/2019028224
LC ebook record available at https://lccn.loc.gov/2019028225

ISBN 13: 978-1-03-208655-2 (pbk)
ISBN 13: 978-1-138-04828-7 (hbk)

Typeset in Bembo
by codeMantra

# Contents

Contents

# Figures

# Tables

# Acknowledgements

The editors would like to thank Annabelle Harris, Matt Shobbrook, and the team at Routledge for their support and enthusiasm in bringing this project to fruition. Many thanks, of course, also to the authors for their contributions and cooperation. We are also grateful for the assistance of Jonas Marggraf at the University of Muenster.

# Abbreviations

| | |
|---|---|
| AR5: | IPCCs Fifth Assessment Report |
| ARC: | Alliance of Religion and Conservation |
| BCE: | Before the Current Era |
| BCOM: | Bloomberg Commodity Index |
| BECCS: | Bioenergy with Carbon Capture and Storage |
| CCS: | Carbon Capture and Storage |
| CDM: | Clean Development Mechanism |
| CFCs: | Chlorofluorocarbons |
| CIFs: | Commodity Index Funds |
| CITES: | Convention on International Trade in Endangered Species of Wild Fauna and Flora |
| CMCD: | Comisión Mundial sobre Cultura y Desarrollo |
| CME: | Chicago Mercantile Exchange |
| $CO_2$: | Carbon Dioxide |
| CSA: | Community-supported agriculture |
| CSD: | Commission on Sustainable Development |
| DA: | Development Alternatives |
| DDT: | Dichlorodiphenyltrichloroethane |
| ETFs: | Exchange Traded Funds |
| ETS: | Emissions Trading Scheme |
| EU: | European Union |
| FAO: | Food and Agricultural Organization of the United Nations |
| FOEE: | Friends of the Earth Europe |
| FSC: | Forest Stewardship Council |
| GDP: | Gross Domestic Product |
| GFN: | Global Footprint Network |
| GSG: | Global Sustainability Governance |
| G20: | Group of Twenty |
| G77: | Group of Seventy-Seven |
| GCSI: | Standard and Poor's Goldman Sachs Commodity Index |
| GHG: | Greenhouse Gas |
| $GtCO_2$/year: | Giga Tonnes of Carbon Dioxide Per Year |
| IAM: | Integrated Assessment Model |
| IBAI: | Index Based Agricultural Insurance |
| ICC: | International Code Council |
| ICLEI: | International Consortium of Local Environmental Initiatives |

IEA:            International Energy Agency
ILO:            International Labour Organization
IMF:            International Monetary Fund
IPAT:           Impact = Population × Affluence × Technology
IPBES:          Intergovernmental Panel on Biodiversity and Ecosystem Services
IPCC:           Intergovernmental Panel on Climate Change
ISDR:           International Strategy for Disaster Reduction
IHGS:           International Union of Geological Sciences
IR:             International Relations
LTA:            Land Transport Authority
MDGs:           Millennium Development Goals
NETs:           Negative Emissions Technologies
NGOs:           Non-governmental Organisations
OECD:           Organisation for Economic Cooperation and Development
OTC:            Over the Counter
PRI:            Principles for Responsible Investment
REDD:           Reducing Emissions from Deforestation and Forest Degradation
REIT:           Real Estate Investment Trust
SD:             Sustainable Development
SDGs:           Sustainable Development Goals
SENPLADES:      Secretaría Nacional de Planificación y Desarrollo (Ecuador)
SR15:           IPCC Special Report on Global Warming of 1.5°C
SRI:            Socially Responsible Investing
SSPs:           Shared Socioeconomic Pathways
TNC:            Transnational Corporation
TOD:            Trinity of Despair
UN:             United Nations
UNCHE:          United Nations Conference on the Human Environment
UNCED:          United Nations Conference on Environment and Development
UNCCD:          United Nations Convention to Combat Desertification
UNDP:           United Nations Development Programme
UNEP:           United Nations Environment Programme
UNESCO:         United Nations Educational, Scientific and Cultural Organization
UNFCCC:         United Nations Framework Convention on Climate Change
UNHCFG:         United Nations High Commissioner for Future Generations
VAT:            Value Added Tax
WCED:           World Commission on Environment and Development
WHO:            World Health Organization
WTO:            World Trade Organization
WWF:            World Wide Fund for Nature

# Contributors

**Samuel Alexander** is a lecturer and researcher at the University of Melbourne, Australia, teaching a course called "Consumerism and the Growth Economy: Critical Interdisciplinary Perspectives" as part of the Master of Environment. He is also co-director of the Simplicity Institute and a research fellow with the Melbourne Sustainable Society Institute.

**Melissa Aronczyk** is an Associate Professor in the School of Communication and Information at Rutgers University. With Robert Brulle and Maria I. Espinoza, she is writing a history of the American public relations industry and its impact on environmentalism.

**Adrián E. Beling** has a PhD in Sociology (Humboldt University of Berlin and Alberto Hurtado University, Chile) and MA in Global Studies. He has a degree in economics and business (Universidad Nacional de Cuyo, Argentina) and later studied social sciences in Germany (Freiburg University), India (Jawaharlal Nehru University), and Argentina (Latin American Faculty of Social Sciences/ FLACSO). Dr Beling is Associate Researcher to the Global Studies Programme at FLACSO Argentina, and an international speaker and academic advisor on social-ecological transformation. He is a founding member of the digital academic journal and blog *Alternautas* and member of its Editorial Board. His research and teaching topics revolve around global environmental change and socio-economic transformations, southern epistemologies, sustainability governance, globalisation, public sphere, civil society, and religion.

**Magdalena Bexell** is Associate Professor in the Department of Political Science at Lund University, Sweden. Her research concerns legitimacy and responsibility in global governance as well as national politics of sustainable development with a special focus on the 2030 Agenda and its Sustainable Development Goals. Prior publications include articles in *Journal of Human Development and Capabilities, Forum for Development Studies, Globalizations, International Feminist Journal of Politics*, and the edited volumes *Global Governance, Legitimacy and Legitimation* and *Democracy and Public-Private Partnerships in Global Governance* (with Ulrika Mörth).

**Ingolfur Blühdorn** is a political sociologist and socio-political theorist. He is Professor for Social Sustainability and Head of the Institute for Social Change and Sustainability (IGN) at the University for Economics and Business in Vienna. His work explores the legacy of the emancipatory social movements since the early 1970s, their participatory revolution and the transformation of emancipatory politics over the past five decades.

**Robert Brulle** is a Visiting Professor of Environment and Society at Brown University in Providence, RI, and a Professor of Sociology and Environmental Science at Drexel University in Philadelphia. His research focuses on US environmental politics, critical theory, and the political and cultural dynamics of climate change. He is the author of over one hundred publications in these areas.

**Jennifer Clapp** is a Canada Research Chair in Global Food Security and Sustainability and Professor in the School of Environment, Resources and Sustainability at the University of Waterloo, Canada. She has published widely on the global governance of problems that arise at the intersection of the global economy, the environment, and food security. Her books include *Speculative Harvests: Financialization, Food, and Agriculture* (with S. Ryan Isakson, Fernwood Press, 2018), *Food,* 2nd Edition (Polity, 2016), *Hunger in the Balance: The New Politics of International Food Aid* (Cornell University Press, 2012), *Paths to a Green World: The Political Economy of the Global Environment,* 2nd Edition (with Peter Dauvergne, MIT Press, 2011), and *Corporate Power in Global Agrifood Governance* (co-edited with Doris Fuchs, MIT Press, 2009).

**Diana Coole** is Emerita Professor of Political and Social Theory at Birkbeck, University of London. She has published extensively in the fields of critical theory, phenomenology, and gender. In the last decade, following a major Leverhulme Trust research award 2010–2013, she has paid particular attention to challenges of environmental and demographic change. Her work here includes *The New Materialisms. Ontology, Agency, and Politics* (Duke 2010, co-edited with Samantha Frost) and *Should We Control World Population?* (Polity 2018), as well as recent articles on the population question.

**Rico Defila** holds a law degree and is deputy leader of the Research Group Inter-/Transdisciplinarity and senior researcher at the Program Man-Society-Environment (MGU), Department of Environmental Sciences, University of Basel, Switzerland. Until 2014, he was senior researcher and head of planning and operations at the Interdisciplinary Centre for General Ecology (IKAÖ), University of Bern. His research interests include theory and methodology of inter- and transdisciplinary research and teaching, structural organisation of interdisciplinary academic units, good life and sustainable development, and sustainable consumption. https://mgu.unibas.ch/en/team/rico-defila/.

**Michael Deflorian** is a Research and Teaching Associate at the Institute for Social Change and Sustainability (IGN) at Vienna University of Economics and Business. His research areas include lifeworld-centred activism, environmental politics, and theories of the late-modern society. He is especially interested in how a "politics of the everyday" is both made possible and complicated by social change.

**Antonietta Di Giulio** holds a PhD in philosophy and is leader of the Research Group Inter-/Transdisciplinarity and senior researcher at the Program Man-Society-Environment (MGU), Department of Environmental Sciences, University of Basel, Switzerland. Until 2014, she was senior researcher and lecturer at the Interdisciplinary Centre for General Ecology (IKAÖ), University of Bern. Her research interests include theory and methodology of inter- and transdisciplinary research and teaching, good life and sustainable development, education for sustainable development, and sustainable consumption. https://mgu.unibas.ch/en/team/antonietta-di-giulio/.

**Doris Fuchs** is Professor of International Relations and Sustainable Development and Speaker of the Center for Interdisciplinary Sustainability Research at the University of Muenster in Germany. Her research focuses on (the potential for and barriers to) sustainability governance with a special focus on consumption, corporate power, the role of religion/religious actors, and financialisation. She is particularly interested in questions of justice, responsibility, and democratic legitimacy. Empirically, she focuses on agrifood, climate, and energy policy in particular. She is a political economist by training. Her research has been published in numerous journals and volumes. https://www.uni-muenster.de/Fuchs/en/mitarbeitende/fuchs.html.

**Andrea K. Gerlak** is an Associate Professor in the School of Geography and Development and Associate Research Professor at the Udall Center for Studies in Public Policy. Her research and teaching focus on institutions and environmental governance. She examines cooperation and conflict around water, including questions of institutional change and adaptation to climate change in rivers basins, and human rights and equity issues in water governance. Gerlak is a senior research fellow with the Earth System Governance Project and serves as a co-editor for *the Journal of Environmental Policy and Planning*. She holds a B.A. in political science from the University of Nevada, Las Vegas and a Ph.D. in political science from the University of Arizona.

**Katharina Glaab** is Associate Professor of Global Change and International Relations at the Department of International Development and Environment Studies at the Norwegian University of Life Sciences. She received her PhD from the University of Muenster. Her fields of research are global environmental politics and International Relations theory.

**Tobias Gumbert** is Research Fellow at the Institute of Political Science, University of Muenster/Germany. He has studied Political Science, Sociology, Anthropology, and Sinology in Muenster and Beijing and is currently completing his PhD on the role of responsibility in global environmental governance. In his research, he focuses on issues of power, knowledge, and environmental limits particularly in the fields of global food and waste governance.

**Anders Hayden** is Associate Professor in the Department of Political Science at Dalhousie University in Halifax, Nova Scotia, with an emphasis on environmental politics. He is particularly interested in the evolving balance between efforts to promote ecological modernisation ("green growth") and sufficiency-based challenges to the endless growth of production and consumption. He has written on efforts to promote "green growth" in Canada, Britain, and the European Union. His interest in the sufficiency approach has included examination of policies and initiatives to reduce hours of work as well as research on Bhutan, a country that has established Gross National Happiness, rather than Gross National Product, as its overriding goal. He is currently involved in research on the political and policy impacts of alternative measures of wellbeing and prosperity ("beyond GDP" measurement). He is the author of two books: *When Green Growth Is Not Enough: Climate Change, Ecological Modernization, and Sufficiency* (McGill-Queen's University Press, 2014) and *Sharing the Work, Sparing the Planet: Work Time, Consumption & Ecology* (Zed Books/Between the Lines, 1999).

**Kerryn Higgs** is an Australian historian and author who, in examining human societies and systems, has focussed on questions of planetary boundaries, and on limits to economic growth and material footprint. She published *Collision Course: Endless Growth on a Finite Planet* (MIT Press) in 2014. The book explores the resistance to ideas about limits, the elevation of growth as the central objective of policy-makers, and the mounting influence of corporate-funded think tanks dedicated to the propagation of neoliberal principles and to the denial of health and environmental dangers – from the effects of tobacco to global warming. She completed her PhD with the School of Geography and Environmental Studies at the University of Tasmania, where she is now a University Associate with the School of Social Sciences. She is an Associate Member of the Club of Rome.

**Cristina Yumie Aoki Inoue** is Professor of International Relations and planet politics at the University of Brasília. She was a lead author to the First National Assessment Report of the Brazilian Panel on Climate Change (PBMC) in 2013 and is currently a member of the Scientific Steering Committee of the Earth System Governance (ESG) Project.

**Agni Kalfagianni** is Associate Professor of Transnational Sustainability Governance at the Copernicus Institute of Sustainable Development, Utrecht University. She specialises in the effectiveness, legitimacy, and ethical and justice considerations of private and transnational forms of governance in the sustainability domain. She is member of the Scientific Steering Committee of the international Earth System Governance project and co-founder of the international Planetary Justice Taskforce. Agni is (co)Editor-in-Chief of the *Global Environmental Governance* book series by Routledge, Associate Editor of the *Earth System Governance* journal, and member of the Editorial Board of *International Environmental Agreements: Politics, Law and Economics* and *Agriculture and Human Values* journals.

**Naomi Krogman** is the Dean of the Faculty of Environment at Simon Fraser University in Burnaby, British Columbia. Her research in recent years has been on sustainable consumption and sustainability scholarship and education. Her current research interests are in climate change disruption implications for humanitarian aid.

**Richard Lane** is a Postdoctoral Researcher at the Copernicus Institute of Sustainable Development, Utrecht University. His research focuses on environmental governance in the age of the Anthropocene. He investigates particularly the contemporary coherence and incoherence of decarbonisation programmes across scales, and the history of environmental and economic governance during great acceleration.

**Steffen Lange** is an economist, researching how economies can be transformed to be socially just and environmentally sustainable. He has studied economics at Maastricht University, the Pontifical Catholic University of Chile, the University of Göttingen and wrote his PhD at the University of Hamburg. Currently, he is a postdoctoral researcher at the Institute for Ecological Economy Research in Berlin. In his research, he focuses on environmental economics and macroeconomics, in particular the social and environmental effects of digitalisation, macroeconomic rebound effects and post-growth economies. Parallel to his academic work, he engages in social movements and civil society organisations, to foster a social and ecological transformation.

**Jörgen Larsson** is a sociologist and has a position as an Assistant Professor in sustainable consumption patterns at Chalmers University of Technology. His research interest is related to how structural dimensions (norms, prices, etc.) are shaping how we use our time and money, as well as how structurally constrained choices result in outcomes regarding individual wellbeing and ecological footprints. He puts specific emphasis on consumption areas with large greenhouse gas emission (GHG), including air travel and food habits. Currently he is engaged in research on vacationing habits and on aviation climate policy instruments.

**Peter Lawrence** is a Senior Lecturer, and co-convenor of the Climate Justice Network, Faculty of Law, University of Tasmania, Australia. He has published extensively in the field of international environmental law. Peter is a former diplomat with the Australian Department of Foreign Affairs and Trade.

**Karen Litfin** is Professor of Political Science at University of Washington, and author of *Ozone Discourses: Science and Politics in Global Environmental Cooperation* (Columbia University Press, 1994) and *Ecovillages: Lessons for Sustainable Community* (Polity, 2014).

**Erik Lundberg** is a researcher and lecturer at the School of Economics, Business and Law at the University of Gothenburg. His research interest follows mainly two paths, consumer behaviour and value creation, and he is often applying a sustainability lens to these fields. He pursues his interests in the empirical fields of tourism, events, and time use. He is also the head of the multidisciplinary Centre for Tourism at the University of Gothenburg.

**Michael Maniates** is Professor of Social Science and the founding Head of Studies of Environmental Studies at Yale-NUS College in Singapore. He holds a BS in Conservation and Resource Studies and MA and PhD in Energy and Resources, all from the University of California. His recent work focuses on systems of sustainable consumption and production, social innovation for a low-growth world, and approaches to teaching for turbulence in undergraduate settings.

**Ayşem Mert** is Associate Senior Lecturer at the Department of Political Science and Director of the International Masters Programme in Environmental Social Science at Stockholm University. Her research explores the discourses of radical democracy, critiques of development and global narratives, and policies of environmental governance. She is the author of *Environmental Governance through Partnerships: A Discourse Theoretical Study* (2015, Edward Elgar) and published various articles on environmental politics and governance in *Environmental Values*, *Journal of Environmental Policy and Planning*, and *Global Policy* among others.

**John M. Meyer** is Professor and Chair of the Department of Politics at Humboldt State University, on California's North Coast. He also serves in interdisciplinary programs on Environmental Studies and Environment & Community (MA). He is author of the award-winning *Engaging the Everyday: Environmental Social Criticism and the Resonance Dilemma* (MIT, 2015), and co-editor of *The Oxford Handbook of Environmental Political Theory* (Oxford, 2016), and *The Greening of Everyday Life: Challenging Practices, Imagining Possibilities* (Oxford, 2016). He is also an editor of the journal *Environmental Politics*.

**Jonas Nässén** is a Senior Researcher (docent) at the Division of Physical Resource Theory, Chalmers University of Technology. His main research focuses on sustainable consumption including time-series analyses and decomposition analyses of GHG emissions from consumption, studies of the relationships between technological and behavioural change, and policy studies of energy use and mobility.

**Chukwumerije Okereke** is the Director of the Development Futures Institute (DFI) at the Alex Ekwueme Federal University, Ndufu-Alike Ikwo, Ebonyi State, Nigeria. He is also visiting professor of Environment and Development at University of Reading, UK and Senior Academic Visitor at Oxford University Centre for the Environment.

**Lennart Olsson** is Professor of Geography at Lund University, was the founding Director of LUCSUS 2000–17. Current research focuses on the politics of climate change in the context of poverty, food insecurity and ill-health in sub-Saharan Africa, and on the transition from annual monocultures to perennial polycultures in agriculture. He was Coordinating Lead Author for the chapter on Livelihoods and Poverty in IPCC's 5th Assessment Report 2011–14 and for the chapter on Land Degradation in the special IPCC report on Climate Change and Land (SRCCL), 2017–19.

**Luigi Pellizzoni** is Professor in Sociology of the Environment at the University of Pisa, Italy. His research interests intersect three fields: risk, uncertainty, environmental change, and sustainability; impacts of scientific advancement and technological innovation; conflict, participation, and the transformation of governance. Recent publications include *Ontological Politics in a Disposable World: The New Mastery of Nature* (Routledge, 2016); "The Ethical Government of Science and Innovation," in D. Tyfield et al. (eds.) *Routledge Handbook of the Political Economy of Science* (2017).

**Dirk Philipsen** teaches economic history at Duke University's Sanford School of Public Policy. He also serves as Senior Fellow at the Kenan Institute for Ethics, co-director of the Duke Sustainability Engagement Certificate, Fellow at the Royal Society of Arts, and founding associate of the Wellbeing Economy Alliance. His research and writing are focused on sustainability, wellbeing economics, and the history of capitalism. His latest work is published by Princeton University Press under the title *The Little Big Number – How GDP Came to Rule the World, And What to Do about It* (2015/2017). His current work is focused on alternative economic performance indicators, the nature and logic of economic growth, and the moral imperative of decoupling material throughput from human development.

**Thomas Princen** researches and teaches about natural resources at the University of Michigan's School for Environment and Sustainability. In his writing, he aims to develop a language of sustainability and positive transition, in his teaching and speaking to promote active learning and engagement. Princen is the author of *The Logic of Sufficiency and Treading Softly: Paths to Ecological Order*, lead editor of *Ending the Fossil Fuel Era and Confronting Consumption*, and co-editor of *The Localization Reader: Adapting to the Coming Downshift*, all published by MIT Press. His current project is "Fire and Flood: A Politics of Urgent Transition."

**Ítalo Sant' Anna Resende** has a bachelor's degree on Social Communication from Centro Universitário de Brasília (UniCEUB), a master's degree in Business Administration from Universidade de Brasília (UnB), and is currently a PhD candidate in International Relations also at UnB. He taught Social Communication and Business Administration courses at Instituto de Educação Superior de Brasília (IESB), Centro Universitário de Brasília (UniCEUB), and Universidade de Brasília (UnB).

**Jonathan Rutherford** is a Research Fellow at the Simplicity Institute and completing his Masters of Environment at the University of Melbourne, Australia. He is currently co-editing, with Samuel Alexander, the collected works of "simpler way" advocate, Ted Trainer.

**Thais Lemos Ribeiro** is a PhD Candidate in International Relations, University of Brasília (IREL/UnB). She was a contributing author to the First National Assessment Report of the Brazilian Panel on Climate Change (PBMC) in 2013 and is one of the authors of the book *International System of Conservative Hegemony: Global Governance and Democracy in Climate Change Crisis Era*, written in Portuguese.

**Michelle Scobie,** PhD, LLB, LEC, is a lecturer and researcher at the Institute of International Relations at The University of the West Indies (UWI), St. Augustine and Co-Editor of the Caribbean Journal of International Relations and Diplomacy. She has practiced as an attorney at law in Trinidad and Tobago and Venezuela. She was the first corporate secretary of the Trinidad and Tobago Heritage and Stabilisation Fund. She is a member of the Caribbean Studies Association, the International Studies Association, the University of the West Indies Oceans Governance Network, the Earth System Governance global research alliance, the Future Earth Ocean Knowledge Action Network and the International Studies Association. Her research areas include international law, international environmental law and developing states' perspectives on global and regional environmental governance. Her most recent publication is the book: *Global Environmental Governance and Small States: Architectures and Agency in the Caribbean.*

**Phoebe Stephens** is a PhD Candidate at the University of Waterloo and a Pierre Elliott Trudeau Foundation Scholar. Her research is funded by the Social Sciences and Humanities Research Council of Canada and explores the role of social finance in sustainable food transitions. She has published articles in the journals *Globalizations* and *Canadian Food Studies.* Prior to academia Phoebe worked in sustainability reporting and on campaign strategy for Oxfam International.

**Julien Vanhulst** has a PhD in Environmental Studies (Université Libre de Bruxelles/ULB) and a PhD in Sociology at Alberto Hurtado University (Chile), MA in Environmental Science and Management (ULB) and Sociologist (ULB). He is currently faculty member at the School of Sociology of the Universidad Católica del Maule (UCM), researcher at the Centre for Urban-Territorial Studies (CEUT – UCM), and external researcher to the Centre of Sustainable Development Studies (ULB). He is co-founder and member of the Editorial Board of the digital academic journal *Alternautas.* He has specialised in sustainability governance, alternative economies, territory and geography of socio-environmental conflicts, and the sociology of science.

# Introduction

## Critical and transformative perspectives on global sustainability governance

*Anders Hayden, Doris Fuchs, and Agni Kalfagianni*

---

The contemporary world is characterised by systemic sustainability challenges that for the first time in human history have the potential to endanger the fate of humanity. Scholars warn that we are crossing critical "planetary boundaries" (Rockström et al. 2009; Steffen et al. 2015), and have introduced novel concepts such as the "Anthropocene" to convey the idea that human beings have become such a major force that they have changed the way the entire earth system operates (Crutzen 2002). The world that is emerging, with more frequent and intense heatwaves, droughts, wildfires, extreme storms, and floods – a planet that environmental activist and author Bill McKibben (2010) called "Eaarth" – is in many ways less hospitable to human beings, more volatile, less predictable, and seemingly angrier. There may still be opportunities to avoid the worst impacts of climate change by limiting warming to 1.5°C – a task that would require, according to the Intergovernmental Panel on Climate Change (IPCC 2018), "rapid, far-reaching and unprecedented changes in all aspects of society," however. The IPCC's counterpart that examines biodiversity issues – the Inter-governmental Science-Policy Platform on Biodiversity and Ecosystem Services – has warned of an accelerating extinction crisis, with one million species threatened and potentially grave impacts on human well-being (IPBES 2019). These ecological challenges exist in a context of significant and growing global economic and social inequality (Alvaredo et al. 2018) – one estimate, provided by Oxfam, is that the 26 richest people on the planet have the same wealth as the poorest 50% of humanity, some 3.8 billion people (Lawson et al. 2019).

Such challenges have awakened demands for a much more substantial political response to the sustainability crisis than has been evident so far. Young people, inspired by the example of Greta Thunberg, have, through their school strike movement, taken their elders, including political leaders, to task for inaction in the face of climate change. Political activists, supported by a new generation of elected officials, have put a Green New Deal – which seeks to combine a radical acceleration of decarbonisation with the promotion of economic, social, and environmental justice – on the agenda in the United States, and similar ideas have spread to several other countries. An "extinction rebellion" movement has emerged, while calls to declare a climate emergency – in some cases, combined with recognition of a biodiversity emergency – have been taken up by a number of national and sub-national governments. Indigenous people in both the global North and South have become increasingly vocal in their resistance to extractivist projects and policies, with parallels in the many examples of the "environmentalism of the poor," in which local people resist the destruction or enclosure of natural resources that they depend on for their livelihoods. Meanwhile, the rapid decline in the cost of renewable energy technologies is quickly changing calculations of what is economically and politically feasible – although some observers caution against over-reliance on technological solutions alone and have greater hope in the growing questioning of a consumerist vision of the good life and the dominant growth-based vision of progress.

The challenges of achieving much stronger sustainability governance in the socio-political sphere are accompanied by the need for scholars to rethink dominant ideas about how to achieve sustainability. Transformative perspectives on sustainability require critical inquiry that goes beyond simple fixes and nudging. Critical inquiry demands that we do not take the current world order as a given and try to find solutions within already established economic, social, and political parameters. Rather, we have to inquire into these parameters in order to better understand how this particular world came about and why, and what its material and ideational structures imply for the potential and characteristics of a sustainability transformation. In other words, global sustainability governance cannot perform the role of steering human societies towards "environmental and social sustainability" without questioning the broader, deeper causes underlying the major sustainability challenges that we are experiencing and without debating who defines and benefits from the proposed solutions.

In this context, we must not forget that sustainability governance is not value-free. Every decision we have to make as individuals and as societies involves choices that rest on or invoke norms and values. At any particular point in time there are multiple alternative courses of action one could consider as a way forward. For example, feeding 10 billion people could be done both by intensifying agriculture and by shifting diets. Likewise, addressing climate change could be done – at least in theory – both by geoengineering technologies and by curbing emissions. And so on. Any alternative course of action entails not only the possibility of solving the problem but also the question of how the problem will be solved and who will be the likely winners or losers from this solution. Accordingly, sustainability research and researchers have the duty to clarify the alternatives, their normative underpinnings, and the related implications. As Cox (1981, 128) famously argued "Theory is always *for* someone and *for* some purpose" and so is sustainability research. This Handbook employing a critical perspective shows that as scholars and as educators we need not only ask the tough questions but also shed light on the alternative ways of identifying the roots of the problems, the pathways proposed to address them, and their underlying ethical assumptions, which are often implicit.

Until now, there is no comprehensive volume that covers global sustainability governance from a critical transformative perspective. Although individual scholars critically reflect upon consumerist values and culture, warn about disempowerment of people, and are disturbed about increasing inequalities and the erosion of democracy, such work has not yet been collectively presented in a single volume on sustainability governance. This Handbook aims to address this gap by providing a state-of-the-art review of core debates and contributions that offer a more normative, critical, and transformative approach to global sustainability governance. Given the interdisciplinary nature of sustainability, furthermore, it draws on a range of perspectives, including political science, sociology, economics, philosophy, and law.

The Handbook contains four main sections. Part I reviews prominent conceptual and analytical lenses critical of mainstream approaches to sustainability governance. Part II provides a review of ethical debates and normative principles related to global sustainability governance. In Part III, key challenges to sustainability governance are reviewed. In Part IV, we turn to transformative approaches to sustainability governance. Finally, the concluding chapter reflects on the chapters' contributions to these themes and adds a discussion of further questions such as the sustainability of sustainability research and challenges to communicating about sustainability. The following pages provide a more detailed outline of these sections.

## Part I: conceptual lenses

The concepts of power and legitimacy are an essential starting point for the analysis of (sustainability) governance. Magdalena Bexell argues that power and legitimacy are inextricably linked. Legitimacy makes power appropriate in the eyes of the governed, but material and ideational power structures shape what is considered legitimate at the same time. In order to study the dynamics between power and legitimacy in a fruitful manner for global sustainability governance, Bexell argues against oversimplified dichotomies between normative and empirical-sociological lenses of inquiry. Instead she proposes a combined normative-sociological perspective consisting of three steps: (a) the empirical identification of how power structures make particular legitimacy beliefs more important than others, (b) the normative interrogation of those legitimacy beliefs, and (c) the normative engagement with substantive and conflicting norms of sustainability, including justice, sufficiency, and development itself. The chapter argues that transformative global sustainability governance research calls for empirically oriented social scientists to reflect on the normative ramifications of their work and for normative oriented political theorists not to neglect the empirical grounds for their argumentation.

Ingolfur Blühdorn and Michael Deflorian provide an "unorthodox interpretation" of environmental governance. Motivated by the puzzling proliferation of new forms of collaborative governance arrangements despite criticisms regarding the latter's effectiveness and democratic legitimacy, they argue that it is precisely these qualities that make them flourish. Environmental governance, they argue, is explicitly designed not to disrupt the established order while simultaneously satisfying the preferences, needs, and dilemmas of contemporary consumer societies. Accordingly, they propose the lens of *performance* to analyse and understand modern forms of environmental governance and contemporary eco-politics more generally. The authors argue that such a lens can help unmask deceptive strategies allegedly pursuing radical transformative action and enable the creation of authentic eco-politics.

A key issue for scholars of sustainability governance is understanding why efforts to promote sustainability have been so limited to date. While many observers point to the everyday concerns (e.g. for jobs, economic and physical security, family, friends, and home) of people as major obstacles to a sustainability transformation, John Meyer argues that such concerns also offer opportunities for social change, but too little has been done to connect sustainability to them. The central challenge, in his view, is resonance: making sustainability and climate action resonate with everyday life. Indeed, he sees opportunities in the fact that many and perhaps most people in affluent consumer societies do not experience their lives as the best of all possible worlds – living, for example with the consequences of environmental injustice, a severe time crunch, or the un-freedom of automobile dependence – which creates possibilities for a politics that simultaneously promotes sustainability and enhances everyday life.

Tobias Gumbert undertakes the challenge of introducing the notions of materiality and non-human agency into environmental sustainability governance. He argues that the Anthropocene raises questions regarding both the radical separation between humans and "nature" as an object of human governance as well as the fixed boundaries of the social sphere and the biosphere. Given the ontological shift signified by the interchangeability between human and nature, it becomes imperative to move beyond dualisms, for example Nature/Culture, in order to examine and transform sustainability governance. In this context, Gumbert proposes that critical scholars need to engage with the concept of assemblages – which includes non-linearity, complexity, and reflexivity – as an analytical lens. Through the example of the non-human agency of waste, Gumbert makes the relatively abstract concept of assemblages accessible, while also calling for more in-depth empirical case studies to uncover the many connections and agentic qualities of non-humans.

Cristina Yumie Aoki Inoue, Thais Lemos Ribeiro, and Ítalo Sant' Anna Resende challenge the liberal-institutionalist, state-centric and positivist perspective that dominates the study of global sustainability governance. They argue that such a perspective and the nature/society divide that prevails with it are inadequate for handling the planetary socio-environmental crisis of the Anthropocene. Instead, we need a new political imagination and different analytical lenses in global sustainability governance studies. Inoue et al. propose "worlding" as such an alternative. Worlding, in their view, recognises many ways of being and experimenting with different worlds with important ontological and epistemological implications. On the basis of five different non-Western cosmovisions, the authors demonstrate the relevance of worlding in advancing new ways of producing knowledge, understanding and transforming global sustainability governance approaches.

## Part II: ethics, principles, and debates

The idea of justice is fundamental to understandings of sustainability. Yet justice has different interpretations making it unclear what it should entail exactly. In their chapter, Agni Kalfagianni, Andrea Gerlak, Lennart Olsson, and Michelle Scobie discuss how we can systematically compare and evaluate justice claims and/or demands for justice in global sustainability governance research in view of the lack of consensus regarding what justice *is*. They propose that any approach to justice needs to clarify the subjects, principles, and mechanisms of (fostering) justice, and the consequences of just societies. Most fundamentally, they argue that critical sustainability scholars need to embrace future generations and the non-human world as subjects of justice; advance an encompassing understanding of principles of justice; move beyond market-based approaches to address injustice; and be sensitive to the contextual conditions in which justice is operationalised and implemented.

Existing institutions "seriously under-represent the interests of future generations," notes Peter Lawrence. This is a fundamental problem for sustainability governance, which, after all, is largely about ensuring intergenerational equity. Such issues are growing more acute as the ecological crisis deepens. Strong reasons thus exist to establish institutions to represent future generations, argues Lawrence. He puts forward an argument for how future generations, who cannot authorise anyone to act on their behalf, can nevertheless be legitimately represented in decisions today. He outlines the normative basis for such representation, emphasising intergenerational justice (requiring the protection of human dignity and human rights), and also examines the proposal to establish a UN Commissioner for Future Generations and the issues related to it. Lawrence concludes that new institutions could help in a modest way to highlight the interests of future generations and incorporate those interests into decision-making.

What it means to live well within planetary boundaries is one of the most fundamental questions for transformative approaches to sustainability governance. Antonietta Di Giulio and Rico Defila offer one approach to defining the good life and needs in the context of sustainability, which involves developing a list of universal "protected needs" that all individuals should have the right to satisfy, along with procedures to adapt these needs to different cultural contexts. Combining theoretical accounts of "needs" with empirical evidence from cross-national studies and a structured dialogue with an interdisciplinary group of scholars, the authors discuss what could serve as a protected list of needs, thus substantiating a good life in the context of sustainability. The chapter concludes with a discussion of the consequences of putting the good life centre stage for sustainability governance.

A different approach to the question of "good living" or "living well" is found in *buen vivir* (BV), a transformation discourse that emerged in the Andean-Amazonian region of Latin America. Julien Vanhulst and Adrián Beling examine this "retro-progressive utopia" that recovers indigenous traditions, connects them to currents of contemporary critical thought, and looks towards emancipatory, socio-ecologically sustainable futures. BV has different variations, but there is a common emphasis on harmony with oneself (identity), with society (equity), and with nature (sustainability). It challenges what it considers the false universality of Eurocentric conceptions of modernity and aims to make space for "pluriversal" approaches, while also seeking alternatives to the Euro-Atlantic development model and to the conventional sustainability governance concepts of "sustainable development," "ecological modernisation," and "green growth." While critical of the limits of statist BV experiments in Bolivia and Ecuador, Vanhulst, and Beling emphasise BV's transformative potential, calling it a "unique living laboratory for social-ecological transformation" and an opportunity to "test a more promising approach to global sustainability governance [...that is] capable of a future."

Luigi Pellizzoni examines the notion of responsibility in sustainability governance. He traces the concept's roots back to ancient Greece and Rome and its ever changing meaning since then. In the contemporary world, responsibility has come to be associated with imputability (the attribution of an action to someone as its actual author) and answerability (the reasons an agent behaved in a certain way and the presence of somebody to whom the agent is deemed accountable). However, the risk and uncertainty that pertain to environmental change complicate responsibility as they make it extremely difficult to determine both a causal chain of events and accountability. In the context of neoliberal politics, this has resulted in pre-emption replacing precaution as the main principle of governing under uncertainty and the market becoming the chief institution allocating responsibility. Yet, Pellizzoni also observes the emergence of new social norms as manifested in alternative forms of community organisation, for example. These may help reconnect action and outcome as long as they make no ambiguous equations between affect and effect, or naïve assumptions about non-dominative orientations resulting from the vanishing threshold between the human and the non-human, the technical and the natural.

While environmental politics has been a particularly secular field, at least on the surface, the search for new approaches and transformations in cultural practices leads some scholars to look to religion. In her chapter addressing the role of religion in sustainability governance and religious actors' potential contribution to sustainable development, Katharina Glaab argues that "one should take seriously the claim that there is a religious answer to the global ecological crisis." Religions provide ethical frameworks with the potential to change views on the human-nature relationship and motivate environmentally friendly actions. In addition to their ability to influence political discourses and bring about normative change, religious actors are playing an important role as environmental-political actors, as seen, for example, in their contribution to the fossil-fuel divestment movement. While critics have rejected religion as irrational or having anti-ecological elements, Glaab maintains that religion deserves its place as an increasingly important topic in sustainability politics.

The question of "how much is enough?" – both having enough to live well and not consuming so much that it is ecologically excessive – is increasingly important for sustainability governance. Anders Hayden argues that "sufficiency" deserves to be a core organising principle for societies facing the need to live within planetary limits. He argues that sufficiency is a key element of a balanced ecological strategy – indeed it is a concept present in many chapters in this Handbook – but it has been neglected by dominant "green growth" and ecological modernisation strategies that focus on efficiency and green technologies. The chapter considers the significant obstacles to sufficiency in contemporary societies, reasons why it nevertheless persists in environmental debates (and is indeed increasingly relevant), potential ways to incorporate it into policy, and examples where it has made inroads in the policy sphere. The idea of sufficiency draws attention to the real possibilities of achieving well-being in less materially intensive ways, although Hayden cautions that it needs to be complemented by other approaches given the scale of the challenge in meeting human needs within planetary boundaries.

## Part III: key challenges

One of the main challenges facing sustainability governance is North-South inequity. Chukwumerjie Okereke demonstrates historically that every significant environmental summit, multilateral agreement, or global environmental institution has been severely challenged by issues related to North-South inequity and justice. He argues that the way problems are framed in present day environmental cooperation, the attribution of responsibility, the solutions offered and the processes of decision-making show very little commitment to addressing inequity. This, in turn, seriously jeopardises the chances of achieving sustainable development. In consequence, more radical interrogations of the basic structure of international society and of patterns of social relations between the North and South are urgently needed, in Okereke's view. This implies that questions of environmental justice must move to the forefront of sustainability governance and not be treated as an optional add-on.

While the gap between the global North and South greatly complicates sustainability governance, critical scholars also challenge the conventional means to close that gap. The dominant approach to development focused on the growth of Gross Domestic Product (GDP) is the subject of the chapter by Kerryn Higgs, who questions the persistent emphasis on "making the cake bigger" in light of its failure to distribute wealth equitably and its collision course with planetary limits. Although the development discourse has evolved from the end of the Second World War to today's Sustainable Development Goals, there remain significant contradictions between the focus on economic growth and environmental goals. Higgs concludes by highlighting the need for development alternatives that focus on sustainably meeting the needs of the rural masses rather than the current growth model that enriches urban elites and expands the middle class at great environmental cost.

One aspect of the unsustainability of the existing economic model is the tendency to treat all resources as something to be "mined" – that is, permanently used up – even renewable resources that could be used in more sustainable ways. Thomas Princen argues that in modern, industrial consumerist societies, mining and sustaining have been conflated – "it's all growth, all wealth formation, all progress" – and indeed the entire economy has come to resemble a mining operation based on extracting irreversibly and moving on. He argues that a transition is needed to a regenerative economy that prevents mining practices from being applied to renewable resources, shifts from an emphasis on growth to finding "sufficient wealth in both mining and sustaining," and in which the limited mining practices that remain are subordinate to and support sustaining practices.

Practices that erode the ecological foundations of contemporary societies are also connected to the processes of the financial economy. Jennifer Clapp and Phoebe Stephens examine the implications for environmental sustainability of financialisation in the neoliberal era – i.e., the growing importance of financial actors, institutions, and motives in guiding economic decisions – which has resulted in a proliferation of new financial instruments linked to natural resources and environmental change. The authors explain how specific financial instruments transform elements of nature (such as land, water, carbon, and weather) into assets that generate returns to investors, and emphasise that such commodification has real-world impacts that can undermine sustainability. However, the distancing between the trading of financial products and impacts on the ground obscures those links. If governance of these processes is to be strengthened, the authors argue, scholars need to play a key role in shining a light on the dynamics at play and in improving our understanding of them.

Sustainability governance would be a considerable challenge even if all major political actors shared core objectives such as significantly reducing greenhouse gas emissions, but there is the additional hurdle of overcoming powerful, organised opposition. Robert Brulle and Melissa Aronczyk examine the phenomenon of environmental counter-movements by looking at the United States, where organised opposition has had considerable "success" in thwarting concerted climate action. The authors trace the history of this movement, which builds on past efforts to resist governmental regulation, promote neoliberalism, increase the influence of American conservativism, and use public relations techniques to advance corporate interests. They show how a sophisticated system of organised opposition to climate action has resulted from long-term efforts to build an "intellectual and ideological infrastructure" capable of turning ideas into policy proposals in the medium term and facilitating political action in the short term. (Those active in movements for sustainability and equity might perhaps think of doing something similar.) The authors argue that more effective strategies are needed to counter the counter-movement; an essential starting point is a deeper understanding of that movement.

Many contributors to this volume see an additional problem even among those who accept climate/environmental science and the need for policy action: an excessive faith in technological solutions. For Samuel Alexander and Thomas Rutherford, techno-optimism is the "belief that science and technology will be able to solve the major social and environmental problems of our times, without fundamentally rethinking the structure or goals of our growth-based economies or the nature of Western-style, affluent lifestyles." Their critique of techno-optimism draws on evidence of the limits to date of efforts to decouple economic growth from environmental impacts. The authors conclude that the degree of decoupling required is too great to have any confidence that technological solutions can save the dominant growth paradigm. They argue that efforts to improve technology and efficiency must be complemented by an ethic of sufficiency, and that alternative economic models are needed – themes examined in depth in chapters in Parts II and IV.

The need to go beyond technological solutions is also reflected in the chapter by Naomi Krogman, who writes that "[a]t the core of all of our environmental problems is consumption." She sees a need to transform consumption patterns in more-developed countries, as well as in medium-developed countries. While her chapter focuses on consumer values and consumption, she cautions that the emphasis on material wealth and status that leads to overconsumption is supported and driven by larger structural factors. As such, scholars should not only focus on values and actions at the individual level, but also consider questions of power and structural change. Krogman examines a number of forces in society that help explain why people consume so much despite increased environmentalism, as well as some hopeful trends towards new consumer values that will require collective political action and supportive governance to achieve a greater impact.

The issues that ought to be the object of sustainability governance are a matter of some debate; one of the most contentious topics is population. While acknowledging the gender, race, and class-based controversies that have made managing fertility a taboo topic, Diana Coole argues that scholars concerned with planetary boundaries need to put the population issue back on the sustainability governance agenda. She maintains that stabilising the global population at a sustainable level requires active policy support rather than a laissez-faire approach (or for that matter pro-natalist policies in several high-income nations that aim to increase fertility). However, if a new approach to population is to be seen as legitimate, it will be necessary to devise effective and ethical policies compatible with human rights and reproductive choice.

## Part IV: transformative approaches

Part IV turns to the question of how to transform sustainability governance (although chapters in other sections also include their own potentially transformative proposals). It begins with Michael Maniates throwing an intellectual Molotov cocktail in the direction of the prominent idea that individual acts of green consumption and lifestyle change will, when combined with similar acts by millions of others, lead to social and ecological transformation. This amounts to "magical thinking," which is leading "environmentally concerned publics into cul-de-sacs of political irrelevance." The necessary transformation, he argues, will only come through political mobilisation of the more ecologically minded segment of the population, which, in turn, requires moving beyond the debilitating and ineffective narratives focused on individual consumer action.

The idea of transformation driven by a mobilised segment of the public assumes a set of democratic institutions in which people can exercise their "citizen muscles," but what type of democracy is needed to address the new challenges of the Anthropocene? Aysem Mert examines this question, arguing that the institutions of the earlier Holocene cannot provide adequate answers for the new era – i.e. democracy in the Anthropocene "cannot be more of the same." The Anthropocene – and the uncertainties and insecurities it is generating – is both a threat to democratic governance and, Mert argues, an opportunity to re-imagine and re-invigorate democracy through a deconstruction of traditions, a reflexive approach to science and decision-making, and democratic experimentation.

If overconsumption is a core driver of ecological degradation, then how can consumption be organised and governed to produce sustainable outcomes that also allow for high levels of well-being? Doris Fuchs argues that "consumption corridors" provide a promising transformative approach in this respect, allowing the pursuit of a good life for all within planetary boundaries while making consumption and its role with respect to both a core concern. Such corridors encompass the space between minimum consumption levels needed to satisfy one's protected needs and maximum consumption levels not to be overstepped in order not to hurt others' chances to do so. The chapter situates its argument in the context of political debates about (limits on) freedom and enquires into pathways towards the development of corridors as well as associated supportive structural changes. Most fundamentally, it suggests that the revisiting and reorganisation of consumption entailed in the idea of consumption corridors provide an opportunity to integrate the pursuit of well-being and justice in a world of limits and guidance in navigating present and future ecological and social crises.

Dirk Philipsen's chapter continues the focus on well-being and links it back to the question of growth by calling for a move beyond GDP – both as a measure of prosperity and a wider economic paradigm rooted in a growth-centred capitalism. He discusses the history of

GDP, its emergence in response to the challenges of the Great Depression, its contribution to the Allied War effort, and growing importance in the post-war era – but argues that this outdated measure ("Grandpa's Definition of Progress") is an obstacle to sustainability governance as it fails to measure whether economic output is sustainable, equitable, and delivering greater well-being. He also examines the many "beyond-GDP" alternatives, their promise, and the obstacles they face, concluding that to be effective, they need to do more than offer improvements to the existing economic system – they need to question the very logic of that system.

One key factor holding back more ambitious environmental policy is the widespread belief that economic growth must be prioritised if key social challenges, such as the creation of adequate employment and the maintenance of economic stability, are to be successfully addressed. With such concerns in mind, Lange considers how a post-growth economy could function and generate high levels of social welfare, which, if achievable, would enable stricter environmental regulation. He analyses the macro-level requirements for a sustainable zero-growth economy, as seen from the perspective of neo-classical, Keynesian, and Marxian theories. He concludes that it is not enough to "get the prices right," as many neoclassical economists would argue; substantial change to economic institutions would be necessary such as moving to employee-owned enterprises to curb the drive for capital accumulation and work-time reduction that keeps pace with the rate of labour productivity growth.

The idea of work-time reduction as a strategy to achieve sustainable livelihoods is examined further by Larsson, Nässén, and Lundberg. They highlight the potential of shorter working hours to limit consumption volumes and – in combination with technical improvements in eco-efficiency – deliver environmental benefits, while generating a higher quality of life through less stressed working lives and more leisure time. In addition to drawing on their comprehensive review of existing studies, they base these conclusions on a survey of municipal employees in Gothenburg, Sweden, who gained the right to choose part-time work – a right that the authors believe could be an important step towards the wider adoption of new, more sustainable work-time norms.

Decarbonisation "as a key goal of global sustainability governance will be transformational in either its successes or its failures," argues Richard Lane. Failure to decarbonise the world's economy is projected to bring major disruptions to life on Earth resulting from ocean warming, coral bleaching, and food systems breakdown, among other impacts. But to succeed, decarbonisation needs a different global governance approach than currently pursued. Today, Lane argues, decarbonisation governance fails because it is deeply incoherent as highlighted by the problem of absolute decoupling of economic output from greenhouse gas emissions and the reliance on Negative Emissions Technologies. These incoherencies both result from, and are maintained by, a series of exclusionary processes: the exclusion of (certain) people; the exclusion of nature and particularly climate change itself; and the exclusion of systemic change. Lane posits that decarbonisation can and should be reconstructed as a transformative locus of climate governance that is inclusive and coherent, but this requires a commitment to emancipatory politics.

Karen Liftin problematises the local as a desirable level of transformative politics. She warns against an oversimplified understanding of localism as *the* solution to current environmental and social challenges. Indeed, localism entails not only ecological and solidaristic voices but also libertarian and populist agendas. Investigating the promise of localism in practice, Liftin contends that the greatest power of localism is its ability to inspire agency, collective action, and innovation in the face of potentially overwhelming complexity. However, given the difficulties in disentangling ourselves from the global, localist approaches should move beyond – and not only against – globalism to reach their full potential. This is captured, Litfin argues, in the notion of organic globalism, a world of locally based but globally networked citizens' initiatives.

## Towards critical sustainability governance

The intensification of ecological crises that is evident today represents a very significant threat; however, in the spirit of Gramsci's pessimism of the intellect and optimism of the will, one can also see opportunities. At a time when stronger sustainability governance is urgently needed, there are opportunities to re-evaluate our understandings of concepts and issues including democracy, justice, legitimacy, the humanity-nature relationship, and the drivers of ecological degradation and necessary socio-political responses to it – and most fundamentally to re-imagine social and ecological futures, with greater emphasis on equity and a new vision of how to live well within planetary limits. Critical scholars of sustainability governance have a key role to play in highlighting the role of power, inequity, and exploitation, and in pointing towards real transformative possibilities, which we hope and believe the contributions to this Handbook will make clear.

## References

Alvaredo, Facundo, Lucas Chancel, Thomas Piketty, Emmanuel Saez, and Gabriel Zucman. 2018. *World Inequality Report 2018.* World Inequality Lab. https://wir2018.wid.world/files/download/wir2018-full-report-english.pdf.

Cox, Robert W. 1981. "Social Forces, States and World Orders: Beyond International Relations Theory." *Millennium* 10 (2): 126–55. doi: 10.1177/03058298810100020501.

Crutzen, Paul J. 2002. "Geology of Mankind." *Nature* 415 (6867): 23. doi: 10.1038/415023a.

IPBES. 2019. *IPBES Global Assessment Summary for Policymakers.* Bonn: Intergovernmental Science-Policy Platform on Biodiversity and Ecosystem Services. www.ipbes.net/news/ipbes-global-assessment-summary-policymakers-pdf.

IPCC. 2018. *Global Warming of 1.5°C: Summary for Policymakers.* Geneva: Intergovernmental Panel on Climate Change.

Lawson, Max, Man-Kwun Chan, Francesca Rhodes, Anam Parvez Butt, Anna Marriott, Ellen Ehmke, Didier Jacobs, Julie Seghers, Jaime Atienza, and Rebecca Gowland. 2019. *Public Good or Private Wealth?* Oxfam. doi: 10.21201/2019.3651.

McKibben, Bill. 2010. *Eaarth: Making a Life on a Tough New Planet.* New York: Times Books.

Rockström, Johan, Will Steffen, Kevin Noone, Asa Persson, F. Stuart Chapin, Eric F. Lambin, Timothy M. Lenton, et al. 2009. "A Safe Operating Space for Humanity." *Nature* 461 (7263): 472–75. doi: 10.1038/461472a.

Steffen, Will, Katherine Richardson, Johan Rockström, Sarah E. Cornell, Ingo Fetzer, Elena M. Bennett, Reinette Biggs, et al. 2015. "Planetary Boundaries: Guiding Human Development on a Changing Planet." *Science* 347 (6223): 1259855. doi: 10.1126/science.1259855.

# Part 1
# Conceptual lenses

# 1

# Power and legitimacy

*Magdalena Bexell*

## Introduction

Global sustainability governance involves the exercise of political power through rule-setting and policy-making in issue domains where domestic rules have proven insufficient. Political legitimacy means that such exercise of power and authority conforms to one or several sources of appropriate rule. Such sources are, for instance, democracy, effective problem-solving, moral authority, or expert-based knowledge. The decision on whether or not power exercise is indeed legitimate can be made by researchers or by those subject to an institution's political authority. Among researchers, normative debates on the (lack of) democratic qualities of global governance have earlier predominated the study of such governance. Today, the empirical study of legitimacy perceptions and processes of legitimation is rapidly evolving. Studying legitimacy is important for several reasons. Legitimacy is required for global sustainability governance to be effective in addressing cross-border sustainability challenges. Without legitimacy, governance attempts are either likely to have less impact or to depend on coercive measures. In the absence of enforcement measures, legitimacy is central to strengthening compliance with globally agreed rules on sustainability. Moreover, studying political legitimacy allows researchers to provide critical and constructive insights into broader societal debates on power, authority, and rule beyond the nation state.

This chapter aims to show in what ways legitimacy is a useful conceptual lens for envisioning transformative approaches to the study of global sustainability governance (see the volume's Introduction). While the chapter does not in itself provide a substantive proposal for such an approach (see instead Part IV of the volume), it contributes conceptual groundwork on which such approaches can build. The chapter proceeds as follows. In the next section, in order to ground the ensuing discussion in explicit attention to power, I identify critical questions for legitimacy research based on two different understandings of power in global governance. Second, I distinguish between normative and empirical-sociological approaches to the study of political legitimacy with regard to concerns those approaches raise for the study of global sustainability governance. The first approach revolves around standards of assessment of such governance. The latter is preoccupied with studying legitimacy perceptions, sources, and processes. Finally, I shall argue that transformative approaches to the study of global sustainability governance ought to employ a combination of both lenses. Such a combination can result in visions that are aspirational yet grounded in systematic empirical knowledge on power relations and legitimacy perceptions. I briefly refer to the consultations on the United Nations' (UN) Sustainable Development Goals (SDGs) in order to illustrate steps involved in combining the two lenses.

## Power and legitimacy

The relationship between power and political legitimacy is intriguing. Legitimacy is the glue that links authority and power while it also reflects and might even reinforce power relationships (Hurd 2007). At the same time as legitimacy is a desired attribute of a governing body, already Inis Claude (1966) pointed out that legitimacy may reinforce existing power relationships. This means that by making power appropriate in the eyes of those who are governed by it, legitimacy makes rulers more secure in their possession of power. Legitimacy is, in brief, both a source of power and a constraint on power. Institutional arrangements for the organisation of power embody legitimating principles that establish how power is obtained and the limits within which it can be exercised (Beetham 2012: 123). Arguably, the power-legitimacy relationship is further complicated by the recognition that power works in and through many forms in global governance (see e.g. Barnett and Duvall 2004; Brassett and Tsingou 2011). This section identifies critical topics for legitimacy research based on contrasting understandings of power in global governance.

### An agent-centred view of power

A first view of power conceptualises it as an asset or attribute that actors can possess more or less of in relation to other actors or in relation to affecting outcomes. Such agent-centred power can reside with individuals as well as with political institutions. It can be based on material resources, formal positions, moral credibility, expertise, coercive measures, or on several other sources. While those sources can be studied empirically, it remains methodologically challenging to assess the relative power of different actors. From this agent-centred view follows questions on the legitimacy of specific governance arrangements in light of power relations between their members. This view of power draws attention to inequality between powerful and less powerful states in intergovernmental organisations, asking questions on participation, accountability, and representation. It also asks who is impacted and whose interests are supported by decisions taken in the organisations of global governance. Such institutions are not neutral problem-solving arenas but sites of power and struggle between rule-makers and rule-takers, with ensuing legitimacy deficits and delegitimation attempts. The agent-centred view also directs interest towards those who are subject to the political authority of global governance institutions. Asking about those groups' legitimacy beliefs is key to better knowledge on the dynamics of (de)legitimation processes (Bexell and Jönsson 2018).

Along this understanding, a key concern for research on sustainability governance is how processes of globalisation impact the distribution of power across actors. Power is broadly dispersed across actors in the realm of sustainable development, making it harder than before for policy-makers and governments to impact outcomes (Meadowcroft 2007). Instead, global sustainability governance is characterised by fragmented regulative power. This steers research questions towards the relative influence of state and nonstate actors on sustainability regulations beyond the nation state, asking what interests are promoted when nonstate actors gain more rule-setting power (Brulle and Aronczyk this volume; Fuchs et al. 2010). Such nonstate actors range from private companies to philanthropic foundations to large civil society organisations. In particular, the political power of global companies has increased in global governance compared to that of labour organisations and civil society (Fuchs 2013). Debates on global public-private partnerships such as the UN Global Compact have been lively in this field. Partnerships are subject to critique stating that companies have obtained too much influence on UN affairs and that partnerships do not deliver on their promises (e.g. Sethi and Schepers 2014).

Finally, an agent-centred view of power also underpins debates on the role of legitimacy for compliance by those subject to sustainability rules in global governance. In contrast to domestic political institutions, global governance organisations have few coercive measures available to obtain compliance with their rules and policy decisions (Hurd 1999). Legitimacy is presumed to generate compliance that is not dependent on monitoring or on coercion by hegemonic states. Beliefs in the legitimacy of a political rule increase the likelihood of rule compliance without coercive measures involved. In the case of global sustainability governance where strong enforcement measures are usually lacking, legitimacy is therefore key to rule compliance by states and nonstate actors.

## A structural view of power

In a structurally oriented understanding, power is not reducible to an attribute of specific actors but considered constitutive of the subjects of global governance (Barnett and Duvall 2004). This understanding of power raises other questions for legitimacy research on global sustainability governance than mentioned earlier. It is less concerned with pinpointing legitimacy beliefs of particular actors or legitimation strategies of governing bodies. A view of power as structural emphasises either material circumstances (e.g. capitalist accumulation) or discursive-ideational conditions (e.g. gender norms). Those allocate different privileges to different actor positions (cf. Fuchs 2013). In a materialist-oriented critical vein, neo-Gramscian theories of world politics focus on how ideological knowledge works to legitimate a capitalist order and its embedded neoliberal institutions and dominant powers (Mittelman 2016). Critical political economy accounts pay particular interest to scrutinising governance attempts at solving environmental problems through a market-based logic. For example, Matthew Paterson (2010) points to tensions between a search for legitimacy and a search for capitalist accumulation in the case of private climate governance. Their increased structural material power has allowed global companies to expand their rule-setting impact in global governance (Fuchs 2013). Key research questions in this vein concern how capitalism legitimates certain approaches to sustainability, how profit motives influence public-private partnerships, and how "consumer power" impacts market-based sustainability schemes.

Approaches that conceptualise power in "productive" terms contain a more diffuse view of where power resides and do not regard power as zero-sum. How power and knowledge interact is at the heart of study in such approaches. They explore how certain domains of truth become predominant and how those, in turn, enable certain practices and modes of governance. Ideational structures of meaning and knowledge define what becomes viewed as possible and impossible fields of governance. These structures thereby set the limits of thought and action. What is it that should be sustained? For whose benefit and how? Studies in this vein explore, for instance, how gender norms are constructed in global sustainable development discourses. Emma A. Foster argues that the *Agenda 21*, adopted at the UN Rio Earth Summit in 1992, reproduced binary gender categories and privileged heterosexual norms in a way that still influences the UN Environment Programme (Foster 2011). Others identify processes of "responsibilisation" through which responsibility for environmental matters is assigned to individuals rather than societal institutions. This is legitimised through subtle forms of ideational power like ideas on "green consumption." Researchers caution that individualised responsibility might reduce attention to political solutions to institutional problems of collective action (Soneryd and Uggla 2015).

In conclusion, material and ideational structures of power set the parameters within which legitimation processes play out. Those structures shape (de)legitimation strategies as well as legitimacy perceptions of individuals. The agent-centred and structural views of power are not necessarily mutually exclusive but, as demonstrated earlier, give rise to different concerns related to political legitimacy. In practice, researchers can be informed by a combined agent-structure approach to power in the study of legitimacy. Mark Suchman's influential definition of legitimacy emphasises that perceptions on the appropriateness of a given entity are shaped within a system of norms, values, beliefs, and definitions (1995: 574). Likewise, Jan Aart Scholte (2018) and Steven Bernstein (2011) underline that prevailing norms of contemporary world politics exert structural power in the sense that they shape legitimacy claims of global governance organisations. The next section goes further into different ways of employing legitimacy as a conceptual lens.

## Legitimacy as conceptual lens

Legitimacy is a useful conceptual lens for the study of global sustainability governance in two main ways, one being normative and the other empirical-sociological. I below introduce these two analytical uses of the concept and point to their respective promises for sustainability research.

### A normative lens

Through a normative lens, political legitimacy means that the exercise of political authority lives up to a standard of appropriateness theoretically defined by the scholar. This line of research debates what principles are most appropriate in scholarly assessments of governing bodies. The outcome of such assessment does not necessarily correspond to the beliefs of those subject to the authority of the governing body at hand. Much scholarly debate in this vein has revolved around how to theoretically conceive of democracy beyond the nation state. To what extent ought democratic theory be revised in light of the lack of a global "demos"? What model of democracy beyond the nation state is most convincing in the realm of sustainability governance? Answers are provided in normative theories of global democracy. Those range from cosmopolitan democracy to democratic intergovernmentalism with deliberative democracy and global stakeholder democracy in between (see Bäckstrand 2011 for further detail). They end up in very different assessments of how democracy fares in global governance. Clearly, the choice of normative "yardstick" determines to what extent global sustainability governance is perceived to suffer from a lack of democratic legitimacy. Yet, there is hardly any yardstick according to which contemporary global sustainability governance can be deemed democratic in terms of a holistic system of political rule. Both public and private forms of environmental governance display great weaknesses regarding democratic legitimacy, reinforced by large resources asymmetries among their participants (Fuchs et al. 2010).

On the other hand, literature that engages with individual democratic values sometimes ends up in more positive assessments, finding settings where such values are at play. Such literature often concerns new hybrid forms of governance. For instance, Bäckstrand et al. (2010) argue that new modes of deliberative governance have in some instances strengthened representation and deliberation in environmental politics. Yet they should best be understood as a piecemeal complement to state-based representative democracy, the authors posit. With regard to accountability, Kuyper and Bäckstrand (2016) map how focal points of nonstate constituency groups of the United Nations Framework Convention on Climate Change (UNFCCC) created mechanisms to remain accountable to the organisations they represented. Studies showing the presence of accountability mechanisms or deliberative stakeholder settings do not, however, warrant conclusions on "democracy" in the sense of a political system of rule. The term democracy should be applied with caution in assessments of global governance as it is not additive in the sense that accountability plus deliberation equates more of democracy (Erman 2010). Rather, when our concern is with power and political legitimacy, a key contribution of normative theories of democratic global governance is to provide leverage for a critique of power and for imagining future transformation.

A normative lens also usefully identifies and explores possible tensions between sources of political legitimacy. The distinction between input/procedural/democratic sources and output/effective problem-solving sources of legitimacy has been formative for such research (Scharpf 1999). In the realm of sustainability, there is no lack of studies demonstrating the acute need for effective global responses to collective action problems. This raises difficult questions on possible hierarchies between sources of legitimacy, if matters come to a head. The relationship between output legitimacy and input/democratic legitimacy is a key normative concern in literature on environmental politics (see Blühdorn and Deflorian this volume; Fuchs et al. 2010). For instance, against what sources of legitimacy should one assess the Forest Stewardship Council (FSC)? Is it more important that FSC certification results in sustainable forestry or that a broad range of stakeholders participates in deciding on FSC certification standards? Ideally, the two sources reinforce each other.

In the case of political legitimacy, a third potential legitimacy source resides in the substance of a rule or decision. Such substantive legitimacy means that the content of policies or rules is congruent with societally shared broader purposes at a given point in time. In other words, this is what Beetham (2012: 123) calls socially accepted beliefs about "the proper ends and standards of government." In this case, the appropriateness of the substantive content of policies is the key source of political legitimacy. This is regardless of how the decision on what policy was made (input legitimacy), or how effective the policy might be in solving problems (output legitimacy) (cf. Hurrell 2005). In the case of sustainability, the normative study of substantive legitimacy taps into academic debates on global justice, equality, and ecology. Normative proposals on such substantive concepts are needed when engaging in debates on the legitimacy of decisions and rules on sustainability. The contemporary study of global sustainability governance would clearly benefit from greater engagement between normative theorists dealing with such substantive debates, and those studying political legitimacy in global governance.

The normative lens, to wrap up and put it simply, provides yardsticks against which to make critical assessments of contemporary governance and suggest alternative visions. In particular, because our focus is on political legitimacy, normative democratic theory offers a basis for pluralistic transformative critique of how power operates in and through the actors and structures of global sustainability governance. Its theoretical diversity ensures that assumptions and visions do not stay unquestioned but become subject of continued debate and scrutiny.

## An empirical-sociological lens

Through a second research lens, legitimacy resides in perceptions about appropriate rule among those who are formally subject to a political institution or otherwise affected by its policies. There is a growing International Relations literature that does not take an explicit normative stance as point of departure for studying legitimacy. Instead, this so-called socio-logical, empirically oriented approach asks questions on perceptions of legitimacy, sources of legitimacy beliefs, and processes of (de)legitimation. This includes explanatory questions on why legitimacy perceptions vary across societal groups or institutions (e.g. Tallberg and Zürn 2019). The process-oriented study of legitimation explores self-legitimation attempts by global governance organisations (Gronau 2016; Gronau and Schmidtke 2016; Steffek 2003) as well as by nonstate-based governance bodies (Bernstein 2011; Dingwerth 2017). It also concerns delegitimation, meaning attempts at contesting the authority of global governance bodies and impacting legitimacy beliefs in negative direction (e.g. Brulle and Aronczyk this volume; Gregoratti and Uhlin 2018). A process-oriented perspective fruitfully shows that legitimacy is not static but intersubjectively shaped through claims-making and contestation.

The empirical study of legitimacy benefits from established research methods such as process-tracing, elite and citizen surveys, participant observation, as well as discourse anal-ysis and other text-analytical methods, depending on the research question at hand. The public communication material of global governance organisations and their critics pro-vides rich ground for exploring self-legitimation claims and delegitimation attempts. That material can be analysed both through quantitative and qualitative methods. Interviews contribute equally important complementary insights into processes and strategies involved in (de)legitimation processes. Yet, the evolving empirical study of legitimation is replete with methodological challenges. For instance, how should one draw the line between what is and is not a legitimation practice? Is all public communication by an international organisation to be understood in terms of legitimation attempts? Or should there be explicit intentionality on the part of agents of legitimation in order for a claim to qualify as a legitimation claim? A narrow conceptualisation of legitimation is convincing on theoretical grounds but harder to study empirically. Moreover, the knowledge of certain international organisations among citizens is low. Does it make sense to study legitimacy beliefs in those cases? How can we de-termine which factors impact legitimacy beliefs? These and other methodological difficulties are not unique for the empirical study of legitimacy, yet need careful consideration by any researcher exploring this topic.

Despite methodological challenges, the sociological study of legitimacy opens up import-ant areas for further research. A key contribution of this lens for this chapter's concern with power and political legitimacy is to ask in whose eyes global governance organisations enjoy or lack legitimacy. Whose legitimacy beliefs are important in the eyes of global sustainability governance bodies? The study of legitimacy has moved beyond a state-centric focus and now includes legitimacy perceptions among both elites and citizens as well as among represen-tatives of different societal spheres: politics, civil society, business, academics, and other ex-perts (Symons 2011; Zaum 2013). The territorially based conception of political constituency works reasonably well for intergovernmental organisations where governments and citizens can be considered constituencies through chains of political institutions. The concept of constituency loses its territorial connotations when applied to nonstate private governance where participants answer to a non-territorial base of support (Bexell and Jönsson 2018). Better systematic knowledge on how, when, and why organisations target different groups would unravel the degree of importance organisations attach to those groups. Ultimately, who is recognised as an audience of legitimation is itself a power-imbued contested question (Bexell 2014; cf. Meine 2016).

In conclusion, the empirical-sociological lens advances knowledge on legitimacy beliefs related to global sustainability bodies and how such beliefs vary between groups. It unravels strategies that organisations use when attempting to influence legitimacy beliefs. This lens can also empirically assess descriptive elements of normative legitimacy theories. Thereby it subjects those theories to critique and refinement. For instance, does broad deliberation in effect make decision-making slow? How do legitimacy beliefs of marginalised groups differ from those of civil society elites or political elites? Transformative visions of global sustainability governance should build on results from empirical legitimacy studies in order to address concerns of those whose voices might not count in the eyes of global governance elites. Next, I discuss how a combination of empirical and normative approaches can nurture the study of political legitimacy in the case of global sustainability governance.

## A normative-sociological lens

### A richer approach to the study of legitimacy

The distinction between normative and empirical approaches to the study of legitimacy is invoked by scholars of global governance in order to delimit and clarify their purpose of inquiry. The validity of the distinction itself is at the same time an object of debate, aligning with broader methodological questions related to the relationship between ideal and nonideal political theory (cf. Beetham 2013). Many studies on legitimacy contain elements of both approaches. This is the case when researchers elaborate normative principles of legitimate rule that guide empirical study of the extent to which particular sustainability governance bodies live up to those principles (e.g. Agné et al. 2015; Dingwerth 2007; Fuchs et al. 2010; Kuyper and Bäckstrand 2016). Indeed, the sociological lens is always based on (implicit) normative assumptions and vice versa. For research guided by transformative aspirations in line with the present volume, an explicitly normative-sociological approach is a fruitful way forward (cf. Agné 2018; Brassett and Tsingou 2011).

Next, I sketch three tentative steps that can guide a combination of the two lenses. I illustrate those steps drawing on the consultations conducted by the UN between 2012 and 2015 on the SDGs, part of the UN 2030 Agenda. Those consultations were unprecedented in terms of outreach to citizens on the part of the UN (Fox and Stoett 2016; Kamau 2018; Sénit et al. 2017). Several consultation processes were conducted in parallel. The UN held national consultations in 88 developing countries as well as thematic consultations across the world. UN-led national consultations aimed to reach groups traditionally excluded from this kind of outreach such as street children, indigenous people, homeless, and people living with HIV (Fox and Stoett 2016: 563). Another key element was web-based provisions for input (the MY World Survey). Such surveys received about 7.5 million answers from individuals stating what should be prioritised in the new set of goals. These processes were transparent insofar as an unprecedented amount of material documenting the consultations was posted online for public access (Bexell 2015). Moreover, civil society and business organised their own consultations independent of UN-led processes. Eventually, this narrowed down into an intergovernmental decision-making phase where nonstate actors were represented through the UN Major Groups system (Kamau 2018; Kanie et al. 2017). This was concluded with the formal adoption of the SDGs by the 193 member states of the UN General Assembly in September 2015.

## First step: unveiling power relations

A first possible step in a normative-sociological approach to legitimacy in global sustainability governance is to empirically explore power relations of the case in focus. A researcher's understanding of power, in terms of the views outlined earlier, impacts where the searchlight is put. As pointed out by Jan Aart Scholte, the choice of empirical focus in legitimacy research is in this respect normatively infused (Scholte 2018). An agent-centred view of power zooms in on the degree of inclusion of and influence by different actors in the SDG consultations. Without doubt, compared to the creation of the Millennium Development Goals (ending 2015), the SDG consultations contained much more inclusive deliberation (Kamau 2018). Yet, recent studies have unveiled power imbalances that permeated this UN-driven legitimation process. Sénit et al. (2017) found that a lack of human and financial resources negatively impacted inclusiveness of civil society organisations during the intergovernmental Open Working Group negotiations on the SDGs (Sénit et al. 2017). The largest proportion of post-2015 proposals in the "World We Want" consultations came from global collaboration bodies or from actors based in countries that are highly placed on the Human Development Index, not low- or middle-income countries (Bergh and Couturier 2013). The exclusion of those lacking Internet access, as well as the short time frames of the post-2015 consultations, left a lot of agenda-setting power in the hands of those compiling the reports from the consultations (Bexell 2015). These were mainly representatives of the convening UN bodies. Input from business was channelled through the UN Global Compact. This promoted representation of large businesses, certain key companies and individuals, often headquartered in Europe or North America. One study shows that certain industry sectors and certain individual companies were overrepresented in the processes that allowed for business input on post-2015 goals (Pingeot 2014). In the end, the 2030 Agenda was crafted through intergovernmental political negotiations where contention between developing and developed countries came to fore (Cummings et al. 2018, 736; Kamau 2018, 229–236).

For their part, structurally oriented views of power reinforce that the SDGs are a product of political negotiations and power relations, as can be expected of any global agenda. In this view, one can go further into questions on knowledge production, on prevailing sustainability discourses, as well as on the role of material structures such as the global political economy. Clearly, prevailing deeper norms of global politics shape legitimacy perceptions and legitimation processes (Scholte 2018). A view of power as ideational and productive privileges questions on what constitutes legitimate knowledge on sustainable development for a global agenda. Here, key documents agreed on during and after the SDG consultations can be examined. What tensions and contradictions can be observed in those documents, as a result of their political character (cf. Bexell and Jönsson 2017)? What views on sustainable development made it into the UN General Assembly's 2030 Agenda outcome document *Transforming Our World*? Critical scholars have argued that the 2030 Agenda does not challenge the ways in which inequalities in wealth and power are produced nationally and globally. Last-minute changes in the negotiation texts on the 2030 Agenda weakened the responsibilities of rich countries (Esquivel 2016). One study argues that consultations and negotiations on *Transforming Our World* privileged knowledge discourses of rich country governments, ignoring local knowledge and pluralist-participatory knowledge discourses (Cummings et al. 2018). In sum, the first step engages with the study of whose legitimacy beliefs matter to global governance institutions and how broader structures of power play out in this regard.

## Second step: legitimacy in whose eyes?

A second step is to normatively engage with the question "whose legitimacy beliefs should matter?". This can take its point of departure in the former step's findings on the empirical question of "whose legitimacy beliefs matter?" (cf. Bernstein 2018). Moreover, researchers need to consider carefully whose legitimacy beliefs should be the object of study. That choice puts the spotlight on certain groups at the expense of others. For example, on what grounds is the selection of respondents made when surveys and interviews are conducted? In addition, an empirically oriented researcher might also on some occasions need to justify his or her selection of legitimacy sources. For instance, beliefs among a majority group of citizens of a particular state that a political institution effectively oppresses subordinate groups would carry little normative relevance for this field (Agné 2018). Such matters are important if the empirical-sociological study of legitimacy is to be relevant to normative debates on legitimacy.

The answer to whose legitimacy beliefs should matter will look different depending on theoretical starting point. For this step, theories on democracy in global governance provide a fruitful basis of normative engagement (cf. Bäckstrand 2011; Mert this volume). Such theories emphasise different groups as key holders of legitimacy beliefs. Two examples illustrate this difference. Theories on democratic intergovernmentalism accord most weight to government elites and the constituencies who are formally bound by the rules of intergovernmental organisations. In this view, legitimacy deficits arise if democratic chains between government elites and citizens are weak domestically. The legitimacy beliefs of constituencies are considered more important than the beliefs of those who are not formally bound by the rules. In the case of the SDGs, this theory would pay most attention to legitimacy beliefs of government elites and citizens. In contrast, theories of stakeholder democracy are centred around the all-affected principle of political legitimacy. This principle implies that the legitimacy beliefs of those who are most affected by particular sustainability policies count heavier than the beliefs of those not affected. In the SDG case, this means that the legitimacy beliefs of those who are most affected by unsustainable practices would matter most when deciding on global sustainability goals. Exactly who those stakeholders are would of course differ between SDGs. Clearly, stakeholder theory is more demanding than democratic intergovernmentalism in terms of deciding whose legitimacy beliefs count.

The SDGs constitute a challenging case for considering whose legitimacy beliefs ought to matter. This is due to their broad scope, long time horizon, and their status as a political agreement among all UN member states. In this light, how could one think about whose legitimacy beliefs on the SDGs ought to matter the most, bearing in mind that this chapter is concerned with *political* legitimacy? The key responsibility for realising the SDGs arguably resides with governments. Therefore, a tentative answer is that the most important legitimacy beliefs are those of national political elites and of marginalised citizen groups. Such elites hold the power over national political institutions, able to decide on change. For their part, marginalised citizen groups are the ones most affected by (a lack of) SDG implementation. Identifying differences in legitimacy beliefs related to the SDGs among those two groups within the same country shows to what extent different societal positions shape views on the SDGs. Arguing for the key relationship between power yielders and their weakest constituencies is also a way to emphasise the politics of sustainability, played out through political institutions. This implies to caution against depoliticisation of sustainability (cf. Soneryd and Uggla 2015). Depoliticisation means that sustainability becomes viewed as a matter for experts, bureaucrats, or the market. This diminishes the room for political contestation based on democratic representation and political accountability.

### Third step: substantive norm conflicts

The third step involves both normative and empirical elements. Here one should engage with substantive issues in order to identify and explore norm conflicts on the basis of previous steps. This step needs normative engagement with substantive propositions on, for instance, justice (Kalfagianni this volume), sufficiency (Hayden this volume), growth and development (Higgs this volume), depending on the issue at hand. The substantive norms of sustainability are manifold and often come into tension with each other. This is particularly the case when filtered through political institutions and political agreements. Consultations and negotiations on the SDGs bear witness of this (Kamau et al. 2018). Again, a key source of legitimacy is the actual substance of sustainability agreements (e.g. environmental protection, reduced emissions, equality in education). Yet, global political agreements on sustainability matters are replete with norm conflicts. The 2030 Agenda outcome document in which the SDGs are detailed also emphasises norms of state sovereignty and national self-determination, leaving extensive scope of interpretation to individual states in light of national circumstances (Bexell and Jönsson 2017). Additional norm conflicts appear when such agreements are to be implemented at the national level. Spelling out norm conflicts and how they impact legitimacy perceptions and delegitimation attempts is central in this step.

An important contribution of transformative approaches is to provide normative visions that can serve as a basis of critique of the substance of intergovernmental political agreements. Beyond critique, such approaches can show that current normative contradictions need not be insurmountable. Moreover, they serve pluralistic purposes of showing there are many ways in which a sustainable future can be imagined. In sum, taken together, the three earlier steps outline key tasks for transformative thinking based on a normative-sociological approach to the study of power and legitimacy. Needless to say, these steps need further elaboration and will look different depending on the issue in focus.

## Concluding remarks

This chapter has argued for the merits of using legitimacy as a conceptual lens underpinning transformative thinking on global sustainability governance that puts power relations of different kinds centre stage. The concept of legitimacy enables both critique and the elaboration of aspirational approaches to the study of such governance. Constructing substantive transformative approaches invites careful consideration of the balance between, on the one hand, knowledge on actually existing legitimacy beliefs, legitimacy sources, and processes of (de)legitimation, and, on the other hand, normative visions of a what a more legitimate condition of global sustainability governance would look like. Empirically oriented researchers should not hesitate to spell out the normative ramifications of their findings, against the backdrop of democratic legitimacy sources or in light of the need for effective political responses to major sustainability challenges. Neither should political theorists working on normative legitimacy theory neglect to ground empirical elements of their arguments on empirical findings on legitimacy beliefs, processes, and strategies.

## Acknowledgement

This chapter builds on research conducted within the research program *Legitimacy in Global Governance*, funded by Riksbankens Jubileumsfond (The Swedish Foundation for Humanities and Social Sciences) and within the project *Realising the Post-2015 Sustainable Development Goals: Whose Responsibility?*, funded by the Swedish Research Council Formas.

# References

Agné, Hans (2018). 'Legitimacy in Global Governance Research: How Normative or Sociological Should It Be?'. In Tallberg, Jonas, Bäckstrand, Karin and Scholte, Jan A. (eds) *Legitimacy in Global Governance. Sources, Processes and Consequences*, pp. 21–34. Oxford: Oxford University Press.

Agné, Hans, Dellmuth, Lisa M. and Tallberg, Jonas (2015). 'Does Stakeholder Involvement Foster Democratic Legitimacy in International Organisations? An Empirical Assessment of a Normative Theory'. *Review of International Organisations* 10, 4: 465–488.

Bäckstrand, Karin (2011). 'The Democratic Legitimacy of Global Governance After Copenhagen'. In Dryzek, John, Norgaard, Richard and Schlosberg, David (eds) *The Oxford Handbook of Climate Change and Society*, pp. 669–685. Oxford: Oxford University Press.

Bäckstrand, Karin, Khan, Jamil, Kronsell, Annica and Lövbrand, Eva (2010). 'Environmental Politics After the Deliberative Turn'. In Bäckstrand, Karin et al. (eds) *Environmental Politics and Deliberative Democracy: Examining the Promise of New Modes of Governance*, pp. 217–234. Cheltenham: Edward Elgar.

Barnett, Michael and Duvall, Raymond (2004). 'Power in Global Governance'. In Barnett, Michael and Duvall, Raymond (eds) *Power in Global Governance*, pp. 1–32. Cambridge: Cambridge University Press.

Beetham, David (2012). 'Political Legitimacy'. In Amenta, Ewdin et al. (eds) *The Wiley-Blackwell Companion to Political Sociology*, pp. 120–129. Oxford: Blackwell Publishing.

Beetham, David (2013). *The Legitimation of Power*. 2nd ed. Basingstoke: Palgrave Macmillan.

Bergh, Gina and Couturier, Jonathan (2013). *The Post-2015 Agenda: Analysis of Current Proposals in Special Areas*. Background briefing prepared for Overseas Development Institute and UN Foundation, New York.

Bernstein, Steven (2011). 'Legitimacy in Intergovernmental and Non-State Global Governance'. *Review of International Political Economy* 18, 1: 17–51.

Bernstein, Steven (2018). 'Challenges in the Empirical Study of Global Governance Legitimacy'. In Tallberg, Jonas, Bäckstrand, Karin and Scholte, Jan A. (eds) *Legitimacy in Global Governance. Sources, Processes and Consequences*, pp. 190–200. Oxford: Oxford University Press.

Bexell, Magdalena (2014). 'Global Governance, Legitimacy and (De)Legitimation'. *Globalizations* 11, 3: 289–299.

Bexell, Magdalena (2015). 'The Post-2015 Consultations: Fig Leaf Policy or Test Bed for Innovation?'. In Debiel, Tobias et al. (eds) *Global Trends 2015. Governance in a Complex World*. Bonn and Duisburg: Centre for Global Cooperation Research, University of Duisburg-Essen.

Bexell, Magdalena and Jönsson, Kristina (2017). 'Responsibility and the United Nations' Sustainable Development Goals'. *Forum for Development Studies* 44, 1: 13–29.

Bexell, Magdalena and Jönsson, Kristina (2018). 'Audiences of (De)legitimation in Global Governance'. In Tallberg, Jonas, Bäckstrand, Karin and Scholte, Jan A. (eds) *Legitimacy in Global Governance. Sources, Processes and Consequences*, pp. 120–133. Oxford: Oxford University Press.

Brassett, James and Tsingou, Eleni (2011). 'The Politics of Legitimate Global Governance'. *Review of International Political Economy* 18, 1: 1–16.

Claude, Inis (1966). 'Collective Legitimization as a Political Function of the United Nations'. *International Organization* 20, 3: 367–379.

Cummings, Sarah, Regeer, Barbara, Haan, Leah, Zweekhorst, Marjolein and Bunders, Joske (2018). 'Critical Discourse Analysis of Perspectives on Knowledge and the Knowledge Society Within the Sustainable Development Goals'. *Development Policy Review* 36, 6: 727–742.

Dingwerth, Klaus (2007). *The New Transnationalism: Transnational Governance and Democratic Legitimacy*. Basingstoke: Palgrave Macmillan.

Dingwerth, Klaus (2017). 'Field Recognition and the State Prerogative: Why Democratic Legitimation Recedes in Private Transnational Sustainability Regulation'. *Politics and Governance* 5, 1: 75–84.

Erman, Eva (2010). 'Why Adding Democratic Values Is Not Enough for Global Democracy'. In Erman, Eva and Uhlin, Anders (eds) *Legitimacy Beyond the State? Re-Examining the Democratic Credentials of Transnational Actors*, pp. 173–193. Houndmills: Palgrave Macmillan.

Esquivel, Valeria (2016). 'Power and the Sustainable Development Goals: A Feminist Analysis'. *Gender & Development* 24, 1: 9–23.

Foster, Emma A. (2011). 'Sustainable Development: Problematising Normative Constructions of Gender within Global Environmental Governmentality', *Globalizations* 8, 2: 135–149.

Fox, Oliver and Stoett, Peter (2016). 'Citizen Participation in the UN Sustainable Development Goals Consultation Process: Toward Global Democratic Governance?'. *Global Governance* 22, 4: 555–574.

Fuchs, Doris (2013). 'Theorizing the Power of Global Companies'. In Mikler, John (ed) *The Handbook of Global Companies*, pp. 79–95. Malden: Wiley-Blackwell.

Fuchs, Doris, Kalfagianni, Agni and Sattelberger, Julia (2010). 'Democratic Legitimacy of Transnational Corporations in Global Governance'. In Erman, Eva and Uhlin, Anders (eds) *Legitimacy Beyond the State? Re-Examining the Democratic Credentials of Transnational Actors*, pp. 41–63. Houndmills: Palgrave Macmillan.

Gregoratti, Catia and Uhlin, Anders (2018). 'Civil Society Protest and the (De)legitimation of Global Governance Institutions'. In Tallberg, Jonas, Bäckstrand, Karin and Scholte, Jan A. (eds) *Legitimacy in Global Governance. Sources, Processes and Consequences*, pp. 135–150. Oxford: Oxford University Press.

Gronau, Jennifer (2016). 'Signaling Legitimacy: Self-Legitimation by the G8 and the G20 in Times of Competitive Multilateralism'. *World Political Science* 12, 1: 107–145.

Gronau, Jennifer and Schmidtke, Henning (2016). 'The Quest for Legitimacy in World Politics – International Institutions' Legitimation Strategies'. *Review of International Studies* 42, 3: 535–557.

Hurd, Ian (1999). 'Legitimacy and Authority in International Politics'. *International Organization* 53, 2: 379–408.

Hurd, Ian (2007). *After Anarchy: Legitimacy and Power in the United Nations Security Council*. Princeton, NJ: Princeton University Press.

Hurrell, Andrew (2005). 'Legitimacy and the Use of Force: Can the Circle be Squared?'. *Review of International Studies* 31: 15–32.

Kamau, Macharia, Chasek, Pamela and O'Connor, David (2018). *Transforming Multilateral Diplomacy. The Inside Story of the Sustainable Development Goals*. London and New York: Routledge.

Kanie, Norichika et al. (2017). 'Introduction: Global Governance through Goal Setting'. In Kanie, Norichika and Biermann, Frank (eds) *Governing through Goals. Sustainable Development Goals as Governance Innovation*, pp. 1–27. Cambridge, MA: MIT Press.

Kuyper, Jonathan W. and Bäckstrand, Karin (2016). 'Accountability and Representation: Non-State Actors in UN Climate Diplomacy'. *Global Environmental Politics* 16, 2: 61–81.

Meadowcroft, James (2007). 'Who Is in Charge Here? Governance for Sustainable Development in a Complex World'. *Journal of Environmental Policy & Planning* 9, 3–4: 299–314.

Meine, Anna (2016). 'Debating Legitimacy Transnationally'. *Global Discourse* 3, 6: 330–346.

Mittelman, James (2016). 'Repositioning in Global Governance: Horizontal and Vertical Shifts Amid Pliable Neoliberalism'. *Third World Quarterly* 37, 4: 665–681.

Paterson, Matthew (2010). 'Legitimation and Accumulation in Climate Change Governance'. *New Political Economy* 15, 3: 345–368.

Pingeot, Lou (2014). *Corporate Influence in the Post-2015 Process*. Bonn: Global Policy Forum. Available at www.globalpolicy.org/images/pdfs/GPFEurope/Corporate_influence_in_the_Post-2015_process_web.pdf (last accessed 20 February 2018).

Scharpf, Fritz (1999). *Governing in Europe. Effective and Democratic?* Oxford: Oxford University Press.

Scholte, Jan Aart (2018). 'Social Structure and Global Governance Legitimacy'. In Tallberg, Jonas, Bäckstrand, Karin and Scholte, Jan A. (eds) *Legitimacy in Global Governance. Sources, Processes and Consequences*, pp. 76–98. Oxford: Oxford University Press.

Sénit, Carole-Anne, Biermann, Frank and Kalfagianni, Agni (2017). 'The Representativeness of Global Deliberation: A Critical Assessment of Civil Society Consultations for Sustainable Development'. *Global Policy* 8, 1: 62–72.

Sethi, S. Prakash and Schepers, Donald H. (2014). 'United Nations Global Compact: The Promise-Performance Gap'. *Journal of Business Ethics* 122: 193–208.

Soneryd, Linda and Uggla, Ylva (2015). 'Green Governmentality and Responsibilization: New Forms of Governance and Responses to "Consumer Responsibility"'. *Environmental Politics* 24, 6: 913–931.

Steffek, Jens (2003). 'The Legitimation of International Governance: A Discourse Approach'. *European Journal of International Relations* 9, 2: 249–275.

Suchman, Marc C. (1995). 'Managing Legitimacy: Strategic and Institutional Approaches'. *The Academy of Management Review* 20, 3: 571–610.

Symons, Jonathan (2011). 'The Legitimation of International Organisations: Examining the Identity of the Communities that Grant Legitimacy'. *Review of International Studies* 37, 5: 2557–2583.

Tallberg, Jonas and Zürn, Michael (2019). The legitimacy and legitimation of international organizations: introduction and framework. *The Review of International Organizations*. https://doi.org/10.1007/s11558-018-9330-7.

Zaum, Dominik (ed.) (2013). *Legitimating International Organisations*. Oxford: Oxford University Press.

# 2

# Environmental governance as performance

*Ingolfur Blühdorn and Michael Deflorian*

## The success and failure of *environmental governance*

In advanced industrial societies, centralised top-down forms of environmental policy-making have been supplemented, sometimes even replaced, by decentralised, flexible, and participatory network approaches. In addition to state agencies they engage a number of actors, including scientific experts, market actors, civil society organisations, and other stakeholders (Bäckstrand et al. 2010). These new modes of collaborative governance are commonly presented as more democratic than traditional approaches. They are said to take into account that governments are no longer the only political actor and source of authority, that in an increasingly complex world environmental problems have multifaceted causes and implications which can only be addressed through constructive collaboration of diverse stakeholders (Pierre and Peters 2000), and that contemporary citizens are more determined than ever to move beyond sheer *protesting* to actually *changing* and *impacting* (Schlosberg and Coles 2016). Furthermore, modern environmental governance is also said to increase the *efficiency* and *effectiveness* of environmental policy-making. It is widely assumed that these cooperative, consensual, and voluntary approaches help to reduce conflicts of interest and to engage even those actors which might otherwise oppose environmental policies or obstruct their effective implementation (Dietz and Stern 2008; Newig and Fritsch 2009).

However, the proliferation of new collaborative modes of environmental governance has also attracted strong criticism. In terms of their democratic quality, it has been argued that they are neither really inclusive nor egalitarian because in most cases governments determine who qualifies as a stakeholder and is admitted into the policy network (Davies 2011). And as governments commonly set the agenda and the rules of engagement, new forms of governance have also been described as *post-democratic* and *post-political*: they are set up to facilitate consensus and, therefore, systematically eclipse all matters of fundamental disagreement and potentially irreconcilable conflict (Swyngedouw 2005). They tightly restrict the boundaries of what can be negotiated as well as the terms of negotiation. Moreover, issues are framed only in ways which allow for *pragmatic* and *viable* solutions that may be implemented within the realm of the currently possible. Thus, rather than genuinely empowering citizens, participatory procedures often only co-opt them, in order to mobilise them as an additional resource for the legitimation and stabilisation of the established order (Boezeman et al. 2014).

As regards their ecological problem solving capacity, it has been noted that flexible actor networks, citizen empowerment, stakeholder engagement, **and** the co-production of knowledge can, indeed, help to devise policies that are acceptable to all stakeholders involved, especially on the local and regional level. Yet, they seem unable to deliver the kind of socio-ecological transformation that scientists and environmental movements are urgently asking for (IPCC 2018; Rockström 2015). This is not least, perhaps, because modern forms of decentralised and participatory governance are explicitly designed not to disrupt the established order. In fact, in the recent literature, democratic procedures and the *democratic legitimation imperative* are, increasingly, seen as a part of the problem rather than the solution to the multiple sustainability crises (Blühdorn 2019; Mitchell 2011). Unsurprisingly, therefore, there is, once again, rising "interest in non-democratic approaches to environmentalism" (Chen and Lees 2018: 2; Beeson 2010).

So, the new modes of environmental governance seem unable to fulfil expectations in terms of both their *democratic performance* (ability to deliver to specifically democratic needs) and their *systemic performance* (practical problem solving capacity) (Roller 2005). Why, then, have they, nevertheless, become so prominent? Exactly what is their appeal for modern consumer societies? What do they deliver? For anyone subscribing to the scientific consensus that environmental and climate change require much more effective counter-action than has been taken so far, these questions must be a priority concern.

In what follows, we explore the argument that these new modes of environmental governance might have become so prevalent because they correspond very closely to the particular preferences, needs, and dilemmas of contemporary consumer societies. If measured by the democratic expectations and eco-political demands of the emancipatory social movements of the past few decades, the performance of these new collaborative forms of governance may, indeed, be found wanting. But if assessed from the perspective of these contemporary needs and dilemmas, they do actually perform exceptionally well: in particular, they provide contemporary consumer societies with a practical policy mechanism that helps them to reconcile the widely perceived seriousness and urgency of socio-environmental problems with their ever more visible inability and unwillingness to deviate from their established societal order, patterns of self-realisation, and logic of development.

Thus, we are proposing that the new forms of governance have become so prominent not *although* they do not perform in terms of a structural transformation of modern societies, but precisely *because* they do not. At the same time, we argue, they are uniquely suited to the articulation and experience – i.e. the performance – of genuine commitment to achieving such change. This unorthodox interpretation of environmental governance sheds light on a dimension which neither the mainstream environmental policy literature nor the critical governance literature touches upon. In order to develop this argument, we will first discuss different notions of performance. Section "Climate summitry, green markets, and niche movements" then investigates specific examples of environmental governance. It aims to demonstrate how the careful distinction between different notions of performance facilitates a nuanced interpretation of actual practices of environmental governance. Section "The multiple disability of contemporary eco-politics" sheds more light on the particular dilemmas to which new forms of environmental governance arguably respond. The concluding section reflects on the potentials and limitations of interpreting environmental governance as performance in this particular sense.

## Varieties of performance

The suggestion that new forms of environmental governance might usefully be interpreted as a kind of performance may well conjure up swift moralising condemnations of environmental governance as *merely* performative rather than substantive, genuinely committed and effective. Such narratives are simplistic. In order to avoid them, a more differentiated understanding of performance is essential. To begin with, a distinction may be made between performance as measuring *fitness for purpose* or *ability to deliver* to expectations and performance in the theatrical sense, i.e. as *presentation, display*, or *enactment*. For a nuanced assessment of environmental governance, both these meanings are relevant. The first one focuses on the output (effectiveness) of the policy process, and also covers the efficiency of delivery, i.e. the relationship between the necessary input and delivered outputs. As new forms of environmental governance are said to deliver to both democratic objectives and in terms of practical solutions to environmental problems, their fitness for purpose can be assessed – as signalled earlier – for both their democratic and their systemic performance.

As regards performance in the theatrical sense, there again are different ways in which the term may be used. The first refers to strategies of *deception and manipulation*, normally by power elites, who are making false promises and take forms of action that are known to be inadequate for their declared purpose but serve the interests of decision-makers. This understanding of performance implies a moral judgement based on the assumption that decision-makers consciously act against the public interest. This kind of deceptive action is often described as window-dressing and fake, and criticised as *symbolic politics* (Edelman 1971), as opposed to genuine, authentic, and effective politics. The second understanding of performance in a theatrical sense does not have moral overtones but recognises that in environmental politics, as elsewhere, political goals, visions, and ideals (e.g. to protect a healthy environment or the integrity of eco-systems) are often abstract and intangible and cannot easily, and immediately, be translated into practical policies. Performative action then may help to signal commitment, to make an abstract goal more imaginable, and to generate and maintain political momentum for its longer-term pursuit. This, too, is a kind of symbolic politics, but rather than being manipulative, deceptive, or immoral, it is *prefigurative and anticipatory* (Blatter 2009; Yates 2014), and often indispensable in order to forge a political consensus and to encourage cooperation between diverse actors.

Third, performance in the theatrical sense may also respond to conditions where commitments are serious and authentic but cannot be implemented, be it for structural reasons or because individual actors and societies at large have multiple commitments that are all equally genuine but cannot be fulfilled at the same time. In such cases, performative action may help to cope with complexities, paradoxes, and dilemmas that *cannot be resolved* but still *must be addressed*. As in the case of symbolic politics that helps to visualise an abstract goal, this form of performative action is not intended to avoid any more effective and moral – yet inconvenient – alternative. In fact, in either scenario, such alternatives are simply not available. Also, in either case, performative action is not based on the unequal social distribution of power. Nevertheless, this third kind of theatrical performance does entail an element of deception or illusioning. But in contrast to the malicious deception of citizens by self-interested elites, such practices might be described as voluntary *self-deception or self-illusioning*. In order to distinguish this kind of performative action from *symbolic* politics (in both its deceptive and the anticipatory variety), it has been conceptualised as *simulative* politics (Blühdorn 2007a). In conditions of high societal complexity, in particular, which often leave individuals disoriented and expose them to paradoxical demands, practices of simulative politics perform a social reality, or discursively construct societal self-descriptions, which help individuals – and society at large – to make sense of their paradoxical experience and manage irresolvable dilemmas (Blühdorn 2007b) (Figure 2.1).

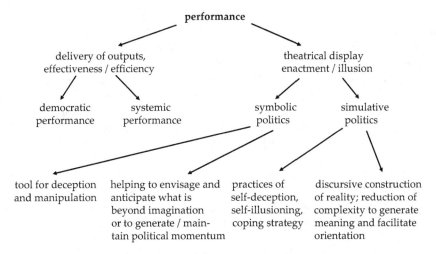

*Figure 2.1* Meanings of *performance*

This careful distinction between different understandings of performance provides the conceptual tools required to move beyond both the simplistic celebration of new forms of environmental governance as a promising strategy for increasing the legitimacy and effectiveness of environmental policy and their equally simplistic rejection as post-democratic and eco-politically ineffective. Undeniably, there are examples of new forms of environmental governance that actually *can* deliver effective solutions for particular problems. In other instances, there are good reasons to criticise prevalent arrangements of environmental governance as eco-political window-dressing and as obstructing more effective measures for the sake of special interests. But for explaining the particular appeal of these arrangements more sophisticated tools are required.

Arguably, the third reading of performance (*simulative politics*) is the most constructive conceptual tool and analytical perspective; at least, it has the greatest potential to shed an innovative light on new forms of environmental governance and contemporary eco-politics more generally. It suggests that these new forms of environmental governance are important practices facilitating the *politics of unsustainability* (Blühdorn 2011, 2013): they allow maximum space for the preservation of what significant majorities in modern consumer societies commonly refer to as their non-negotiable values, their freedom, and their way of life; and, at the same time, they provide optimal opportunities for the articulation and experience of deeply felt socio-ecological values and commitments. Thus, new forms of environmental governance may be said to deliver in that they address the dilemma that liberal consumer societies are lacking the will and ability to resolve their multiple sustainability crisis but fully acknowledge that this crisis is more real and urgent than ever before.

## Climate summitry, green markets, and niche movements

Important examples for contemporary environmental governance include international climate summits such as the *United Nations Climate Change Conferences* (a), the involvement of market actors into environmental policy-making (b), and the official recognition of civil society initiatives and niche movements as pioneers, drivers, and laboratories of societal change towards sustainability (c). They all demonstrate how modern environmental governance has moved, vertically and horizontally, beyond conventional state-centred government. Refer-ring to different levels of policy-making, these examples help to illustrate the significance and analytical power of investigating contemporary forms of environmental governance through the lens of performance.

(a) International climate politics under the roof of the 1992 United Nations Framework Convention on Climate Change is one of the most prominent examples of international en-vironmental governance. In terms of performance as the *delivery of policy outputs*, the annual *Conferences of Parties* (COP) are regarded as unique opportunities for far-reaching politi-cal agreements because these gatherings are prepared and attended by international leaders, scientific experts, climate activists, and a wide range of non-governmental organisations (NGOs), all protesting their commitment to fast, coordinated, and effective climate action. At the same time, however, such summits are also commonly portrayed as performance in the *theatrical sense*. They may be interpreted as *symbolic politics* because the 1.5°C or 2°C limit itself has acquired the status of a political symbol that helps to visualise something that is highly abstract: a relatively safe level of change in global average temperatures compared to pre-industrial levels. In addition, the regular climate summits help to maintain the interna-tional diplomatic infrastructure and to uphold the political momentum, even when factual progress is difficult to achieve. In these respects, international climate summitry is symbolic politics in the *prefigurative* sense (Bäckstrand and Lövbrand 2016). Yet, it may also be inter-preted as symbolic politics in the *deceptive and manipulative* sense (Death 2011). In fact, UN climate summits have been criticised as professionally staged media events and costly and in-effective public relations exercises, providing international lobby groups with an exceptional opportunity to exert their influence and world leaders with a stage to make big promises which they neither can nor perhaps even want to fulfil (Carter et al. 2011).

The 2015 Paris Agreement, for example, that fails to specify any legally binding pol-icy measures but was still presented to the world as a major climate-political achievement (Droege 2016), may be seen to corroborate the view that international climate summitry is, more than anything, a strategic tool, which economic and political elites use to mollify the concerns of the international public while closely protecting their respective interests (Spash 2016). Yet, this interpretation fails to take into account that the negotiating parties may well be very genuinely committed to reaching an effective agreement, but find themselves caught up in the dilemma that it is extremely difficult to translate the abstract 2°C target into nationally acceptable and internationally interlocking action plans (Mahapatra and Ratha 2016). Furthermore, all national governments have to accommodate a range of different state goals, which are all categorical, yet often mutually incommensurable. The interpretation of international climate summitry in terms of *simulative politics* takes account of these dilemmas. From this point of view, these high profile events may be understood as an opportunity for policy-makers to articulate their genuine and serious commitment to the goal of environ-mental sustainability, and provide evidence that they are pursuing this goal at the highest possible level and in cooperation with the widest possible range of stakeholders – while at the same time avoiding, or at least postponing, any detrimental implications for their other, equally serious commitments. The same applies to the wide range of other actors attending such summits. And it even extends to those non-participants who join into celebrating out-comes such as the Paris Treaty as a major eco-political achievement – relieved that a solution has been found that does not directly affect their personal value preferences and lifestyle choices. Seen through the lens of simulative politics, their intentions do not appear immoral or malicious. Instead, these actors and audiences are experiencing inescapable dilemmas and have to cope with unmanageable paradoxes.

(b) Close cooperation with market actors, ranging from large international corporations right down to individual consumers, is another pillar of modern environmental governance. Traditionally, the interests of these actors were seen as diametrically opposed to those of environmentalists. In recent years, however, major corporations have been working closely with state agencies, expert bodies, and environmental NGOs to reduce the environmental impact of production, distribution, and consumption processes. At the other end of the market, these new corporate efforts are mirrored by a shift in consumer choices, in particular milieus, towards environmentally and ethically oriented products (Connolly and Prothero 2008; see also Krogman, this volume; Micheletti 2013). Looking from the perspective of problem solving, self-monitoring, Corporate Social Responsibility (CSR) and Corporate Environmental Responsibility (CER) schemes and the related reporting requirements may help companies to reorient their unsustainable practices and enable other stakeholders to assess the environmental credentials of companies (Gunningham 2009). They potentially empower consumers to exert some market pressure. And NGOs can provide evidence that, in addition to mere protesting, they also engage in devising constructive solutions to eco-political challenges (Camilleri 2017; Fontana 2018). Hence, the engagement of market actors in networks of collaborative governance may be seen to have the potential to reconfigure the established logic of consumer capitalism.

The lens of symbolic politics, in contrast, suggests that the engagement of market actors performs mainly in the sense of facilitating business strategies of deception. On the one hand, their involvement may, of course, be symbolic in the prefigurative sense: it may help to envisage the goal of a comprehensive transformation of society and subtly create the conditions for this project (WBGU 2011). On the other hand, however, CER and CSR schemes are often only a tool that companies use to greenwash their public image (Delmas and Burbano 2011). Since they are a marketing strategy, they are designed to further increase sales and consumption and, thus, they sustain rather than overcome the destructive principles of capital accumulation, social inequality, commodification, and so forth (Banerjee 2008).

Green forms of consumption, in turn, are symbolic in that they tend to focus on very particular products. They may anticipate a much more thorough transformation of established consumption practices (Emerich 2011), but they may also just be a means of articulating a particular self-understanding, social status, or lifestyle. In particular, green consumerism is rarely about consuming less or even abandoning the shopping and consumer culture (see also Fuchs, this volume). Instead, neoliberal governments are keen to open up new markets and stimulate *green growth* (Fuchs and Lorek 2005), while trying to offload their eco-political obligations to *responsible consumers* (Hobson 2013; Soneryd and Uggla 2015). And supposedly responsible, green consumer choices may well be counter-productive, for example, if they merely ennoble ecologically damaging practices (e.g. voluntary carbon payment for air tickets) or if environmentalism incentivises additional purchases (e.g. e-bikes) or premature product replacement (e.g. motor cars) (Hertwich 2005; Murray 2013). Hence, just as CER and CSR never unhinge the logic of capitalism, green consumerism rarely suspends the logic of mass consumption and resource overuse (Seyfang 2005).

The lens of simulative politics, however, adds another layer to the understanding of both corporate environmentalism and green consumerism. It suggests that these two dimensions of modern environmental governance have become so important because they enable businesses and consumers alike to manage the diverse and often conflicting pressures they find themselves confronted with (see also Maniates, this volume). CER and CSR schemes help businesses to respond simultaneously to civil society critiques, innovation pressures, government agendas, and shareholder interests, by offering multi-layered narratives which each stakeholder group may read selectively from their particular perspective. Green consumerism, in turn, helps modern citizens to hold on to their established patterns of self-realisation, further pursue their lifestyle preferences and articulate their individual identities (Longo et al. 2017; Soron 2010). More specifically, it allows them to act and experience themselves not only as individualised consumers, but also as socially and ecologically oriented and responsible citizens (Micheletti 2013). Thus, corporate environmentalism and green consumerism both respond to the problems of highly differentiated selves and societies.

(c) In addition to the engagement of market actors, civil society initiatives and local niche movements, too, have acquired an important role in modern environmental governance. They have been identified as *pioneers of change*, who might make a substantial contribution to the great transformation to sustainability (WBGU 2011). Community-supported agriculture, renewable energy cooperatives, alternative housing collectives, repair cafés, or food- and tool-sharing platforms flourish in a realm beyond both the market and the state. They can serve as laboratories for experimental social practices and societal change. Thus, they are assumed to substantially increase the democratic performance and also boost the problem solving capacity of local communities and society at large (Schlosberg and Coles 2016; Seyfang and Longhurst 2015). And beyond their effects in the immediate present, they also perform in the sense that they symbolically anticipate a radically different future society. Indeed, this prefigurative element is central to the transformative potential of civil society-driven initiatives. As the hegemony of market-liberal thinking renders it ever more difficult to envisage an alternative to the consumer capitalist order of unsustainability, these niche movements help to imagine and subtly implement alternative practices and socio-ecological relations (Wright 2013).

There are observers, however, who have noted that practitioners within niche movements often remain rather ambivalent about really abandoning the established practices and lifestyles of unsustainability. In fact, activists often refrain from adopting a critical or antagonist stance towards the existing political-economic regime and conceive of their alternative practices not as a radical political project but as a playful effort *to do something good* (Schlosberg and Coles 2016). This view is corroborated by studies that compare the environmental impact of different social milieus. They have revealed that members of the post-materialist, critical-creative milieus – which provide an important reservoir for alternative niche movements (Grossmann and Creamer 2016) – often remain strongly attached to mobility practices, communication technologies, or living patterns that prevail in the societal mainstream (Moser and Kleinhückelkotten 2017). Thus, engagement in niche movements and experiments with alternative practices do not necessarily signal much commitment to a profound transformation of the prevailing order of unsustainability, but may simply be an ingredient of a particular personal image or lifestyle.

From a traditional critical point of view, such ambivalence may then be read as inconsistency, false posturing, fake and performance, signalling dishonesty or a lack of genuine eco-social commitment (symbolic politics). Yet, if viewed from the perspective of simulative politics, this simultaneity of seemingly incompatible practices and value-orientations may be interpreted as a performative strategy to deal with the complexities and cope with the irresolvable contradictions of advanced modern societies. From this perspective, the engagement in niche movements might be interpreted as the attempt of individuals to recuperate an autonomous space for expressing their environmental and democratic values. In these arenas they can experience some degree of sovereignty with regard to everyday needs such as food, energy, or clothing, while fully acknowledging that in other respects their everyday conduct may be much less self-determined – and sustainable. Hence, the engagement in a food cooperative or alternative housing project may well articulate a serious pledge to a sustainable way of living (Certomà 2016; Schlosberg and Coles 2016), but rather than anticipating a full and consistent transformation of the personal – and then societal – way of life, it may be an experiential strategy for the management of irresolvable dilemmas and contradictions.

## The multiple disability of contemporary eco-politics

Further discussion of these dilemmas and contradictions helps to clarify in what respects new forms of governance may be said to correspond very closely to the particular conditions, preferences, and needs of contemporary consumer societies. The failure of contemporary consumer societies to achieve a socio-ecological transformation is commonly explained with reference to the institutionalised interests and the overwhelming power of economic elites who effectively block any substantial deviation from the established order and pattern of development (e.g. Swyngedouw 2005, 2009; see also Brulle and Aronczyk, this volume). While capturing significant parts of the truth, this line of argument ignores that contemporary consumer democracies are caught up in the trilemma that (a) despite the wealth of scientific information, the normative foundations for transformative action are becoming ever more uncertain and unreliable; (b) despite unprecedented levels of environmental awareness and commitment, there is neither a political actor who could effectively drive and coordinate transformative action, nor a promising political strategy; and (c) in contemporary consumer societies prevailing notions of freedom, identity, self-realisation, and a *good life* are ever more incompatible with ideals of environmental integrity, social justice, and democracy.

(a) As processes of modernisation have rendered contemporary societies ever more complex, giving rise to an ever increasing number of perspectives onto reality and to competing views of what ought to be sustained, for whom, for which reasons and so forth, the normative foundations of environmental politics became increasingly uncertain (Pellizzoni, this volume). Also, the growing wealth of scientific knowledge and information has triggered disorientation at least as much as they were politically mobilising and enabling (Longo et al. 2017). In addition, the acceleration of societal change, the unpredictability of societal development, and the complexity of international relations have a paralysing effect. And when neoliberals and right-wing populists then started to amplify alleged scientific disagreement, for example, about anthropogenic climate change, discredited the public media as *fake news*, and replaced public deliberation and rational argument by fabricated fears and *alternative facts*, it became even more impossible to achieve and maintain agreement about eco-political problems, priorities, objectives, and strategies.

(b) As regards the driver and strategy for the socio-ecological transformation of contemporary societies, environmental movements had always been sceptical of the state, which they thought was pre-occupied with the reproduction of power and too closely entangled with business interests. Hence, environmentalists preferred to rely on civil society's capacity for self-organisation, and believed that the thorough democratisation of every dimension of societal affairs would be a promising strategy to achieve social justice and equality, secure the integrity of the natural environment, and guarantee a *good life* for all. Yet, in increasingly complex societies, democratic procedures proved, in many respects, unconducive to the attainment of ecological goals, for example, because they are long winded, focused on short-term electoral returns, based on the principle of compromise, and rather limited in terms of their geographical reach (e.g. Blühdorn 2014). In line with a notable decline of public confidence in democracy more generally, environmentalists, too, began to suspect that there might, in fact, be an underlying "complicity" between democracy and unsustainability

(Eckersley 2017): after all, the demands of ever larger parts of (international) society for a *good life* and their fair share of societal wealth are, and have always been, a powerful driver of ever more economic growth and socio-ecological exploitation (Hausknost 2019; Mitchell 2011). And with right-wing populist movements powerfully demanding less stringent environmental regulations and laws (Lockwood 2018), civil society and democratic approaches may be becoming even less suitable for a profound socio-ecological transformation.

(c) As regards their prevailing notions of freedom, self-realisation, and a fulfilled life, modern consumer societies have acquired value preferences, aspirations, and lifestyles that are categorically incompatible with the principles of sustainability. In fact, with the pluralisation and flexibilisation of traditional notions of subjectivity and identity, with consumerism having emerged as the primary mode of self-realisation and self-expression, with the massive geographical expansion of individual lifeworlds (mobility, product sourcing, waste disposal), and the steady acceleration of innovation and consumption cycles, unsustainability has in many ways become a constitutive principle of modern identity, lifestyles, and society (Blühdorn 2019). And despite all narratives of sustainability and a socio-ecological transformation, contemporary individuals, and societies at large, are incrementally abandoning established social and ecological imperatives (e.g. social equality, solidarity, and redistribution; protection of natural habitats, biodiversity, and the planet's atmosphere), which from the perspective of contemporary value preferences and necessities seem unacceptably restrictive (Blühdorn 2014, 2017).

So advanced modern societies are confronted with the dilemma that radical transformative action may seem more urgently required than ever, but the normative foundations for such an agenda are more uncertain than ever, there is no primary actor nor a promising political strategy for a significant transformation, and the prevailing and sacrosanct notions of freedom, self-determination, and self-realisation are firmly based on the principle of unsustainability. It is exactly these three dilemmas, to which contemporary forms of environmental governance respond when they perform in the sense of *simulative politics*: with the absence of reliable eco-political norms they deal by delegating the issue to *self-regulating* entities, *responsible* consumers, and their voluntary self-commitment, thereby alleviating the problems governments and policy-makers have with defining, politically legitimating, and executing particular laws and standards. The lack of a promising actor and strategy for transformative change is addressed by setting up new policy networks, which are charged with the task to negotiate and implement an inclusive social contract for sustainability. And concerning the value preferences of contemporary individuals and consumer society at large, these new forms of governance deliver in that they allow for the articulation and experience of eco-social commitments and, at the same time, also for the desire to maintain the established principles of unsustainability. In none of these respects, modern arrangements of environmental governance actually *resolve* the underlying problems, but in all of them, they *address* them and respond to the particular needs of modern consumer societies. They are a performative response to the multi-dimensional disability of contemporary eco-politics and to the unwillingness of liberal consumer societies to depart from their established values and order.

## Potentials and pitfalls of interpreting environmental governance as performance

Thus, the lens of performance can make a major contribution to the analysis and understanding of modern forms of environmental governance and contemporary eco-politics more generally. It reveals aspects which neither the celebration of these new decentralised practices as a promising strategy for the great transformation to sustainability can capture, nor the critique of environmental governance as being *merely performative* rather than genuinely committed and effective. The latter aims to unmask the deceptive strategies of elites and thus create new spaces for the *authentic* kind of eco-politics which scientists, activists, and eco-movements demand – and supposedly want. Yet, the analysis has revealed that in advanced modern societies things are more complex than this narrative suggests. It shows that the governance approach to environmental policy-making has become so prominent, not least because it corresponds very closely to the particular concerns, preferences, and dilemmas of contemporary consumer democracies. For anyone wondering why modern forms of environmental governance have become so prominent, and looking for a more nuanced explanation of modern consumer societies' *politics of unsustainability*, this analysis through the lens of performance will provide valuable insights.

A predictable objection to this interpretation of environmental governance as performance concerns its political use-value, be it for official policy-makers or social movement activists. The primary emphasis of this conceptual model is, indeed, on *understanding* rather than *changing* the eco-political conduct of liberal consumer societies. It does not directly give rise to suggestions for alternative policy approaches which might perform *better* – not least because it remains itself caught up in the normative dilemma it diagnoses. Accordingly, it confines itself to investigating how modern consumer societies organise their environmental politics and why they might be doing so. Of course, policy-makers expect, more than anything, practical policy recommendations. And activists may complain that the analysis undertaken here not only undermines the efforts of those who seek to unmask the *merely performative* strategies of the elites, but actually subjects these well-intentioned critics to the meta-critique that their discourse, too, may have to be read as the performance of particular values and commitments.

This concern needs to be taken seriously. Yet, if we hold on to the belief that a radical socio-ecological transformation is urgently required, it is essential to explore why the new forms of governance, which increasingly substitute for traditional-style interventionist politics, do not perform in that regard – and, vice versa, in which respects they *do* actually perform very well. Moreover, it is essential to demonstrate that the limitations of modern forms of environmental governance cannot be overcome by involving yet another stakeholder or adding yet another layer of deliberation. Although the analysis in terms of performance may, at first sight, appear politically unconstructive or even disabling, the first essential step towards a successful socio-ecological transformation of modern liberal consumer democracies necessarily has to be the critical investigation of the prevailing practices that, for the time being, contribute to sustaining the established order of unsustainability.

## References

Bäckstrand, Karin, Jamil Khan, Annica Kronsell, and Eva Lövbrand. 2010. "The Promise of New Modes of Environmental Governance." In *Environmental Politics and Deliberative Democracy. Examining the Promise of New Modes of Governance*, edited by Karin Bäckstrand, Jamil Khan, Annica Kronsell, and Eva Lövbrand, 3–7. Cheltenham/Northampton: Edward Elgar Publishing.

Bäckstrand, Karin, and Eva Lövbrand. 2016. "The Road to Paris: Contending Climate Governance Discourses in the Post-Copenhagen Era." *Journal of Environmental Policy & Planning* 10: 1–19.

Banerjee, Subhabrata B. 2008. "Corporate Social Responsibility: The Good, the Bad and the Ugly." *Critical Sociology* 34 (1): 51–79.

Beeson, Mark. 2010. "The Coming of Environmental Authoritarianism." *Environmental Politics* 19 (2): 276–294.

Blatter, Joachim. 2009. "Performing Symbolic Politics and International Environmental Regulation: Tracing and Theorizing a Causal Mechanism beyond Regime Theory." *Global Environmental Politics* 9 (4): 81–110.

Blühdorn, Ingolfur. 2007a. "Sustaining the Unsustainable: Symbolic Politics and the Politics of Simulation", *Environmental Politics* 16 (2): 251–275.

Blühdorn, Ingolfur. 2007b. "Self-Description, Self-Deception, Simulation: A Systems-theoretical Perspective on Contemporary Discourses of Radical Change." *Social Movement Studies* 6 (1): 1–20.

Blühdorn, Ingolfur. 2011. "The Politics of Unsustainability: COP15, Post-Ecologism and the Ecological Paradox." *Organization & Environment* 24 (1): 34–53.

Blühdorn, Ingolfur. 2013. "The Governance of Unsustainability: Ecology and Democracy after the Post-Democratic Turn." *Environmental Politics* 22 (1): 16–36.

Blühdorn, Ingolfur. 2014. "Post-Ecologist Governmentality: Post-Democracy, Post-Politics and the Politics of Unsustainability." In *The Post-Political and Its Discontents: Spaces of Depoliticisation, Spectres of Radical Politics*, edited by Japhy Wilson and Erik Swyngedouw, 146–166. Edinburgh: Edinburgh University Press.

Blühdorn, Ingolfur. 2017. "Post-Capitalism, Post-Growth, Post-Consumerism? Eco-political Hopes beyond Sustainability." *Global Discourse* 7 (1): 42–61.

Blühdorn, Ingolfur. 2019. "The Legitimation Crisis of Democracy. Emancipatory Politics, the Environmental State and the Glass Ceiling to Socio-Ecological Transformation." *Environmental Politics* (forthcoming).

Boezeman, Daan, Martinus Vink, Pieter Leroy, and Willem Halffman. 2014. "Participation under a Spell of Instrumentalization? Reflections on Action Research in an Entrenched Climate Adaptation Policy Process. *Critical Policy Studies* 8 (4): 407–426.

Camilleri, Mark Anthony. 2017. *Corporate Sustainability, Social Responsibility and Environmental Management*. Cham: Springer International Publishing.

Carter, Chris, Stewart Clegg, and Nils Wåhlin. 2011. "When Science Meets Strategic Realpolitik: The Case of the Copenhagen UN Climate Change Summit." *Critical Perspectives on Accounting* 22 (7): 682–697.

Certomà, Chiara. 2016. *Postenvironmentalism. A Material Semiotic Perspective on Living Spaces*. New York: Palgrave Macmillan.

Chen, Geoffrey C., and Charles Lees. 2018. "The New, Green, Urbanization in China: Between Authoritarian Environmentalism and Decentralization." *Chinese Political Science Review* 3 (2): 212–231.

Connolly, John, and Andrea Prothero. 2008. "Green Consumption. Life-Politics, Risk and Contradictions." *Journal of Consumer Culture* 8 (1): 117–145.

Davies, Jonathan. 2011. *Challenging Governance Theory. From Network to Hegemony*. Bristol: Bristol Policy Press.

Death, Carl. 2011. "Summit Theatre. Exemplary Governmentality and Environmental Diplomacy in Johannesburg and Copenhagen." *Environmental Politics* 20 (1): 1–19.

Delmas, Magali A., and Vanessa Cuerel Burbano. 2011. "The Drivers of Greenwashing". *California Management Review* 54 (1): 34–87.

Droege, Susanne. 2016. "The Paris Agreement 2015. Turning Point for the International Climate Regime." Working Paper. Stiftung Wissenschaft und Politik. www.ssoar.info/ssoar/handle/document/46462.

Eckersley, Robyn. 2017. "Geopolitan Democracy in the Anthropocene." *Political Studies* 65 (4): 983–999.

Edelman, Murray. 1971. *Politics as Symbolic Action. Mass Arousal and Quiescence*. Chicago: Markham.

Emerich, Monica. 2011. *The Gospel of Sustainability. Media, Market, and LOHAS*. Urbana: University of Illinois Press.

Fontana, Enrico. 2018. "Corporate Social Responsibility as Stakeholder Engagement: Firm-NGO Collaboration in Sweden." *Corporate Social Responsibility and Environmental Management* 25 (4): 327–338.

Fuchs, Doris, and Sylvia Lorek. 2005. "Sustainable Consumption Governance: A History of Promises and Failures." *Journal of Consumer Policy* 28 (3): 261–288.

Grossmann, Mena, and Emily Creamer. 2016. "Assessing Diversity and Inclusivity within the Transition Movement. An Urban Case Study." *Environmental Politics* 26 (1): 1–22.

Gunningham, Neil (ed.). 2009. *Corporate Environmental Responsibility*. London: Ashgate.

Hausknost, Daniel. 2019. "The Environmental State and the Glass Ceiling to Transformation." *Environmental Politics* (forthcoming).

Hertwich, Edward. 2005. "Consumption and the Rebound Effect: An Industrial Ecology Perspective." *Journal of Industrial Ecology* 9 (1–2): 8 –98.

Hobson, Kersty. 2013. "On the Making of the Environmental Citizen." *Environmental Politics* 22 (1): 56–72.

IPCC (Intergovernmental Panel on Climate Change). 2018. Summary for Policymakers of IPCC Special Report on Global Warming of 1.5°C Approved by Governments. www.ipcc.ch/2018/10/08/summary-for-policymakers-of-ipcc-special-report-on-global-warming-of-1-5c-approved-by-governments/.

Lockwood, Matthew. 2018. "Right-wing Populism and the Climate Change Agenda. Exploring the Linkages." *Environmental Politics* 27 (4): 712–732.

Longo, Cristina, Avi Shankar, and Peter Nuttal. 2017. "'It's Not Easy Living a Sustainable Lifestyle': How Greater Knowledge Leads to Dilemmas, Tensions and Paralysis." *Journal of Business Ethics* 154 (3): 759–779.

Mahapatra, S. Kumar, and Keshab C. Ratha. 2016. "The 21st Conference of the Parties Climate Summit in Paris: A Slippery Slope." *Journal of International Development* 28 (16): 991–996.

Micheletti, Michele. 2013. *Political Virtue and Shopping: Individuals, Consumerism, and Collective Action.* New York: Palgrave Macmillan.

Mitchell, Timothy. 2011. *Carbon Democracy: Political Power in the Age of Oil.* New York: Verso.

Moser, Stephanie, and Silke Kleinhückelkotten. 2017. "Good Intents, but Low Impacts: Diverging Importance of Motivational and Socioeconomic Determinants Explaining Pro-Environmental Behaviours, Energy Use, and Carbon Footprint." *Environment and Behaviour* 50 (6): 626–656.

Murray, Cameron. 2013. "What If Consumers Decided to All Go Green? Environmental Rebound Effects from Consumption Decisions." *Energy Policy* 54 (c): 240–256.

Newig, Jens, and Oliver Fritsch. 2009. "Environmental Governance: Participatory, Multi-Level – And Effective?" *Environmental Policy and Governance* 19 (3): 19–214.

Pierre, Jon, and B. Guy Peters. 2000. *Governance, Politics and the State.* Basingstoke: Macmillan.

Rockström, Johan. 2015. *Bounding the Planetary Future. Why We Need a Great Transition.* www.greattransition.org/publication/bounding-the-planetary-future-why-we-need-a-great-transition.

Roller, Edeltraud. 2005. *The Performance of Democracies. Political Institutions and Public Policies.* Oxford: Oxford University Press.

Schlosberg, David, and Romand Coles. 2016. "The New Environmentalism of Everyday Life. Sustainability, Material Flows and Movements." *Contemporary Political Theory* 15 (2): 160–181.

Seyfang, Gill. 2005. "Shopping for Sustainability: Can Sustainable Consumption Promote Ecological Citizenship?" *Environmental Politics* 14 (2): 290–306.

Seyfang, Gill, and Noel Longhurst. 2015. "What Influences the Diffusion of Grassroots Innovations for Sustainability? Investigating Community Currency Niches." *Technology Analysis & Strategic Management* 28 (1): 1–23.

Soneryd, Linda, and Ylva Uggla. 2015. "Green Governmentality and Responsibilization: New Forms of Governance and Responses to 'Consumer Responsibility'". *Environmental Politics* 24 (6): 913–931.

Soron, Dennis. 2010. "Sustainability, Self-Identity and the Sociology of Consumption." *Sustainable Development* 18 (3): 172–181.

Spash, Clive L. 2016. "This Changes Nothing: The Paris Agreement to Ignore Reality." *Globalizations* 13 (6): 928–933.

Swyngedouw, Erik. 2005. "Governance Innovation and the Citizen: The Janus Face of Governance-beyond-the-State." *Urban Studies* 42 (11): 1991–2006.

Swyngedouw, Erik. 2009. "The Antinomies of the Post-Political City: In Search of a Democratic Politics of Environmental Production." *International Journal of Urban and Regional Research* 33 (3): 601–620.

Thomas, Dietz and Paul C. Stern (eds). 2008. *Public Participation in Environmental Assessment and Decision-Making.* Panel on Public Participation in Environmental Assessment and Decision Making. Washington: National Research Council.

WBGU (German Advisory Council on Global Change). 2011. *World in Transition. A Social Contract for Sustainability.* Flagship Report. www.wbgu.de/fileadmin/user_upload/wbgu.de/templates/dateien/veroeffentlichungen/hauptgutachten/jg2011/wbgu_jg2011_en.pdf.

Wright, Erik O. 2010. *Envisioning Real Utopias.* London, New York: Verso.

Yates, Luke. 2014. "Rethinking Prefiguration: Alternatives, Micropolitics and Goals in Social Movements." *Social Movement Studies* 14 (1): 1–21.

# 3

# Engaging the everyday
## Sustainability as resonance

*John M. Meyer*

## Introduction

The question of how people might foster a more sustainable future requires us to better understand why efforts to promote such possibilities have so far been anaemic. This demands that we grapple with strategic considerations about how social change is best achieved. In an effort to do so here, I distinguish between familiar diagnoses that point to the everyday concerns of the masses of people in affluent, post-industrial societies of the Global North as a key obstacle to needed social change and others that suggest that these concerns might instead – or also – present opportunities for such change.

I don't intend to offer a definitive rejection of the former sort of diagnosis in favour of the latter in this chapter. The pursuit of sustainability is such a "wicked" problem that no simple or single answer can possibly solve it. Yet I do suggest that there are characteristics of familiar diagnoses that make it harder rather than easier to pursue a more sustainable society. I will then argue that there are disparate threads of both academic discourse and movement activism in recent years that orient us towards a renewed engagement with everyday life. Drawn together, these can help us delineate an alternative diagnosis that cultivates a more expansive imaginary and so offers greater possibilities to promote social change democratically.

## Two ways of thinking about the challenge of sustainability

In broad brush-strokes, one influential way of thinking about the challenge of sustainability goes like this: *People are in denial. Many are too selfish, or greedy, or apathetic to make the changes that addressing the climate crisis or pursuing sustainability would require of them or even to accept the idea that they are needed. Others seem to be sympathetic, but are simply too preoccupied with daily life to even know what is required, nonetheless act on it. We must get people to "think globally."* A characterisation such as this one is likely to be familiar to readers of this chapter; it can be found in numerous works by both academics and activists. Even those who don't defend such claims explicitly often echo this lament in more informal settings among colleagues.

It's important to acknowledge, here, that "denial" does not mean the same thing throughout this discourse. Certainly, there are those who deny the scientifically established consensus of the occurrence of anthropogenic climate change. These "denialists" – backed up by a network of right-wing funders, corporate interests, think-tanks, and media organisations that have been characterised as the "climate change denial machine" (Dunlap and McCright 2011; see also Brulle and Aronczyk, this volume) – have an outsized influence on policy-making in the United States and a handful of other countries. Yet even in the United States, public opinion surveys consistently show that such views are held by only a minority of citizens (Leiserowitz et al. 2017). Yet when climate activist and author Bill McKibben argues that "we're losing the fight … because, most of all, we remain in denial about the peril that human civilization is in," this sense of being "in denial" need not – and often does not – entail a denial of the science itself (McKibben 2012). Instead, it is tied to a disconnect between the abstract cognition of climate change and the experiential realities of personal and political life, as well as the extent of social change required to address climate change meaningfully (for nuanced accounts of this sense of denial, see Norgaard 2011; Anfinson 2018; cf. Maniates' discussion of magical thinking in this volume).

What is to be done? If the people are diagnosed as the problem, then this way of thinking leads many to conclude that overcoming denial in its various forms will require strategies that will shake people out of their apathy: we must shout louder to be heard over the din, we must emphasise the consensus of the world's best scientists, we must make them see that catastrophe is imminent.

It leads others to dismiss the prospect of democratic social change that will advance sustainability, concluding that popular attitudes are most likely to sustain unsustainability (Blühdorn and Deflorian, this volume). This has led some who accept this sort of conclusion to go on to reject democratic strategies and argue that only authoritarian powers might be able to advance sustainability (e.g., Shearman and Smith 2007; Beeson 2016).

It is clear that much of the blame, as diagnosed by this way of thinking, is placed on the people. But who is the "we" who must address this problem? This is less commonly articulated but no less important. "We" appear to be a minority – an enlightened elite, by virtue of having transcended the ignorance, apathy, or greed of the masses. "We" are informed by science and expertise, motivated by morality and the common good. To be sure, "we" are not perfect (after all, "we" fly too much) but at least "we" care.

In this way of thinking, concerns about climate change and the pursuit of sustainability require both individuals and societies to overcome the material self-interest that foster denial in all its forms. This self-interest is reflected in a focus on jobs, economic and physical security, and time-consuming everyday matters connected with children, family, friends, home, and other aspects of our lives that preoccupy most of us. Countless public opinion polls in the United States and other societies identify issues related to these quotidian matters as top political priorities, while – despite some recent upward movement – action on climate change and environmental concern linger much lower on the list (Guber 2003; Pew Research Center for the People and the Press 2018). Only if "we" can move these to the top of the priorities list, the thinking goes, will meaningful action follow. In earlier writing, I've called this characterisation a "resonance dilemma:" the challenges of climate change and sustainability don't seem to resonate deeply with most people's everyday lives, yet public pressure for change will only emerge if or when it does (Meyer 2015). Within this frame of reference, I argue, the resonance dilemma is unlikely to be resolved, because few are truly willing or able to overcome their everyday concerns.

One of the most influential social scientific efforts to understand public opinion in the past half century has been political scientist Ronald Inglehart's analysis of the World Values Survey. This has led to the development of his "postmaterialist values thesis," which contrasts individuals preoccupied with material, economic concerns and those who are said to have transcended this preoccupation to be motivated by "postmaterialist" values focused on quality of life, self-expression, and self-fulfilment, which are higher up on Maslow's hierarchy of needs (Inglehart 1990). Although environmental concern is only one instance of this thesis, it captures – in often precise and revealing language – a way of understanding the basis for environmental concern as values-oriented and non-materialist (Meyer 2015, chapter 3). This thesis echoes in a wide variety of contemporary environmental claims, like the ones mentioned earlier.

The danger in applying Inglehart's thesis to environmental and – even more so – to sustainability concerns is that it presumes that these are subjective values that tend to be activated only once more fundamental and material needs are met. Certainly there is an influential image of an environmental activist in many affluent post-industrial societies – especially those who focus on protecting and recreating in majestic wild places – which is consistent with this presumption. Yet – despite its familiarity – the generalised notion that a concern for climate, environment, or sustainability is a product of relative comfort, education, or affluence does not bear scrutiny. There is more than one sort of motivation for such concern. In many cases, it is rooted in a concern for livelihood and material vulnerability or the survival of a way of life (Dunlap and York 2008; Kim and Wolinsky-Nahmias 2014). Even Inglehart has acknowledged this (Inglehart 1995). More to the point, by characterising concerns as at odds with, or requiring us to transcend, everyday ones we reinforce a flawed dichotomy between post-material values and material needs.

In sum, if one perceives sustainability as the sort of concern that is likely to be embraced by those whose more quotidian and material needs have already been met, then we will expect it to be a preoccupation limited to the relatively affluent middle- and upper-classes and those who have access to education and resources to live a comfortable life. This perception is then reinforced by (some) people who fit this image proclaiming that we are in denial about climate change and must change our ways by prioritising values that require us to turn away from everyday concerns and preoccupations. This, I have argued earlier, is a recipe for paternalism that has too often been followed by prominent environmental organisations and activists (Meyer 2008).

What, then, is an alternative way of thinking about the challenge of sustainability? The general contours are already coming into focus. It would engage with existing everyday concerns, rather than seek to transcend them. To the extent that these concerns are understood as "material" ones, then it will be a materialist rather than post-materialist approach, though labelling it as such will require us to think more expansively about what a materialist approach entails. Its starting point, in other words, is where people are at now, rather than where one might wish them to be in the future. Rather than exhorting others to begin to "think globally," those cultivating a sustainability transition must first "think locally."[1] Rather than diagnosing the central challenge as denial, here, we would focus on *resonance*: how can we rethink our understanding of sustainability – or of action to address climate change – in a manner that resonates *with* everyday life, rather than pursuing the Sisyphean task of trying to overcome it? How might we challenge social structures and institutions that serve as obstacles to a resonant vision of sustainability?

The point here is emphatically not to simply repackage sustainability as consistent with every popular attitude or belief. That would achieve little more than the legitimation of the status quo – what has been called "sustaining the unsustainable" (Blühdorn 2007, 2016). It is, however, to recognise that any social transformation will require public support and that a strategy that begins with a respect for local knowledge and popular concerns is more likely to succeed than one that begins with the message that these must be dismissed or overcome.

All too often, public arguments for change are countered by the claim that people in post-industrial societies are unwilling to sacrifice anything to address environmental challenges (cf. Maniates and Meyer 2010). As one scholar puts this pithily, we should not "expect a sudden moral epiphany that clashes brutally with contemporary lifestyles … it is unlikely that citizens abandon their smartphones in order to embrace the charms of a more embedded rural lifestyle" (Arias-Maldonado 2012, 118). Another asserts that

> [n]ot only do the already rich people in industrialized countries value their current lifestyle … [but] billions in the developing world also aspire for a decent material living standard … these global aspirations to modernity make it impossible to win popular support for any radical green programme.
>
> *(Karlsson 2012, 464; cf. Nordhaus and Shellenberger 2007)*

There's a relevant insight in these claims that ought not to be casually dismissed. Yet the premise that seems to underlie this sort of retort is that people in post-industrial societies live in something like the best of all possible worlds now, and so there is little motivation for most people here to alter "contemporary lifestyles." Similarly, it is assumed here that for those outside these societies to achieve a "decent material living standard" requires mimicking the particular development trajectory of the Global North. These assumptions are overstated and overgeneralised.

Reflecting upon everyday practices can help us to identify those spaces where these assumptions are least persuasive and so arguments for change most likely to resonate. For example, Christopher Buck identifies contemporary anxieties about a time-crunch as such a space (2012); he, as well as Jörgen Larsson and Jonas Nässén in this volume, argues for work-time reduction as a strategic response. Similarly, I have argued that reimagining automobility in contemporary society must begin by recognising the distinctive ways such a society enables a sense of individual freedom. Otherwise, ideas for reducing cars and driving will rightly be resisted as a paternalistic threat to this freedom. Yet the relationship between automobility and freedom is not unidirectional. Attention to freedom also enables greater clarity about the many ways in which dependence upon cars also constrains freedom. In this way, we can reorient our analysis towards changes to our practices and the built environment that would reduce this dependence. Here, the primary goal might be to enable greater freedom *from* automobility. For drivers, this could mean freedom from expensive car and insurance payments, from being stuck in traffic, from needing to drive for daily provisioning and childcare responsibilities. For those who can't drive or don't have access to a car, freedom from automobility can have a somewhat different meaning, including: greater freedom from reliance on others, from unsafe streets, from social isolation. A more complex and practice-based account of the relationship between cars and freedom does not determine answers, but it does open up possibilities that are otherwise all-too-readily dismissed as infeasible (Meyer 2015, chapter 6).

Engaging in material practices is something we all do every day. Reflecting upon them, in the pursuit of sustainability, is something we can and ought to do far more often. It would be a mistake, however, to imagine that by doing so we might fix upon a singular, determinate, end. It is precisely when our material practices lead to differences that a *politics* of sustainability emerges.

## Scholarly tributaries that flow into a view of sustainability as engagement with everyday life

Elements of a resonant view of sustainability that engages with everyday life can be found in a range of activist and academic discourses and literatures. Rather than begin from an abstract or technocratic orientation inaccessible to non-experts, these attend to experiences and practices familiar to many. Rather than privileging certain (i.e. post-materialist) values or (i.e. wild) places as paradigmatically "about" environmental sustainability, these open up the question of why people are motivated to care or to act and what spaces or places are vital to protect. Rather than projecting a class, gender, or racial profile onto sustainable practices, these are regarded as – at least in part – a product of how "environment" and "sustainability" come to be understood in particular social contexts. Rather than a narrow focus on ethical consumerism and individual behavioural change, these focus on large-scale practices that lie at the intersection of individuals and social structures. As a result of some or all of the mentioned earlier, the framing of sustainability can turn away from the recitation of a litany of foreboding catastrophes meant to overcome denial and towards the diverse, sometimes elusive and fragmentary, but nonetheless real possibilities for engaging the everyday.

One integral tributary here is the expansive scholarship and activism around environmental justice (EJ). While activism in poor and marginalised communities addressing threats to their health, livelihood, and vitality has deep roots in many places, the explicit framing of this as a matter of EJ emerged first among African-American anti-toxics organisers in the 1990s (First National People of Color Environmental Leadership Summit 1991; Bullard 1993; Schlosberg 1999). Crucial, for our purposes here, has been the emphasis on the importance and consequence of re-conceptualising the "environment" as – at least in part – "where we live, work, and play" (Gottlieb 1993; Novotny 2000), and the focus on race, gender, and class as central to differences in how injustices are experienced and how individuals and communities respond to these experiences.

EJ discourse has expanded in several directions in recent years (for a more detailed account, see Schlosberg 2013). First, this discourse has been increasingly used in other national contexts and to highlight commonalities among movements and constituencies to preserve and promote sustainable livelihoods in the Global South – what Ramachandra Guha and Joan Martinez-Alier have termed the "environmentalism of the poor" (Guha and Martinez-Alier 1997; Martinez-Alier 2016). Second, it has been used to address the experiences of other social groups, including people with disabilities (Ray and Sibara 2017; Salkeld 2017). Third, it has expanded beyond the protesting of injustice and location of environmental harms to the promotion of what Julian Agyeman and his collaborators have termed "just sustainabilities" – a pluralistic framework for addressing inclusive community spaces, alternative bases for livelihood and economic production, and environmental quality together (Agyeman 2013; McLaren and Agyeman 2015). This framework is also reflected in a variety of other contemporary movements and discourses, including food justice, energy justice, climate justice, and – more inclusively – the notion of "just transitions" (Akuno and Nangwaya 2017; Heffron and McCauley 2018; Gottlieb and Joshi 2010).

In a complementary fashion, Giovanna Di Chiro and Alyssa Battistoni have both highlighted the Marxist-feminist concept of "social reproduction" as a lens for understanding how inequality and differential vulnerability to the effects of climate change, toxic pollution, and other hazards can impact and impair the ability of many to provide for their daily lives and the lives of others in their households and communities. Gender is central to this dynamic of social reproduction and these authors make it clear that environmental injustice can neither be fully understood nor adequately challenged without attention to it (Di Chiro 2008; Battistoni 2017; see also Krauss 1994; Mellor 2017). While this growing diversity of EJ and feminist work may seem to make it harder than ever to characterise succinctly, it seems clear that the animating premise is that people's everyday lives and experiences are the foundation upon which the pursuit of sustainability must be based.

A second tributary that has buoyed this engagement with everyday life is interdisciplinary work advancing a "new materialism." While it should be clear that EJ work is already attentive to both social and biophysical materiality, this wide-ranging body of work pushes back against the dichotomy between human agency and vitality on the one hand and "dead" matter on the other hand. Scholars including Bruno Latour, Jane Bennett, Stacy Alaimo, Noortje Marres, and many others have drawn our attention to the inescapably complex webs of humans and non-humans within which action and agency exist (Latour 2007; Alaimo 2010; Bennett 2010; Marres 2012; for valuable overviews and anthologies, see Braun and Whatmore 2010 and Coole and Frost 2010).

Similarly, Maria Kaika, Fiona Allon, and others' close attention to metabolic flows in and out of homes and cities brings to the forefront elements of everyday life often literally hidden beneath the surface: those that enable many people to heat our homes, flush our waste, cook our meals, and draw a bath (Kaika 2005; Allon 2016). David Schlosberg and Romand Coles also argue for the tracing of material flows – of food, energy, and other commodity chains – as the basis for emergent social movements that seek to reconfigure these. This new materialist work operates in many different registers, but what is most notable and relevant here is its renewed confidence in our ability to attend to the material world without collapsing into reductionist and deterministic frameworks that have often been associated with materialist philosophies (Schlosberg and Coles 2016).

The third tributary is reflected in what has been termed social practice theory. Here, Elizabeth Shove, Gert Spaargaren, Alan Warde, and others have crafted a framework for the empirical examination of everyday practices such as bathing, heating and cooling, and shopping and cooking (Warde 2005; Spaargaren 2011; Shove et al. 2012). Following on earlier theoretical work, they position these practices as mid-level phenomena that cannot be properly understood as simply individual-level choices (Reckwitz 2002; see also Giddens 1984; Schatzki 2001). At the same time, they avoid the abstraction and disempowerment that can result from thinking of these practices as determined by distant structures that individuals have no ability to influence. By examining practices in this way, they lend greater empirical specificity and insight to the ways in which everyday practice can be understood and the relation of theory to such practices.

While the three tributaries described here are interdisciplinary, it is unfortunately the case that there is relatively limited interaction among those involved in each. Yet all can contribute to an agenda of scholarship and activism that seeks to promote the engagement with the everyday as a basis for fostering greater sustainability. Rather than conceptualising the material world in contrast to social values or activities, practice theory recognises them as thoroughly enmeshed with each other. Echoing an insight of new materialism, but in a more empirical vein, practices are understood as the appropriate level of inquiry precisely because they bring together an engagement with the material world and careful attention to values, interpretations, and activities. An engagement with practices, in turn, demands attention to their social location and positionality of their practitioners – are they more widespread among urban, suburban, or rural dwellers, for instance; among the poor and marginalised, or among privileged elites; are practices gendered, do they assume certain bodily abilities, are they racialised? Empirical and normative concerns are intermingled, here.

What I hope has become clear is that engaging the everyday is in no sense an ideology, a political position, or a singular scholarly agenda. In its inescapable plurality, it is probably best understood as a disposition or orientation, one that begins with a recognition and respect for the embodied experiences of people who constitute a society in which a sustainability transition is even possible. It is only upon the basis of this recognition and respect that we can begin to imagine what the pursuit of sustainability might mean in practice and it is only upon this basis that we might imagine what strategies might resonate and allow for movement in this direction.

## Conclusion: engaging the everyday as insider criticism

I have argued elsewhere that strategic questions about social change explored in this chapter are best advanced by immanent or engaged social critics who boldly challenge the status quo, yet deeply respect both the values and the concerns of those they wish to persuade. In this way, such critics break down a dichotomy between "reformers" who seek only modest, incremental change within the existing structures of society and "radicals" who reject such structures and advocate for systemic transformation from what seems to be an Archimedean point disconnected from society (Meyer 2015, 5–8).

The likely success of the engaged critic depends crucially upon their ability to speak in a manner that resonates with citizens, while simultaneously arguing for extensive, meaningful change from the status quo. With the proponents of practice theory, engaging the everyday focuses us on the intermediate level of analysis at which much change happens (Geels et al. 2015). These critics engage pragmatically with existing concerns, while it also offers critical distance from the status quo and a reimagining of what is possible. This is also what is so promising about much of the work to promote just sustainabilities. It draws our attention to the diversity of spaces and opportunities for sustainability strategies that resonate with the everyday concerns of many. It has transformative potential precisely because for many if not most of us, our everyday lives do not now even approximate the best of all possible worlds. Recognising this requires truly listening and respecting diverse voices, especially those that are least privileged and most marginalised. Val Plumwood has called this a reduction of "remoteness" that deepens democracy by cultivating local knowledge and experiences of those least heard (Plumwood 1998). In an age where "populism" has come into wide use as a label for authoritarian forms of exclusion and hierarchy, engaging the everyday can foster a more expansive political imagination by allowing us to envision the pursuit of sustainability as grounded in a vision of popular, political action.

## Note

1 Thanks to Frank Baber and Bob Bartlett for suggesting this as the message embedded in my analysis (Baber and Bartlett 2016).

## References

Agyeman, Julian. 2013. *Introducing Just Sustainabilities: Policy, Planning and Practice*. London: Zed Books.
Akuno, Kali, and Ajamu Nangwaya, eds. 2017. *Jackson Rising: The Struggle for Economic Democracy and Black Self-Determination in Jackson, Mississippi*. Daraja Press.
Alaimo, Stacy. 2010. *Bodily Natures: Science, Environment, and the Material Self*. Bloomington: Indiana University Press.
Allon, Fiona. 2016. "The Household as Infrastructure: The Politics and Porosity of Dwelling in a Time of Environmental Emergency." In *The Greening of Everyday Life: Challenging Practices, Imagining Possibilities*, edited by John M. Meyer and Jens M. Kersten, 47–64. Oxford: Oxford University Press.
Anfinson, Kellan. 2018. "How to Tell the Truth about Climate Change." *Environmental Politics* 27 (2): 209–27. doi: 10.1080/09644016.2017.1413723.
Arias-Maldonado, Manuel. 2012. *Real Green: Sustainability after the End of Nature*. Farnham, Surrey: Ashgate.

Baber, Walter F., and Robert V. Bartlett. 2016. "Critical Dialogue – Engaging the Everyday: Environmental Social Criticism and the Resonance Dilemma. By Meyer John M. Cambridge, MA: The MIT Press, 2015. 253p. $24.00." *Perspectives on Politics* 14 (1): 168–70. doi: 10.1017/S1537592715003394.

Battistoni, Alyssa. 2017. "Living, Not Just Surviving." *Jacobin*, August 15, 2017. http://jacobinmag.com/2017/08/living-not-just-surviving.

Beeson, Mark. 2016. "Environmental Authoritarianism and China." In *Oxford Handbook of Environmental Political Theory*, edited by Teena Gabrielson, Cheryl Hall, John M. Meyer, and David Schlosberg, 520–32. Oxford: Oxford University Press.

Bennett, Jane. 2010. *Vibrant Matter a Political Ecology of Things*. Durham: Duke University Press.

Blühdorn, Ingolfur. 2007. "Sustaining the Unsustainable: Symbolic Politics and the Politics of Simulation." *Environmental Politics* 16 (2): 251–75. doi: 10.1080/09644010701211759.

———. 2016. "Sustainability – Post-Sustainability – Unsustainability." In *Oxford Handbook of Environmental Political Theory*, edited by Teena Gabrielson, Cheryl Hall, John M. Meyer, and David Schlosberg, 259–73. Oxford: Oxford University Press.

Braun, Bruce, and Sarah Whatmore, eds. 2010. *Political Matter: Technoscience, Democracy, and Public Life*. Minneapolis: University of Minnesota Press.

Buck, Christopher D. 2012. "Post-Environmentalism: An Internal Critique." *Environmental Politics*, September, 1–18. doi: 10.1080/09644016.2012.712793.

Bullard, Robert D. 1993. *Confronting Environmental Racism: Voices from the Grassroots*. Boston: South End Press.

Coole, Diana, and Samantha Frost, eds. 2010. *New Materialisms: Ontology, Agency, and Politics*. Durham: Duke University Press.

Di Chiro, Giovanna. 2008. "Living Environmentalisms: Coalition Politics, Social Reproduction, and Environmental Justice." *Environmental Politics* 17 (2): 276–98. doi: 10.1080/09644010801936230.

Dunlap, Riley E., and Aaron M. McCright. 2011. "Organized Climate Change Denial." In *Oxford Handbook of Climate Change and Society*, edited by John S. Dryzek, Richard B. Norgaard, and David Schlosberg, 144–60. Oxford: Oxford University Press.

Dunlap, Riley E., and Richard York. 2008. "The Globalization of Environmental Concern and the Limits of the Postmaterialist Values Explanation: Evidence from Four Multinational Surveys." *Sociological Quarterly* 49 (3): 529–63.

First National People of Color Environmental Leadership Summit. 1991. "Principles of Environmental Justice." www.ejnet.org/ej/principles.html.

Geels, Frank W., Andy McMeekin, Josephine Mylan, and Dale Southerton. 2015. "A Critical Appraisal of Sustainable Consumption and Production Research: The Reformist, Revolutionary and Reconfiguration Positions." *Global Environmental Change* 34 (September): 1–12. doi: 10.1016/j.gloenvcha.2015.04.013.

Giddens, Anthony. 1984. *The Constitution of Society: Introduction of the Theory of Structuration*. Berkeley: University of California Press.

Gottlieb, Robert. 1993. *Forcing the Spring: The Transformation of the American Environmental Movement*. Washington: Island Press.

Gottlieb, Robert, and Anupama Joshi. 2010. "Food Justice." *Dissent Magazine*. Accessed March 15, 2018. www.dissentmagazine.org/online_articles/food-justice.

Guber, Deborah L. 2003. *The Grassroots of a Green Revolution: Polling America on the Environment*. Cambridge, MA: MIT Press.

Guha, Ramachandra, and Juan Martinez-Alier. 1997. *Varieties of Environmentalism: Essays North and South*. London: Earthscan.

Heffron, Raphael J., and Darren McCauley. 2018. "What Is the 'Just Transition'?" *Geoforum* 88 (January): 74–77. doi: 10.1016/j.geoforum.2017.11.016.

Inglehart, Ronald. 1990. *Culture Shift in Advanced Industrial Society*. Princeton: Princeton University Press.

———. 1995. "Public Support for Environmental Protection: Objective Problems and Subjective Values in 43 Societies." *PS: Political Science and Politics* 28 (1): 57–72.

Kaika, Maria. 2005. *City of Flows: Modernity, Nature, and the City*. New York: Routledge.

Karlsson, Rasmus. 2012. "Individual Guilt or Collective Progressive Action? Challenging the Strategic Potential of Environmental Citizenship Theory." *Environmental Values* 21 (4): 459–74. doi: 10.3197/096327112X13466893628102.

Kim, So Young, and Yael Wolinsky-Nahmias. 2014. "Cross-National Public Opinion on Climate Change: The Effects of Affluence and Vulnerability." *Global Environmental Politics* 14 (1): 79–106. doi: 10.1162/GLEP_a_00215.

Krauss, Celene. 1994. "Women of Color on the Front Line." In *Unequal Protection: Environmental Justice and Communities of Color*, edited by Robert D. Bullard, 256–71. San Francisco: Sierra Club Books.

Latour, Bruno. 2007. *Reassembling the Social: An Introduction to Actor-Network-Theory*. Cambridge: Cambridge University Press.

Leiserowitz, Anthony, E. Maibach, S. Rosenthal, M. Cutler, and J. Kotcher. 2017. *Politics and Global Warming, October 2017*. New Haven: Yale Program on Climate Change Communication.

Maniates, Michael, and John M. Meyer, eds. 2010. *The Environmental Politics of Sacrifice*. Cambridge, MA: MIT Press.

Marres, Noortje. 2012. *Material Participation: Technology, the Environment and Everyday Publics*. Basingstoke, UK: Palgrave Macmillan.

Martinez-Alier, Joan. 2016. "Global Environmental Justice and the Environmentalism of the Poor." In *Oxford Handbook of Environmental Political Theory*, edited by Teena Gabrielson, Cheryl Hall, John M. Meyer, and David Schlosberg, 547–62. Oxford: Oxford University Press.

McKibben, Bill. 2012. "Global Warming's Terrifying New Math – Rolling Stone." *Rolling Stone*, July 12, 2012. www.rollingstone.com/politics/news/global-warmings-terrifying-new-math-20120719.

McLaren, Duncan, and Julian Agyeman. 2015. *Sharing Cities: A Case for Truly Smart and Sustainable Cities*. Cambridge, MA: MIT Press.

Mellor, Mary. 2017. "Ecofeminist Political Economy." In *Routledge Handbook of Gender and Environment*, edited by Sherilyn MacGregor, 86–100. Abingdon: Routledge.

Meyer, John M. 2008. "Populism, Paternalism, and the State of Environmentalism in the U.S." *Environmental Politics* 17 (2): 219–36.

———. 2015. *Engaging the Everyday: Environmental Social Criticism and the Resonance Dilemma*. Cambridge, MA: MIT Press.

Nordhaus, Ted, and Michael Shellenberger. 2007. *Break Through: From the Death of Environmentalism to the Politics of Possibility*. Boston: Houghton Mifflin.

Norgaard, Kari Marie. 2011. *Living in Denial: Climate Change, Emotions, and Everyday Life*. Cambridge, MA: MIT Press.

Novotny, Patrick. 2000. *Where We Live, Work, and Play: The Environmental Justice Movement and the Struggle for a New Environmentalism*. Westport, CT: Praeger Publishers.

Pew Research Center for the People and the Press, Suite 800 Washington, and DC 20036 USA202-419-4300 | Main202-419-4349 | Fax202-419-4372 | Media Inquiries. 2018. "Economic Issues Decline among Public's Policy Priorities." *Pew Research Center for the People and the Press* (blog). January 25, 2018. www.people-press.org/2018/01/25/economic-issues-decline-among-publics-policy-priorities/.

Plumwood, Val. 1998. "Inequality, Ecojustice and Ecological Rationality." *Ecotheology* 5 and 6: 185–218.

Ray, Sarah Jaquette, and Jay Sibara. 2017. *Disability Studies and the Environmental Humanities: Toward an Eco-Crip Theory*. Lincoln: University of Nebraska Press.

Reckwitz, Andreas. 2002. "Toward a Theory of Social Practices: A Development in Culturalist Theorizing." *European Journal of Social Theory* 5 (2): 243–63. doi: 10.1177/13684310222225432.

Salkeld, Deborah Fenney. 2017. "Environmental Citizenship and Disability Equality: The Need for an Inclusive Approach." *Environmental Politics* 0 (0): 1–22. doi: 10.1080/09644016.2017.1413726.

Schatzki, Theodore. 2001. *The Practice Turn in Contemporary Theory*. Milton Park: Routledge.

Schlosberg, David. 1999. *Environmental Justice and the New Pluralism*. New York: Oxford University Press.

———. 2013. "Theorizing Environmental Justice: The Expanding Sphere of a Discourse." *Environmental Politics* 22 (1): 37–55.

Schlosberg, David, and Romand Coles. 2016. "The New Environmentalism of Everyday Life: Sustainability, Material Flows and Movements." *Contemporary Political Theory* 15 (2): 160–81. doi: 10.1057/cpt.2015.34.

Shearman, David and Joseph Smith. 2007. "The Climate Change Challenge and the Failure of Democracy." Westport, CT: Praeger Publishers.

Shove, Elizabeth, Mika Pantzar, and Matt Watson. 2012. *The Dynamics of Social Practice: Everyday Life and How It Changes*. Thousand Oaks, CA: Sage.

Spaargaren, Gert. 2011. "Theories of Practices: Agency, Technology, and Culture." *Global Environmental Change* 21 (3): 813–22. doi: 10.1016/j.gloenvcha.2011.03.010.

Warde, Alan. 2005. "Consumption and Theories of Practice." *Journal of Consumer Culture* 5 (2): 131–53. doi: 10.1177/1469540505053090.

# 4

# Materiality and nonhuman agency

*Tobias Gumbert*

## Introduction

Let us address the most fundamental issues causing reluctance and restraint, among many practitioners and academics alike, to adopt thinking in terms of the nonhuman head-on: how do we even begin to grasp agentic qualities of material things, biological organisms, or whole ecosystems? How can broad-spectrum systemic herbicides or piles of electronic waste be anything other than objects that need to be politically regulated? Why are increasingly more academics from various disciplines so keen on decentring the human subject in environmental affairs, in times when we should arguably look even more to marginalised and disenfranchised groups and the dynamics of power relations? Or in other words: what's the whole point? It is the first and foremost aim of this chapter to clarify many of the admittedly rather abstract conceptions that scholars of various disciplines put forth to make sense of the changing times we live in that prompt us to rethink our firmly held beliefs about humanity, nature, and their interrelation. It is the claim of many of these authors that the certainty about who or what we should understand as subjects and objects of ethical, legal, or political action has been suspended. It is argued that "as humans meddle more vigorously in natural processes and thus become more materially, if not yet ethically, responsible for outcomes" (Coole and Frost 2010, 16), we need to cultivate political practices that do not just add environmental concerns as a new layer to the management and regulation of behaviour and everyday concerns. In Bruno Latour's terms, it's about ecologising rather than modernising: "It is very much a question of considering everything differently, but this 'everything' cannot be subsumed under the expression Nature, and this difference does not reduce to the importation of naturalistic knowledge into human quarrels." (Latour 1998, 35). We will see how various authors have worked towards making this difference knowledgeable by taking leave from established notions of human agency and materiality.

For environmental governance scholarship, it is almost by definition difficult to recognise the material world as anything other than the canvas on which problems of environmental degradation and global environmental change unfold. When key theoretical debates in the field speak of "the changing nature of agency," "new agency," and "the move towards a more holistic approach to studying governance at different levels"

(Pattberg and Widerberg 2015, 689), the nonhuman is usually omitted. Novel discussions on agency in environmental governance refer instead to certification organisations, environmental consultancies, and big corporations. Change is here a result of dynamic human actor-networks, and governance failure often the outcome of misguided, unsuccessful negotiations of utility-seeking public and private authorities. Our moving into the Anthropocene, however, questions the radical separation in human manager and "nature" as object of management and the fixed boundaries of the social sphere and the biosphere (see also Mert this volume). The result is an ontological shift that casts both spheres as "existentially indistinguishable" (Burke et al. 2016, 510). It is this new geological epoch in which, it is claimed, humanity itself has become a "force of nature" (Nicholson and Jinnah 2016, 3) altering the functioning of the Earth system and causing "global shifts of an unprecedented scale and speed" (Hamilton et al. 2015, 5). While a number of authors speak of the Capitalocene, the Anthrobscene or the Plantationoscene (e.g. Parikka 2014; Haraway 2016) to "make clear 'who' and what practices are responsible" (Chandler et al. 2018, 199) for our current situation, it is especially the new attention to materiality and nonhuman agency that re-orients political thinking:

> Agencies are swapping properties in surprising ways: we become unable to recognize either the old ideas of freedom and humanism, or those of matter and reductionism. Responsibilities [...] are now redistributed.
>
> *(Latour 2016, 167)*

To govern in the Anthropocene entails to recognise the complex environment in which human action and decision-making takes place, i.e. not only relations of power and inequality among human agents and communities (e.g. Bexell this volume, Okereke this volume) but also interactions with animal communities or even material things and processes (such as technologies). Many authors who argue in favour of recognising nonhuman agency share the common starting point, that, against the backdrop of the Anthropocene, we need to critically challenge our normative foundations and reset our operating system. For the most hopeful among them, a radical reformulation of our thinking vis-à-vis nonhumans and materiality is the hard reboot desperately needed.

In the next section, the chapter gives an account of how scholars try to make sense of nonhuman agency and materiality in the social sciences by drawing on fundamental ethical debates and very stable, firmly held beliefs about governing "ecological objects." This section is followed by an overview of contemporary theoretical and methodological approaches that aim to think and represent the relation of humans to the nonhuman realm not in terms of separation and hierarchical domination, but as entanglement and mutual dependence, thereby looking explicitly beyond mainstream approaches. Subsequently, current research on the politics and governance of waste is presented to illustrate how a resetting of our conceptual tools in regard to human/nonhuman relations can be a transformative project. Finally, the chapter concludes by arguing for broadening political recognition towards nonhumans in order to organise political responsibility and build strong institutions towards an ecologically safe and socially just future.

## What is nonhuman agency?

Agency, in its most basic understanding and use, is the capacity to act, resonating with notions of motivation, will, intentionality, choice, freedom, and creativity (Emirbayer and Mische 1998, 962). Accounts of human agency are regularly related back to Immanuel Kant, whose philosophy split reality into two opposing orders: the normative, the realm of freedom and autonomous moral beings, and the conditional, the realm of necessity and material surroundings (ibid., 965). Building on this dichotomous distinction, to grasp agency analytically in its different dimensions and effects has been to conceptualise it in relation to structure. Accordingly, various traditions either downplay the effect of structure on agency (rational choice, phenomenology), emphasise it (discourse theory, the concept of habitus), or try to level the importance of both (theories of social practice). Similarly, a number of different bodies of knowledge, such as pragmatist philosophy (John Dewey, George Herbert Mead), explicitly go beyond the dichotomy of normative and instrumental action in arguing that all objects and purposes are socially constructed (and therefore only "make sense" in the context of shared meanings and values attributed within social interactions). However, going beyond the human in accessing the various dimensions of agency has in the long history of thinking on this topic rarely been an issue. When scholars speak of the agency of nonhumans, agency does not only refer to things acting on their own accord (as animal species, bacteria, fungi of course do), but emphasises the fact that humans and nonhumans are relationally embedded in networks (or assemblages) and influence each other's capacity to act (and by extension, to have effects on the environment). But what exactly does the term *nonhuman* refer to? Nonhuman interests figure indirectly in decisions about biodiversity, nature conversation, and ecosystem management (Ellis 2016, 515). The term is applied from particular animal species and plant varieties to ecological wholes such as lotic ecosystems (springs, rivers) that include interactions amongst various organisms and abiotic (non-living) physical and chemical interactions. Apart from these understandings that position the nonhuman as being part of the natural environment, the nonhuman category can essentially be extended to include *everything-other-than-human*, e.g. artificial intelligence, material things, physical infrastructures, or their material remainders (waste). In this context it is, however, important to clarify if the term *nonhuman* refers largely to lively matter (Tsing 2015, Haraway 2016), or if it includes non-living matter (Latour 2004, Bennett 2010; Jane Bennett also considers the inorganic "vitality" of things, transcending the binary of living/non-living to some degree).

Nonhuman forces can also more generally be understood as inhuman or posthuman, i.e. they can be denied "human" characteristics such as dignity and worth, morality and rationality, or refer to something that is beyond being human such as the enhancement of human (intellectual, physical) capabilities and the elimination of negative traits. For Cary Wolfe (2010), the idea that the autonomous liberal subject can be expanded by technological mastery aims to intensify humanism rather than going beyond it – it is a transhumanism. When thinking in posthuman terms, the emphasis should be on "posthuman*ist*, in the sense that it opposes the fantasies of disembodiment and autonomy, inherited from humanism itself" (Wolfe 2010, xv). While humanism would undeniably produce good commitments on issues such as cruelty towards animals, it reproduces at the same time "the very kind of normative subjectivity – a specific concept of the human – that grounds discrimination against nonhuman animals […] in the first place" (ibid., xvii). This split, which implies dominance, hierarchy, and superiority for humanist values, is further exemplified in the work of Giorgio Agamben and the politics of "life itself," where he draws on the Aristotelian idea that life is made up of *zoê* (zoology, zoophilic) and *bios* (biology) (Agamben 1998). As yet another qualitative distinction of Western reason, *zoê* relates to the "being-aliveness of the subject," the body as well as animal life in general, whereas *bios* stands for intelligent life, the mind, and is in its radical demarcation from *zoē* "almost holy" (Braidotti 2010, 206–207). Braidotti goes on:

> *Zoē* is always second best, and the idea of life carrying on independent of, even regardless of, and at times in spite of rational control is the dubious privilege attributed to the nonhumans. These cover all of the animal kingdoms as well as the classical "others" of metaphysically based visions of the subject, namely the sexual other (woman) and the ethnic other (the native). In the old regime this used to be called "Nature".
>
> *(ibid.)*

Nature is traditionally seen as the domain of material things and materiality in general, the inanimate objects which passively crowd the world of human subjects as background context. Jane Bennett refers to this thinking as "parsing the world into passive matter (it) and vibrant life (us)" (Khan 2009, 92). This understanding rests on a particular view of nature as a mechanism of material parts, as "defining materiality as deterministic and devoid of spirit" (Bennett 2006, 216), which is part of the process of disenchanting and rationalising the world in the context of modernity. Apart from this Weberian root of materialism, the concepts of commodification and fetishisation developed and emphasised by Marx explain how its contemporary understanding became a hegemonic frame (Bennett 2006). The rapidly developing paradigm of "new materialism" (Coole and Frost 2010) argues instead for an ontologised agency of things and matter, dissipating ontotheological binaries and challenging "the human of traditional humanism [and the] [...] human/nonhuman hierarchy of value and power" (Washick and Wingrove 2015, 64). Authors like Jane Bennett develop such an alternative account of materialism from a very different philosophical tradition, applying the thought of Epicurus, Spinoza, Leibniz, Nietzsche, Thoreau, Bergson, Deleuze and Guattari, and others to elucidate that in the history of thinking on materiality, Marx's version, which governs our common sense, is not without its counterparts (see for example Bennett 2010). The starting point of "new materialism" is therefore the "conceptualization of matter as interconnected assemblages of human and nonhuman, living and nonliving, and hence the self as transcorporeal – never existing in isolation from this broader network" (Meyer 2015, 64; see also Meyer this volume). It recognises humanity's embeddedness in, or humans as the very "stuff" of, the material world. Following Latour, material objects are not "matters of fact," imbued with meaning that would be transparent or determined, but characterised as "matters of concern," understood as a "path to respectful recognition of the materiality of things, which requires that we not misrepresent their import by presuming that their meaning is fixed, singular, and, thus, apolitical" (ibid., 72).

The recognition of nonhumans presents scholars with different challenges. By adopting a nonhuman agency perspective, researchers have to be very careful not to reproduce dominant dichotomies and pit *real* human interests against *abstract* nonhuman interests. For example, if we follow the assumption that modern instrumental reason, i.e. to conceive of nonhuman nature as a means to human ends, has become a universal human project of mastering nature (Biro 2016), then strategies to save endangered species are not a remedy against such instrumental reason (while e.g. providing nonhuman entities with limited legal rights may undoubtedly act as a potential counterbalance, see Mert this volume). Generally, these strategies reify the separation of *human* or *social* from *natural*. As Elisabeth Ellis rightly points out, in most environmental conflicts, political cleavages run between "a minority of short-term 'extractors' [and] a majority of 'sustainers,' or people with medium-term interests in the sustainable provision of things like clean air and water" (Ellis 2016, 514). If we were to differentiate broadly between human and nonhuman interests, we would "unwittingly adopt the ideological lenses of an unrealistic minority position" (ibid.). Analysing nonhuman agency is therefore never separate from human power struggles – it rather enhances our understanding of the mechanisms that mask and distort powerful human interests. That is why many notions of nonhuman agency break with the idea that nature would signify a gigantic whole, governed by the harmonious coexistence of diversity. Against this powerful idea which relates back to romanticism, nature must be seen for what it is, as a "turbulent field in which various and variable materialities collide, congeal, morph and disintegrate" (Khan 2009, 94). This picture of nature as a turbulent field, where agencies are distributed, constantly mixing and swapping, and blame cannot be assigned on the basis of singular cause and effect chains, is perhaps the prime blueprint for understanding the Anthropocene. That is not to say that we should deny the fact that the advent of the Anthropocene is, fundamentally, a consequence of Capitalist exploitation built on terrifying historical injustices. But from the point of view of "new materialism," to deny the agentic qualities of nonhumans would give even more power to the guiding principles of ecological modernisation, managerialism, and techno-scientific promises of salvation.

What are the consequences of the decentring of humanist ideas in environmental philosophy and politics? It is important to note that arguing for the recognition of nonhuman agency in the context of sustainable development does not diminish or neglect the adherence to fundamental human values and principles such as intra- and intergenerational justice (see Lawrence this volume), leading a good life (in terms of quality of life) or satisfying human needs (see Di Giulio and Defila this volume). Rather, a posthuman ethics (Braidotti 2011) emphasises the human as part of diverse networks of distributed agencies, which is at its core an ethics of interdependence and belonging. Humanity is no longer seen as a homogeneous, universal category, with nature as its radical other, but redefined as a "companion species" (Haraway 2007). This postanthropocentric perspective envisions the human subject as embodied and embedded, i.e. as constitutively relational, in constant exchange with and dependent on lively matter (Braidotti 2011) as well as directly influenced in its interaction with non-living matter. This leads to two related conclusions that drive such a posthuman ethics: first, if the perspective of separateness and instrumentalisation towards nonhumans is not revised, the human self continues to misrecognise its own existence, and by that its role and responsibility in the Anthropocene – with potentially (even more) severe consequences. And second, if the dualism of human and nonhuman worlds is to be rejected, and their ontological interdependence to be embraced, then it is necessary to reflect human–nonhuman relations through new ways of knowing, representing, and story-telling, e.g. by extending socio-political categories such as justice or structural discrimination (i.e. concepts of the *humanities*) to nonhuman entities (Latour 1998, Tsing 2013, Haraway 2016, see also Mert this volume). Because after all, as Anna Tsing reminds us, "[i]f social means 'made in entangling relations with significant others,' clearly living beings other than humans are fully social – with or without humans." (Tsing 2013, 27).

In short, nonhuman agency denotes a radical departure from traditional and historically long-standing ideas of human-nature relations. It urges us to critically engage with the domination paradigm that subjugates nature to human will, and not just for moral reasons. The consequences of nature's agentic qualities are experienced and *felt* by human communities everywhere. In the words of Ulrich Beck, climate change has become "an agent of metamorphosis," rendering what was unthinkable yesterday as real and possible today (Beck 2016, 4). Especially in ontological terms, many firmly held beliefs need to be revisited, not just within scholarly work, but also explicitly on the level of everyday experience. A postanthropocentric perspective emphasises the need to improve humanity's self-awareness and reflexivity in the Anthropocene as a necessary first step towards engaging *Nature* differently. The next section introduces selected concepts that assist in achieving this goal and try to capture this complexity theoretically and empirically.

## From hybrids to assemblages

Critical scholars aim to right the wrongs of the duality of culture and nature through different theoretical and methodological means. The theoretical approaches gathered here share the common goal to develop remedies against the fallacies of humanist and modernist thinking that allegedly got us into the current socio-ecological predicament. For the French philosopher Bruno Latour, the notion of the separation of humans and nature is an illusory product of modernity, since both nature and culture can only be defined by reference to each other. They form a single concept that has been split into two parts: "They were born together, as inseparable as Siamese twins who hug or hit each other without ceasing to belong to the same body." (Latour 2017, 15). Latour argues that the notions of objective nature and subjective values are always intertwined and entangled in political controversies (as in genetic engineering or nuclear energy), a fact which he illustrates with the typography "Nature/Culture" to escape the self-evidence of these categories (Latour 2017). With reference to Donna Haraway (2016), he suggests the term "worlding" (see also Inoue this volume) or "world making" for this dualism to indicate the multiplicity of existences and their respective modes of existence, which are intermingled beings of Nature/Culture. The central term with which Latour tries to capture the production of these beings is hybridisation. Hybrids are networks of generative capacities, acting in multiple ways – they are neither pure nature nor pure culture. The hole in the ozone layer is, for example, a hybrid, a network of chlorofluorocarbons (CFCs), air masses, consumers, industrial plants, environmentalists, etc. On the ontological level, it represents an intertwined whole that cannot be separated into individual components.

According to Latour, the objects of nature have been scientifically, that is, objectively and technically, examined by the natural sciences, and are indisputable matters of fact, while society, with its different interests and values, is characterised as a sphere of controversy and must therefore be politically represented. For Latour, what is problematic in this context is the lack of political interpretations of these hybrids, which are assigned to the realm of nature and determined by hard sciences. This leads him to the diagnosis of a paradoxical effect: the denial of hybridisation, the constant production of Nature/Culture, leads to its rapid expansion, since all moral and political limits of assessing this transformation have been suspended (Latour 2004). Latour discusses these conflicting tendencies with the concepts of purification and translation. The work of purification is carried out constantly by science and politics; it separates society and nature according to strict criteria when a blending is immanent. The work of translation, on the other hand, signifies the ever-increasing, systematic mixture of nature and society in the process of technological and industrial production (hybridisation), which transfers agency to actants (term that comprises human and nonhuman actors) who consequently multiply and spread. Taken together, the process of purification, as a whole ensemble of material practices and discourses, renders invisible and uncontrollable what is being accomplished in the process of translation: the unrestricted proliferation of hybrids.

The political conflicts surrounding the governance of the systemic herbicide glyphosate in the European Union are a prime example of this, because the political problem is apparent: since the technical production of hybrids is based on objective scientific knowledge, it does not have to be democratised and subjected to the political influence of citizens. Alleged natural facts are thus strategically de-politicised, they are not debatable and negotiable, and arguments that are not framed in the language of the natural sciences are rendered "unreasonable." Simultaneously, these discursive practices cement the rule of experts, who claim privileged access to objective knowledge. As a consequence, with regard to the legitimacy of different social and political values, positions that are not "purified," that is, arguments rejecting the separation of nature and society, tend not to be recognised and therefore do not materialise in practical political matters.

Analysing these processes is methodologically challenging, since the inclusion of nonhumans into environmental political research demands a specific framework that redefines their ontological status in relation to human actors. Among academics, the concept of "assemblage" is often prominently applied to make visible the diverse effects these interactions can produce (see e.g. Tsing 2015, who also convincingly relates "assemblage" to "world making"). By drawing on the work of Deleuze and Guattari (1987), assemblage can be described as an "*ad hoc* grouping of an ontologically diverse range of actants, of vital materialities of various sorts" (Khan 2009, 92). The concept draws on developments in the non-linear sciences and concepts such as open systems, complexity, emergence, and non-linear dynamics (Acuto and Curtis 2014, 4). Latour and Deleuze in fact share many ontological suppositions to "disentangle social processes from the constraints of modernist thinking, recharting the geography of the social as embedded in endless connections among 'actants,' that is things people and ideas that shape that very geography" (ibid., 5).

In mainstream governance approaches, the effects of a governance architecture, whether in the form of a hierarchical, market, or networked structure, are assessed as the causal results of stakeholder interactions. The effects generated by an assemblage are on the contrary understood as "emergent properties," in the sense that distinct webs of overlapping and interpenetrating agencies of humans and nonhumans "make something happen" that cannot be reduced to the sum of each materiality involved. For example, the effects of natural disasters are not due to objective forces of nature that have to be anticipated and managed by human security and risk governance structures. Seen as an assemblage, a distinctive history of formation becomes visible that does not reduce outcomes to governance failure caused by particular human agencies (e.g. emergency services), nor points apologetically to the unforeseeable consequences of nonhuman agencies (e.g. floodings due to spontaneous weather extremes). Rather, within assemblages, agency is distributed between such heterogeneous elements as individual persons, bodies of knowledge, technologies, measuring tools, physical infrastructures, degrading soils, biological organisms, weather conditions, sea level rise, etc. The dichotomy of human/nonhuman is resolved by paying attention to the constantly evolving interaction of these elements.

Working with assemblages is therefore challenging, since the concept seems structural, yet "the intent in its aesthetic uses is precisely to undermine such ideas of structure" (Marcus and Saka 2006, 102, ct. in Bueger 2014, 60). Christian Bueger provides some excellent "rules of thumb" for coming to terms with a number of concepts such as "multiplicity, relations, practice, ordering, expressivity and territoriality" (Bueger 2014, 65) that are discussed in the context of assemblage thinking. He suggests that the researcher should treat apparent wholes as puzzles for empirical research, and binaries and dualisms such as human/nonhuman as distinctions that need explanations. Bueger describes the vocabulary of assemblage as "voluntarily poor" in order not to limit research to an a priori focus, which is why attention to mundane activities is needed (which also implies an ethnographic gaze) as well as the crucial realisation that assemblages are "real-time enactments," constantly changing (ibid., 65–67). In this sense, we can view assemblage thinking as "anti-methodologist," since every attempt at structuring and ordering is in itself an act to be explained. Finally, and importantly, the researcher herself is part of an assemblage, as there is no possibility of an outside view, which presents normative challenges. As Jane Bennett reminds us:

> It is ultimately a matter of political judgment what is more needed today: should we acknowledge the distributive quality of agency to address the power of human-nonhuman assemblages and to resist a politics of blame? Or should we persist with a strategic understatement of material agency in the hopes of enhancing the accountability of specific humans?
>
> *(Bennett 2010, 38)*

All in all, the normative implications of assemblage thinking can be said to comprise a "recovery of the dignity of material objects, as pushing towards a more ecological sensibility" (Acuto and Curtis 2014, 14). Beyond a mere structuration devise of heterogeneous elements, assemblage thinking can be, with its project of making visible excluded agencies, deconstruction and scientific non-conformity, a deeply political endeavour pushing towards pluralism in environmental politics. The challenges for empirical research not withstanding, many scholars have provided good examples how to think along the categories of nonhuman agency and materiality about environmental problems and make a strong case why it is important and relevant. The following section illustrates such thinking on the topic of human relations with wasted materials.

## The nonhuman agency of waste

In the context of environmental governance, research on waste constitutes a burgeoning field of research. As economic growth curves in almost every sphere of production have grown exponentially since the 1950s, so have the material remainders and residues of global production processes: nuclear waste, electronic waste, plastic waste, food waste, and many other waste streams create transnational, yet place-specific ecological and social problems that need to be addressed by governing institutions. Garbage, trash, and waste have been employed as lenses to explore various fields of study, e.g. environmental politics, capitalism, modernity, risk, or governance, to name just a few (Moore 2012, 2). The question that is addressed in all of these accounts is what waste is and how, why, and to whom it matters (ibid., see also Gregson and Crang 2010).

Waste lends itself formidably to study the agency of nonhumans. Among very diverse conceptualisations, we find waste as "hazard, object of management, commodity, resource, archive, filth, fetish, risk, disorder, matter out of place, governable object, abject and actant" (Moore 2012, 2). The international community has produced many legal frameworks to define and identify when a thing becomes waste, when it must be considered hazardous (see Gillespie 2015), and yet, depending on the circumstances and on the observing subject, things "shimmer back and forth" (Bennett 2010, 4) between various identities and conceptualisations. Seeing a thing as debris, litter or waste also provokes different effects than considering an object as still "being of use." The object of waste itself "symbolizes an idea of improper use, and therefore operates within a more or less moral economy of the right, the good, the proper, their opposites and all values in between" (Scanlan 2005, 22). How humans act in relation to waste is therefore irreducibly connected to meanings (and the practices of "meaning-making"), emotions, and moral codes that change according to the type of waste, geographical location, socio-cultural attitudes, political stakeholders, legal frameworks, and other variables. For many waste scholars, the impossibility to distinguish subjective values from objective facts is apparent in this case, and this is why concepts like hybridity or assemblage are applied to model and understand this inherent complexity.

Analyses in this field often start with rendering visible dominant dichotomies that govern particular human relations to waste. In the process of wasting (as well as through calls to recycle more), we see how humans actively establish mastery over a passive nature and simultaneously reproduce their separation from it: the identities of both, human and nonhuman nature, are fixed. For Gay Hawkins, this "dualistic thinking inhibits any serious consideration of the specificities of waste and our relations with it" (Hawkins 2006, 9). Since waste can only be a bad thing in this framework, it has to be eliminated to keep the identity of each category intact. The productive, generative capacities of waste can only be activated through economic revalorisation processes (recycled material, waste-to-energy). Apart from this human work on the object, it is only destructive. Drawing on Latour, Hawkins argues that our understanding of waste changes if we think of it as part of the proliferation of hybrids. Waste is not inherently bad, but rather subject to relations:

> What is rubbish in one context is perfectly useful in another. Different classifications, valuing regimes, practices, and uses, enhance or elaborate different material qualities in things and persons – actively producing the distinctions between what will count as natural or cultural, a wasted thing or a valued object.
>
> *(ibid., 10)*

In "the politics of bottled water," Hawkins shows how plastic bottles exhibit a vital force connecting material agency and political agency (Hawkins 2009). Bottled water industries imbue drinking water with ethical and biological values, producing a 'will to health' (ibid., 183). Biopolitical discourses utilise medical and hygiene narratives that problematise living and invite subjects to engage in self-reflection and to reform their everyday habits. They implicitly devalue tap water, rendering it ordinary and suspect. Other companies, such as producers of water filters, allude to the material origins of bottle water by showing humans in ads, literally, drinking oil. Using the "material presence as political disturbance […], [o]nce the bottle's material substance is revealed as oil it becomes capable of generating a new assemblage which foregrounds the specific logics of unsustainable production and an over packaged world." (ibid., 187). How matter is represented in its being and its effects on other agents can have diverse political ramifications. Hawkins reminds us that these different materialities are present in the bottle at all times; it is not just a question of discursive framing, but a real "material multiplicity" enacted or suppressed in different assemblages capable of making or destroying markets. Recognising these materialities can (for example in the case of banning plastic bags) at the very least "shift the discourse […] away from an all-too-easy rejection of wasteful consumerism and toward meaningful engagement with the roles and responsibilities associated with these material practices" (Meyer 2015, 70).

Importantly, though, besides being attentive to the multiple meanings and effects of waste, we also see how the human self is governed in its interactions with waste. Within the moral economy of waste, recycling, composting, and other means of revaluing are cast as practices of virtue, negotiating activism, righteousness, resentment, and guilt of human agents. On the municipal level in the United Kingdom, Gregson (2009) argues that recycling is best understood as an assemblage of policy, households, green bags, mundane washing practices, sustainability principles, and various materials that creates "uncertainty, indecision and inappropriate action replicated across thousands of households" (64) while giving rise to citizen practices of environmental self-regulation. The causes of waste relating to productivist logics and growth imperatives are thereby gradually removed from sight.

On the global level, research on waste has focused on its multiple mobilities, its dynamism and shifting geographies, the impacts of processes of decomposition and degradation, its organic and inorganic vitality, and problematic presence (Davies 2012, Lepawsky 2015). By way of mapping flows and following things, we see in the case of global e-waste governance that due to a lack of political regulation of these hybrids, consumers are now carrying responsibility to purchase certified products, the economies of different regions in the South that rely on e-waste processing are excluded from negotiating certification proposals, and the question of how these materials are produced in the first place (toxicity, longevity) is omitted; it is solely about what to do with end-of-pipe waste (Pickren 2013). Understanding waste however in its "fundamental ontological indeterminacy" and as a "phenomenon of non-coherence" (being subject to differing knowledge from place to place, and jurisdiction to jurisdiction) renders visible the multiple effects generated and suggests differing strategies to gain control over them (Lepawsky 2018).

In sum, thinking waste as part of socio-natural relations presupposes the recognition of waste's otherness beyond repulsive, nausea-inducing object and manageable resource. Sarah Moore suggests with reference to Slavoj Žižek to think of waste as "that which objects, that which disturbs the smooth running of things" (Žižek 2006, 17), creating indeterminacy and disrupting very stable social norms (Moore 2012, 2). Since the predominant part of anthropogenic waste will haunt this planet for centuries and (in the case of radioactive waste) millennia to come, finding new ways to acknowledge, interact, and live with waste (instead of hiding, burying, burning, or shipping it out of sight) may be simultaneously an experiment in cultivating a different ecological politics.

## Conclusion

It has been argued that the recognition of nonhuman agency and materiality in environmental politics reforms our thinking on the socio-ecological transformations that are under way and the question of who counts as a political subject in environmental affairs alike. The departure from mainstream orientations of sustainability governance, i.e. questions of authority, effectiveness, and power of some over other human agents to which the material world is staged as a passive background, signals the need for the acknowledgement of human relations with nonhumans and materiality as the basis for cultivating new linkages. These links, whether understood as hybrids or assemblages, are only visible through attention to material things, their conditions of production, transformation through consumption and disposal, as well as their emergent properties. A large interdisciplinary research interest lies in the socio-ecological transformations of our time and how the separation in human agency and material passivity links to them, while showing that both categories constantly transcend each other. Empirically, we already begin to see the benefits of these research approaches applied in various disciplines. Human work is not only done on inanimate objects – nonhumans are represented politically and through ads and campaigns and "given voice" through social movements in order to highlight and enact a range of different issues, from structures of inequality and unsustainability to local processes of community-building. Similarly, if the notion of waste as an essentialised object of negativity is abandoned, we begin to see the multiple links and meanings that adhere to it. The clear dividing line between human subjects and nonhuman/material objects is challenged when calls to find solutions to environmental problems do not demand extended human mastery over nature, but start to press for paying more attention to the generative capacities of assemblages.

That being said, the theoretical concepts and debates discussed here can be seen as off-putting due to their level of abstraction and philosophy-heavy concepts. Making these ideas more accessible and demonstrating their worth by more in-depth empirical case studies is a steep, but nevertheless important challenge. It starts with broadening research on power, justice, recognition, and responsibility to include nonhumans and materiality in ways that seem odd at first, but might prompt us to rephrase some of our own research questions. Following authors like Deleuze, Latour, and Bennett, this must also be understood as an ethical endeavour:

> It is, I think, the 'responsibility' of humans to pay attention to the effects of the assemblages in which we find ourselves participating, and then work experimentally to alter the machine so as to minimize or compensate for the suffering it manufactures.
>
> *(Khan 2009, 93)*

If we reject the many dualisms inherent in our political thinking tools (subject/object, agency/structure, mind/body, materiality/discursivity), there is room to discover many connections and agentic qualities of nonhumans and "things" we have not witnessed earlier. Research on environmental governance in the Anthropocene can hardly afford to bypass these opportunities.

# References

Acuto, M. and S. Curtis. 2014. 'Assemblage Thinking and International Relations'. In *Reassembing International Theory: Assemblage Thinking and International Relations*, eds Acuto and Curtis. New York: Palgrave Macmillan, pp. 1–15.

Agamben, G. 1998. *Homo Sacer: Sovereign Power and Bare Life*. Stanford: Stanford University Press.

Beck, U. 2016. *The Metamorphosis of the World*. Cambridge: Polity Press.

Bennett, J. 2006. 'Modernity and Its Critics.' In *The Oxford Handbook of Political Theory*, eds Dryzek, Honig and Phillips. Oxford: Oxford University Press, pp. 211–224.

Bennett, J. 2010. *Vibrant Matter. A Political Ecology of Things*. Durham/London: Duke University Press.

Biro, A. 2016. 'Human Nature, Non-Human Nature, and Needs. Environmental Political Theory and Critical Theory.' In *The Oxford Handbook of Environmental Political Theory*, eds Gabrielson, Hall, Meyer and Schlosberg. Oxford: Oxford University Press, pp. 89–102.

Braidotti, R. 2010. 'The Politics of "Life Itself" and New Ways of Dying'. In *New Materialisms. Ontology, Agency and Politics*, eds Coole and Frost. Durham/London: Duke University Press, pp. 201–218.

Braidotti, R. 2011. *Nomadic Subjects: Embodiment and Sexual Difference in Contemporary Feminist Theory*. New York: Columbia University Press.

Bueger, C. 2014. 'Thinking Assemblages Methodologically: Some Rules of Thumb'. In *Reassembling International Theory: Assemblage Thinking and International Relations*, eds Acuto and Curtis. New York: Palgrave Macmillan, pp. 58–66.

Burke, A., S. Fishel, A. Mitchell, S. Dalby and D.J. Levine. 2016. 'Planet Politics: A Manifesto from the End of IR.' *Millennium* 44(3): 499–523.

Chandler, D., E. Cudworth and S. Hobden. 2018. 'Anthropocene, Capitalocene and Liberal Cosmopolitan IR: A Response to Burke et al.'s 'Planet Politics'.' *Millennium* 46(2): 190–208.

Coole, D. and S. Frost. 2010. 'Introducing the New Materialisms'. In *New Materialisms. Ontology, Agency and Politics*, eds Coole and Frost. Durham/London: Duke University Press, pp. 1–43.

Davies, A.R. 2012. 'Geography and the Matter of Waste Mobilities.' *Transactions* 37: 191–196.

Deleuze, G. and F. Guattari. 1987. *A Thousand Plateaus*. London/New York: Continuum.

Ellis, E. 2016. 'Democracy as Constraint and Possibility for Environmental Action.' In *The Oxford Handbook of Environmental Political Theory*, eds Gabrielson, Hall, Meyer and Schlosberg. Oxford: Oxford University Press, pp. 505–519.

Emirbayer, M. and A. Mische. 1998. 'What Is Agency?' *American Journal of Sociology* 103(4): 962–1023.

Gillespie, A. 2015. *Waste Policy. International Regulation, Comparative and Contextual Perspectives*. Cheltenham/UK: Edward Elgar.

Gregson, N. 2009. 'Recycling as Policy and Assemblage.' *Geography* 94(1): 61–65.

Gregson, N. and M. Crang. 2010. 'Materiality and Waste: Inorganic Vitality in a Networked World'. *Environment and Planning A* 42: 1026–1032.

Hamilton, C., C. Bonneuil and F. Gemenne. 2015. 'Thinking the Anthropocene.' In *The Anthropocene and the Global Environmental Crisis*, eds Hamilton, Gemenne and Bonneuil. New York: Routledge, pp. 1–14.

Haraway, D. 2007. *When Species Meet*. Minneapolis: University of Minnesota Press.

Haraway, D. 2016. *Staying with the Trouble. Making Kin in the Chthulucene*. Durham/London: Duke University Press.

Hawkins, G. 2006. *The Ethics of Waste. How We Relate to Rubbish*. Oxford: Rowman and Littlefield.

Hawkins, G. 2009. 'The Politics of Bottled Water: Assembling Bottled Water as Brand, Waste and Oil.' *Journal of Cultural Economy* 2: 183–195.

Khan, G. 2009. 'Agency, Nature and Emergent Properties: An Interview with Jane Bennett'. *Contemporary Political Theory* 8(1): 90–105.

Latour, B. 1998. 'To Modernize or to Ecologize? That's the Question'. In *Remaking Reality: Nature at the Millennium*, eds Castree and Willems-Braun. London/New York: Routledge, pp. 221–242.

Latour, B. 2004. *Politics of Nature: How to Bring the Sciences into Democracy*. Cambridge/MA: Harvard University Press.

Latour, B. 2016. 'Sharing Responsibility: Farewell to the Sublime.' In *Reset modernity!*, ed Latour. Cambridge/London: MIT Press.

Latour, B. 2017. *Facing Gaia. Eight Lectures on the New Climate Regime*. Cambridge: Polity Press.

Lepawsky, J. 2015. 'The Changing Geography of Global Trade in Electronic Discards: Time to Rethink the e-Waste Problem.' *The Geographical Journal* 181(2): 147–159.

Lepawsky, J. 2018. *Reassembling Rubbish. Worlding Electronic Waste*. Cambridge/London: MIT Press.

Marcus, G.E. and E. Saka. 2006. 'Assemblage'. *Theory, Culture & Society* 23(2–3): 101–106.

Meyer, J. 2015. *Engaging the Everyday. Environmental Criticism and the Resonance Dilemma*. Cambridge/London: MIT Press.

Moore, S.A. 2012. 'Garbage Matters: Concepts in New Geographies of Waste. *Progress in Human Geography* 36(6): 1–20.

Nicholson, S. and S. Jinnah. 2016. 'Living on a New Earth'. In *New Earth Politics. Essays from the Anthropocene*, eds Nicholson and Jinnah. Cambridge/London: MIT Press, pp. 1–16.

Parikka, J. 2014. *The Anthrobscene*. Minneapolis: University of Minnesota Press.

Pattberg, P. and O. Widerberg. 2015. 'Theorising Global Environmental Governance: Key Findings and Future Questions'. *Millennium* 43(2): 684–705.

Pickren, G. 2013. 'Political Ecologies of Electronic Waste: Uncertainty and Legitimacy in the Governance of e-Waste Geographies.' *Environment and Planning A* 46(2): 26–45.

Scanlan, J. 2005. *On Garbage*. London: Reaktion Books Ltd.

Tsing, A. 2013. 'More-than-Human Sociality: A Call for Critical Description'. In *Anthropology and Nature*, ed Hastrup. New York/London: Routledge.

Tsing, A. 2015. *The Mushroom at the End of the World. On the Possibility of Life in Capitalist Ruins*. Princeton/Oxford: Princeton University Press, pp. 27–42.

Washick, B. and E. Wingrove. 2015. 'Politics that Matter: Thinking about Power and Justice with the New Materialists.' *Contemporary Political Theory* 14: 63–89.

Wolfe, C. 2010. *What Is Posthumanism?* Minneapolis/London: University of Minnesota Press.

Žižek, S. 2006. *The Parallax View*. Cambridge/London: MIT Press.

# 5

# Worlding global sustainability governance

Cristina Yumie Aoki Inoue, Thais Lemos Ribeiro,
and Ítalo Sant' Anna Resende

## Introduction

To face the current planetary socio-environmental crisis, Burke et al. (2016) claim that we need a new political imagination in which "humans, animals, ecologies, biosphere are all together," and to adapt or to create new institutions. Beyond that, we need to develop "a politics to nurture worlds for all humans and species co-living in the biosphere" (Burke et al. 2016, 2).

In this chapter, we will explore why worlding is necessary to global sustainability governance (GSG) studies and will provide a few examples of how to "world" them. GSG has predominantly been studied from a liberal-institutionalist perspective, from an international scale, based on a state-centric and top-down ontology (Barnett and Duvall 2005; Jordan et al. 2015; Jang et al. 2016), and with positivist epistemology and methodologies. Consequently, as in other modern perspectives, the nature/society divide has prevailed in most GSG approaches, definitions, and practices. We argue that in the Anthropocene, the relationship between human societies and "nature" must be reconceived and one way forward could be the "many worlds-one planet" perspective.

Worlding means that we are always making worlds and, consequently, recognising multiple worlds, different knowledge systems, and notions of nature is a starting point for inquiry (Tickner and Weaver 2009, 9; Inoue and Moreira 2016, 2). Many worlds mean more than tolerating different perspectives, but recognising many ways of being and experimenting with different worlds (Querejazu 2016, 3). Such recognition has ontological and epistemological implications. It entails examining and situating (contextualising) our own worldviews, concepts, theories, and opening space to other "reals," in the sense that there are many different realities,[1] which are simply other worlds, ontologically speaking, as well as acknowledging the ways through which we construct and validate knowledge in different times and places (Tickner and Weaver 2009, 11).

According to Latour (1993), modern Western knowledge is trapped in a Cartesian bind that consists of great separations or divides. Such separations include, for example, the binaries human/non-human, mind/body, animate/inanimate, nature/society, and aim to represent a singular reality. The separation of nature and society, in particular, is a central tenet of modernity on which global environmental politics has been founded. However, modernity's particular way of representing reality is not universally shared, as many cultures around the world make little distinction between humans, plants, spirits, and other entities (Blaney and Tickner 2017).

This modern common sense is also challenged by what Burke et al. (2016, 4–12) call a new kind of power represented by the Anthropocene, in which nature is being reshaped by social actions, and human activity and nature are bound together in a complex and indistinguishable manner. Hence, categories and methodologies are being challenged, and there is a demand for a new global political project in which social justice is inextricably linked together with environmental justice, the end of human-caused extinctions, climate change mitigation, among other necessary actions (see also Kalfagianni et al., this volume). This means an effort towards new practices, new ideas, new stories, and new myths in the worlds we have created (Burke et al. 2016), and the inclusion of actors, who usually do not have voice, in processes of knowledge co-production.

Worlding and the recognition of the pluriverse are ways of making alternatives and engaging in these new dynamics. In this sense, the chapter will present five different cosmovisions from non-Western cultures that we consider as "worlds."[2] However, we do not have the ambition to present them comprehensively but to focus on how these visions challenge these divisions and separations and may contribute to the development of a more pluralistic and inclusive approach to environmental governance. We understand GSG as the many collective efforts by states and non-state actors to identify, understand, and address various present and future sustainability issues, considering many worlds in one planet, and the interdependence among ecological, social justice, and economic dimensions and the novel challenges presented by the Anthropocene. This chapter is divided into three parts: first, we discuss the implications of the Anthropocene for GSG studies; second we present worldism and the pluriverse as alternative approaches, and finally, we present different worldviews that could contribute to worlding GSG.

## The Anthropocene and limitations of conventional approaches

Most social science accounts start from the assumption, framed in the Holocene, that nature is a stable scenario. The nature-society dichotomy, state-centrism, and the focus on global institutions, power, interests, and inter-state bargain highlight specific questions, commitments, and worldviews, which do not properly respond to today's planetary reality and socio-environmental challenges. Such limitations reflect the way that GSG has been conceived and practiced. Burke et al. (2016) and Pereira (2017) contest this stability assumption and argue that the Anthropocene calls for new approaches.

Though a contested notion (Harrington 2016; Chandler et al. 2017), the Anthropocene indicates that the stability experienced during the Holocene no longer holds because human activities are substantially altering the planet and its natural systems. Population mobility and growth and changes in the demographic structure, high level of energy and resource use, growing consumerism, urbanisation, globalisation of transport and communication systems, decoupling of financial from production economy are large-scale socio-economic changes that characterise the post Second World War great acceleration (Steffen et al. 2015). Humans can now be considered the main altering force on the planet and of its natural systems, with evidence of large-scale impacts on Earth systems.

However, the Anthropocene is not just an accumulation of environmental change effects, but a new world reality, with complex relations of life and non-life, between humans, non-humans, things, materials, sociopolitical, and biophysical elements (Burke et al. 2016;

Harrington 2016; Pereira 2017). Therefore, it becomes difficult to make a clear distinction between human and non-human realms, and to keep the dichotomy between nature and society (Rudy and White 2014; Wapner 2014; Steffen et al. 2015; Hamilton 2016; Pereira 2017).

Nature is a social, geographical, and historical construct, a repository of meaning, which is framed in relation to human experience and subject to change, but not in a unilateral relationship: humans and nature are co-produced social constructions that can change over time. As a consequence, traditional categories of environmental politics may not adequately respond to present social/nature challenges (Wapner 2014; Harrington 2016; Inoue and Moreira 2016) and the Anthropocene can be understood as a new way of apprehending reality, which orients political subjectivity and rationality (Hamilton 2016).

One of these new challenges is how to consider the agency of non-humans (Harrington 2016; Gumbert this volume) and to look for a new understanding of the world in the Anthropocene, based on the notion of the social, geographical, and historical co-production of humans and nature.

Another challenge is the tendency of some approaches to the Anthropocene that consider humans as a unified category and as if all humans were equally responsible for the current socio-environmental crisis (Harrington 2016; Chandler et al. 2017). Top-down and allegedly universal solutions, based on this kind of generalisation, have a potential to be authoritarian and exclusionist. Bäckstrand (2006), working on the "stakeholder democracy" model for environmental governance, presents a less state-centric and multi-layered notion of global governance, with informal, participatory, non-electoral, and non-territorial references, which includes marginalised groups, like women and indigenous peoples. In this model, a more *bottom-up* participatory and deliberative environmental governance has the potential to generate more effective and legitimate collective problem-solving, an argument based on the concepts of *input* legitimacy, which is the quality of decision-making processes regarding participation (openness to public scrutiny and inclusion of different stakeholders' interests) and *output* legitimacy – effectiveness or problem-solving capacity of the governance system (Bäckstrand 2006, 473).

Many worlds mean that on a single planet – the Earth – there is a multiplicity of worlds that intersect, overlap, and conflict, and which are co-constituents and co-vulnerable. To think of many worlds on one planet opens up the possibility of new research avenues and has the potential to bring new solutions for GSG and challenges. Scientists around the world have been working with local populations and indigenous peoples in joint research, local development, biodiversity conservation, and other projects (Inoue 2007; Kassam 2009; Berkes 2012; Whyte 2013; Swift and Cock 2015; Athayde et al. 2016). Examples include the cooperation between scientists and Skolt Sami people on a biodiversity project to restore salmon in Finland and to gather information about insects as an indicator of changes in the Arctic, or the use of aborigine fire stick farming as a fire control practice in Australia (Robbins 2018).

However, global governance scholars have paid less attention to other ways of knowing. Inoue and Moreira (2016, 8) and Burke et al. (2016, 20–21) consider that it is better to think of multiple worlds (an Earth-worldly politics) and many natures to meet the new challenges brought by the Anthropocene, because this recognition can contribute to a new understanding of GSG, with broad participation and legitimacy.

## Worldism and the pluriverse as alternative approaches to governance

Querejazu (2016) considers the pluriverse as an ontological starting point, based on the incommensurability of different ways of being and living in the world – many worlds exist on their own and are interrelated. This means that reality is not a universe made of different realities, but it is per se a plurality, or a pluriverse. This perspective opens up the possibility of one to assess and to acknowledge multiple realities that stand ontologically as other "reals."

Worlding GSG means recognising that different ontologies ask for different epistemologies and methodologies. These ontologies and epistemologies allow one to overcome restraining categories and traditional concepts of international politics (e.g. power, security, sovereignty, state), and, as a consequence, give new insights about how to theorise the global. They can help us to understand how some concepts, like global governance, have become established, what mechanisms have naturalised them, and how these concepts influence the choice for issues or non-issues in research and policy development. Besides, they open possibilities to improve the international environmental agenda, and to assume a politically emancipatory position (Tickner and Weaver 2009, 3 and, 18). Despite that fact, many scholars still relegate different ontologies, such as those of indigenous peoples, to the realm of myths, legends, and beliefs (Querejazu 2016, 4–11).

For many people the human, the natural, and the spiritual worlds are interconnected and coexist in time and space. Therefore, the political is much more than what is conventionally considered, since it implies, for example, the recognition that non-humans (living and non-living) can also have political agency, rights, or responsibilities. Querejazu (2016) complements this perspective, including "subjects" who are neither human nor things – spiritual entities (meaning nonmaterial or transcendent, like nature forces), for reality is the result of intersubjective practices that are compatible with the existence of different kinds of subjects.[3]

In the Anthropocene, the Earth systems are starting to alter politics and have power on their own. Several indigenous peoples around the planet consider animals, plants, rocks, mountains, rivers, and so as sentient, with life and spirit (Inoue and Moreira 2016) and, in this sense, have rights and responsibilities, and so do human beings towards them. Some national states like Ecuador and Bolivia have recognised the rights of nature.

It is interesting to observe that while indigenous and traditional ways of knowing can call our attention to forms of non-human agency, with political, moral, and legal implications, the fast technological development in several areas is also calling our attention to non-human agency and power. Harrington (2016, 13) presents examples of how the agency of "things, humans and non-humans, configure the practices and understandings of war, diplomacy, security and the economy," as algorithms in global economic relations, that can be seen as having independent agency, as they alter the conditions of human possibilities. Fast technological and artificial intelligence developments in many fronts (nanotechnology, biological engineering, genetics, facial and voice recognition, etc.), and the merger of human intelligence and technology point to this direction as well. Consider, for example, the possibility of a superhuman intelligence with no biological body, and how its agency can change today's practices and understandings and build a new reality. Thus, to speak of non-human agency and the end of nature-society dichotomy is not a religious understanding of politics, but a view that changes our relation to things and to the more-than-human and, as such, can have moral, legal, and political implications.

In this sense, perhaps, worldism can be a better way to respond to present complexities and challenges, because it implies communication and negotiation across difference, not incommensurability. Worldism is defined by Ling (2014, 13–37) as an analytical portal that encompasses different traditions and cultures. Ling (2014) recognises multiple worlds with modes of thinking, doing, being, and relating that intersect with the *Westphalian World*.[4] Worldism is different from the cosmopolitan version of the *Westphalian world* because it does not put forward isolated understandings of different modes of thinking departing from the *Westphalian* worldview, but it acknowledges the existence and the role of multiple worlds in international politics.

Different from the pluriverse, Ling (2014, 15) argues that Multiple Worlds and the West-phalian World contradict and complement each other, using the image of the Dao, in which there is communication and negotiation across difference, and not incommensurability. In Daoist dialogics, as the yin and the yang, you and I constitute each other (*I am in you, you in me*). The result is the recognition of relations and contributions of multiple worlds, in a co-constituted and engaged perspective. The *Westphalian* worldview constitutes and is con-stituted by the other worldviews, bringing a more accurate portrayal of world politics. In this sense, a worldist perspective in GSG can better respond to the present complexities of global life by bringing to the fore actors from many worlds that are usually not recognised.

For Jang et al. (2015, 2), a multiple-actors governance configuration can broaden the scope of policy solutions, but it can also increase fragmentation and segmentation of rule-making and rule implementing, resulting in competition that can lead either to paralysis or innovative solutions. There is indeed a risk of paralysis. The dialectic and hybrid approach of worldism, however, questions dualities and complements competition with continuities and connections.

The benefits of broadening the scope of policy solutions and encouraging novel solutions, with the development and diffusion of different normative standards and new knowledge that can inform unilateral and multilateral behaviour and practices (Underdal 2008; Robbins 2018), may outweigh the risks of fragmentation, segmentation, and competition. Scholars have ar-gued that indigenous knowledge can improve the implementation of global policies at the local level, for instance, in climate change adaptation or reducing emissions from deforestation and forest degradation (REDD) (Moreira and Baniwa 2011; Schroeder and González 2019) as well as by linking local, indigenous, and scientific knowledge systems in processes of knowl-edge co-production, for example, within the Intergovernmental Science-Policy Platform on Biodiversity and Ecosystem Services (IPBES).[5] Then, the future of global governance rests not only on the liberal paradigm of world politics (Jang et al. 2015) but also on many worldviews and cosmologies that are also concerned with the future of our planet, natures, and societies.

So, worlding in GSG is needed to:

a   interrogate mainstream ontology and epistemology of GSG and consider other knowl-edge systems and ways of being, in parity with Western epistemologies. This proposition may offer a more realistic portrayal of world affairs and invigorate research about global environmental issues with new methodological developments;

b   question the nature-society divide, which makes it difficult to cope with the An-thropocene and other sustainability challenges, opening the possibility of considering non-human agency and bringing different worldviews that overcome this divide, like indigenous knowledge (Inoue and Moreira 2016);

c   present new ways to overcome the limitations of the Westphalian theorisation about the world, such as the subject-object divide and the top-down formulations, especially in the Anthropocene;

d   consider humanity as non-homogeneous, made of different actors and agents with di-verse worldviews, which are equally important for making an effective, socially and environmentally just, and legitimate GSG.

## Examples of perspectives that a worldist approach could draw from

When Western approaches to GSG acknowledge non-Western perspectives merely as objects of discourse rather than as an essential source of theory building, they abdicate from a more pluralistic perspective, in which the existence of many worlds is taken into consideration. As different cultures have different views on the relations between humans and nature, each one of them could potentially provide relevant insights when it comes to GSG.

In our view, these attempts that are in search of new ontologies, ways of knowing, and being in the world coincide with the concept of worldism. Non-Western knowledge systems illustrate how GSG can be "worlded," as different worldviews can enrich and contribute to establishing new principles, norms, and rules that would be more adequate to how we relate to the new power in the Anthropocene proposed by Burke et al. (2016).

Using Ling's worldist approach, Inoue and Moreira (2016) develop the understanding of many worlds as many knowledge systems in epistemological parity, to which many different notions of nature correspond. One way to "world" GSG is to recognise and connect different knowledge systems in processes of knowledge co-production in which all the actors are equally agents, and research problems, methods, and validation procedures are jointly defined (Tengö et al. 2014). This can help one to better understand and find solutions for socio-environmental problems, as illustrated in the example of cooperation between scientists and Skolt Sami people in Finland or between scientists, non-governmental organisations (NGOs), and indigenous peoples in Xingu (Brazilian Amazon) with respect to bees, pollination, pollinators, and food production (Biodiversity and Ecosystem Services Network 2019).

Five perspectives are presented because of their richness and relevance regarding their views on humans, cosmos, and nature, and because they present perspectives from distinct geographies and ethnicities. It is not our intention to generalise these worldviews, as there are many differences within and among them, but to call attention to the diversity of views that exist beyond the "West" and to their potential to enrich GSG approaches. The first two perspectives are from South American indigenous cosmovisions: Amazonian and Andean-Kichwa[6] peoples; the third one refers to Islamic traditions as revealed by the Quran (the sacred scripture of Islam) and the teachings of the prophet Muhammad; the fourth reveals the cosmovision of the Oromo people, the largest ethnic group in Ethiopia and in the wider "Horn of Africa;" and the fifth refers do the Daoist perspective as shown in the Daodejing, a classic of Chinese philosophical literature.

## Amazonian and Andean indigenous voices

Amazonian and other South American indigenous cosmovisions – especially those of Andean peoples – can contribute to the development of more pluralistic environmental approaches. Relevant aspects of Andean indigenous cultures regarding environmentalism are addressed with special care in the chapter by Vanhulst and Beling (this volume).

Several Amazonian and Andean indigenous peoples believe that subjects can embody objects and objects can embody subjects (Querejazu 2016, 9). This shaman's ability to embody other beings and to transit between worlds can be explained by these beings' common origins. For many Amazonian peoples, what we call "nature" comes from humanity or from culture. In their mythical cosmovisions (Ashaninka/Campa, Yanomami, Yawanawa, Aikewara) humans are empirically prior to the world: a kind of primordial humanity existed as the only substrate or matter from which the world was formed. Portions of this humanity were turned into animals, plants, other living beings, meteorological phenomena, and parts of the cosmos (stars, the moon, and so on), spontaneously or through the action of a demiurge. The portion that did not transform into something else is the historical or present humanity (Danowski and Viveiros de Castro 2017, 91–92).

Danowski and Viveiros de Castro (2017, 90–94) call this period of time the first Anthropocene. Unlike modern myths, animals and *"nature"* are humanity's future, not the past; and humans are the seed, the soil, the background, and not an innovation or the apex of creation.

By rejecting anthropocentrism, the Yanomami and other Amazonian peoples also reject the nature-society dichotomy. They see themselves as just one more among the various peoples and entities of the forest – such as animals, spirits, other beings associated with the land and even dreams. For them, the forest is a living entity and endowed with intrinsic value, not only as a repository of biodiversity or carbon stocks. Animals and other living beings are considered other kinds of "folks" and "peoples." As a consequence, they can be considered political entities, who live in their societies, in their realities (their "reals"). The natural world, or the "world," is an interconnected "multiplicity of multiplicities" (Danowski and Viveiros de Castro 2017, 97).

In the same direction, several Andean cosmovisions are based on relational ontologies. These ontologies present different forms of interaction and recognise that truth is not about uniqueness, or singularity. According to them, the connections between the factual, the divine, the social, and nature are paramount, and reality cannot be separated from the observer. Consequently, "nature" needs to be liberated from the condition of a mere object which belongs to human beings, since all beings have the same ontologic value – which does not mean they are identical, but that they have the same importance and deserve to be protected. This would be achieved by a major political effort in the sense of recognising nature as a subject of rights (Acosta 2016).[7]

Andean relational cosmovisions consider the state or condition of being relational as the most important principle of reality. Other principles are correspondence, complementarity, and reciprocity. Correspondence implies that elements are correlated in a balanced duality; complementarity implies that opposites complete and cannot exist without each other; and reciprocity is the fundamental idea of justice which permeates the relations between natural, human, spiritual, and cosmic (Querejazu 2016). A particularly interesting aspect of *"Pacha Mama"* is its relationship with the concept of pluralism. The deity has different roles: it can both be a generous entity who provides food and water; or a cruel one when she does not receive what she expects from humans. Actually, more than a deity, *"Pacha Mama"* could be considered a connection between three worlds (heaven, earth, and hell). Their roles do not follow fixed Manichean interpretations, for they can be both good and evil. It shows that in the Andean pluralistic cosmovision sacred supernatural entities are political actors with agency as well as interests (Acosta 2016).

## Islamic worldview

Islamic culture(s) and their teachings also provide valuable insights regarding the relationship between humans and nature. According to Kamla, Gallhofer and Haslam (2006), the environment is central in Islam, and both the Quran and the words of the Prophet carry concepts and principles that reflect the way Muslims relate to nature. The colour green – as used by many Muslim countries on their national flags – is considered the most blessed of all colours and represents in a deep sense the importance of nature for Muslims. According to the authors, there are key guiding principles that are fundamental to understand the relationship between Islam and nature.

The first principle – "Unity of God" (*"Tawheed"*) – implies unity and equality concerning both the worship of God and the partnership among all beings in terms of appreciation of interdependence and interconnectedness. It also implies that there is an equilibrium ruling the natural world, and that all of God's creations coexist in harmony or balance. According to the "*Tawheed*," people are part of the environment – different from the idea of separations and divisions proposed by Western and positivist perspectives – and humans deserve no more respect than the rest of Creation. All beings are equally important when it comes to maintaining balance (Kamla, Gallhofer and Haslam 2006).

The second principle is trusteeship. Although people are considered part of the environment and deserve the same respect as other beings, this principle implies a special role for humans in relation to the environment, for they are considered guardians (or "*Khalifah*") and have the duty to look after the self and others, including non-humans. Being a trustee implies that besides acting as guardians, Muslims must also cultivate the environment – but in a way that does not affect the equilibrium. They should cultivate it to the highest point consistent with sustainability (Kamla, Gallhofer and Haslam 2006).

The third principle gravitates around the idea of community ("*Umma*"), and emphasises social justice, social welfare, and the countering of oppression, implying that concerns about people are also relevant for the environment. Since people are part of nature, important for the equilibrium, and have the responsibility of being guardians of nature, respecting people also means respecting the environment.

The fourth principle refers to the idea of holism in a way that encompasses the future. Holism, in this sense, addresses the duty to ensure the wellbeing of future generations and could be considered the Islamic principle most closely related to the idea of sustainable development, since it emphasises the negativity of waste. The Prophet, for example, demanded Muslims not to waste water even when they are washing before a prayer. In this sense, today's consumption patterns could be considered incompatible with the holistic perspective, since they are potentially dangerous to the environment. The Quran itself condemns any sort of abuse of the Earth, for it compromises the wellbeing of future generations (Kamla, Gallhofer and Haslam 2006).

The last principle is about appreciating the beauty of nature, which provides the inspiration humans need to act as guardians of the environment. The splendour of nature is mentioned in many parts of the Quran, and the beauty of animals is considered a sign of God's existence. Muslims are encouraged to contemplate and meditate on the beauty of nature, so that their connection to it and their will to protect it remain strong (Kamla, Gallhofer and Haslam 2006).

### An African cosmovision – the Oromo

African cosmovisions concerning the relationship between humans and nature are not homogeneous. Given the impossibility to address multiple African cosmovisions in a single chapter, we focus on the Oromo People, since – according to Kelbessa (2005) – the critical examination of Oromo worldviews suggests that some Oromo groups have developed strong indigenous environmental ethics. They are the largest ethnic group in Ethiopia (more than 30% of the country's population) and also inhabit parts of Kenya and Somalia.

Oromo ethics regarding nature have two sides: one is material and pragmatic, while the other one is spiritual and moral. The material and pragmatic side is related to the fact that the Oromo people protect their environment for utilitarian reasons. They are aware that, if their environment deteriorates, their very existence will be in danger, as well as the lives of future generations. They pay special attention to changes in nature: the cycles, the seasons, the movement of the stars, the behaviour of wild and domestic animals, the condition of trees, and many other signs that could potentially reveal practical problems for their subsistence in the present and in the future (Kelbessa 2005).

The more spiritual and moral side is related to principles implicit in Oromo thought and practice. Besides being a means for their subsistence, the land is also a gift from God ("*Waaqa*"). Therefore, it has inherent value, despite any utilitarian considerations. Since the land is a gift and "*Waaqa*" is its guardian, no one is free to destroy nature in order to satisfy particular needs or see the land as something to be owned. Humans are simply friends, beneficiaries or users, and do not have the right to treat it as a commodity and dispose of it as they please. It should be stressed, however, that although "*Waaqa*" is an important influence when it comes to handling the environment, religious beliefs are not a necessary condition for ethical behaviour among Oromo farmers and peasants. Activities such as tilling the land, dealing with animals, and planting trees have their own moral codes, regardless of one's particular beliefs.

By means of both pragmatism and moral codes – religious or not – Oromo cosmovision implies a responsible attitude towards nature, and the essence of this vision is the partnership between humans and the environment. Once again, unlike Western and positivist approaches, humans are perceived as part of nature (Kelbessa 2005).

The Oromo political system (the "*Gadaa*" system) is also a source of insights, since it addresses the need to protect the rights of both human and non-human species. The Oromo care not only about animals and plants that are economically exploitable, for they consider that all species have inherent value. Some wild animals and groves are even considered sacred and have symbolic meanings; and domestic animals should not be mistreated, according to their principles (Kelbessa 2005).

Their concept of "*saffuu*" is also a relevant one for the understanding of the Oromo relationship with nature and what differentiates humans from other animals. It serves as a moral code that guides human behaviour in multiple kinds of situations, for it refers to mutual relationships in the cosmic order. As Kelbessa (2005) points out, "*saffuu*" is respecting one another and respecting one's own "*Ayyaana*" (spirit) and that of others. According to the Oromo, "*saffuu*" is "*ulfina*" (respect). We need to show respect to our father, mother, aunt, uncle, and our mother Earth. Since "*saffuu*" guides people's activities, it regulates the exploitation of natural resources.

## Daoism

Daoism provides relevant contributions regarding environmental ethics, and there is relatively abundant literature on the application of Daoist perspectives to discussions about the environment. Such contributions could be summarised in three main aspects, as proposed by Lai (2014): anti-anthropocentrism; opposition to human separateness and other dualisms; and holism and integrity.

Anti-anthropocentrism is a major theme in many debates on environmental philosophy and relates to the idea that human beings are unwilling or unable to act according to moral considerations towards nature, being responsible for the degradation of the environment and the extinction of species. The "*Daodejing*" – a Daoist classic often cited for its insights regarding the nature of reality and the relations between individual things and beings – provides interesting insights about the matter (Lai 2014).

First, it provides an inclusive theory which opposes the idea of human priority and implies that the relationship between humans and nature is characterised by interdependence and unity instead of dependence and separation. Second, it criticises the Confucian proposal of creating a human cultural identity which would be different and separate from all other beings and nature. Third, it promotes the idea of transcendence of the human condition and opposes that of imposition of human values on non-human existence. Fourth, as it rejects separation, it also rejects both dualism and the perspective that humans deserve absolute priority. Fifth, it recognises the integrity of all individuals – human and non-human – and seeks to promote the wellbeing of all (Lai 2014).

The anthropocentric perspective, often linked to a dualistic framework, can potentially discourage ethical behaviour, since it implies that, by harming nature, humans would not be negatively affecting themselves. Such a perspective could be damaging not only for the environment but also for humans. Also problematic is the fact that the dualities, or dichotomies, oversimplify many aspects of reality, especially those related to the connections between humans and the environment, and could restrict humans' capacity to properly evaluate ethical issues, for they present such connections as simple trade-offs (Lai 2014).

According to Lai (2014), holism is not universally accepted among environmental philosophers, because certain versions of environmental holism neglect the needs of individuals. Holism in Daoist philosophy, however, maintains a sense of individual integrity and assumes that the whole is neither just the sum of its parts nor independent from them. Since it is not considered an end in itself, its integrity and stability are valued because such conditions are fundamental to ensure the wellbeing of its parts, no matter if they are humans or non-humans.

Although holism in the *"Daodejing"* implies harmony and balance between the realisation of individual excellence within a context of interdependence and mutual enrichment, Daoist environmental ethics are not about the satisfaction of all parties. They are about achieving a "maximally coherent and superlative state of affairs" (Lai 2014, 189). This would be a state in which the whole is considered to be more than simply the sum of its parts, in the sense that not all individuals or groups may always achieve their desired outcomes, for interdependencies must be considered, as well as the need to negotiate and to compromise (Lai 2014).

## Conclusion

GSG has been traditionally based on the ideas of independence and separation of humans from nature and on the Holocene's stability patterns, which no longer hold. The severity of the current socio-environmental crisis and the evidence that planetary boundaries have been trespassed indicate that GSG studies need new frameworks or a new political imagination, as argued by Burke et al. (2016), to address non-human agency, the challenges presented by a new kind of power in the Anthropocene, and the need for a new global political project.

Based on this assumption, the chapter has tried to answer two main questions: why worlding is necessary to GSG studies and how to do it. Both the pluriverse and worldism are approaches to recognise that we live in many worlds, many natures, but on one planet (Ling 2014; Burke et al. 2016; Inoue and Moreira 2016; Querejazu 2016). We emphasised worldism as a way to respond to present complexities and challenges, because it implies communication and negotiation across difference, not incommensurability.

Therefore, GSG can be re-conceived and reframed by an Earth-worldly politics, based on the understanding of a single planet inhabited by multiple worlds, many worldviews, many knowledges, and many natures, in which the relationship between humans and nature is a social, geographical, and historical co-production (Burke et al. 2016, 20–21; Inoue and Moreira 2016, 8). This understanding has ontological, epistemological, and methodological implications for how GSG has been studied and practiced, with potentially more just, legitimate, and democratic practices.

The many worlds-one planet research agenda is a work in progress. The five cosmovisions offer some examples of how to world GSG, with ontological and normative implications.

In common, they all reject the human-nature dichotomy. When Andean and Oromo cosmovisions consider that all beings have the same ontological value, or all have an inherent value, they put forward matters of equality and justice that should be enforced in GSG. Also, Andean indigenous peoples consider that reality cannot be separated from the observer, which can reinforce mind-world monism – the world, or the "real," is not an independent part of knowledge – and, as consequence, challenge mainstream GSG research (Jackson 2011). Amazonian peoples consider that everything has a human origin, so there is no nature, but a society of societies (Danowski and Viveiros de Castro 2017). The Yanomamis, for example, consider the forest as the world. In this sense, politics is about everything (humans and non-humans, or former humans) in constant interaction, and the relations within the forest-world are considered a delicate balance.

The Islamic view can be considered the most anthropocentric of all the cosmovisions presented here, as it puts humans as the guardians of nature. However, it is based on trusteeship, justice, and intergenerational concerns, values that should be reinforced in GSG. Daoism is a valuable approach to consider complexities, and the maximally and superlative state of affairs offers ground to reconsider bargaining and interdependence, for all stakeholders in GSG should negotiate and compromise, bearing in mind that the wellbeing of all is not equivalent to the satisfaction of individual desired outcomes.

Finally, a GSG research agenda based on worldist and/or the pluriverse perspectives can, perhaps, be advanced in three ways. The first is to answer why is it necessary to recognise the existence of many worlds and to think about the possibilities of knowledge co-production among different worlds. The second is empirical research to look for alternative ways to understand today's main socio-environmental challenges and the normative claims that come with it. The third is to learn how to co-produce knowledge, listening and interacting with agents, whose voices have not been heard, to build new theories and practices of GSG. Such a move, perhaps, could respond to the call by Burke et al. (2016, 2) to develop "new practices, new ideas, stories, and myths (…)" in order to face Earth politics in a way that "humans, animals, ecologies, biosphere are all together."

## Notes

1  For Escobar (2016, 22), looking at reality as the "pluriverse" in contrast to the "universe" means acknowledging that there are multiple realities.
2  The pluriverse and worldism are not the same as multiculturalism. According to Querejazu (2016, 8), multiculturalism works within Western ontology, in a process of dominance. Different beliefs and imaginations are allowed in a general framework that does not question Western ontological premises. Therefore, in this work, cosmovision should not be read as a synonym for culture.
3  The role of non-humans in indigenous perspectives is addressed in more detail in section "Examples of perspectives that a worldist approach could draw from" of this chapter. For more about the Andean cosmovision, see also Vanhulst and Beling (this volume).
4  Ling (2014) considers the Westphalian world as state-centric and with an exclusionist version of civilisation, meaning that there is a hierarchy between world understandings – the Western rule with the upper position – and no possibility of mutual learning, reciprocity, respect, adaptation, connections, or continuities.
5  See Biodiversity and Ecosystem Services Network (2019) and IPBES (n.d.).
6  The expression Quechua is sometimes used as a synonym for Kichwa, but actually each word stands for different languages spoken in different South American regions, although they are very similar and share common origins. The language used as a reference to incorporate "Sumak Kawsay" in the Ecuadorian Constitution is Kichwa, the language spoken by most indigenous peoples living in Ecuador.
7  Ecuador was the first country to explicitly recognise the existence of Nature's Rights in its 2008 Constitution, including the right to be fully restored in case of damage or degradation.

# References

Acosta, Alberto. 2016. *O Bem Viver: uma oportunidade para imaginar outros mundos*. São Paulo: Editora Elefante.

Athayde, Simone et al., 2016. "Engaging Indigenous and Academic Knowledge on Bees in the Amazon: Implications for Environmental Management and Transdisciplinary Research". *Journal of Ethnobiology and Ethnomedicine*, 12 no. 1: 26.

Bäckstrand, Karin. 2006. "Democratizing Global Environmental Governance? Stakeholder Democracy after the World Summit on Sustainable Development". *European Journal of International Relations*, 12 no. 4: 467–498.

Barnett, Michael and Duvall, Raymond. 2005. "Power in Global Governance". In *Power in Global Governance*, Michael Barnett and Raymond Duvall, 1–32. New York: Cambridge University Press.

Berkes, Fikret. 2012. *Sacred Ecology*, 3rd ed. New York and Oxon, UK: Routledge.

Biodiversity and Ecosystem Services Network. 2019. "Dialogue Across Indigenous, Local and Scientific Knowledge Systems Reflecting on the IPBES Assessment on Pollinators, Pollination and Food Production", 20 January 2019 to 25 January 2019, Chiang Mai University, Thailand. www.besnet.world/event/dialogue-across-indigenous-local-and-scientific-knowledge-systems-reflecting-ipbes-assessment. Accessed on February 4, 2019.

Blaney, David L. and Tickner, Arlene B. 2017. "Worlding, Ontological Politics and the Possibility of a Decolonial IR". *Millennium: Journal of International Studies*, 45 no. 3: 1–19.

Burke, Anthony et al. 2016. "Planet Politics: A Manifesto from the End of IR". *Millennium Journal of International Studies*, 44 no. 3: 499–523. First Published April 7, 2016. doi: 10.1177/0305829816636674.

Chandler, David et al. 2017. "Anthropocene, Capitalocene and Liberal Cosmopolitan IR: A Response to Burke et al.'s 'Planet Politics'". *Millennium: Journal of International Studies*, 46 no. 2: 190–208. To link to this article: doi: 10.1177/0305829817715247.

Danowski, Déborah and Viveiros de Castro, Eduardo. 2017. *Há mundo por vir? Ensaios sobre os medos e os fins*. 2ª Edição Desterro [Florianópolis]: Instituto Socioambiental.

Escobar, Arturo. 2016. "Thinking-feeling with the Earth: Territorial Struggles and the Ontological Dimension of the Epistemologies of the South". *Revista de Antropología Iberoamericana* 11 no. 1: 11–32.

Hamilton, Scott. 2016. "The Measure of all Things? The Anthropocene as a Global Biopolitics of Carbon". *European Journal of International Relations*, 24 no. 1: 33–57.

Harrington, Cameron. 2016. "The Ends of the World: International Relations and the Anthropocene". *Millennium Journal of International Studies*, 44 no. 3: 478–498.

Inoue, Cristina Yumie A. 2007. *Regime Global de Biodiversidade e o Caso Mamirauá*. Brasília: Editora Universidade de Brasília.

Inoue, Cristina Yumie A. and Moreira, Paula Franco. 2016. "Many Worlds, Many Nature(s), One Planet: Indigenous Knowledge in the Anthropocene". *Revista Brasileira de Política Internacional* 59 no. 2: e009.

IPBES Science and Policy for People and Nature. n.d. "New IPBES Assessment Begins: Making the Many Values of Nature Count". Accessed on February 4, 2019. www.ipbes.net/news/new-ipbes-assessment-begins-making-many-values-nature-count.

Jackson, Patrick T. 2011. *The Conduct of Inquiry in International Relations. Philosophy of Science and Its Implications for the Study of World Politics*. New York: Routledge.

Jang, Jinseop et al. 2016. "Global Governance: Present and Future." *Palgrave Communications*, 2 Article number: 15045.

Jordan, Andrew J. et al. 2015. "Emergence of Polycentric Climate Governance and Its Future Prospects". *Nature Climate Change*, 5: 977–982.

Kamla, Rania, Gallhofer, Sonja and Haslam, Jim. 2006. "Islam, Nature and Accounting: Islamic Principles and the Notion of Accounting for the Environment". *Accounting Forum*, 30 no. 3: 245–265.

Kassam, Karim-Aly S. 2009. *Biocultural Diversity and Indigenous Ways of Knowing: Human Ecology in the Arctic*. Calgary: University of Calgary Press, open access. https://press.ucalgary.ca/books/9781552382530. Accessed on February 6, 2019.

Kelbessa, Workineh. 2005. "The Rehabilitation of Indigenous Environmental Ethics in Africa." *Diogenes*, 52 no. 3: 17–34.

Lai, Karyn L. 2014. "Conceptual Foundations for Environmental Ethics: A Daoist Perspective". In *Environmental Philosophy in Asian Traditions of Thought*, J. B. Callicott and J. McCae, 173–195. Albany: State University of New York Press.

Latour, Bruno. 1993. *We Have Never Been Modern*. Cambridge: Harvard University Press.

Ling, Lily H. M. 2014. *The Dao of World Politics. Towards a Post-Westphalian Worldist International Relations*. London and New York: Routledge.

Moreira, Paula Franco and Baniwa, D. 2011. "Linking Local Knowledge of Indigenous Peoples to Global Climate Knowledge. A Preliminary Assessment of Climate Change Impacts and Needs for Adaptation from Baniwa Based on the Içana River, Brazilian Amazon – A Case Study." Presentation at Proceedings of the Expert Workshop on Indigenous Peoples, Marginalized Populations and Climate Change. Vulnerability, Adaptation and Traditional Knowledge, Mexico City, Mexico, 19–21 July 2011. https://unesdoc.unesco.org/in/rest/annotationSVC/DownloadWatermarkedAttachment/attach_import_c8568f33-b235-4b4e-8ab7-caaea4eb923a?_=265874eng.pdf.

Pereira, Joana Castro. 2017. "The Limitations of IR Theory Regarding the Environment: Lessons from the Anthropocene." *Revista Brasileira de Política Internacional* 60 no. 1: e018.

Querejazu, Amaya. 2016. "Encountering the Pluriverse: Looking for Alternatives in Other Worlds." *Revista Brasileira de Política Internacional* 59 no. 2: e007.

Robbins, Jim. 2018. "Native Knowledge: What Ecologists Are Learning from Indigenous People." *Yale Environment 360*. April 26, 2018. https://e360.yale.edu/features/native-knowledge-what-ecologists-are-learning-from-indigenous-people.

Schroeder, Heike and González, Nídia C. 2019, March. "Bridging Knowledge Divides: The Case of Indigenous Ontologies of Territoriality and REDD+." *Forest Policy and Economics*, 100: 198–120. doi: 10.1016/j.forpol.2018.12.010.

Steffen, Will et al. 2015. "The Trajectory of Anthropocene: The Great Acceleration." *The Anthropocene Review* 2 no. 1: 81–98.

Swift, Peter and Cock, Andrew. 2015. "Traditional Khmer Systems of Forest Management". *Journal of the Royal Asiatic Society* 25 no. 1: 153–173.

Tengö, Maria et al. 2014. "Connecting Diverse Knowledge Systems for Enhanced Ecosystem Governance: The Multiple Evidence Base Approach." *AMBIO*, 43 no. 5: 579–591.

Tickner, Arlene, and Waever, Ole. 2009. "Introduction. Geocultural Epistemologies". In *International Relations Scholarship around the World. Worlding beyond the West*, organized by Arlene Tickner and Ole Waever, 2–31. London and New York: Routledge.

Underdal, Arild. 2008 "Determining the Causal Significance of Institutions. Accomplishments and Challenges. In *Institutions and Environmental Change. Principal Findings, Applications and Research Frontiers*, organized by Oran R. Young, Leslie A. King and Heike Schroeder, 49–78. Cambridge: MIT Press.

Wapner, Paul. 2014. "The Changing Nature of Nature: Environmental Politics in the Anthropocene." *Global Environmental Politics* 14 no. 4: 36–54.

Whyte, Kyle Powis. 2013. "On the Role of Traditional Ecological Knowledge as a Collaborative Concept: A Philosophical Study." *Ecological Process* 2 no. 7. doi: 10.1186/2192-1709-2-7.

# Part 2
# Ethics, principles, and debates

# 6

# Justice[1]

*Agni Kalfagianni, Andrea K. Gerlak, Lennart Olsson, and Michelle Scobie*

## Introduction

Questions of justice are becoming central political discourses in a world characterised by growing inequality. Currently, governments and intergovernmental organisations formulate goals and set priorities for action with the aim to address justice issues on a global scale. For example, three of the recently adopted Sustainable Development Goals (SDGs) focus explicitly on reducing inequalities within and across countries (Goal 10), promoting gender equality (Goal 5), and peace and justice (Goal 16). In addition, private actors like businesses and civil society organisations (e.g. the Ethical Trading Initiative and the Fairtrade Labelling Organization) are creating institutions that rely on market forces to generate "fair" distribution of environmental or social goods. Likewise, activist and grassroots networks such as Global Justice Now are engaged with justice concerns around the globe.

Despite these efforts to address inequalities in the world today, the very concept of justice is elusive and means different things to different people. Justice is typically associated with allocation, or "the process of allocating or sharing out something" (Oxford Dictionary). Yet, justice as a broader term is also understood in its colloquial use as "the quality of being fair and reasonable" (Oxford Dictionary) as "fairness in the way people are dealt with" (Cambridge Dictionary) or as "the principle of fairness that like cases should be treated alike" (Collins Dictionary). Also, different disciplines tend to refer to justice differently. While economists emphasise allocation, lawyers adopt a language of rights, resource analysts of access, political scientists of fairness in representation and access, and sociologists of social justice (Biermann et al. 2009; Gupta and Lebel 2010). Likewise, political philosophers have developed different theories of justice. For example, in the socialist tradition the central requirement for justice is culminated in the phrase: "from each according to his ability, to each according to his needs" (Marx 1875). In contrast, liberal egalitarian theorists would stress "fair distribution" of various goods and resources resulting from the fairness of procedures (e.g. Rawls 1971, 1999). Capabilities theorists would propose to evaluate how these goods or resources are transformed into the capacity of individuals to function in lives of their own choosing (e.g. Sen 1999, 2009; Nussbaum 2000). Cosmopolitan theorists would extend liberal egalitarian concerns at the global level (e.g. Beitz 1999; Caney 2005; Beck 2006). And others would deny the possibility of global justice entirely (e.g. Nagel 2005).

In view of this complexity, this chapter asks: how can we systematically compare and evaluate justice claims or demands for justice in global sustainability governance research? We propose that a justice lens would need to clarify the subjects, principles, mechanisms and instruments for justice, as well as the consequences of just societies. We discuss these various elements in some greater detail below.

## The subjects of justice

According to political philosopher Thomas Nagel "we do not live in a just world" (Nagel 2005). The fact that this proposition is uncontroversial does not mean that the concept of justice is uncontested nor that it is clear what it should entail. But before examining what a system of justice covers, any approach to justice needs to clarify who the subjects of justice – or else the units of moral concern – are. For the purpose of global sustainability governance research, we find it useful to conceptualise the subjects of justice in four dimensions (Jerneck et al. 2011): intergenerational (between generations; see also Lawrence this volume), international (between states and regions; see also Okereke this volume), intersectional (between groups/categories in society), and interspecies (between humans and other species).

### Justice between generations

Intergenerational justice is core to environmental concerns for both natural and social reasons (Gündling 1990; Weiss 1990). The inertia of many natural systems and phenomena is one obvious reason why intergenerational considerations are essential. For example, greenhouse gases are persistent over several generations, while the atmosphere responding to these gases interacts with oceans and icecaps operating at time scales of decades and millennia. Extraction of finite resources like oil, coal, or minerals is fundamentally a matter of intergenerational justice. The generation of long-lived hazardous materials, such as nuclear waste where one generation reaps the benefits of nuclear power while hundreds of generations will live with the waste and the potential use of nuclear arms also raises intergenerational issues. Finally, irreversible processes, such as extinction of species or permanent depletion of resources, are of intergenerational importance. Accordingly, intergenerational justice brings to the fore questions of responsibility (Birnbacher 2006; Pellizzoni this volume) of present generations to the future together with questions of representation and voice. We provide some examples of the latter in the section of "mechanisms for justice" below, but for a more extended discussion see Lawrence (this volume).

### Justice between states and regions

International justice has a long tradition of research and scholarship in global sustainability governance (Toth 2013), often from the point of view of international relations and international law. As Okereke (this volume) highlights there has been no significant environmental summit, multilateral agreement, or global environmental institution that has not been severely challenged by issues of North-South inequity and justice since the 1972 United Nations Conference on the Human Environment (UNCHE), when sustainability was first introduced as a focus for international governance. Many of our most pressing environmental challenges have explicit and implicit international implications and drivers. Climate change, in particular, is to a large extent caused by industrialised countries over many decades (and thereby seen as an intergenerational justice issue) but will have much more severe negative impacts on countries in the global South (Field et al. 2014). Many of the policies and mechanisms for addressing climate change are initiated by the global North but with significant implications for energy security and energy resilience for people in the global South (Scobie 2019). Many value chains benefitting affluent societies are based on exploitation of natural resources and repression of people elsewhere, for example, phosphorous for agriculture and rare earth minerals for computers and telephones. Similar observations extend to loss of biodiversity, marine pollution, overfishing, or depletion of water resources (Scobie 2013).

## Justice between groups in society

Intersectional justice relates to expressing the multiple dimensions and modalities of social relations and subject formations we belong to (McCall 2005). In sustainability governance, intersectional justice can be understood in relation to multiple deprivations at context-specific intersections of age, class, caste, (dis)ability, gender, indigeneity, and race. Examples of intersectional (in)justice are rife in regards to climate change impacts as well as impacts of climate change policies (Kaijser and Kronsell 2014; Olsson et al. 2014). However, intersectional justice also has an emancipatory perspective and is used by activist groups who want to overcome a more narrow-focused politics addressing singular justice concerns and move towards a more strategic and relational vision of environmental and social issues. Examples of this include women activists who link environmental and feminist concerns (Di Chiro 2008), alliances between voices who are usually marginalised in the political agenda (Kaijser and Kronsell 2014), and more just and equitable planning, preparedness, response, and recovery activities in response to environmental disasters (Ryder 2017).

## Justice between species

Environmental ethics approaches emphasise the need to extend justice considerations beyond humans towards non-human species and the natural world (Donaldson and Kymlicka 2011; Benton 2018; see also Gumbert, this volume). Such an ecocentric view on justice requires moving away from an instrumentalist perspective of the non-human world (e.g. as one that provides satisfaction of basic needs and human flourishing) towards one that acknowledges the independent intrinsic value of nature. This would require recognising and respecting other ways of life as equal to humans. Similar to intergenerational justice, however, this approach is also contested. The main criticisms include the convincingness of moral arguments of equality among species (Cooper 2018), our limits to fully understand and know what is good for other species (Soper 2018), and the appropriateness of a discourse of justice for capturing humans' ethical relationships to other species (Hay 2018). As a way around these challenges, some scholars argue for interspecies justice from the perspective of stewardship, care (Hay 2018), and solidarity (Hayward 2018).

## Principles of justice

Principles of justice are moral propositions that serve as the foundation of all actions and institutions that aim to achieve just outcomes. Below we identify three main principles of relevance to the current literature of environmental justice and global sustainability governance research.

## Distribution

Distribution as a principle of justice evaluates how and to what end a just society allocates the costs and benefits of social cooperation (Rawls 1971). This perspective emphasises that justice fundamentally concerns the basic structure of society and how this defines and regulates social, economic, and environmental equality and inequality. For sustainability governance, distributive justice would pay attention to the institutions that are responsible for distributing such costs and benefits across different generations, among jurisdictions and other regions, and among different groups in societies worldwide. There is no widespread consensus on what is considered just distribution, however (Luterbacher and Sprinz 2001). To illustrate, utilitarians accept as just the distribution that, on average, produces more benefits than costs. Scholars in the liberal egalitarian tradition, in contrast, adopt a (global) "difference principle" whereby inequality in the distribution of costs and benefits is acceptable as long as this benefits the least advantaged members of society (Beitz 1979; Caney 2001, 2005; Moellendorf 2002). Environmental justice advocates focus on access to resources, environmental rights and duties, and the fairness of their distribution among geographical, ethnic, economic, social, and other communities (Bullard 1994). Still others advocate a needs-based minimum floor principle whereby basic needs should be satisfied first before any distribution is considered (Brock 2009). The plurality of distributive justice principles invites sustainability scholars to clarify and unravel the principles that underline the multiple governance processes in which decisions regarding "who gets what and why" are being negotiated and disputed.

## Recognition

Recognition as a principle of justice contends that, if a group or individual lacks recognition in the social or political structures within a society, it will contribute to maldistribution (Young 1990; Fraser 1997, 2001). Lack of recognition occurs when people are devalued, dominated, or disrespected due to their identity or status. Recognition and distribution are two distinct experiences of justice, but are intrinsically linked. Misrecognition manifests in the structures, practices, rules norms, and language. In turn, it is within this context that the maldistribution is instigated (Fraser 1997). Recognition can be achieved when individuals are free of physical threats, offered complete and equal political rights, and have distinguishing cultural traditions free from various forms of disparagement (Honneth 2001). This perspective of justice invites scholars to contemplate who is being recognised or misrecognised as a subject of justice in global sustainability governance processes.

## Representation

Representation or procedural justice describes democratic, fair, and equitable processes in decision-making (Schlosberg 2007). It demands that all groups, especially those most affected, are fully provided the opportunity to participate in the decision-making process, and the decision-making should be shared. It also requires that all (affected) actors participate in an impartial way and ensure full disclosure of the content of information, how it is provided, if it is provided in a timely manner, and to whom it is given, so as to facilitate effective participation. In other words, representation emphasises the importance of the political process through which existing injustices in distribution and recognition can be addressed (Young 1990). For global sustainability governance research, representation requires evaluating, for instance, the democratic character of the processes through which decisions affect the distribution of environmental costs and benefits, as well as the economic costs and benefits of proposed solutions. It further entails questioning who are considered and recognised as legitimate participants and beneficiaries of cooperation along with those who are not considered like certain nation-states, social groups, different generations, and non-human species.

## Mechanisms and instruments for justice

How – if at all – do sustainability governance institutions organise themselves to address injustice? Below we discuss existing and potential mechanisms and instruments to foster justice with varying demands and ambitions which reflect and inform global sustainability governance research.

### Legal rights

In international forums, legal and human rights are seen as one path to advance equity claims of disadvantaged and underserved peoples and nations. The Common but Differentiated Responsibility (CBDR) and capability principle, for instance, is the main legal justice norm addressing North-South injustices. It has enabled countries to maintain cooperation in climate, biodiversity, and other environmental concerns even though there remain important differences over how to interpret and operationalise CBDR in practice (see Okereke this volume).

Human rights to water are considered to have enormous mobilising potential and may help redress the imbalance between the have and have-nots in water allocation and use (Sultana and Loftus 2012). In those countries that have institutionalised the human right to water in their constitutions or national legislation, it may serve as a moral articulation and as a basis for legal challenges, even if there are limitations in terms of implementation (Gerlak and Wilder 2012; Baer and Gerlak 2018). Among other things, access to systems of implementation and justice at national and international levels is needed to ensure implementation of those rights for the poorest and most vulnerable (Gupta and Lebel 2010), and proper recognition is afforded. An example of combining the human right to water with the environmental right to water, for instance, is South Africa's Free Basic Water policy, which institutionalised the idea that water used for necessities is free, but above that level it needs to be paid for (Muller 2008).

Today also a variety of institutional and legal efforts to represent future generations can be identified (see also Lawrence, this volume). At the national level, a number of countries have made progress towards institutionally recognising future generations such as the Canadian Commissioner of the Environment and Sustainable Development, the Finnish Committee for the Future, the Parliamentary Advisory Council for Sustainable Development in Germany, and the Parliamentary Commissioner for Future Generations and Deputy Ombudsperson for Future Generations in Hungary. Internationally, although proposals for a United Nations High Commissioner for Future Generations (UNHCFG) have not yet materialised, the concept of intergenerational equity has arguably influenced the decision to include a more ambitious temperature goal of 1.5°C (United Nations Joint Framework Initiative on Children Youth and Climate Change 2010).

Another application of legal rights is to non-human entities. Although this idea is not entirely new (Salmond 1947; Stone 1972) it has only recently begun to be implemented (O'Donnell and Talbot-Jones 2018). Examples include the granting of constitutional rights to nature in Ecuador in 2008 (Constitution of the Republic of Ecuador 2008, articles 71–74), the creation of legal rights for nature in Bolivia in 2010 ("the Law of Mother Earth"), and the attribution of legal status of persons to rivers in 2017 (the Whanganui River in New Zealand, and the Ganges and Yamuna rivers in India). Although it is too soon to evaluate the effects of such legal approaches, scholars underline their importance as complementary to existing legal frameworks to address complex cultural, environmental, and economic issues to the extent that there is the financial and institutional capacity to support them (O'Donnell and Talbot-Jones 2018).

## Democratic processes

Legitimate and transparent democratic processes permit societies and communities to choose equitable policies to address environmental problems (Biermann et al. 2012) and is part of procedural justice. Justice can be achieved through public participation in decision-making, by empowering communities, and seeking equitable distribution (Anand 2004). Governance architectures (Biermann et al. 2009) have the potential to challenge injustices if adequately inclusive in their construction and provide systems for rule making, monitoring, and enforcement in matters of the global commons (Andersson and Agrawal 2011; see however Blühdorn and Deflorian, this volume, on challenges to this potential). If not, these architectures risk locking in existing injustices. In this context, democratic governance systems that seek to distribute power amongst citizens in ways that curtail the power of any single individual or interest group can potentially reduce inequalities amongst individuals and groups.

Accordingly, mobilising the agency of local communities, indigenous peoples, and non-governmental organisations to help shift towards more mutual learning and capacity building approaches at different governance levels is considered to be a key part of promoting democracy in sustainability governance (Dryzek and Stevenson 2011), and illustrates representative or procedural justice (see, however, Litfin, this volume, for a critical view on initiatives limited to the local level). New alternative discourses and social movements are often necessary to promote a re-allocation of resources and shift to more just and equitable patterns of use (Gupta and Lebel 2010). The widespread anti-privatisation movement around water in Latin America over the past two decades illustrates the power of social movements to protect marginalised populations and reverse neoliberal water reforms at national and local levels (Bustamante et al. 2012). The debates supporting privatisation of marine genetic resources obtained from areas beyond national jurisdiction via patenting show that justice challenges continue (Scovazzi 2016). Increasingly, climate justice activists and movements are also relying on the local experience of increasing vulnerability to climate change and adaptive responses to climate change helping to shift beyond traditional distributive justice approaches to addressing injustice (Schlosberg 2013). Simultaneously, deliberative democracy processes are supposed to facilitate the inclusion of the voices of those not present such as future generations and the non-human world (Dryzek and Stevenson 2011).

## Economic tools

Intergenerational justice is at the core of macro-economic analyses and policy advice through the practice of discounting the future. Mainstream economic models often discount the welfare of future generations by adding the utilities of all people of the current generation, and adding the weighted (by a discount rate) utilities of all people of future generations. The choice of discount rate explains why the Stern Review (using a very low discount rate of 0.1%) concluded that early mitigation of climate change is a priority in order to achieve a stabilisation of the climate at about 2°C (Stern 2007), while William Nordhaus (using a discount rate of 6%) concluded that the social optimum of mitigation is a mere 11% below business as usual (Nordhaus 1992).

Economic tools typically focus on distributive justice. For example, some scholars advocate stronger financial support for poorer countries, through direct support payments for climate change mitigation and adaptation programmes based on international agreements or through international market mechanisms, like global emissions markets (Biermann et al. 2010) or more recently through payments for loss and damage caused by climate change through the United Nations Framework Convention on Climate Change (Darren 2016). Carefully designed and monitored market mechanisms for climate change mitigation and technology transfers can help to address inequalities between industrialised and developing countries (Dryzek and Stevenson 2011; Rode et al. 2015). However, fossil fuel subsidies, long seen as a tool of redistributive justice in some developing states, are being removed with pressure from climate lobbies, and the challenge is to find more efficient redistributive mechanisms to help the energy poor (Scobie 2017).

Internationally, labelling strategies and financial instruments like tradable certificates or taxes have been promoted to better inform consumers, producers, and institutions about water usage and ultimately, to shift the financial burden to customers of water-intensive products (Hoff 2009). However, private standards and certification schemes are also criticised as mechanisms for delivering justice and have been questioned in terms of delivering environmental benefits (see Ebeling and Yasué 2009). First, standards are products and extensions of broader political and socio-economic structures, such as the neoliberalisation of policies worldwide, designed to strengthen market independence from government regulation while providing market alternatives rather than addressing structural inequalities and injustices. Research underlines the use of such instruments to open up new spaces for material control, social legitimation, political power, and environmental management, particularly for transnational corporations (Kalfagianni 2014). Second, in contrast to command and control regulation that nominally forces all to abide by a set of constraints, voluntary economic tools enable economic actors to pay compensation to continue environmentally damaging behaviour (Guthman 2007). As a result, environmental and social sustainability becomes subject to the highest bidder (Busch 2011) or lacks strong enough governance with highly uneven effects (Ebling and Yasué 2009).

## Consequences of just societies

Several of the key conceptual elements discussed earlier are also part of the global sustainable development agenda and the UN 2030 SDGs. As mentioned earlier, Goal 5 aims to *"Achieve gender equality and empower all women and girls."* Goal 10 aims to *"Reduce inequality within and among countries."* Goal 16 is designed to *"Promote peaceful and inclusive societies for sustainable development, provide access to justice for all and build effective, accountable and inclusive institutions at all levels."* These SDGs depend upon humanity being able to achieve intergenerational, international, intersectional, and interspecies justice (Linklater 1999; Levy and Scobie 2015; Stevens and Kanie 2016; Zhang et al. 2016). Further, the SDGs related to food and nutritional security (Patrick, Syme, and Horwitz 2014; Shingirai and Happy 2017), health, education, job security, fair access to land, sustainable agriculture (van Bommel et al. 2016), fair access to genetic resources, sustainable resource use (Jaeckel, Gjerde, and Ardron 2017), fair access to technology and to energy (Wolf et al. 2016) and fair and sustainable systems of trade (Scobie 2013) all face the challenge of being dependent on or creating robust justice systems (Levy and Scobie 2015).

The governance of complex global economic and environmental systems requires strong social, economic, and legal institutions that are multilevel and hybrid in terms of the actors, sectors, spaces, and forms of relationships (Boehmelt, Koubi, and Bernauer 2014; Kalfagianni 2014; Ramos 2015; Nunes, Rajão, and Soares-Filho 2016; Stratoudakis et al. 2016). They should also be context-specific, flexible, participatory, representative, inclusive, accountable (Donald and Way 2016), transparent, resilient, and effective (Bracking 2015; van Bommel et al. 2016).

Just and non-discriminatory legal and regulatory systems and institutional frameworks can serve to reduce human suffering. They can resolve conflicts and are indispensable for promoting and maintaining peaceful societies, the fair distribution of environmental rights to goods and services (Griggs et al. 2014; Paloniemi et al. 2015; Scovazzi 2016), risks (Thaler and Hartmann 2016), duties (including positive duties of care or negative duties to refrain from harming the environment) (Barrett 2011; Duus-Otterström and Jagers 2012; Saarinen 2013; Norstrom et al. 2014; Asmelash 2015), and preserving delicate physical environmental systems including genetic, historical, and cultural assets.

Finally, religious and ethical worldviews (Dash 2014; Esquivel and Mallimaci 2017; see also Glaab this volume) have the potential to form the overarching delivery framework and contexts for partnerships for ending poverty and inequality (Feygina 2013), for sustainable financing, capacity building, technology sharing and transfer, and for quick responses to environmental shocks and crisis situations at national or local scales. Indeed, religious and ethical worldviews are often the drivers of global solidarity but also of subsidiarity, supporting local and community-scaled initiatives and solutions that are outside centralised political governance and that directly redress economic and environmental inequity and poverty. This is crucial as subsidiarity is especially important in sustainability governance in that it favours solution brokering wherever possible at local scales and in areas that are directly impacted by improvements.

## Conclusion and looking forward

Experts contend that justice needs to become an explicit topic of academic research (Klinsky et al. 2017), as no institutional framework for sustainable development will be effective and legitimate in the long run unless it has justice and fairness concerns at its core (Pritchard 1969; Adger et al. 2005; Bierman et al. 2012). In this chapter, we argued that a justice lens for global sustainability governance research would need to clarify the subjects, principles, mechanisms and instruments for justice, and consequences of just societies.

Scholars of critical sustainability governance, in particular, need to embrace a broad range of subjects of justice including future generations and the non-human world. In addition, such scholarship needs to advance an encompassing understanding of principles of justice. Distributive, procedural, and recognition principles can both reinforce and undermine each other. For example, fair procedures cannot materialise unless actors are recognised as subjects worthy of participation and representation in them. Likewise, fair distribution can be undermined by unfair procedures when deciding what is at stake and who should get what and why. Regarding mechanisms of justice, these are currently both underdeveloped and tend to rely too much on voluntary economic instruments. We are sceptical that reliance on such instruments alone will deliver the more progressive and encompassing forms of justice that are necessary today. Instead, we find encouraging the legal and institutional innovations specifically towards the recognition of subjects of justice across generations and across species.

However, we also wish to underline that contextual conditions matter for justice. Indeed, the way justice is operationalised and the hierarchy of and relationships between principles and instruments to be applied depends on the nature of the injustices and of the limited resources that are to be apportioned. A one size fits all model in applying principles and instruments may result in injustices. Scholars of critical sustainability governance would need to include historical and future, economic, political, social, and environmental contexts and the impact of these on human and non-human actors in any analysis of the way justice should be operationalised for fair sustainability outcomes.

To conclude, in order to advance a research agenda on justice, we propose that future research should take into account the following aspects.

First, there is a need for more interdisciplinary approaches to better understand the outcomes associated with mechanisms to advance justice and how best to understand success in achieving it. The integration of theoretical constructs and methodologies from diverse disciplines can allow us to better understand the most fruitful governance pathways for addressing injustices in distribution, recognition, or representation of the various subjects of justice.

Second, we argue that it is necessary to examine which new demands for justice, fairness, and allocation are emerging in a world where "planetary boundaries" are being crossed (Rockström et al. 2009; Steffen et al. 2015), where the Anthropocene (Crutzen 2006) confronts us with our ability to change the entire earth system, and where global inequalities are ever increasing.

Finally, determining which types of steering have been helpful and not helpful to channel personal, regional, national, and global worldviews towards more sustainable approaches to environmental rights and duties is of the utmost importance. Sustainable outcomes are a consequence of the willingness of peoples and groups to "live more simply so that others may simply live," but achieving this is anything but simple. Global sustainability governance science can help discover the triggers, drivers, and types of agency that can be scaled up to support just outcomes, making both intellectual and policy contributions towards a more sustainable world.

## Note

1 A version of this chapter is available online as part of the Earth System Governance Science and Implementation Plan 2018: www.earthsystemgovernance.org/wp-content/uploads/2018/11/Earth-System-Governance-Science-Plan-2018.pdf.

## References

Adger, W. N., K. Brown and M. Hulme. (2005). Redefining global environmental change (editorial). *Global Environmental Change: Human and Policy Dimensions* 15(1): 1–4.

Anand, R. (2004). *International Environmental Justice: A North-South Dimension.* Aldershot: Ashgate Publishing.

Andersson, K. and A. Agrawal. (2011). Inequalities, institutions, and forest commons. *Global Environmental Change* 21(3): 866–875.

Asmelash, H. B. (2015). Energy subsidies and WTO dispute settlement: Why only renewable energy subsidies are challenged. *Journal of International Economic Law* 18(2): 261–285.

Baer, M. and A. K. Gerlak. (2015). Implementing the human right to water and sanitation: A study of global and local discourses. *Third World Quarterly* 36(8): 1527–1545.

Barrett, S. (2011). Rethinking climate change governance and its relationship to the world trading system. *World Economy,* 34(11): 1863–1882.

Beck, U. (2006). *Cosmopolitan Vision.* Cambridge: Polity Press.

Beitz, C. (1979). *Political Theory and International Relations.* Princeton, NJ: Princeton University Press.

Benton, T. (2018, 2nd Edition). Ecology, community and justice. In: T. Hayward and J. O'Neil (eds.). *Justice, Property and the Environment: Social and Legal Perspectives*. London: Routledge.

Biermann, F., M. M. Betsill, J. Gupta, N. Kanie, L. Lebel, D. Liverman, H. Schroeder and B. Siebenhüner with contributions from K. Conca, L. da Costa Ferreira, B. Desai, S. Tay and R. Zondervan. (2009). *Earth System Governance: People, Places and the Planet. Science and Implementation Plan of the Earth System Governance Project*. Earth System Governance Report 1, IHDP Report 20. Bonn, IHDP: The Earth System Governance Project.

Biermann, F., et al. (2012). Navigating the Anthropocene: Improving earth system governance. *Science* 335(6074): 1306–1307.

Birnbacher, D. (2006). Responsibility for future generations: Scope and limits. In: J. C. Tremmel (ed.). *Handbook of Intergenerational Justice*. Cheltenham: Edward Elgar.

Boehmelt, T., V. Koubi and T. Bernauer (2014). Civil society participation in global governance: Insights from climate politics. *European Journal of Political Research* 53(1): 18–36. doi: 10.1111/1475-6765.12016

Bracking, S. (2015). The anti-politics of climate finance: The creation and performativity of the Green Climate Fund. *Antipode* 47(2): 281–302. doi: 10.1111/anti.12123

Brock, G. (2009). *Global Justice: A Cosmopolitan Account*. Oxford: Oxford University Press.

Busch, L. (2011). The private governance of food: Equitable exchange or bizarre bazaar? *Agriculture and Human Values* 28(3): 345–352.

Bustamante, R., C. Crespo and A. M. Walnycki. (2012). Seeing through the concept of water as a human right in Bolivia. In: F. Sultana and A. Loftus (eds.). *The Right to Water: Politics, Governance and Social Struggles*. London: Routledge, pp. 223–240.

Caney, S. (2001). International distributive justice. *Political Studies* 49(5): 974–997.

Caney, S. (2005). *Justice Beyond Borders*. Oxford: Oxford University Press.

Cooper, D. E. (2018). Justice, consistency and 'non-human' ethics. In: T. Hayward and J. O'Neil (eds.). *Justice, Property and the Environment: Social and Legal Perspectives*. London: Routledge.

Crutzen, P. J. (2006). The "Anthropocene". In: E. Ehlers and T. Kraft (eds.). *Earth System Science in the Anthropocene*. Berlin: Springer, pp. 13–18.

Dash, A. (2014). The moral basis of sustainable society: The Gandhian concept of ecological citizenship. *International Review of Sociology*, 1–11. doi: 10.1080/03906701.2014.894343

Di Chiro, G. (2008). Living environmentalisms: Coalition politics, social reproduction and environmental justice. *Environmental Politics* 17(2): 276–298.

Donald, K. and S. A. Way. (2016). Accountability for the sustainable development goals: A lost opportunity? *Ethics & International Affairs* 30(2): 201–213.

Donaldson, S. and W. Kymlicka. (2011). *Zoopolis: A Political Theory of Animal Rights*. Oxford: Oxford University Press.

Dryzek, J. S. and H. Stevenson. (2011). Global democracy and earth system governance. *Ecological Economics* 70: 1865–1874.

Duus-Otterstrom, G. and S. C. Jagers. (2012). Identifying burdens of coping with climate change: A typology of the duties of climate justice. *Global Environmental Change-Human and Policy Dimensions* 22(3): 746–753. doi: 10.1016/j.gloenvcha.2012.04.005

Ebeling, J. and M. Yasué. (2009). The Effectiveness of market-based conservation in the tropics: Forest certification in Ecuador and Bolivia. *Journal of Environmental Management* 90(2): 1145–1153.

ESG (Earth System Governance). (2009). Science and Implementation Plan of the Earth System Governance Project, IHDP Report No. 20.

Esquivel, J. C. and F. Mallimaci. (2017). Religión, medioambiente y desarrollo sustentable: la integralidad en la cosmología católica. [Religion, environment and sustainable development: Integrality in Catholic cosmology]. *Revista de Estudios Sociales* 60: 72–86.

Feygina, I. (2013). Social justice and the human–environment relationship: Common systemic, ideological, and psychological roots and processes. *Social Justice Research* 26(3): 363–381. doi: 10.1007/s11211-013-0189-8

Fraser, N. (1997). *Justice Interruptus: Critical Reflections on the 'Postsocialist' Condition*. New York: Routledge.

Fraser, N. (2001). Recognition without ethics? *Theory, Culture, and Society* 18: 21–42.

Gerlak, A. K., M. Baer and P. Lopes. (2018). Taking stock of human right to water. *International Journal of Water Governance* 6: 108–134.

Gerlak, A. K. and M. Wilder. (2012). Exploring the textured landscape of water insecurity and the human right to water. *Environment: Science and Policy for Sustainable Development* 54(2): 4–17.

Griggs, D., M. S. Smith, J. Rockstrom, M. C. Ohman, O. Gaffney, G. Glaser, … P. Shyamsundar. (2014). An integrated framework for sustainable development goals. *Ecology and Society* 19(4): 24.

Gündling, L. (1990). Our responsibility to future generations. *American Journal of International Law* 84(1): 207–212.

Gupta, J. and L. Lebel. (2010). Access and allocation in earth system governance: Water and climate change compared. *International Environ Agreements* 10: 377–395.

Guthman, J. (2007). The Polanyian way? Voluntary food labels as neoliberal governance. *Antipode* 39(3): 456–478.

Hayward, T. (2018). Interspecies solidarity: Care operated upon justice. In: T. Hayward and J. O'Neil (eds.). *Justice, Property and the Environment: Social and Legal Perspectives.* London: Routledge.

Hoff, H. (2009). Global water resources and their management. *Current Opinion in Environmental Sustainability* 1: 141–147.

Honneth, A. (2001). Recognition or redistribution: Changing perspectives on the moral order of society. *Theory, Culture, and Society* 18(2–3): 43–55.

Jaeckel, A., K. M. Gjerde and J. A. Ardron. (2017). Conserving the common heritage of humankind – Options for the deep-seabed mining regime. *Marine Policy* 78: 150–157. doi: 10.1016/j. marpol.2017.01.019

Jerneck, A., L. Olsson, B. Ness, S. Anderberg, M. Baier, E. Clark, … J. Persson. (2011). Structuring sustainability science. *Sustainability Science* 6(1): 69–82.

Kaijser, A. and A. Kronsell. (2014). Climate change through the lens of intersectionality. *Environmental Politics*, 23(3): 417–433.

Kalfagianni, A. (2014). Addressing the global sustainability challenge: The potential and pitfalls of private governance from the perspective of human capabilities. *Journal of Business Ethics* 122(2): 307–320.

Klinsky, S., J. T. Roberts, S. Huq, C. Okereke, P. Newell, P. Dauvergne, K. O'Brien, H. Schroeder, P. Tschakert, J. Clapp, M. Keck, F. Biermann, D. Liverman, J. Gupta, A. Rahman, D. Messner, D. Pellow and S. Bauee. (2017). Why equity is fundamental in climate change. *Global Environmental Change* 44: 170–173.

Levy, M. and M. Scobie. (2015). Goal 16: Promote peaceful and inclusive societies for sustainable development, provide access to justice for all and build effective, accountable and inclusive institutions at all levels. In International Council for Science and International Social Science Council (ed.), Review of Targets for the Sustainable Development Goals: The Science Perspective (pp. 73–76). International Council for Science.

Linklater, A. (1999). The evolving spheres of international justice. *International Affairs* 75(3): 473–482.

Luterbacher, U. and D. Sprinz. (2001). *International Relations and Global Climate Change.* Cambridge: MIT Press.

Marx, K. (1875). Critique of the Gotha Program. Retrieved from www.marxists.org/archive/marx/ works/download/Marx_Critque_of_the_Gotha_Programme.pdf

McCall, L. (2005). The complexity of intersectionality. *Signs: Journal of Women in Culture and Society* 30(3): 1771–1800.

Moellendorf, Darrel. (2002). *Cosmopolitan Justice.* Colorado: Westview Press.

Muller, M. (2008). Free Basic Water—A sustainable instrument for a sustainable future in South Africa. *Environment and Urbanization* 20(1): 67–87.

Nagel, T. (2005). The problem of global justice. *Philosophy & Public Affairs* 33(2): 113–147.

Nordhaus, W. D. (1992). An optimal transition path for controlling greenhouse gases. *Science* 258(5086): 1315–1319.

Norstrom, A. V., A. Dannenberg, G. McCarney, M. Milkoreit, F. Diekert, G. Engstrom, … M. Sjostedt. (2014). Three necessary conditions for establishing effective Sustainable Development Goals in the Anthropocene. *Ecology and Society* 19(3): 8. Doi: 10.5751/es-06602-190308

Nunes, F., R. Rajao and B. Soares. (2016). Boundary work in climate policy making in Brazil: Reflections from the frontlines of the science-policy interface. *Environmental Science & Policy* 59: 85–92.

Nussbaum, M. C. (2000). Aristotle, politics and human capabilities: A response to Antony Arneson, Charlesworth and Mulgan. *Ethics* 111(1): 102–128.

O'Donnell, E. and J. Talbot-Jones. (2018). Creating legal rights for rivers: Lessons from Australia, New Zealand and India. *Ecology and Society* 23(1): 7.

Olsson, L., M. Opondo, P. Tschakert, A. Agrawal, S. H. Eriksen, S. Ma, … S. A. Zakieldeen. (2014). Livelihoods and poverty. In: C. B. Field, V. R. Barros, D. J. Dokken, K. J. Mach, M. D. Mastrandrea, T. E. Bilir, M. Chatterjee, K. L. Ebi, Y. O. Estrada, R. C. Genova, B. Girma, E. S.

Kissel, A. N. Levy, S. MacCracken, P. R. Mastrandrea and L. L. White (eds.). *Climate Change 2014: Impacts, Adaptation, and Vulnerability. Part A: Global and Sectoral Aspects. Contribution of Working Group II to the Fifth Assessment Report of the IPCC*, Cambridge, UK, and New York, NY: Cambridge University Press, pp. 793–832.

Paloniemi, R., E. Apostolopoulou, J. Cent, D. Bormpoudakis, A. Scott, M. Grodzińska-Jurczak, M., ... J. D. Pantis. (2015). Public Participation and environmental justice in biodiversity governance in Finland, Greece, Poland and the UK. *Environmental Policy and Governance* 25(5): 330–342.

Patrick, M. J., G. J. Syme and P. Horwitz. (2014). How reframing a water management issue across scales and levels impacts on perceptions of justice and injustice. *Journal of Hydrology* 519, 2475–2482.

Pritchard, R. D. (1969). Equity theory: A review and critique. *Organizational Behavior and Human Performance* 4(2): 176–211.

Ramos, H. (2015). Mapping the field of environmental justice: Redistribution, recognition and representation in ENGO press advocacy. *Canadian Journal of Sociology-Cahiers Canadiens De Sociologie* 40(3): 355–375.

Rawls, J. (1971). *A Theory of Justice*. Oxford: Oxford University Press.

Rawls, J. (1999). *The Law of Peoples*. Cambridge, MA: Harvard University Press.

Rockström, J., W. Steffen, K. Noone, et al. (2009). A safe operating space for humanity. *Nature* 461: 472–475.

Rode, J., E. Gómez-Baggethun and T. Krause. (2015). Motivation crowding by economic incentives in conservation policy: A review of the empirical evidence. *Ecological Economics* 117: 270–282.

Ryder, S. S. (2017). A bridge to challenging environmental inequality: Intersectionality, environmental justice, and disaster vulnerability. *Social Thought & Research* 34: 85–115.

Saarinen, J. (2013). Critical sustainability: Setting the limits to growth and responsibility in tourism. *Sustainability* 6(1): 1–17.

Schlosberg, D. (2007). *Defining Environmental Justice*. Oxford: Oxford University Press.

Schlosberg, D. (2013). Theorising environmental justice: The expanding sphere of a discourse. *Environmental Politics* 22(1): 37–55.

Scobie, M. (2013). Climate regulation: Implications for trade competitiveness in Caribbean States. In: L. F. Walter, F. Mannke, R. Mohee, V. Schulte and D. Surroop (eds.). *Climate-Smart Technologies Integrating Renewable Energy and Energy Efficiency in Mitigation and Adaptation Responses*, Berlin: Springer Berlin Heidelberg, pp. 33–49.

Scovazzi, T. (2016). The negotiations for a binding instrument on the conservation and sustainable use of marine biological diversity beyond national jurisdiction. *Marine Policy* 70: 188–191.

Sen, A. K. (1999). *Development as Freedom*. New York: Knopf Press.

Sen, A. K. (2009). *The Idea of Justice*. London: Penguin Books.

Shingirai, S. M. and M. T. Happy. (2017). Climate change: A threat towards achieving 'Sustainable Development Goal number two' (end hunger, achieve food security and improved nutrition and promote sustainable agriculture) in South Africa. *Jàmbá: Journal of Disaster Risk Studies* 9(1): e1–e6.

Soper, K. (2018). Human needs and natural relations: The dilemmas of ecology. In: T. Hayward and J. O'Neil (eds.). *Justice, Property and the Environment: Social and Legal Perspectives*. London: Routledge.

Steffen, W., K. Richardson, J. Rockström, et al. (2015). Planetary boundaries: Guiding human development on a changing planet. *Science* 347(6223): 736–746.

Stern, N. (2007). *The Economics of Climate Change: The Stern Review*. Cambridge: Cambridge University Press.

Stevens, C. and N. Kanie. (2016). The transformative potential of the Sustainable Development Goals (SDGs). *International Environmental Agreements-Politics Law and Economics* 16(3): 393–396. doi: 10.1007/s10784-016-9324-y

Stratoudakis, Y., P. McConney, J. Duncan, A. Ghofar, N. Gitonga, K. S. Mohamed, ... L. Bourillon. (2016). Fisheries certification in the developing world: Locks and keys or square pegs in round holes? *Fisheries Research* 182: 39–49.

Sultana, F. and A. Loftus (eds.). (2012). *The Right to Water: Politics, Governance and Social Struggles*. London: Routledge.

Thaler, T. and T. Hartmann. (2016). Justice and flood risk management: Reflecting on different approaches to distribute and allocate flood risk management in Europe. *Natural Hazards* 83(1): 129–147.

Tóth, F. L. (2013). *Fair Weather: Equity Concerns in Climate Change*. London: Routledge.

United Nations Joint Framework Initiative on Children Youth and Climate Change. (2010). Youth Participation in the UNFCCC Negotiation Process. *UNFCCC*. Retrieved from http://unfccc.int/cc_inet/files/cc_inet/information_pool/application/pdf/unfccc_youthparticipation.pdf

van Bommel, S., C. Blackmore, N. Foster and J. de Vries. (2016). Performing and orchestrating governance learning for systemic transformation in practice for climate change adaptation. *Outlook on Agriculture* 45(4): 231–237.

Weiss, E. B. (1990). Our rights and obligations to future generations for the environment. *American Journal of International Law* 84(1): 198–207.

Wolf, F., D. Surroop, A. Singh and W. Leal. (2016). Energy access and security strategies in Small Island Developing States. *Energy Policy* 98: 663–673.

Young, I. (1990). *Justice and the Politics of Difference*. Princeton, NJ: Princeton University Press.

Zhang, Q., C. Prouty, J. B. Zimmerman and J. R. Mihelcic. (2016). Research: More than target 6.3: A systems approach to rethinking Sustainable Development Goals in a resource-scarce world. *Engineering* 2: 481–489.

# 7

# Representation of future generations

*Peter Lawrence*[1]

## Introduction

Future generations cannot speak. But their interests can be represented. Indeed over recent decades the international community created institutions which included amongst their goals, representation of future generations' interests through sustainable development-related objectives, and even proposals for a United Nations High Commissioner for Future Generations (UNHCFG). There are compelling reasons for such institutions, given the current global ecological crisis. Scientists have pointed out that we have now entered the Anthropocene, an era in which human beings have permanently altered the earth's climatic and other ecological systems (Steffen et al. 2011). This has led to calls for international institutions that represent the interests of future generations and the ecological systems upon which human beings are dependent (Dryzek 2015: 938). In this chapter the term "future generations" is used to refer to people not yet born. While strictly outside the scope of this chapter many of the same justifications for representing future generations also apply to young people, especially those below voting age.

Representation of the interests of future generations is strongly linked to intergenerational justice or equity which has been integral to concepts of sustainability and sustainable development which entered international discourse over recent decades. An influential formulation of sustainable development is that found in the Brundtland Report (1987: 43), which defined it as: "development that meets the needs of the present generations without compromising the ability of future generations to meet their own needs," thus including a concept of intergenerational justice. However, this definition leaves ambiguous the content of the obligation towards future generations (Page 2006: 91), and is consistent with a number of competing concepts of intergenerational justice (Gosseries 2005). It has also been strongly criticised for favouring economic growth at the cost of preservation of ecological systems (Bosselmann 2008). In spite of the inherent ambiguity in the Brundtland Report concept, there is no doubt that intergenerational justice lies at the heart of the concept of sustainable development, and many notions of sustainability. Therefore, consideration of institutions to implement sustainability involves careful consideration of the normative underpinnings of institutions claiming to represent future generations, as this constitutes one important vehicle for realising intergenerational justice.

The normative basis for such an approach, however, throws up complex issues. Can assumptions validly be made about the interests of future generations, without arrogantly projecting current preferences onto future generations? Do such proposals entail a flawed assumption that future generations possess rights? More fundamentally, does it make sense to say that a contemporary institution can *represent* future generations or their interests? Does it make sense to apply democratic legitimacy criteria to such institutions at the international level, given ongoing disagreement as to whether democratic legitimacy norms can be applied at this level?

This chapter argues that there are good answers to these questions and that representation of future generations in global institutions can play a modest role in furthering intergenerational justice and sustainability. To make the analysis more concrete, these normative issues are illustrated by drawing on proposals over recent years for a UNHCFG. This case study elucidates issues relating to the applicability of democratic legitimacy criteria in relation to institutions to represent future generations more generally. Issues include what mandate such institutions should have; should human rights, sustainability, or both be at the core of the relevant mandate?

Some caveats. Improved inclusiveness in decision-making processes by giving future generations a voice is necessary but not sufficient to ensuring intergenerational justice. This is because injustice – both intergenerational and intra-generational – rests on structural socio-economic inequalities that need to be systematically addressed (Schlossberg 2007: 28). These inequalities extend to the global economic system which must be reformed in order to build the trust between countries of the North and South required to strengthen the global climate regime (Okereke, this volume).

Furthermore, intergenerational justice cannot be addressed in isolation from intra-generational justice. Environmental issues need to be tackled at the same time as issues of poverty. As stated by Konrad Ott, "Extending the justice perspective to future generations would remain an abstract exercise if it were to ignore the many different existential needs and environmental problems facing poor people currently living on the planet" (Ott 2017: 20). Difficult issues have arisen as to how intergenerational and intra-generational objectives should be balanced, particularly where they come into conflict. It is unnecessary for the purpose of this chapter to take a stand on this issue, because the argument made here is that international sustainable governance institutions can play an important role in highlighting the interests of future generations without deciding one way or another in terms of how this balance should be struck. The latter is best left to governments or international legal frameworks negotiated by governments.

This chapter is structured as follows: first the normative basis for institutions that purport to represent future generations is explored (section "Normative basis"). Rehfeld's theory of representation is presented and tested to see whether it is coherent in the intergenerational context (section "Concept of representation"). Intergenerational justice is argued as providing a convincing basis for institutions representing future generations, resting on an assumption that current institutions massively underrepresent the interests of future generations (section "Intergenerational justice as a normative basis"). A further argument based on democratic legitimacy is sketched and objections to these arguments are addressed (section "A democracy argument – extending the 'demos' into the future"). Section "International institutions to represent future generations" examines institutional possibilities for representation of future generations at the international level while not examining the detailed design of such institutions and their likely effectiveness. The history of such institutions is sketched (section "Some history") before presenting a case study of a proposed UNHCFG (section "A proposed UN commissioner for future generations").

Future prospects and the need for more research in this field conclude the chapter (section "Conclusions and future research").

## Normative basis

Any normative grounding of institutions to represent future generations necessarily entails a concept of representation, to which we now turn.

### Concept of representation

There is no one concept of representation in political philosophy. This chapter proceeds on the basis of the concept of representation developed by Rehfeld (2006). Rehfeld's theory has the advantage that it defines representation in a manner which keeps it separate from the concept of democracy, and related concepts of democratic legitimacy. According to Rehfeld, "representation" involves a claim by a "representative" to be acting on behalf of a person or thing being represented (the "representee") in relation to a particular function, which is accepted by a particular audience (Rehfeld 2006: 5). Applying this concept to a person X who purports to represent future generations, in relation to decisions taken within institution Y, this case will be one of "representation," provided that the members of Y accept that X can speak on behalf of future generations.[2] Alternative – and narrower – concepts of representation rely on the representative being an *agent* of the person represented, and instructing them or providing them with a mandate (Pitkin 1967: 209). This concept clearly does not work in relation to future generations.

Superficially, there would appear to be a serious difficulty here in applying Rehfeld's theory, in that future generations cannot recognise a claim made by an institution purporting to act in their interests. However, recall that under Rehfeld's approach, representation is not dependent upon a representative being authorised by the person or thing being represented. Thus, if we consider the audience before which a particular claim is made as comprising contemporary institutions – rather than future generations – this difficulty vanishes. Thus, for example, it would be consistent with Rehfeld's concept of representation to have a UNHCFG with a mandate to represent the interests of future generations in the UN climate change negotiations, provided, first, that the "audience" – the states involved in the negotiations – accepted this claim, and, second, that the rules of procedure of the negotiations allowed for such representation. Acceptance of the claim should thus be in accordance with the procedures and practices of the particular international institution (i.e. consensus and, failing this, whatever majority applies to the particular body).[3] Whether such representation *ought* to occur is a separate question further discussed below.

Moreover, while applying Rehfeld's concept of representation to future generations may, at first blush, seem rather novel, at the national level we already clearly have institutions such as guardianship, which involve representation, without the person being represented expressing their will or preferences. Furthermore, at the national level ombudsman/commissioners for future generations have been established over recent years (Beckman and Uggla 2016).

In summary then, it is clear that it is coherent and consistent with Rehfeld's concept of representation to have institutions at the international level that represent future generations. Having established that it is coherent to talk about the representation of future generations by international institutions, the next question is whether such representation ought to occur, and to this issue we now turn.

## Intergenerational justice as a normative basis

In the discussion that follows, we include within the notion of institutions, not just bricks and mortar, but also international law and practices. Conceived in this way, it seems clear that such institutions seriously underrepresent the interests of future generations. To cite one example, the Paris Climate Agreement establishes a goal of keeping mean temperature increases "well below 2°C" and if possible below 1.5°C.[4] But current pledges by governments to implement their so-called "nationally determined contributions" will – if added together – lead to global warming in the order of 3.5–4°C with serious impacts for humanity, ecological systems, and future generations (Robiou du Pont 2016).

A key premise of the argument presented here is that international institutions ought – whatever other functions they have – to promote human rights essential to human dignity. This premise rests upon the acceptance by the international community of human rights as flowing from the inherent value of all human beings reflected in the notion of human dignity.

The value of human dignity as a universal value has been reflected by the international community's acceptance of the Universal Declaration of Human Rights which in article 1 proclaims that: "All human beings are born free and equal in dignity and rights." Article 1 accepts the premise that human dignity is an inherent criterion of human existence. Flowing from the value of human dignity as a universal value is the proposition that all human beings regardless of when and where they are born possess basic human rights essential for this dignity (Beyleveld et al. 2015: 549). Thus human dignity and human rights extend both spatially and temporally into the future.

Human dignity and human rights, in turn, play a role as building blocks in intergenerational justice. While there is no single theory of intergenerational justice (Gosseries 2001), a minimal obligation of intergenerational justice would require that current generations convey to future generations a global ecological system – including climate system – capable of providing a minimal subsistence level of existence for future generations, where this is defined as protection of core human rights to life health and subsistence (Caney 2009).

If we accept that existing international institutions underrepresent the interests of future generations, this provides a powerful rationale for establishing institutions that purport to represent future generations. Such representation potentially functions as a vehicle for redressing the current imbalance, increases the likelihood of meeting the requirements of intergenerational justice, and can assist in implementing sustainability. This minimal sufficientarian content of intergenerational justice (e.g. Meyer and Roser 2009, 219) is sufficient to provide the basis for an argument that international institutions should – whatever other functions they have – promote justice including intergenerational justice. This, in turn, provides a rationale for international institutions to represent the interests of future generations in order to counterbalance the existing underrepresentation of their interests.[5]

The argument made so far, however, involves an implicit assumption that such representational institutions are potentially effective in fulfilling such a role. It could be contended that a more effective response would be to factor-in the interests of future generations into mainstream existing institutions, rather than establishing separate institutions purporting to represent future generations. But given the current failure of this "mainstreaming" approach, the rationale remains strong for separate institutions which, at the very least, highlight the particular interests of future generations. Moreover, these are not alternative strategies, and both "mainstreaming" and separate institutions which highlight the interests of future generations may work in a complementary fashion.

## A democracy argument – extending the "demos" into the future[6]

An alternative basis for the representation of future generations in global governance would rest on an argument from democracy. In summary form, this argument would extend the "demos" (public) into the future to incorporate future generations, who arguably should be represented in decisions affecting their interests. This rests on the so-called "all affected principle" (Goodin 1999).

The democratic ideal is supported not just in philosophical writing, but has received strong endorsement by the international community. The Universal Declaration of Human Rights, adopted by the General Assembly in 1948 proclaimed that "the will of the people shall be the basis of the authority of government" and guaranteeing to everyone rights essential for effective political participation. The UN Millennium Declaration, adopted by the General Assembly in 2000, proclaims that "[d]emocratic and participatory governance based on the will of the people" best ensures "dignity" and a range of human rights including rights to be free from hunger, free of violence, oppression, or injustice (United Nations Millennium Declaration 2000). The UN Human Rights Council (2012) has built on this theme, making an explicit link between legitimacy and democracy, stating that: "…A functioning democracy, strong and accountable institutions, transparent and inclusive decision-making and effective rule of law is essential for a legitimate and effective Government that is respectful of human rights." While these instruments and resolutions are referring to national decision-making, as explained earlier, there is no reason in principle not to extend this principle to decision-making at the international level.

## Objections

This argument, however, rests on controversial assumptions relating to whether the democratic ideal can apply at the international level,[7] and also how it can apply in the context of the Anthropocene (Mert, this volume). Some have argued that the democratic ideal cannot apply at the international level (e.g. Dahl 1999), because many states are not democratic. In addition, it has been pointed out that there is no international organisation through which citizens of various countries can find representation. Moreover, in many cases parliaments play a minimal role in global treaty-making.

In response, it has been pointed out that the democratic ideal is important in relation to international institutions, for the same reason it is at the national level. The democratic ideal aims to ensure that public decision-making is responsive to those affected by the decisions in question. It also aims to minimise the risk of the abuse of public power by accountability mechanisms. And finally democratic decision-making is more likely to ensure just outcomes.[8] While the democratic ideal, when applied to international institutions purporting to represent future generations, requires some modification to meet the requirements of this particular context, its underlying purpose remains the same. In addition, it is important to remember that we are talking about an *ideal* – institutions are not either democratic or non-democratic, but the democratic ideal can provide guidance in terms of their reform. This involves, in turn, some specific democratic legitimacy criteria, which we discuss below.

A further objection to international institutions representing future generations is the claim that this involves contemporaries arrogantly making assumptions about the unknown preferences of future generations. But this objection does not stand up to close scrutiny. We can reasonably assume that future generations will share our basic needs for food, water, and shelter, and a stable climatic system upon which human beings depend (Vanderheiden

2008: 129). Given this reasonable assumption, it makes sense to establish a representative of future generations with a mandate to ensure that contemporary decision-making ensures that these minimum requirements are met in relation to future generations. Giving such an institution a veto over existing decision-making bodies may be objectionable because it would arguably push decision-making too far in the direction of favouring future generations' interests. But an institution that highlights the distinctive interests of future generations and helps ensure that these interests are factored into the difficult balancing between contemporary and future interests would avoid this objection.

## International institutions to represent future generations

### Some history

Institutions aimed at furthering sustainable development – an integral component of which is intergenerational equity – have a long history. Since the early 1990s international institutions designed to promote sustainable development include the Commission on Sustainable Development (CSD).[9] The CSD was established following the 1992 Rio Conference on Environment and Development with a mandate to implement a wide-ranging program contained in the so-called Agenda 21 aimed at achieving sustainable development.

International institutional reform proposals can be categorised in terms of proposals that involve (1) new institutions such as a UNHCFG and (2) proposals that involve reform of existing institutions or programs (Report of the Secretary-General 2013). Recent proposals have also included the convening of a global constitutional convention for future generations (Gardiner 2014), giving greater space to non-governmental organisations (NGOs) in the UN climate negotiating agenda, and more inclusion of NGOs in government delegations (Caney 2014).

### A proposed UN commissioner for future generations

During the lead up to the 2012 Rio +20 UN conference on sustainable development, civil society and a number of governments proposed the establishment of a High Commissioner for future generations who would have the following functions:

a    acting "as an advocate for intergenerational solidarity … across United Nations entities";
b    undertaking research on policy practices;
c    offering advice on the "implementation of existing intergovernmental commitments to enhance the rights … of future generations";
d    offering "support and advice, including to individual member states, on best practices and on policy measures to enhance intergenerational solidarity" (Report of UN Secretary General 2013: para 63).

Under some civil society proposals, a UN Commissioner for future generations would have a mandate to bring intergenerational justice issues to the attention of UN organs, participate in treaty negotiations, and request advisory opinions from the International Court of Justice (Ward 2012: 4). The latter proposal reflected the call by many civil society groups for the establishment of ombudsmen for future generations at both the national and international level. Under these proposals, the function of such a national institution would include not only advocacy for sustainable development and its implementation in policy practices but also include responding to citizen petitions, with the power to conciliate and even initiate litigation (Report of UN Secretary General 2013: para 54). A new international ombudsman was envisaged as providing an overarching support for such national mechanisms.

It is important to emphasise that the recent proposals to create a commissioner for future generations (like the existing CSD) did not involve the transfer of sovereignty away from nation states to an international organisation. The UNHCFG would not be authorised to take decisions but rather seek to exert pressure on states by reminding them of their obligations towards future generations, similar to the role played by the existing UN commissioners for human rights and refugees (Caney 2014: 18). In spite of this, the proposal was seen as too intrusive by a number of developing countries and favouring developed country interests against developing countries and was rejected (Lawrence 2014).

The rejection of the UNHCFG proposal at the Rio Conference demonstrates the need to build trust between the north and south as a precondition for developing international agreement on an international institution to represent future generations, as mentioned earlier. We now turn to consider the democratic legitimacy and mandate of a UNHCFG.

## Democratic legitimacy[10]

While there is no single set of democratic legitimacy criteria, Dingwerth's (2007) criteria capture many elements common to theories in this area. According to Dingwerth, democratic legitimacy is assured through: (1) fair or inclusive representation, where those impacted by a decision are included in the decision-making process; (2) democratic control exercised through accountability and transparency; and (3) deliberation.

If we consider the first criteria of *inclusive representation*, there seems to be an obvious difficulty in that future generations cannot participate, in the sense of expressing preferences or being consulted in decision-making processes. However, if we accept Rehfeld's concept of representation sketched earlier, this objection would not stand. Applying Rehfeld's theory to the particular context of future generations, a proxy institution that purports to represent future generations can act as a representative in contemporary decision-making, provided that (i) we accept that such a proxy can make reasonable assumptions as to the interests of future generations (Eckersley 2000: 117), and (ii) the proxy is accepted as representing future generations in the context of its operations – here the states in relation to which the UNHCFG would be operating. As explained earlier, it is reasonable and feasible for such a body to make assumptions that future generations share our basic interests in the functioning climate system, and basic rights to life, health, and subsistence and factor this into recommendations.

In terms of accountability and transparency, a participatory model of accountability would on the face of it be implausible as future generations cannot hold to account in any direct fashion contemporary decision-makers. However Grant and Keohane (2005: 31) have argued that at the international level, a delegation model of accountability makes more sense. According to this approach, accountability of international institutions can be ensured by states holding a particular institution to the requirements of the particular agreed mandate. This, in turn, rests on an assumption that the mandate, or particular agreed standards, reflects the interests of those involved.

Applied to a UNHCFG this model would suggest that states, and in addition NGOs, could ensure accountability for a UNHCFG by insisting that it complied with the particular standards or mandate in question. The weak point in this approach is that this, in turn, depends on the standards or mandate reflecting sufficiently the interests of future generations. In relation to climate change, and also a host of other global ecological threats, international law frameworks are highly variable in terms of the stringency of the rules involved, their scope, precision, and bindingness (Stephens 2017).

A final criterion is that of deliberation, which features in many contemporary theories of democracy. Put simply, deliberation involves the idea that there be equal consideration of different interests and views. If one considers proxies for future generations being able to articulate the interests of persons not yet born, rather than representing them as such, there seems to be no reason in principle why this criterion could not be met in relation to a UN-HCFG. Dryzek and Niemeyer (2008) have formulated a particular version of deliberation which involves the articulation of discourses rather than positions.[11] Under this approach, deliberative democracy entails the equal consideration of discourses, with discourses, in turn, reflecting the variety of positions involved in decision-making. This approach would be well suited to a proxy model, where the proxy articulates discourses that reflect the interests of future generations.[12] In considering accountability and deliberation, clearly the mandate of the UNHCFG would be crucial, and to this issue we now turn.

## Mandate

Sustainable global governance institutions purporting to represent future generations can potentially take on a variety of forms. Christopher Stone has suggested that one could create such institutions across thematic areas, e.g. having separate commissioners for future generations in relation to the oceans, forests, biodiversity, and climate change (Stone 2010: 104). An advantage of this approach would be that the particular commissioner would carve out a particular area of expertise. A significant disadvantage would be the difficulty in persuading states to invest sufficient resources in funding a number of institutions of this nature.

Broader-based institutions, such as a proposed UNHCFG, have tended to be framed with a wide-ranging mandate (as mentioned earlier), with the idea that the commissioner assesses the extent to which the interests of future generations are being taken into account in policies, standards, and rules across the whole UN system. A range of proposals have been made in terms of principles that could be incorporated in such a mandate to guide a commissioner fulfilling this role. Possible candidates for such principles include concepts of sustainable development, or sustainability, human rights, intergenerational justice or solidarity, and global ecology, or a combination of these concepts.

The concept of sustainable development has the attraction that it already is widely accepted by governments and is reflected in a number of international instruments. However, as mentioned earlier, the concept has been much criticised for favouring economic development at the expense of the environment. Sustainable development is often seen as synonymous with so-called "weak sustainability" which involves the notion that natural capital – in the form of biodiversity, forests fauna, and fauna – is substitutable by artificial capital (Ott 2017: 24). However, the multi-functionality of many ecological systems limits their substitutability (Harte 1995). Weak sustainability can imply that future generations end up with the risk that their basic human rights are not protected and thus fails to satisfy minimal requirements of intergenerational justice.

A UNHCFG could also incorporate into its mandate a minimal subsistence concept of intergenerational justice. As argued previously, as a minimum, it could be argued that intergenerational justice implies an obligation on contemporaries not to impair future generations' core human rights to life, health, and subsistence (Caney 2009: 69). A slim list of core human rights would be sufficient to enable a UNHCFG to put pressure on governments to take action on global ecological threats such as climate change which clearly threatens these rights.

A slim list of rights would also deal with the epistemic objection raised earlier that we cannot make assumptions about the particular values of future generations, including in relation to a wide range of rights such as freedom of expression and political participation. It does seem reasonable to make assumptions that future generations share our concern for the rights to life, health, and subsistence. However, a slim list of rights could be seen as undermining other human rights that are arguably of equal value. Thus, for example, human rights related to nondiscrimination, and democratic political participation could be argued as being of equal importance in ensuring minimal requirements of justice are met.

An objection to these proposals so far is that they are all anthropocentric, ignoring the inherent value of ecological systems. Indeed, a further option would be to include a reference to the integrity of the global ecological system within the mandate of a UNHCFG. Such an approach would ensure that the damage per se which is occurring to global ecological systems is given sufficient weight in the activities of the UNHCFG. Tim Stephens (2017) has argued that the current web of global environmental treaties fails to take into account sufficiently the interconnections between various ecological systems, which calls out for a more integrated systemic approach to international environmental law-making. But concepts of ecological integrity are inherently vague and may be difficult to negotiate, given that states may perceive this as limiting their sovereignty to pursue unshackled economic growth. At the end of the day, failure to incorporate a strong version of sustainability into the mandate of a UNHCFG may not be fatal; given human beings' dependence on a healthy global ecological system, an approach based on core human rights would largely point in the same direction in terms of policy prescriptions.

A general problem with all the principles discussed so far is their inherent vagueness. Arguably, for a UNHCFG to be effective it is necessary to have precise standards against which the action of governments can be measured. These standards can take the form of both precise and binding international treaty rules, but also soft law non-binding standards that are being developed through the UN system (Kim 2016). A UNHCFG could add value to the existing global governance institutions by highlighting in particular the impacts on future generations involved in such standards and norms.

## Conclusions and future research

There is a sound normative argument for the establishment of international institutions that represent future generations as a vehicle for promoting sustainability. The normative basis for this approach rests on an assumption that international institutions ought to promote justice defined as human dignity through the protection of core human rights. This chapter argued that core human rights should be protected both spatially and temporally into the future, thus underpinning a minimal subsistence concept of intergenerational justice based on protection of core human rights. The massive bias against future generations in contemporary rule-making and institutions justifies international institutions with a mandate to represent future generations as a means for redressing this imbalance.

The democratic ideal also justifies such institutions which can provide a voice for future generations that are inevitably impacted by contemporary decision-making; the all affected principle provides the basis for expanding the *demos* to incorporate future generations. This approach rests on consensus instruments reflecting the international community's support for democracy as an ideal, essential for securing protection of human rights.

This chapter further argued that democratic legitimacy criteria should be applied to global institutions purporting to represent future generations and applied criteria developed by Dingwerth (2007) to this end.

A host of further issues remain in relation to such proposals, which to date remain relatively poorly researched.[13] In terms of effectiveness, at the national level, ombudsmen for future generations, or sustainable development commissions, have received little systematic empirical study, in terms of their effectiveness.[14] Further important issues that remain relatively un-researched include how the mandate for such representative institutions at the international level should be framed. This chapter briefly canvassed issues relating to whether a mandate of a UNHCFG should be narrow or broad, and whether it should rest on the notion of sustainability (weak or strong) or on human rights.

A further possible research direction agenda taken up by Lawrence and Köhler (2017) would involve examining the role of international courts as a vehicle for representing future generations. There is scope for further research which addresses these issues in relation to other areas of law and policy-making involving sustainability, including for example law and policy relating to the oceans and biological diversity.

Given the current trajectory of human civilisation and worsening ecological crisis, issues of intergenerational justice will only become more acute. Against this context, international institutions to represent future generations may play a modest role in highlighting the interests of future generations and helping to ensure that these interests are better incorporated into decision-making.

## Acknowledgements

I would like to thank the editors of this Handbook, and Jan McDonald and Nicky van Dijk for their helpful suggestions in relation to this chapter. All errors and omissions remain of course the responsibility of the author.

## Notes

1 Thank you to the editors of this volume, Jan McDonald and Nicky van Dijk for helpful suggestions in relation to this chapter. All errors and omissions remain of course the responsibility of the author.
2 Anja Karnein argues that Rehfeld's approach ignores the fact that while a representative need not necessarily be democratically legitimate, there must be some background norms shared between the represented and the audience. She notes that shared understandings between contemporaries and future generations are inherently problematic, but argues that some broad principles, which future generations would likely agree to if present today, provide a basis for background norms applicable in this context. She argues a principle involving future generations and contemporaries being treated with equal respect would fall within this category. See Karnein (2016) 9–10.
3 See critique of Rehfeld in Karnein (2016).
4 Paris Agreement to the United Nations Framework Convention on Climate Change (adopted 22 April 2016, entered into force 4 November 2016) http://unfccc.int/files/essential_background/convention/application/pdf/english_paris_agreement.pdf (Paris Agreement), article 2 (1).
5 This argument relies on a similar argument made in relation to international tribunals by Lawrence and Köhler (2017).
6 This section relies upon Lawrence in Szabó and Cordonier Segger (2019 forthcoming).
7 Compare, for example, Dahl (1999) with Held (1995).
8 Sen argues on the basis of empirical evidence that democracy plays a significant role in reducing famines and inter-state conflict, thus increasing the likelihood that the requirements of justice will be achieved (Sen 2009: 348).
9 Commission on Sustainable Development (CSD), https://sustainabledevelopment.un.org/csd.html (accessed 18 August 2018).
10 This section relies upon Lawrence in Szabó and Cordonier Segger (2019 forthcoming).
11 Cf. Niemeyer, Simon and Julia Jennstål (2017).

12 However, a weakness of this model is its reliance upon social science experts who must analyse the differing discourses in terms of their positions.

13 With the exception of González-Ricoy and Gosseries (2016).

14 But see Beckman and Uggla (2017).

## References

Beckman, Ludwig and Fredrik Ugglal (2016) 'An ombudsman for future generations' in González-Ricoy, Iñigo and Axel Gosseries (eds), *Institutions for Future Generations*, Oxford University Press, 117–134.

Beyleveld, Deryck, Marcus Düwell and Andreas Spahn (2015) 'Why and How Should We Represent Future Generations in Policymaking?' *Jurisprudence*, 6, 549.

Bosselmann, Klaus (2008) *The Principle of Sustainability: Transforming Law and Governance*, Ashgate Publishing.

Brundtland Report (1987) *Our Common Future*, United Nations World Commission on Environment and Development.

Caney, Simon (2009) 'Climate change, human rights and moral thresholds' in Stephen Humphreys (ed), *Human Rights and Climate Change*, Cambridge University Press.

Caney, Simon (2014) 'Applying the Principle of Intergenerational Equity to the 2015 Multilateral Processes' *Background Paper* commissioned by the Mary Robinson Foundation Climate Justice.

Christiano, Thomas (2011) 'Is Democratic legitimacy possible for international institutions?' in Daniele Archibugi, Mathias Koenig-Archibugi in Daniele Archibugi, Mathias Koenig-Archibugi and Raffaele Marchetti (eds), *Global Democracy: Normative and Empirical Perspectives* Cambridge University Press, 69, 81.

Ciplet, David J., Timmons Roberts and Mizan Khan (2015) *Power in a Warming World, The New Global Politics of Climate Change and the Remaking of Environmental Inequality*, MIT Press.

Dahl, Robert (1999) 'Can international organisations be democratic? A sceptic's view' in Ian Shapiro and Casiono Hacker-Cordon (eds), *Democracy's Edges*, Cambridge University Press.

Dingwerth, Klaus (2007) *The New Transnationalism: Transnational Governance and Its Democratic Legitimacy*, Palgrave Macmillan.

Dryzek, John and Simon Niemeyer (2008) 'Discursive Representation' *American Political Science Review*, 102(4), 481.

Dryzek, John (2015) 'Institutions for the Anthropocene: Governance in a Changing Earth System' *British Journal of Political Science*, 46(4), 937–956.

Eckersley, Robyn (2000) 'Deliberative democracy, ecological representation and risk: Towards a democracy of the affected' in Michael Saward (ed), *Democratic Innovation, Deliberation, Representation and Association*, Routledge.

González-Ricoy, Iñigo and Axel Gosseries (eds) (2016) *Institutions for Future Generations*, Oxford University Press.

Goodin, Robert (2007) 'Enfranchising All Affected Interests, and Its Alternatives' *Philosophy and Public Affairs*, 35(1), 40.

Gosseries, Axel (2001) 'What Do We Owe the Next Generation(s)?' *Loyola of Los Angeles Law Review*, 35, 293.

Gosseries, Axel (2005) 'The Egalitarian Case against Brundtland's Sustainability' *Gaia: Okologische Perspektiven in Natur-, Geistes- und Wirtschaftswissenschaften*, 14(1), 40–46.

Grant, Ruth and Robert Keohane (2005) 'Accountability and Abuses of Power in World Politics' *American Political Science Review*, 99(1), 29.

Harte, Michael J. (1995) 'Ecology, Sustainability, and Environmental Capital' *Ecological Economics*, 15(2), 157–164 cited in Ott (2017: 24).

Held, David (1995) *Democracy and the Global Order*, Polity Press.

Karnein, Anja (2016) 'Can we represent future generations?' in González-Ricoy, Iñigo and Gosseries, Axel (eds), *Institutions for Future Generations*, Oxford University Press, 83.

Kim, Rakhyun E. (2016) 'The Nexus between International Law and the Sustainable Development Goals' *RECIEL*, 25, 15–26.

Lawrence, Peter and Köhler, Lukas (2017) 'Representation of Future Generations through International Climate Litigation: A Normative Framework' *German Yearbook of International Law*, 60, 639–666.

Lawrence, Peter (forthcoming 2019) 'International institutions to represent future generations in the Anthropocene and democratic legitimacy' in Marcel Szabó and Marie-Claire Cordonier Segger (eds), *Intergenerational Equity in Sustainable Development Treaty Implementation*, Cambridge University Press.

Meyer, Lukas H. and Dominic Roser (2009) 'Enough for the future' in Axel Gosseries and Lukas H. Meyer (eds), *Intergenerational Justice*, Oxford University Press, 219–248.

Niemeyer, Simon and Julia Jennstål (2016) 'The deliberative democratic inclusion of future generations' in Iñigo González-Ricoy and Axel Gosseries (eds), *Institutions For Future Generations*, Oxford University Press.

Ott, Konrad (2017) 'Sustainability: Theory and policy' in Bernd Klauer, Reiner Manstetten, Thomas Petersen and Johannees Schiller (eds), *Sustainability and the Art of Long-Term Thinking*, New York, Routledge, 16–30.

Page, Edward (2006) *Climate Change, Justice and Future Generations*, Cheltenham, Edward Elgar.

Pitkin, Hannah (1967) *The Concept of Representation*, University of California Press.

Rehfeld, Andrew (2006) 'Towards a General Theory of Political Representation' *The Journal of Politics*, 68, 1.

Report of the Secretary General (2013) *Intergenerational Solidarity and the Needs of Future Generations*, United Nations Secretary General, UNGA, 68th session item 19(a), UN Doc A/68/322 (15 August 2013).

Robiou du Pont, Yann, M. Louise Jeffery, Johannes Gutschow, Joeri Rogelj, Peter Christoff and Malte Meinshausen (2016) 'Equitable Mitigation to Achieve the Paris Agreement Goals', *Nature Climate Change*, advance online publication. doi:10.1038/nclimate3186.

Schlossberg, David (2007) *Defining Environmental Justice, Theories, Movements, and Nature*, Oxford University Press.

Sen, Amartya (2009) *The Idea of Justice*, Allen Lane.

Steffen, Will, Jacques Grinevald, Paul Crutzen and John McNeill, (2011) 'The Anthropocene: Conceptual and Historical Perspectives, *Philosophical Transactions of the Royal Society A*, 369, 842–867.

Stephens, Tim (2017) 'Reimagining international environmental law in the Anthropocene' in Kotzé, Louis (ed), *Environmental Law and Governance for the Anthropocene*, Hart Publishing.

Stone, Christopher (2010) *Should Trees Have Standing? Law, Morality, and the Environment*, 3rd edn, Oxford University Press.

United Nations Human Rights Council (2012) *Human Rights, Democracy and the Rule of Law*, resolution 19/36 UN GA doc. A/HRC/RES/19/36 (19 April 2012) para 5.

United Nations Millennium Declaration (2000) Resolution adopted by the General Assembly (UN doc.A/RES/55/2 (18 September 2000). www.un.org/millennium/declaration/ares552e.pdf (accessed 21 September 2015) para 6.

Vanderheiden, Steve (2008) *Atmospheric Justice*, Oxford, Oxford University Press.

Ward, Halina (2012) *Committing to the Future We Want: A High Commissioner for Future Generations at Rio+20*, Foundation for Democracy and Sustainable Development, World Future Council. www.fdsd.org/wordpress/wp-content/uploads/ Committing-to-the-future-we-want-main-report.pdf (accessed 15 December 2014).

# 8

# The 'good life' and Protected Needs

*Antonietta Di Giulio and Rico Defila*

## The 'good life' and needs are central to sustainability

Since the onset of the debate about sustainability in the late 1980s, United Nations (UN) documents have used different terms to capture the overarching goal of sustainable development such as meeting peoples' "needs" or "basic needs", achieving a "better life", a "decent life", "quality of life", or "well-being" for all, or providing everyone living now and in the future with a "healthy and productive life" (Di Giulio 2004). In sum, according to the UN, sustainability is about making sure that all human beings in present and future generations have the possibility of satisfying their needs and leading a good life. This has been pointed out by different authors in the past (e.g. Di Giulio 2004; Manstetten 1996; Michaelis 2000; Rauschmayer et al. 2011) and still holds true (see also Fuchs, this volume, and Vanhulst and Beling, this volume). The 'good life' in this context is not about a life being good in a moral or ethical sense but about a life being good in terms of the quality it holds for individuals. This aligns with data showing that organisations committed to sustainability seem to distinguish themselves from other organisations in the significance they attach to four values (Ruesch and Di Giulio 2016). One of these values is the 'good life', the others are intergenerational justice (see Lawrence this volume), intragenerational justice (see Kalfagianni et al. this volume), and the common good.

Both the 'good life' and needs are thus essential to the idea of sustainability and so should also be the point of reference for sustainability governance. This will only be possible if these notions are defined in a way that one can develop suitable policies and assess societal development based on them (see also Higgs, this volume).

This chapter first provides criteria for defining the 'good life' and needs for the context of sustainability[1] and then examines promising approaches to defining these notions based on a literature review. Subsequently, it presents our suggestion to operationalise the 'good life' with nine Protected Needs that should receive special protection within and across societies. The last section then discusses the consequences of focusing on the 'good life' in sustainability governance.

## Criteria for defining the 'good life' and needs for the context of sustainability

Defining the notions of the 'good life' and needs for the context of sustainability means defining them for a specific field of national and global governance. In other words, whether the 'good life' and needs are defined for the context of sustainability or for another context leads to different definitions and has to meet different criteria. One should therefore first ascertain the specific characteristics of the sustainability context and derive criteria from these characteristics.

Sustainability is not a descriptive concept but a normative one, and its normativity applies to the whole of society, from individuals to governments. It is meant to inform governance on both a national and global scale. It cannot be achieved by single nation states alone. Sustainability is a visionary idea, that is, it sets a positive goal for human society and thus for individuals, civil organisations (including economic actors), national governments, and international organisations. Accordingly, quality of life is more than mere physiological survival (the goal of sustainable development cannot be reduced to the goal that people do not die). Sustainability is a long-term concept, demanding policies with a future-oriented perspective, and it proceeds from the assumption that at least some natural and social resources might be finite and impaired in their quality by human actions.

These characteristics lead to five criteria for formulating definitions of the 'good life' and needs in order for these concepts to be suitable to inform sustainability governance:

1   They must be defined in a way that sound governmental responsibilities can be derived from them.
2   They must be defined so as to be applicable to different cultural contexts and to allow for diverse lifestyles and value systems. The definitions thus necessarily presuppose the existence of universals but must avoid privileging some cultures and excluding others.
3   Their definitions must be based on a salutogenic approach[2] rather than proceeding from problems or deficits.
4   Their definitions must allow for adaptations to future developments.
5   Their definitions must make it possible to clearly distinguish the 'good life' and needs from (natural and social) resources and to distinguish legitimate from illegitimate behaviours related to such resources.

One might object that criteria 1 and 2 are, to some extent, mutually exclusive: the 'good life' and needs must be defined as concretely and precisely as possible to serve as a sound basis for policies, but they must, at the same time, be defined as openly as possible to be culturally and historically neutral. This argument, though, does not point to a flaw in the idea of sustainability but to a strength: the idea of sustainability is regulative, that is, its precise meaning must be renegotiated in each generation (similar to freedom or justice). It is therefore neither possible nor necessary to determine exactly what sustainable development looks like once and for all. Rather, this must be tailored to a specific cultural and historical context (see also Vanhulst and Beling, this volume). This also applies to the definitions of the 'good life' and needs. The challenge is to find the golden mean between specification and openness.

One might also object that criterion 4 should be left out because future generations should have as much leeway as possible to specify their needs on their own and because it is not possible to know anything about future generations' needs with certainty. This argument does not stand up to scrutiny: almost every single action (as well as every single omission and inaction) impacts the lives and scope of action of future generations. Thus, although each generation must define its needs for itself, it also must make (and actually does make) assumptions about the needs of future generations (see also Pellizzoni, this volume, about prospective responsibility, and Kalfagianni et al., this volume, about justice and the subjects of justice).

Providing a substantive definition of the 'good life' and needs for the context of sustainability is both a necessity and a challenge, and it entails both negotiating the 'good life' and needs in society and drawing on existing theoretical approaches and empirical evidence. The challenge is augmented by the fact that the literature on quality of life is extensive, scattered among a broad diversity of disciplines, and based on a broad range of approaches. In the next section, we use our criteria to navigate the literature and identify promising approaches to defining the 'good life' and needs for the context of sustainability.

## Approaches to defining the 'good life' and needs

Philosophy offers an overarching perspective for sorting the different approaches to defining the 'good life' in the sense of quality of life, such as subjective well-being, psychological well-being, freedoms, life satisfaction, happiness, needs, and capabilities, thus providing a way to narrow down the field. In philosophy, the 'good life' is a (primarily theoretical) term used to describe what constitutes a fulfilled human life. Such descriptions are by definition positive (criterion 3), supra-individual (criterion 2), and, as far as this is possible, ahistorical (criterion 4).

One can distinguish between objective and subjective philosophical theories of the 'good life'. Objective theories claim to define universally valid elements of a good life that are independent of subjective wishes and individual preferences. By contrast, subjective theories define a good life exclusively in terms of subjectively experienced well-being or happiness (living a good life is solely defined by what makes an individual feel good). Objective theories do not negate the importance of subjectively experienced well-being, but they do not regard it as the only decisive factor in defining a good life.[3] For the context of sustainability, objective theories provide a promising starting point for defining the 'good life'. Subjective theories do not meet criterion 2 since they do not provide universals; nor do they meet criterion 1 since they do not allow one to derive responsibilities for governments or communities, which cannot assume the responsibility of ensuring that each individual actually feels happy (Di Giulio 2008; Di Giulio et al. 2010).

Among objective theories, anthropological approaches are the most promising for defining the 'good life' in the context of sustainability and are thus prevalent in sustainability discourses (e.g. Alkire 2007, 2010; Jackson et al. 2004; Michaelis 2000; Rauschmayer et al. 2011; Robeyns and van der Veen 2007). Anthropological approaches proceed from what their authors assume to be universal characteristics of human beings. They then posit that living a good life equals having the opportunity to develop and realise these characteristics according to an individual's physical and psychological traits, personal values, and preferences. With regard to normativity, anthropological approaches neither dictate a specific lifestyle nor claim to determine what a fulfilled life for a specific individual should be. Instead, they infer the ethical obligation of providing the necessary external conditions for individuals to lead a life they individually perceive as meaningful. These external conditions are derived from the posited universal human characteristics. The duty to provide all human beings with these conditions exists regardless of whether an individual makes use of them or not (e.g. the duty of providing individuals with the opportunity to eat does not entail their obligation to eat). It is rather obvious that an obligation to provide such external conditions can legitimately be imposed on states. That is, anthropological approaches are able to offer definitions of the 'good life' that meet criterion 1 while at the same time complying with criteria 2, 3, and 4.

Anthropological approaches differ with regard to how they conceptualise and describe universal human characteristics. The two most dominant approaches in sustainability discourses are the capability approach and the needs approach. The capability approach (e.g. Nussbaum 1992) argues that all human beings have certain identifiable capabilities and that living a good life means that an individual has the opportunity to choose which of these he or she wants to realise (thus making them into "functionings"). Lists of capabilities have been provided for instance by Burchardt and Vizard (2011), Nussbaum (2006), and Robeyns and van der Veen (2007). The needs approach (e.g. Max-Neef 1991) argues that all human beings have certain urges and that living a good life means having the opportunity to satisfy these urges. According to Jackson et al. (2004), needs are universal motivating drives (similarly, Doyal and Gough 1991); according to Rauschmayer et al. (2011), they are constitutive aspects of human flourishing that require no further justification as a reason to act; and according to O'Neill (2011), they are indispensable, irreducible, and non-substitutable objectives that detrimentally affect people if they are not met (similarly, Baumeister and Leary 1995). Lists of needs have been provided for instance by Max-Neef et al. (1991) and by Doyal and Gough (1991). The capability and needs approaches are not incommensurable, as shown, for example, by Costanza et al.'s (2007) list of needs, which was developed by integrating, among others, Nussbaum's list of capabilities and Max-Neef et al.'s list of human needs. Alkire's comprehensive comparison of approaches (e.g. 2007, 2010) also confirms the commensurability of these two approaches. She herself provides a definition of "human development" that she suggests using as an overarching concept (Alkire 2010, 27) but without providing a list because she is convinced that no such list should be provided (Alkire 2007, 14).

Although the capability and needs approaches are commensurable, it is nevertheless necessary to choose whether to proceed from one or the other in defining the 'good life' for the context of sustainability. There are two reasons to adhere to the concept of needs. First, the language is more in line with that used by the UN and more accessible to a broad public since the concept of needs is easier to understand than that of capabilities. Second, the concept of needs is more suited to meet criterion 5. While the concept of capabilities aims to capture everything that is assumed to be characteristic of human beings, the concept of needs entails the possibility of distinguishing needs from other urges or drives that are assumed to be less decisive to human well-being and thus do not create the same ethical obligation and the same entitlement as needs do (mirrored in the debate by differentiating, for example, between needs and wants, desires, or preferences; see e.g. Soper 2006). This, in turn, allows one to distinguish between legitimate and illegitimate concerns and to make clear that in the context of sustainability, living a good life is not the same as living out all individual preferences.

Adopting a needs approach leads to distinguishing between needs and satisfiers (e.g. Max-Neef et al. 1991; Soper 2006). Satisfiers are the means used to satisfy needs (in the terminology of anthropological approaches, satisfiers are external conditions). The notion of satisfiers covers actions, products, structures, institutions, services, infrastructures that is, both material and non-material means (thus, it opens the possibility of integrating natural and social resources into the equation). Distinguishing needs from satisfiers allows one to classify satisfiers in terms of their potential for actually satisfying needs. Drawing on Max-Neef et al. (1991), Jackson et al. (2004) propose a typology of satisfiers, distinguishing satisfiers according to whether they (a) do indeed satisfy a need, (b) are detrimental to the satisfaction of one or more needs, (c) can potentially fulfil more than one need, or (d) are indispensable for the satisfaction of a need (for similar proposals, see Cruz 2011; Rauschmayer et al. 2011). In contrast to needs that are assumed to be universal and stable, satisfiers are highly dependent on historical and societal contexts and unstable (e.g. Jackson et al. 2004; Soper 2006).

By applying the criteria presented in section "Criteria for defining the 'good life' and needs for the context of sustainability", we were able to narrow down the field of theoretical approaches to the 'good life' first to objective theories, then to anthropological approaches, and, finally, to a needs approach. These criteria lead, in other words, to a clear preference for a needs approach. This adds up to the suggestion that listing the needs all individuals must have the possibility to satisfy is a suitable approach to defining the 'good life' for the context of sustainability. If the 'good life' is meant to inform national and international governance, then such a list must not negate evidence and can thus not be developed by relying on theoretical approaches alone. Rather, empirical findings about human well-being must be considered as well in order to make sure to capture only needs that are decisive to human well-being and to include nothing that has been proven not to be decisive.

Such empirical findings are provided by studies on subjective well-being, psychological well-being, freedoms, life satisfaction, quality of life, happiness, capabilities, human flourishing, and needs (see also Alkire 2007, 2010). Although it would be a naturalistic fallacy to infer normativity from such evidence, such research provides knowledge about, for instance, activities leading to more or less satisfaction in life (e.g. Lewinsohn 2016; White and Dolan 2009), about elements that are crucial for an individual's mental health (e.g. Abbott et al. 2010; Ryan and Deci 2000, 2001; Ryff 1989), or about how factors such as financial resources or access to infrastructure influence well-being in general or in specific life domains (e.g. Aked et al. 2008; Anand et al. 2011; Cummings 1996; Dolan et al. 2008, 2011; Shin 2015; Veenhoven 2008). Finally, an increasing number of national and international reports are providing data about factors beneficial or detrimental to human well-being that can be compared across nations (e.g. in Bhutan, Canada, UK; World Happiness Report, EU-SILC,[4] Happy Planet Index). Such reports can be used to compare life satisfaction in different nations (e.g. Brulé and Veenhoven 2014). To meet criterion 2, and because of the overwhelming number of inquiries and publications, priority should be given to reports presenting international data or comparing national data preferably from different cultural contexts.

There is one last point to consider with regard to defining the 'good life' for the context of sustainability. Against Alkire (2007), we argue that a substantive definition is necessary if the concept of the 'good life' is to have practical policy relevance. And this substantive definition should not be limited to providing titles of needs. Titles offer, first, too much scope for differing interpretations. Second, titles do not provide a rich description of quality of life, that is, they can easily be reduced to expressions of bare survival (thus failing criterion 3). Third, titles do not provide a sufficient guide to policy formation (thus failing criterion 1). It is hence necessary to provide what Soper calls a "thick" theory in contrast to a "thin" theory of the 'good life' even though this increases the potential controversies (Soper 2006; similarly, Doyal and Gough 1991). In the next section, we present the concept of Protected Needs, that is, our proposal for a thick theory of the 'good life' for the context of sustainability.

## Protected Needs: a theory of the 'good life' as a basis for global sustainability governance

Drawing on what has been said previously, we define the 'good life' for the context of sustainability as the opportunity for individuals to satisfy their individual constructs of wanting that belong to a given list of Protected Needs (based on Di Giulio et al. 2012).

Protected Needs are ends in themselves; that is, they are legitimate needs and cannot be contested on ethical grounds. Individuals and communities (encompassing national governments and intergovernmental organisations) have an obligation to provide (social, cultural, economic, environmental, etc.) conditions under which all human beings can – now and in the future – satisfy these needs. We call these needs *Protected* Needs to express this ethical obligation. This obligation does not imply that each and every human being *does* in fact feel the same needs nor that each and every human being *must* feel the same needs, but it does imply that all human beings are entitled to satisfy their individual constructs of wanting that correspond to needs on the list of Protected Needs. We use "individual constructs of wanting" (Di Giulio et al. 2012) to emphasise both that needs are always subjectively experienced by individuals (see also Soper 2006) and that needs depend, in how they are individually delineated and weighted, on social and cultural contexts. Humans have quite different conceptions of how their needs should be satisfied and what a life they value looks like, and they are entitled to live accordingly. That is, they are entitled to satisfy the needs they develop according to their individual preferences, culture, and physical, emotional, and cognitive features – and thus to have the possibility of living a life they value. This individual freedom is limited, though, by the ethical rule of not compromising others' possibility to satisfy Protected Needs. The ethical obligation for communities thus includes the obligation to prevent individuals from impairing others' possibility to satisfy their Protected Needs (see also Fuchs, this volume, about the task of governments to constrain individual freedom).

To distinguish Protected Needs from less legitimate (and thus not equally protected) constructs of wanting, we use the notion of "subjective desires" (Di Giulio et al. 2010, 2012; see Figure 8.1). Subjective desires are individual sensations of wanting that do not correspond to needs on the list of Protected Needs. Satisfying subjective desires is legitimate only as far as it does not prevent other humans from satisfying their Protected Needs. Neither individuals nor states have an obligation to ensure that people can satisfy subjective desires.

Finally, with regard to the conditions with which humans must be provided, we suggest complementing the notion of satisfiers with the element of resources (both natural and social). This allows for the distinction between satisfiers and the resources used to perform actions; to establish and maintain institutions, services, and infrastructures; to manufacture, distribute, use, and dispose of consumer goods; or that are affected by corresponding activities, systems of provision, or systems of disposal (see Figure 8.1). The extent of the use of both satisfiers and resources can be contested on ethical grounds; that is, they should be used only insofar as their use does not prevent others from satisfying Protected Needs. In addition, the extent to which it is legitimate to use resources depends on the (assumed) amount available (now and in the future; see also Fuchs, this volume).

The outlined definition of the 'good life' for the context of sustainability is not useful if it is not accompanied by a justified, comprehensive, and manageable list of Protected Needs. Otherwise it cannot, as argued earlier, guide national and international governance. Based on the reasoning and literature presented earlier, we have developed such a list.

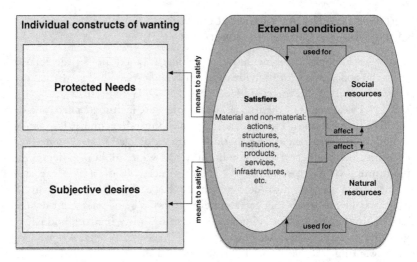

*Figure 8.1* The 'good life' for the context of sustainability – a conceptual framework: both Protected Needs and subjective desires are individual constructs of wanting. Subjective desires do not have the same status of protection, that is, there is no obligation to ensure that people can satisfy subjective desires. Satisfiers and resources are external conditions. While satisfiers are instrumentally linked to needs and desires, natural and social resources are linked to satisfiers and not directly to needs and desires.

## How we developed the list of Protected Needs

In developing the list of Protected Needs, we started with the lists of needs and capabilities noted in section "Approaches to defining the 'good life' and needs". We analysed these lists in three ways. First, we considered whether the single needs (and capabilities) on these lists meet criteria 1, 2, and 5 (see section "Criteria for defining the 'good life' and needs for the context of sustainability"; we did not consider lists that do not meet criteria 3 and 4). Second, we compared the needs (and capabilities) on these lists with empirical findings about human well-being (see section "Approaches to defining the 'good life' and needs"), focusing not on national but on cross-national and cross-cultural studies. Third, we conducted a structured dialogue lasting two days with an interdisciplinary group of German and Swiss scholars from economics, the educational sciences, political science, psychology, sociology, and our own disciplines, law and philosophy.[5] We did this to assess whether the needs (and capabilities) provided by these lists could serve as Protected Needs and thus whether they could be used to delineate the 'good life' for the context of sustainability, although they had not been developed for this context.

An in-depth discussion of all our findings and a comprehensive presentation of the literature we consulted are beyond the scope of this chapter. We can only provide a summary. The major result was that none of the existing lists could serve our purpose because only a few needs (or capabilities) on these lists both met all our criteria and withstood a comparison with empirical findings. The most important reasons for rejecting needs (and capabilities) in existing lists were:

• Criterion 1: Some of the needs are not suitable for deriving ethical obligations for communities and governments although their satisfaction undoubtedly impacts individual well-being. Examples of such needs are being loved (or showing affection) or being healthy. No individual and no government can assume the responsibility of ensuring that someone is loved (or shows affection) or that someone is healthy. Treating such needs as Protected Needs would lead to a nanny state and dictatorial policies.

- Criterion 2: If critically scrutinised, some of the needs lose their potential of being universal because instead of leaving room for a diversity of lifestyles, life choices, and values across or within time and cultures, they privilege specific life choices and/or value systems. Some are even tailored to specific life phases. Examples for such needs are living in a family, having children (reproduction), being religious (spirituality), working, and receiving training or an education. Moreover, some of the needs denote activities such as taking care of others that empirical inquiries have proven are not decisive for the well-being of individuals (in contrast to performing activities that are meaningful to individuals).

- Criterion 5: Instead of needs, existing lists often name satisfiers (or single resources) such as income, standard of living, education, mobility, work and employment, healthcare, and clean water and do not clearly distinguish needs from satisfiers (or resources). Some even entail expressions such as "the highest possible standard of". This does not meet this criterion, nor does it correspond to what is known from empirical inquiries because neither increases in income nor a higher level of education nor other augmentations of material circumstances are (above a specific threshold) decisive with regard to individual well-being. Moreover, including income, education, mobility, and similar satisfiers or resources in definitions of well-being feeds mechanisms of overconsumption by equating positional goods with quality of life and thus equating well-being with the idea of "the more the better" (see also Krogman, this volume).

Finally, with a view to providing a thick theory of the 'good life', existing lists proved to be either too thin because they do not provide much more than titles such as "freedom", which makes them devoid of content, or too specific because they go so far as to list things such as "freedom of assembly and demonstration", making them too comprehensive or arbitrary. In sum, our analysis and discussions showed that it is, for the context of sustainability, justified and necessary to develop a specific list of needs based on needs in existing lists that both meet the five criteria and are justified by empirical inquiries into human well-being. We developed an initial list of nine Protected Needs with descriptions specifying each (in section "The list of Protected Needs" we argue why descriptions are necessary). This list was subjected to qualitative cognitive testing by neutral interviewers to determine whether it was comprehensible and complete and whether respondents would object to any needs (and/or descriptions) on the list. The list was revised based on the results of these interviews with ten respondents (Table 8.1).

In a next step, this list was subjected to a quantitative representative survey in Switzerland ($N = 1059$). It would be beyond the scope of this chapter to present all the results of this survey. Basically, the survey confirms that the Swiss population also views the nine Protected Needs as protected. Respondents have corresponding constructs of wanting regardless of their individual circumstances (age, gender, income, education, employment, etc.). The individually perceived possibility of satisfying the nine needs correlates with general life satisfaction, while the individually attributed importance of the needs does not. If the importance attached to a need is higher than the possibility of satisfying it, the perceived quality of life is impaired. Each of the nine needs is considered generally important to other people, and most respondents think it would be blatantly unjust if others could not satisfy each of the nine needs (these numbers differ for each need, but in sum, 43–62% of respondents think it would be unjust with regard to all people on the world, and 29–45% only with regard to people living in Switzerland). Respondents posit both an individual and a collective responsibility to ensure the possibility that others can satisfy each of the nine needs, but this responsibility decreases the further away that people are in temporal and spatial terms. Individual and collective responsibility correlate.

## The list of Protected Needs

The list of Protected Needs consists of nine needs arranged in three groups (see Figure 8.2 and Table 8.1, left column). These needs denote what individuals must be allowed to want. We choose this rather bulky phrase to emphasise that this list of needs does not entail that individuals must develop a corresponding construct of wanting but that they have to be allowed to do so; and if they do, they are entitled to satisfy it. That is, their development must not be inhibited by oppressive social and/or governmental behaviour but must be facilitated by empowering and conducive behaviour. We claim that these nine needs are universal and therefore suggest using them to operationalise the 'good life' for the context of sustainability.

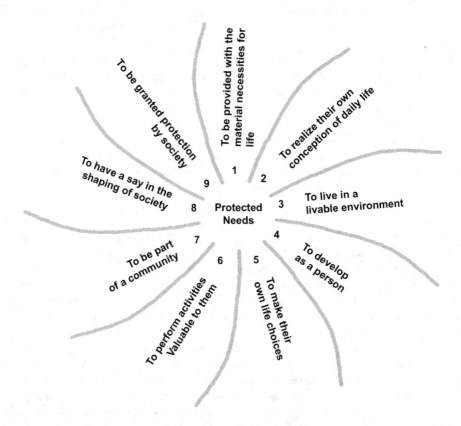

*Figure 8.2* Protected Needs – what individuals must be allowed to want: the list of Protected Needs consists of nine needs arranged in three groups. These needs denote what individuals must be allowed to want. Group 1 focuses upon tangibles, material things (Protected Needs 1–3); group 2 upon the person (Protected Needs 4–6); and group 3 upon community (Protected Needs 7–9).

It goes without saying that these needs do not yet represent a thick theory of the 'good life'. Rather, they have to be further delineated, that is, these needs have to be described. The thick descriptions must outline the possibilities (not the satisfiers!) individuals should be provided with (not implying that individuals must seize these possibilities). To produce such a description of the Protected Needs is quite challenging. While the needs are universal, their description has to be tailored to a specific (cultural and historical) context. But the five criteria used to identify the Protected Needs (see section "Criteria for defining the 'good life' and needs for the context of sustainability") also apply to their description. And the same applies to the criterion of corresponding with empirical evidence, that is, the description of each need must name possibilities that negatively affect an individual's quality of life if the individual develops a corresponding construct of wanting and is not provided with them. In addition, it must be taken care to make sure that the Protected Needs (= name and thick, specified description) are clear enough to be broadly understood, and that their legitimacy, the resulting entitlement, and the responsibility ingrained in them are shared in a society (for instance by means of representative surveys). Ideally this process encompasses participatory elements (see also e.g. Alkire 2007; Robeyns and van der Veen 2007).

We developed such a specification for the cultural context of Switzerland (also covering Germany due to the group of German scholars involved). Each of the nine Protected Needs is described by three specifying elements (see Table 8.1, right column). In producing these elements, we took great care to stick to a universal perspective as much as possible and to avoid smuggling in satisfiers (or resources). This entire process is quite demanding and time consuming. The list of Protected Needs has also been used in qualitative research in Asian cities (Chennai, Metro Manila, Shanghai, Singapore) to explore how green public spaces act as satisfiers with regard to these needs. This research has shown that the effort of producing specified descriptions for the Protected Needs can be eased by taking our thick list as a starting point and identifying differences instead of starting from scratch.

Table 8.1 The nine Protected Needs (left column), specified for the cultural context of Switzerland and Germany (right column)[6]

| Group 1, focusing upon tangibles, material things (Protected Needs 1–3) | |
| --- | --- |
| Need (what individuals must be allowed to want) | Specified description: Individuals should have the possibility ... |
| (1) To be provided with the material necessities for life | ... to feed themselves sufficiently, with variety, and with food that is not detrimental to health. |
| | ... to live in a suitably protected and equipped accommodation, offering privacy and sufficient space and allowing them to realise their idea of living. |
| | ... to care for their bodies with dignity and dress suitably. |
| (2) To realize their own conception of daily life | ... to shape their daily life according to their own ideas. |
| | ... to procure and use the material necessities for life from a diverse range of supply, and to have sufficient means to do so. |
| | ... to move freely in public space. |

*(Continued)*

*Group 1, focusing upon tangibles, material things (Protected Needs 1–3)*

| (3) To live in a livable environment | ... to live in an environment (built and natural) that is not harmful to health and is aesthetically pleasing. |
| | ... to develop a sensorial and emotional relationship with nature. |
| | ... to have access to and be able to move about in diverse natural and cultural landscapes. |

*Group 2, focusing upon the person (Protected Needs 4–6)*

| *Need (what individuals must be allowed to want)* | *Specified description: Individuals should have the possibility ...* |
| --- | --- |
| (4) To develop as a person | ... to develop their potential (knowledge, skills, attitudes, feelings, etc.) and thus their individual identity. |
| | ... to face the challenges of their choice. |
| | ... to freely access reliable information and thus form their own opinion. |
| (5) To make their own life choices | ... to freely decide and act upon the value-orientations they choose to adopt or reject (spirituality, religiosity, ideology, etc.). |
| | ... to set their own life goals and pursue them. |
| | ... to determine how they want to lead their life in terms of intimate relationships, family planning, where to live, etc. |
| (6) To perform activities valuable to them | ... to carry out activities that they consider to be fulfilling (in work and leisure; paid and unpaid). |
| | ... to carry out activities that match their personality and in which they can unfold their potential (in work and leisure; paid and unpaid). |
| | ... to allocate their time for their different activities according to their own preferences and to have time for idleness. |

*Group 3, focusing upon community (Protected Needs 7–9)*

| *Need (what individuals must be allowed to want)* | *Specified description: Individuals should have the possibility ...* |
| --- | --- |
| (7) To be part of a community | ... to maintain social relationships with other people (private, professional, during training, etc.). |
| | ... to take part in cultural activities and celebrations and to participate in associations. |
| | ... to access the cultural and historical heritage of their community. |
| (8) To have a say in the shaping of society | ... to co-determine the affairs of the society in which they live. |
| | ... to take an active stand for concerns and problems (local, national, international) they hold dear. |
| | ... to voice their opinion, by themselves and with others. |
| (9) To be granted protection by society | ... to be protected from public and private violence, from infringements on physical and mental integrity, and from natural hazards. |
| | ... to pursue their goals without discrimination and with equal opportunity, to live in legal certainty, and to be treated with dignity and respect. |
| | ... to be supported in the event of physical or mental impairment, unemployment, poverty, and other impairing conditions. |

## Conclusion: focusing on the 'good life' and needs in sustainability governance

The aim of this chapter was to present how the notions of the 'good life' and of needs should be approached in order to serve as a foundation for sustainability governance. Based on some general considerations and criteria (section "Criteria for defining the 'good life' and needs for the context of sustainability"), we first discussed the literature (section "Approaches to defining the 'good life' and needs") and then presented our own suggestion of a theory of the 'good life' for the context of sustainability (section "Protected Needs: a theory of the 'good life' as a basis for global sustainability governance"). Our proposal comprises of both a list of (universal) Protected Needs and a procedure to adapt the thick descriptions of these Protected Needs for other cultural contexts. Having a list of Protected Needs that is specified for different cultural contexts could provide a basis for coherent international and national sustainability governance. Having such a list is a prerequisite to negotiating need satisfaction, that is, in negotiations about the satisfiers and resources that should be provided on a national and global scale, about the satisfiers and resources that should be safeguarded, and about the limits for individuals to freely choose and implement their individual way of living (see Fuchs, this volume).

But, and this could turn out to be equally important, focusing on the 'good life' and on needs not only in the theory of sustainability but also in sustainability governance requires, first, acknowledging that human beings share many characteristics and aspirations across time and cultures and that it is therefore wrong to think in terms of "us versus them" defined by nations and cultures instead of in terms of "we" defined as humanity. Second, it requires acknowledging the value of each human being and granting all individuals the right to freely choose and implement their own way of living based on a common and shared responsibility limiting this freedom for the sake of the other human beings. It requires, third, acknowledging that ensuring a good life for all human beings now and in the distant future necessitates observing the potential finitude of resources, adopting a precautionary strategy, and engaging in a global debate, informed by a salutogenic approach, about what quality of life is. This, in turn, allows one to draw on a powerful narrative that is appealing to most if not all human beings and thus has the potential to support sustainability governance with a view to necessary, fundamental changes.

## Acknowledgement

We wish to thank, first of all, all scholars that have contributed to our research. All of them are mentioned by name in our paper. Second, we wish to thank the funders that have made our research possible. Finally, we wish to thank the editors for valuable contributions to an earlier version of the paper, and Anthony Mahler for his highly appreciated support in editing the paper.

## Notes

1 Much of what we present here draws on two research projects: first, the research programme "From Knowledge to Action – New Paths towards Sustainable Consumption" (2008–2014) funded by the German Federal Ministry of Education and Research (BMBF) and consisting of ten research groups (involving a

total of 100 researchers from more than 15 disciplines and 80 partners from practice) and an accompanying project with the task of facilitating the development of integrated results (see e.g. Blättel-Mink et al. 2013; Defila et al. 2012, 2014); and second, the project "Sustainable Consumption – Searching for the Right Balance between Boundless Freedom, Hostile Asceticism and Dogmatically Imposed Lifestyles" (2015–2017) funded by the Mercator Foundation Switzerland.

2 Salutogenic approaches are based on "a positive perspective on human life" and aim to investigate the origins of health rather than those of disease and risk (Mittelmark and Bauer 2017).

3 The distinction between subjective and objective theories corresponds, at least to some extent, to the distinction between "hedonistic" and "eudemonic" concepts of the 'good life' in more empirical approaches in disciplines such as psychology, sociology, and economics.

4 European Union Statistics on Income and Living Conditions.

5 The list of Protected Needs was developed in the project "Sustainable Consumption" mentioned in footnote 1. The scholars involved in the structured dialogue were Peter Bartelheimer, Mathias Binswanger, Birgit Blättel-Mink, Rico Defila (project team), Antonietta Di Giulio (project team), Doris Fuchs, Konrad Götz, Gerd Scholl, Ruth Kaufmann-Hayoz (project team), Lisa Lauper (project team), Gerd Michelsen, Martina Schäfer, Michael Stauffacher, Roland Stulz, and Stefan Zundel.

6 The original and thus authoritative version of the list is in German (dating 15 October 2016), authored by Rico Defila and Antonietta Di Giulio. This version has been translated into French by M. I. S. Trend and into English by Antonietta Di Giulio and Rico Defila. Valuable contributions and feedback to the English version have been provided by Manisha Anantharaman, Marlyne Sahakian, Czarina Saloma-Akpedonu, and Anders Hayden.

# References

Abbott R., Ploubidis G. B., Huppert F. A., Kuh D., Croudace T. J. (2010): An evaluation of the precision of measurement of Ryff's psychological well-being scales in a population sample. *Social Indicators Research* 97: 357–373.

Aked J., Marks N., Cordon C., Thompson S. (2008): Five Ways to Wellbeing. A Report Presented to the Foresight Project on Communicating the Evidence Base for Improving People's Well-Being. London: New Economics Foundation.

Alkire S. (2007): Choosing Dimensions: The Capability Approach and Multidimensional Poverty. Paper presented at the CPRC Workshop on Concepts and Methods for Analysing Poverty Dynamics and Chronic Poverty, 23–25 October 2006, University of Manchester. OPHI, working paper 88. Oxford: University of Oxford.

Alkire S. (2010): Human Development: Definitions, Critiques, and Related Concepts. Background paper for the 2010 Human Development Report. OPHI, working paper 36. Oxford: University of Oxford.

Anand P., Krishnakumar J., Tran N. B. (2011): Measuring welfare: Latent variable models for happiness and capabilities in the presence of unobservable heterogeneity. *Journal of Public Economics* 95: 205–215.

Baumeister R. F., Leary M. R. (1995): The need to belong: Desire for interpersonal attachments as a fundamental human motivation. *Psychological Bulletin* 117(3): 497–529.

Blättel-Mink B., Brohmann B., Defila R., Di Giulio A., Fischer D., Fuchs D., Gölz S., Götz K., Homburg A., Kaufmann-Hayoz R., Matthies E., Michelsen G., Schäfer M., Tews K., Wassermann S., Zundel S. (2013): Konsum-Botschaften. Was Forschende für die gesellschaftliche Gestaltung nachhaltigen Konsums empfehlen. Stuttgart: Hirzel Verlag.

Brulé G., Veenhoven R. (2014): Freedom and happiness in nations: Why the Finns are happier than the French. *Psychology of Well-Being: Theory, Research and Practice* 4(17): 1–14.

Burchardt T., Vizard P. (2011): 'Operationalizing' the capability approach as a basis for equality and human rights monitoring in twenty-first-century Britain. *Journal of Human Development and Capabilities* 12(1): 91–119.

Costanza R., Fisher B., Ali S., Beer C., Bond L., Boumans R., Danigelis N. L., Dickinson J., Elliott C., Farley J., Elliott Gayer D., MacDonald G. L., Hudspeth T., Mahoney D., McCahil L., McIntosh B., Reed B., Turab Rizvi S. A., Rizzo D. M., Simpatico T., Snapp R. (2007): Quality of life: An approach integrating opportunities, human needs, and subjective well-being. *Ecological Economics* 61: 267–276.

Cummins R. (1996): The domains of life satisfaction: An attempt to order chaos. *Social Indicators Research* 38(3): 303–328.

Defila R., Di Giulio A., Kaufmann-Hayoz R. (eds.) (2012): The Nature of Sustainable Consumption and How to Achieve it. Results from the Focal Topic "From Knowledge to Action – New Paths towards Sustainable Consumption". München: oekom.

Defila R., Di Giulio A., Kaufmann-Hayoz R. (2014): Sustainable consumption – an unwieldy object of research. *GAIA* S1/2014: 148–157. doi:10.14512/gaia.23.S1.2.

Di Giulio A. (2004): Die Idee der Nachhaltigkeit im Verständnis der Vereinten Nationen – Anspruch, Bedeutung und Schwierigkeiten. Muenster et al.: LIT Verlag.

Di Giulio A. (2008): Ressourcenverbrauch als Bedürfnis? Annäherung an die Bestimmung von Lebensqualität im Kontext einer nachhaltigen Entwicklung. *Wissenschaft & Umwelt INTERDISZIPLINÄR* 11, Energiezukunft: 228–237.

Di Giulio A., Brohmann B., Clausen J., Defila R., Fuchs D., Kaufmann-Hayoz R., Koch A. (2012): Needs and consumption – a conceptual system and its meaning in the context of sustainability. In: Defila R., Di Giulio A., Kaufmann-Hayoz R. (eds.): The Nature of Sustainable Consumption and How to Achieve It. Results from the Focal Topic "From Knowledge to Action – New Paths towards Sustainable Consumption". München: oekom. 45–66.

Di Giulio A., Defila R., Kaufmann-Hayoz R. (2010): Gutes Leben, Bedürfnisse und nachhaltiger Konsum. Nachhaltiger Konsum, Teil 1. *Umweltpsychologie* 14(2/27): 10–29.

Di Giulio A., Fischer D., Schäfer M., Blättel-Mink B. (2014): Conceptualizing sustainable consumption: Toward an integrative framework. *Sustainability: Science, Practice, & Policy (SSPP)* 10(1): 45–61.

Dolan P., Layard R., Metcalfe R. (2011): Measuring Subjective Wellbeing for Public Policy: Recommendations on Measures. London: Centre for Economic Performance.

Dolan P., Peasgood T., White M. P. (2008): Do we really know what makes us happy? A review of the economic literature on the factors associated with subjective well-being. *Journal of Economic Psychology* 29: 94–122.

Doyal L., Gough I. (1991): A Theory of Human Need. New York: Macmillan.

Jackson T., Jager W., Stagl S. (2004): Beyond insatiability – needs theory, consumption and sustainabilty. In: Reisch L., Røpke I. (eds.): The Ecological Economics of Consumption. Cheltenham, Northampton: Edward Elgar. 79–110.

Lewinsohn P. M. (2016): Pleasant Events Schedule.

Manstetten R. (1996): Zukunftsfähigkeit und Zukunftswürdigkeit – Philosophische Bemerkungen zum Konzept der nachhaltigen Entwicklung. *GAIA* 5(6): 291–298.

Max-Neef M. A. (ed.) (1991): Human Scale Development: Conception, Application and Further Reflections. London: Zed Books.

Max-Neef M. A., Elizalde A., Hopenhayn M (1991): Development and human needs. In: Max-Neef M. A. (ed.): Human Scale Development: Conception, Application and Further Reflections. London: Zed Books. 13–54.

Michaelis L. (2000): Ethics of Consumption. Oxford: Oxford Centre for the Environment, Ethics & Society.

Mittelmark M. C., Bauer G. F. (2017): The meanings of salutogenesis. In: Mittelmark M. B., Sagy Sh., Eriksson M., Bauer G. F., Pelikan J. M., Lindström B., Espnes G. A. (eds.): The Handbook of Salutogenesis. Cham: Springer. 7–13. doi:10.1007/978-3-319-04600-6.

Nussbaum M. (1992): Human functioning and social justice: In defense of Aristotelian essentialism. *Political Theory* 20(2): 202–246.

Nussbaum M. (2006): Frontiers of Justice. Disability, Nationality, Species Membership. Cambridge, MA and London: Belknap Press of Harvard University Press.

O'Neill J. (2011): The overshadowing of needs. In: Rauschmayer F., Omann I., Frühmann J. (eds.): Sustainable Development. Capabilities, Needs, and Well-Being. London and New York: Routledge. 25–42.

Rauschmayer F., Omann I., Frühmann J. (2011): Needs, capabilities and quality of life. Refocusing sustainable development. In: Rauschmayer F., Omann I., Frühmann J. (eds.): Sustainable Development. Capabilities, Needs, and Well-Being. London, New York: Routledge. 1–24.

Robeyns I., van der Veen R. J. (2007): Sustainable quality of life: Conceptual analysis for a policy-relevant empirical specification. Bilthoven, Amsterdam: Netherlands Environmental Assessment Agency and University of Amsterdam.

Ruesch Schweizer C., Di Giulio A. (2016): Nachhaltigkeitswerte: Eine Kultur der Nachhaltigkeit? Sekundäranalyse der Daten aus der StabeNE-Erhebung 2008. *SOCIENCE, Journal of Science-Society Interfaces* 1(1): 91–104. http://openjournals.wu.ac.at/ojs/index.php/socience/index.

Ryan R. M., Deci E. L. (2000): Self-determination theory and the facilitation of intrinsic motivation, social development, and well-being. *American Psychologist* 55(1): 68–78.

Ryan R. M., Deci E. L. (2001): On happiness and human potentials: A review of research on hedonic and eudaimonic well-being. *Annual Review of Psychology* 52: 141–166.

Ryff C. D. (1989): Happiness is everything, or is it? Explorations on the meaning of psychological well-being. *Journal of Personality and Social Psychology* 57(6): 1069–1081.

Shin D. C. (2015): How People Perceive and Appraise the Quality of Their Lives: Recent Advances in the Study of Happiness and Wellbeing. Irvine: Center for the Study of Democracy.

Soper K. (2006): Conceptualizing needs in the context of consumer politics. *Journal of Consumer Policy* 29: 355–372.

Veenhoven R. (2008): Sociological theories of subjective well-being. In: Eid M., Larsen R. (eds.): The Science of Subjective Well-Being: A Tribute to Ed Diener. New York: Guilford Publications. 44–61.

White M. P., Dolan P. (2009): Accounting for the richness of daily activities. *Psychological Science* 20(8): 1000–1008.

# 9

# Post-Eurocentric sustainability governance

## Lessons from the Latin American *Buen Vivir* experiment

*Julien Vanhulst and Adrián E. Beling*

## Introduction[1]

The fundamental problem of socio-ecological sustainability[2] has opened a space of disputes among multiple discourses or visions about global sustainability governance that are more or less critical of the currently prevailing system of social organisation. We hereafter refer to this space of deliberations as the "discursive field of social-ecological sustainability" (in line with Dryzek 2005; Hopwood, Mellor, and O'Brien 2005; Sachs 1997). Until now, conservative or mild reformist discourses of sustainability governance (such as ecological modernisation, sustainable development, or green growth)[3] have dominated the field and directed the trajectories of socio-ecological adaptations. Nevertheless, some governance approaches in the discursive field are much more critical of the status quo or incremental reforms and call for a radical "paradigm shift." In light of the poor track-record of sustainability governance over the last decades, the growing evidence of a deepening of the environmental crisis, and a broader global discontent with the status quo, these *"transformation* or *transition discourses"* (Escobar 2011) have gathered momentum at the turn of the new millennium.[4] In contrast with approaches reproductive of the status quo, transition discourses seek to "transform" the political-economic chessboard set by industrial societies (Dryzek 2005), transcending the universalistic prism of the cultural model of the modern West towards a *pluriverse* (Escobar 2011, 2015). A pluriverse can be defined as a plural yet interdependent ecology of knowledges and practices, opening up space for alternative socio-cognitive and normative structures for social action towards socio-ecological sustainability. In its scope and ambition, the Latin American *Buen Vivir* (BV) can be considered one such transformation discourse (Beling et al. 2018; Beling, Gómez, and Vanhulst 2014; Beling and Vanhulst 2016; Vanhulst 2015; Vanhulst and Beling 2013, 2014, 2017).

BV has risen to international policy and academic centre-stage since the early 2000s and spearheaded many of the political discussions and policies in the region throughout the first decade of the century. Its emersion in the domestic political sphere of some Latin American countries, especially Ecuador and Bolivia, which included BV as a regulative principle in their respective national constitutions, had the effect of enlarging the scope of "what matters in the political arena" (Gerlach 2017), thus effectively challenging "politics as usual" (de la Cadena 2010). The roots of BV can be traced back to various indigenous traditions in Latin America, and their intertwining with other currents of critical thought throughout the 20th century (Beling and Vanhulst 2016). In fact, the term "Buen vivir" is a controversial Spanish translation of the Quechua *Sumak Kawsay* and the Aymara *Suma Qamaña*, as well as other similar ideas that can be found in other indigenous cultures. It can be loosely translated into English as "good living," "living well," or, perhaps more accurately, "living a plentiful life."

As a contemporary discourse, BV arises in a particular historical–political juncture at the interface of the *local* – where longstanding indigenous struggles for cultural and material recognition, eventually converged with the disenchantment of the masses with the neoliberal order at the dawn of the century – and the *global*, where the development paradigm[5] is undergoing a severe legitimacy crisis. Indeed, in light of the grave and sustained (if not worsening) social, environmental, and economic challenges of our time, and the lack of suitable answers within the development discourse, BV has often been defined as a (sometimes dialogical, sometimes antithetical) *alternative to development* (Acosta 2015; Gudynas 2014, 2011; Vanhulst and Beling 2014), challenging not only the development worldview but also the socio-cultural matrix of Eurocentric modernity as a whole.

As a result of this *glocal* and synthetic character, BV constitutes a collective and heterogeneous proposal for a new cultural model, a plural vision of the good society that aims at transcending the dualistic (universalistic, materialistic, individualistic) frameworks that have characterised the modern political and cultural imagination, as embodied in modern institutions and values such as industrialism, modern science, capitalism, anthropocentrism, secularism, individualism, liberal freedom, etc. Such institutions and values are challenged by means of (re)introducing notions of the intrinsic value of non-human nature and its symbiotic connection with humans and their societies (Pachamama or "mother Earth"), by stressing the collective as a condition of possibility for the individual, and the non-material dimensions of human wellbeing, among others.

A common element among the diverse versions of BV is a three-dimensional constitutive structure: it is a way of life in harmony with oneself (identity), with society (equity), and with nature (sustainability) (Cubillo-Guevara et al. 2018). As shown below, the diverse strands of the BV discourse varyingly prioritise one of these dimensions over the other two, with, respectively, distinct implications for the way of conceiving sustainability governance (cf. section "Buen Vivir: genealogy and key contents").

The text is structured in two main sections: the first section discusses the emergence of the idea of a "civilizational crisis" and the need to set up an effective mode of governance for socio-ecological sustainability beyond the cultural and institutional frames derived from Eurocentric modernity. The second section presents an analysis of BV as an approach to sustainability governance that opens up a space of disputes around default responses to the socio-ecological predicament, raising fundamental questions about the globally dominant cultural model. The text ends with some reflections on the lessons gained from BV as a cultural and political experiment for a reconceptualisation of global sustainability governance.

## Eurocentric modernity and socio-ecological crisis: the need for transformation

### Eurocentric modernity, critiques, and pluralist perspective

Since its origins, classical social science has made the empirical and conceptual distinction between so-called "modern" and "traditional" societies, one of its central quests – not only with descriptive, but mainly with normative and hierarchising purposes. The conventional understanding of modernity has long been characterised by a social configuration character-ised by monolithic traits, amongst which individualism, secularisation, experimental/posi-tivistic science as the superior way of knowing the world, the differentiation of functions and institutions in society, and the idea of progress as a horizon of perpetual material and moral improvement (Martinelli 2005). These features allegedly came to their highest realisation in Europe and North America, turning this "Euro-Atlantic modernization" into a universal model for all societies, including governance modes.

Over the second half of the 20th century, however, this orthodoxy envisioning a universal path of convergence towards Eurocentric modernity started to be strongly questioned and criticised for its linear, ethnocentric, evolutionist, and colonialist biases (Arnason 1991; Beck 1992; Eisenstadt 2000; Escobar 1995; Therborn 2003; Wagner 2008). These new theories deconstruct the idea of modernity as reduced to the singular "Euro-Atlantic" prism, and thus move away from the Western model as the ideal type of modernity. However, they remain committed to the core promise of modernity, namely to provide the conditions of possibility for individuals and societies to govern themselves as autonomous entities. While the promise of autonomy is hailed, its conditions of possibility are acknowledged to vary according to context-specific, contingent, or structural factors (Wagner 2008). This interpretation of mo-dernity as a perpetually produced and reproduced social order, coupled with a broad recogni-tion of autonomous political projects, draws the contours of the pluralist image of modernity.

Yet within the globally prevailing model of society, shaped after the Eurocentric under-standing of modernity, it is precisely this utopia of autonomy which is widely experienced as elusive in a global development pathway unfolding at an ever accelerating and automatised pace (Rosa 2010). Furthermore, in light of the global ecological crisis, the cultural model of the West becomes a "doomsday model" (Beck 2015); yet societies around the world con-tinue to pursue the Euro-Atlantic path of development regardless, with both "developed" and "developing" societies operating at enormous ecological and social cost. This necessarily brought conventional (sustainability) governance approaches under severe scrutiny.

## The sustainability imperative and the plurality of human cultures

The debates around the socio-ecological sustainability are part of the late-modern project. Since the 1970s, these debates have gradually forged a space of discursive disputation, offering different solutions to the Society/Environment equation. This equation cannot be separated from the tension between the idea of global socio-ecological interdependence (and therefore of unity) and the multiplicity of cultures and development trajectories. In its introduction, the Brundtland Report *Our Common Future* (WCED 1987) recognises the tension between global socio-ecological interdependence and local disparities in socio-economic terms and development trajectories. However, it makes no explicit reference to the cultural dimension of sustainable development. Since the early 1990s, a parallel reflection has begun on the topic of culture and development within United Nations Educational, Scientific and Cultural Or-ganization (UNESCO) (Porcedda and Petit 2013). The report *Our Creative Diversity* (CMCD 1996) emphasises the desirability of cultural diversity and pluralism, and calls for more at-tention to the interactions between culture and development in the search for ecological sustainability (CMCD 1996, Chapter 8). In line with these discussions, and in preparation of the Johannesburg Summit (2002), UNESCO published the *Universal Declaration on Cultural Diversity*, which was reasserted in 2005 with the *Convention on the Protection and Promotion of the Diversity of Cultural Expressions* and, more recently, with the *Declaration of Hangzhou* (2013), which resulted from a UNESCO international conference especially dedicated to the links between culture and sustainable development. Thus, the cultural dimension has progressively become a central element in debates around socio-ecological sustainability.[6] In fact, the multiple historical experiences and worldviews of diverse social groups (which do not necessarily correspond to national borders) have a direct influence on the meanings attached to sustainability and on the responses to the Society/Environment equation (Vanhulst and Zaccai 2016).

## Sustainability "made in Latin-America"

In Latin America, some socio-ecological discourses adopt a critical stance towards Eurocentric modernity (Vanhulst and Zaccai 2016), and align with social movements fighting for social and environmental justice. Thus, in a way, they reflect the debates around cultural diversity and social justice, and the multiple struggles for the recognition of *"saberes otros"* ("other forms of knowledge") marginalised by Eurocentric modernity.

Although there are some common experiences and understandings with the "North," in Latin America, sustainability discourses are directly linked to the historical experience of colonisation, the current situation of the region in the (semi-)periphery of the globalised capitalist economy, and an extractivist economic matrix, with all its social, economic, environmental, and cultural consequences. Conflict is more tangible and, therefore, also thematised more explicitly. For example, the expropriation and exploitation of land on which local communities depended (and still depend today) for their subsistence, the unregulated process of hyper-urbanisation, the brutal exposure to the unrestrained forces of the world market, with their negative socio-economic and environmental externalities, have configured a culturally specific framework for the interpretation of socio-ecological questions, which starkly differs from Northern interpretative schemes or frames (Adams 2001; Brand 2015), as well as a distinctive Latin American environmental ethics (Heyd 2004, 2005).

Another core difference between Latin American and Euro-Atlantic debates around sustainability pertains to a key dimension which is heavily underestimated in the latter, namely: power. While in the North the social and ecological consequences of economic activity are often framed as a material-technical problem, with a language heavily marked by abstract and dry concepts such as "anthropocentrism" or "patterns of production/consumption," in Latin America they are closely linked to relationships of domination, such as capitalist seizure or land-grabbing, post-colonial, patriarchal, or racial ones. Capitalism is not framed merely as a system of production and consumption, but rather as a system of power and domination (Adams 2001; Brand 2015; Heyd 2004, 2005).

While in the North sustainability debates are largely de-territorialised, in Latin America they tend to be firmly anchored in concrete territories, which are understood as vital spaces of struggle, belonging, and autonomy. Particularly strong emphasis is given to the importance of revalorising other forms of knowledge and other epistemologies beyond Cartesian rationality and techno-scientific knowledge: *"saberes otros"* ("other forms of knowledge") condensed in language and in "epistemic dialogues" (Leff 2004), leading to a "democratization of knowledge" characterised by epistemic or "cognitive justice" (de Sousa Santos 2009), or else to an "ecology of knowledges."

The pairing of ecological issues with structures and dynamics of power and domination implies that the search for sustainability futures is simultaneously a search for emancipatory futures. Euro-Atlantic sustainability discourses, in turn, often have a markedly burdensome character, framing the environmental crisis as a problematic by-product of an otherwise desirable societal trajectory. Inspirational narratives of change are thus crowded out by cryptic-technocratic narratives of continuity, with sustainability governance being understood rather as "ecological modernization" without fundamental changes in socio-cultural structures.

Unlike in the European debates on sustainability governance, in Latin America the talk is hardly about consumption or individual behaviour – although such approaches are on the rise on account of the rampant consumerism of the urban elites and the so-called "new middle classes." Discourses about alternative ways of life, however, are conceived of at a rather collective level. BV, for example, deals with the creation and reproduction of integral conditions for socio-ecological reproduction.

## The *Buen Vivir* experiment

### Buen Vivir: *genealogy and key contents*

As a cultural-political experiment in the Andean-Amazonian region around the turn of the century, BV can be viewed as a multi-round clash over political and societal shaping-power among competing visions of desired socio-ecological futures. However, as we have shown elsewhere (Beling and Vanhulst 2016; Beling et al. forthcoming), far from portraying a purely idiosyncratic cultural and political process, the Latin American BV results from a dialectical synthesis between global and local forces, in a context of global political contestation around the prevailing development model. This *glocal* character has placed BV in a global spotlight. Thus, while the role of local subaltern actors (indigenous, socialist, environmentalist groups) has been determinant for the materialisation and global dissemination of the BV discourse (*inside-outward* processes), the role of foreign agents – especially from the fields of development, ecology, and counter-hegemonic global struggles – has been no less important (*outside-inward* processes) in both the co-production and dissemination of BV narratives in international discursive arenas through the institutional civil society and state channels.

At the same time, however, as BV became territorially anchored in the socio-cognitive and cultural landscape and in certain socio-political practices, its content diversified, forking into a range of more or less (dis)similar discourses respectively re-articulated by the successive groups that have adopted and adapted it (Hidalgo-Capitán and Cubillo-Guevara 2014; Le Quang and Vercoutère 2013; Vanhulst 2015; Vanhulst and Beling 2013, 2014). The distinct narratives of BV can be traced back to different phases in the diachronic process of the discursive construction of BV. Three broad phases can be distinguished: *emergence* (or primordial discursive shaping), *institutionalisation* (discursive hybridisation or consensual assemblage), and *dislocation* (or discursive diversification), the latter giving way to three new discursive configurations. This resulted in five different versions of BV (see Cubillo-Guevara et al. 2018 for detailed treatment.): (1) a *primordial BV*, corresponding to the moment of discursive emergence of BV in the Ecuadorian Amazonia and its expansion through the Andean-Amazonian region; (2) a *hybrid BV*, corresponding to the process of political consensus-building and institutionalisation in the Bolivian and Ecuadorian constitutions, as well as with official development plans at the national level (Ministerio de Planificación del Desarrollo 2007; SENPLADES 2007, 2009, 2013); and finally, the phase of discursive dislocation that opens up three new declinations: (3) an *indigenist BV*, which is closely tied to Latin American indigenous movements; (4) a *socialist/statist BV*, associated with the Ecuadorian and Bolivian governments and their academic interlocutors; and (5) a *post-developmentalist BV*, which is advocated by other Latin American social movements

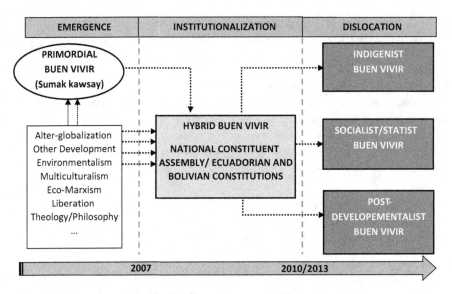

*Figure 9.1* Emergence/institutionalisation/dislocation of *Buen Vivir*
*Source*: Own elaboration (cf. Cubillo-Guevara et al., 2018).

(environmentalist, peasant, etc.) and their intellectual referents (for more detailed elaboration on the genealogy of BV from a discursive and actor-centred and discursive perspectives, respectively, see Beling et al., forthcoming; Beling and Vanhulst 2016; Cubillo-Guevara et al., 2018) (Figure 9.1).

While the primordial BV can be largely viewed as an exercise in formal discursive articulation aimed at conceiving and communicating a native development-horizon, and the hybrid BV largely remains the purely rhetorical expression of a factional negotiation process, it is the post-consensual dislocation or trifurcation that expresses the key cultural and political cleavages around BV.

The indigenist strand tends to privilege the identitary-traditional dimension over the socio-economic equity and environmental aspects, being largely dismissive of the reciprocal local and global entanglements of these three dimensions. Under the banner of decolonisation and self-determination, an essentialist view of indigenous culture is typically played against (an equally reified view of) modernity. While the ecological dimension is implicit in the indigenous cosmologies, political (territorial) and cultural self-determination are envisioned as the royal road towards all these goals. The relationship towards the modern world is thus antagonistic.

The socialist strand, in turn, under the (declared) premise of attending to the immediate and basic needs of the impoverished population (and assuming that the only way to do this is with a rent-capturing and -redistributing state) places the ecological and identitary dimensions in a subordinate position, whereby the relationship among these goals is framed as conflictual. Despite opposing colonial-like economic subordination to Western powers, the relation to (Western-style) modernity is aspirational.

Finally, the post-developmentalist strand sees all dimensions of BV as interdependent or organically interlinked. The various dimensions can thus not be played against each other without jeopardising the whole emancipatory and transformative enterprise: the ecological dimension is central, since all other dimensions "derive cascading from it, as it forces into rethinking the economic system, the cultural parameters of reference, the patterns of social interaction, and the form of political organization" (Belotti 2014, 43–44). Western-style modernity is not rejected wholesale, but rather *provincialised* (Chakrabarty 2000), that is, removed from the place of a (false) universal and understood as a particular spatiotemporal constellation, and thus as a valuable repository of embedded human experience. Cultural creativity and inter-cultural dialogue engaged at eye level are thus prescribed as a means towards emancipation in an interdependent world. Epistemic or "cognitive justice" (de Sousa Santos 2009) is deemed a precondition for global social justice and ecological sustainability.

## Lessons from BV for a renewed conceptualisation of sustainability governance

The cultural effervescence unleashed by the intense social movement activity leading to the so-called "post-neoliberal" or "left turn" in the early 21st century Latin America comes into stark contrast with the *Realpolitik*-flavour of the path followed thereafter by the Correa and Morales administrations in Ecuador and Bolivia, respectively. From the above, it becomes clear that the socialist-statist version of BV has largely forsaken culturally innovative and politically transformative aspirations towards a social-ecological transition, and rather taken a business-as-usual pathway towards conventional development. Rather than a post-capitalist, biocentric political project, as it is rhetorically proclaimed, the statist-socialist BV constitutes an attempt at humanising capitalism and reproducing the liberal vision of society. Moreover, steeply accumulating evidence contradicts the principle of harmony with nature and the effective recognition of rights awarded to subaltern groups – let alone the inclusion of their values and practices in the political sphere. Indeed, of the constitutive elements of BV highlighted in the consensual "hybrid BV" and anchored in national constitutions, legal tools, and in official development plans (harmony with nature, revaluation of marginalised voices, democratic autonomism, and the universal satisfaction of basic needs, in addition to the critique of Western-style development and cultural plurality), only the satisfaction of basic needs has been met to a significant degree, though arguably in a precarious, unsustainable manner, yet at the cost of deepening dependence on international commodity-markets and locking the national economies further into the (neo)extractivist matrix (Beling et al. 2018; Lalander and Cuestas-Caza 2017; Vanhulst 2015).

Statist experiments with BV showcase the limitations of a political revolution without an effective transformation of the material base (Becker 2011; Beling et al. 2018). The BV experiment thus begs the question, we argue, about the cultural and material conditions under which sustainability governance can be effective. Indeed, BV seems to confirm the thesis that premature institutionalisation of a transformative programme in the form of a political party leads into the trap of mere "politicking," i.e. political actors becoming divorced from social and ecological realities and being trapped in the political game (Blühdorn 2007; Latouche 2009). In addition, BV shows that once the state enters what traditionally is the realm of civil society, it tends to monopolise the transformative impulse, weakening other agents (Svampa 2016).

But BV also taught us positive lessons. Indeed, BV appears not only as a product but also as a powerful source of cultural transformation, matching long marginalised voices from the global South with a global momentum for a discursive shift in the imaginary of development and modernity. In this regard, BV can be considered as the first large-scale experiment discursively articulating modern and non-modern ontologies, not only at the ideational level but also in the institutional and material spheres, thus bridging the hitherto unbridgeable: the holistic, relational, circular, mystical world of the Andean ancestral traditions and the dualistic, individualistic, linear, and rational world of development, thereby unleashing a process of cross-fertilisation or cultural learning. The provisional outcome has been variously referred to as a "retro-progressive utopia" (Serrano 1999, cited in Cortez 2011), a "mobilizing illusion" (Fander Falconí), or a "real utopia" (René Ramírez) (Vanhulst 2015, 11).

The scale of this cultural transformation is such that Raúl Zibechi has even proposed dropping the conventional concept of "social movement," as it cannot capture the extension or the depth of the phenomenon of what he calls, instead, "societies in movement" (Zibechi and Nuin 2008): a larger subset of social relations that differ from the hegemonic ones. Social movements are conventionally understood as means towards the end of conquering political power. The concept of "societies in movement," in turn, remits to the idea of social movements as "carriers of a different world" in motion, fighting a struggle with a disarticulating character, subverting institutionalised and naturalised symbolic orders toward a civilisational turn (de Sousa Santos 2009).

In Ecuador and Bolivia, this has translated into political leverage: indeed, the BV debate snowballed societal forces into a critical mass for political change, empowering agents formerly conceived of as mere objects (rather than subjects) of politics. Hence BV generated not only a cultural but also a political democratisation-effect, building-up intermediate political infrastructure and bridging rigid social hierarchies. In order to reach the point of tipping material-economic structures, however, we have elsewhere argued (Beling et al., 2018) that synergic cross-pollinations with other transformative and reformist waves in sustainability governance are required.

At the agency level, we identify three key agents, with particular roles and practices, who acted as enablers in the BV debate: civil society (local social movements and foreign non-governmental organisations (NGOs)), the state, and the intellectual sphere – what we have termed the "triple helix of BV" propelling cultural learning (Vanhulst 2015; Vanhulst and Beling 2014).

Civil society was observed to perform a complex role with at least three clearly distinguishable facets: (1) as philosophical (civilisational critique, articulation of the *Sumak Kawsay* utopia) and programmatic architect of BV (i.e. the Yasuní-ITT[7] initiative, Rights of nature)[8]; (2) as "societies in movement," working as a ferment for the broader "citizen revolution" and the "cultural revolution" in Ecuador and Bolivia, respectively; and (3) as a resistance agent against the perceived co-optation of BV by the state (Beling 2019).

The state, for its part, had a pivotal role in scaling and re-framing BV into a narrative of macro-societal (also global) relevance. The legitimacy and visibility that BV acquired through state sponsorship prompted a global snowball of academic reflections, activist initiatives, and renewed political rhetoric. In addition, the unconventional role adopted by the state as "champion of nature" (rather than merely a maker of environmental policy) breaks open the horizon of expectations about the conceivable role of the state in advancing a social-ecological transformation. The role of the state thus remains a *sine qua non* enabling role in triggering a large-scale learning process – alas not the straightforward type of performativity conventionally expected from the state as empowered space for decision-making, implementation, and enforcement of norms and programs. This favourable assessment of the role of the state, however, is to be understood as the positive by-product of a largely failed implementation attempt, and to be weighed against the above-mentioned negative by-products of monopolising (and degrading) the transformative impulse and damaging the credibility of BV as an alternative to the mainstream sustainability discourse (Beling 2019).

The intellectual sphere, in turn, particularly public scholars close either to social movements or to the governments, was shown to provide substantial inputs to the process of societal reflection, as well as for political mobilisation. However, there were hardly any strategic-analytical inputs to the programmatic debate around the idea of a transformation, as indicated by the virtually total absence of "blueprints" for BV. Rather, these intellectuals engage in one of two (partly competing) enterprises: either in far-reaching deconstructive, philosophical-critical work (chiefly, the diagnosis of "civilizational crisis"), thus playing a stronger role in terms of "unbuilding unsustainability" than in "building sustainability," or, alternatively, in legitimising government agendas (Beling 2019).

## Conclusions

BV represents a novel contribution from Latin America to the global discursive war about sustainability futures. The BV discourse(s), as we have seen, embody certain practices and institutional approaches that offer an alternative socio-cognitive structure and normative orientations for social action towards social-ecological sustainability. It thus offers a widened horizon for the conceptualisation and praxis of sustainability governance.

The foremost substantial contribution of BV is arguably showcasing the need for revising the *frame* within which the prevalent mode of sustainability governance is defined. In this light, BV can be understood as an attempt at redefining the terms of the modern project in a collective and dialectical exercise of retro-progressively expanding the cultural and political imagination, disembedding the modern aspiration towards self-determination from the particular coordinates established by Eurocentric cultural imaginaries and the socio-political and economic institutions emerging from the latter. Indeed, the practical experiment with BV in the Andean-Amazonian region allows us to draw lessons also about the limitations of solo-ventures with a transformative ambition, yet channelled (a-critically) through conventional governance structures.

Within the cultural model of Eurocentric modernity as an overarching framework, "sustainable development" has been conceived of as a new avatar of the same old program of Westernisation, and sustainability governance reduced to "ecological modernization," at best. However, voices have been raised in diverse locations around the world to criticise this universalist aspiration and the cultural reductionism coupled with it. The black-boxed "iron cage" of Eurocentric modernity starts to break open, making space for a pluriversal approach: one that, while acknowledging global socio-ecological interdependence, resorts to cultural diversity for solutions to the imperative of sustainability. This idea finds theoretical anchorage in the pluralist theories of modernity (Arnason 1991; Beck 1992; Eisenstadt 2000; Escobar 1995; Therborn 2003; Wagner 2008), which detach the idea of modernity from the concrete materialisations observed in the evolutionary trajectory of the Euro-Atlantic constellation and its colonial ramifications (including liberalism and industrialism as naturalised forms of politics and economics), acknowledging instead multiple cultural matrixes and multiple possible trajectories for modern (i.e. self-determined) societies (Castoriadis 1999).

The ambiguity derived from the transitional resignification of modernity is mirrored in cultural and political debates around BV, creating a field of inner discursive tension. What we have called the "indigenist strand" of BV rejects Euro-Atlantic heritage and aims at building a wholesale alternative outside it. Oddly enough, the associated approach to (sustainability) governance of this essentialised indigeneity resembles a typically Western post-modern particularism. The *pluriversal* approach of the "post-developmentalist strand" of BV, instead, does not reject Euro-Atlantic heritage but its "false universal" character, and thus seek to "provincialize" it (Chakrabarty 2000). This approach highlights the dialogical and subversive potential of BV for deconstructing the current dominant development model, without wholesale opposing modernity, which is framed here as "multiple" and/or "entangled." Finally, the adoption of BV as an electoral and state-program has created ample room for shallow engagements with (if not outright instrumentalisation of) this otherwise transformative approach from a universalistic, Eurocentric perspective of modernity, giving way to conventional sustainability governance. BV is rhetorically instrumentalised, but deprived, in practice, of its critical quality vis-à-vis the Eurocentric understanding of modernity.

The historical-cultural diffraction of the signification of modernity, on the one hand, and the sorry track-record of narrow sustainability governance approaches shaped after the imaginary of ecological modernisation, on the other hand, require the urgent democratisation and cultural diversification of political discourse, allowing for "ontological openings" in the conceptualisation of global sustainability governance. There is a lively space of debates involving a great variety of alternative conceptualisations and programmatic proposals struggling for recognition and for finding pathways towards institutionalisation. BV is but one of them.

What makes BV distinctive, then? Wherein lies its potential? In our view, BV has a triple performativity underpinning its transformative potential:

First, the "BV hype" has channelled social energies towards the cultural and political recovery, valuation, and activation of forgotten (and at the same time prefigurative) worlds, which allows unsuspected approaches to "modern problems which have no modern solutions" (de Sousa Santos 2010, 39). That is, in the face of the negative socio-ecological consequences of the dominant Western cultural model, it is necessary to de-strand inherited perspectives to find effective solutions through governance approaches drawing on an "ecology" of both forgotten and emergent knowledges. In this sense, BV can be defined as a retro-progressive utopia towards socio-ecologically sustainable futures.

Second, the world looks at Ecuador and Bolivia because BV achieved a breakthrough into the macro-cultural and political sphere (see sections "Buen Vivir: genealogy and key contents" and "Lessons from BV for a renewed conceptualisation of sustainability governance"). All limitations in its practical operationalisation notwithstanding, the mainstreaming of BV has lent this retro-progressive utopia visibility, and maybe even credibility. As a political phenomenon, however, it is a rather exceptional outcome of a contingent *glocal* articulative process (Beling and Vanhulst 2016; Cubillo-Guevara et al. 2018; Beling et al. forthcoming), and hence unlikely to be conceived of in terms of a "model" that would be reproducible in different spatiotemporal coordinates.

Third and finally, the combination of the two earlier features makes BV a worldwide unique living laboratory for the social-ecological transformation. As with any social phenomenon, BV is an open process, a story still in the process of being written. We have elsewhere suggested that if BV is to overcome *Realpolitik*-type of limitations and become a success story in terms of opening pathways to sustainable social-ecological futures, articulations with other both transformative and reformist approaches in the north and in the south need to be explored and realised (Beling et al. 2018; Vanhulst and Beling 2014). Yet there is much more at stake in this story than the fate of a political project in the remote foothills of the Andes: at stake is a singular opportunity to learn and test a more promising approach to global sustainability governance towards viable models of societal organisation that are "capable of a future."

## Acknowledgement

This work is part of the research projects FONDECYT n°11180256 and n°1160186, funded by CONICYT (Chile).

## Notes

1 The present work is part of the research project FONDECYT n°11180256 and n°1160186 funded by CONICYT (Chile).
2 The very existence of any human collectivity entails addressing certain fundamental problems, whose cultural and political resolution makes up the basis of the social fabric (Arnason 1991; Bajoit 2003; Wagner 2008, 2010). Traditionally in social theory, these problems are mainly focused on intra-social and inter-social cohesion (internal order, socialisation, relations between collectivities, etc.). However, during the late 20th century, with the increasing awareness of the impacts of human activities on natural ecosystems, the relation between human societies and their ecological environment became defined as a fundamental problem of collective life in the "global village." The "Great Acceleration" observed since the mid-20th century (McNeill 2001; Steffen et al. 2015) – that is, the generalisation of ecologically unsustainable exponential growth-trends in virtually all areas of human activity – implies that the modern imaginary of control over the natural world gradually began to shatter and be displaced by the idea of constitutive interdependence between society and its natural environment. Thereby, the imperative of socio-ecological sustainability has become a new fundamental problem of collective life on a planetary scale, insofar the future of many forms of life on Earth (including human beings), as well as the balance of bio-geo-chemical systems, depends on the regulation of the social world.
3 All are proposals that redefine positively the opposition between growth economy and the environment in base of techno-scientific recipes to respond to the imperatives of sustainability.
4 By way of example, consider the discourses of Degrowth, Convivialism, Food Sovereignty, Post-Extractivism, Eco-Feminism, the defense of sentient beings and the Rights of Nature, or the growing global Movement for Environmental Justice.
5 The development paradigm (often also referred to as development discourse, worldview, or ideology) can be broadly defined as the vision of a common horizon for all evolutionary trajectories in convergence with that of the industrialised countries, and the pursuit of actively realising such alignment with societies around the globe. Despite its changing avatars in response to various criticisms – from Ethno-development to Human Development (UNDP, Sen, Nussbaum), passing through Human-Scale Development (Max-Neef), and, finally, Sustainable Development – the promise-land of development remained ever-elusive for the majority of human societies. And even those viewed as "developed societies" operate at enormous social and ecological costs, which are externalised to other spatial and temporal coordinates (United Nations Environment Programme 2016).
6 This inclusion of culture in mainstream sustainability debates has been understood in terms of a "fourth pillar" (Hawkes 2001; Nurse 2006; Yencken and Wilkinson 2000) in what would be the "tetrahedron" of sustainable development, formed by an ontologically particularistic understanding of the world as divisible into distinct spheres which need to be articulated or harmonised ex-post: the economy, society, the environment – and now, the sphere of culture. *Buen vivir*, as shown below, radically challenges and deconstructs this functionally differentiated understanding of the world in favour of a holistic model of embeddedness of the economy into society, and of both into the Earth's biosphere.
7 The name of the initiative "Yasuni-ITT" refers to a sector of Yasuni National Park located between the oil exploration quadrants Ishpingo, Tiputini, and Tambococha (ITT).
8 For an in-depth analysis of these iconic civil-society led initiatives, see for example the work of Espinosa on the Rights of Nature (2015) and of Arsel and Avila Angel on the Yasuni-ITT initiative (2012).

## Bibliography

Acosta, Alberto. 2015. "El Buen Vivir como alternativa al desarrollo. Algunas reflexiones económicas y no tan económicas." *Política y Sociedad* 52 (2): 299–330. doi: 10.5209/rev_POSO.2015.v52.n2.45203.

Adams, William Mark. 2001. *Green Development: Environment and Sustainability in the Third World*. Routledge.

Arnason, Johann P. 1991. "Praxis and Action – Mainstream Theories and Marxian Correctives." *Thesis Eleven* 29 (1): 63–81. doi: 10.1177/072551369102900106.

Arsel, Murat, and Natalia Avila Angel. 2012. "'Stating' Nature's Role in Ecuadorian Development Civil Society and the Yasuní-ITT Initiative." *Journal of Developing Societies* 28 (2): 203–227. doi: 10.1177/0169796X12448758.

Bajoit, Guy. 2003. *Le Changement Social : Approche Sociologique Des Sociétés Occidentales Contemporaines.* Armand Colin.

Beck, Ulrich. 1992. *Risk Society: Towards a New Modernity.* SAGE.

———. 2015. *Ecological Politics in an Age of Risk.* UK: Polity Press.

Becker, Marc. 2011. "Correa, Indigenous Movements, and the Writing of a New Constitution in Ecuador." *Latin American Perspectives* 38 (1): 47–62. doi: 10.1177/0094582X10384209.

Beling, Adrián E. 2019. "Unravelling the Making of Real Utopias: Debates on 'Great Transformation' and Buen Vivir as Collective Learning Experiments towards Sustainability." PhD Thesis, Berlin: Humboldt University of Berlin, Alberto Hurtado University (Chile). https://doi.org/10.18452/20387

Beling, Adrián E., Ana Patricia Cubillo-Guevara, Julien Vanhulst, and Antonio Luis Hidalgo-Capitán. forthcoming. "Tracing the Origins and Evolution of Buen Vivir (Good Living): Glocal Genealogy of a Latin-American Utopia." *Latin American Perspectives.*

Beling, Adrián E., Francisca Gómez, and Julien Vanhulst. 2014. "Del Sumak Kawsay Al Buen Vivir: Filosofía Andina Como Base Para Una Racionalidad Ambiental Moderna." In *La Religión En La Sociedad Postsecular. Transformación y Relocalización de Lo Religioso En La Modernidad Tardía*, edited by Carlos Miguel Gómez. Bogotá: Universidad del Rosario – Centro de Estudios Teológicos y de las Religiones (CETRE).

Beling, Adrián E., and Julien Vanhulst. 2016. "Aportes Para Una Genealogía Glocal Del Buen Vivir." *Dossier Economistas Sin Fronteras* 23: 12–17.

Beling, Adrián E., Julien Vanhulst, Federico Demaria, Violeta Rabi, Ana E. Carballo, and Jérôme Pelenc. 2018. "Discursive Synergies for a 'Great Transformation' Towards Sustainability: Pragmatic Contributions to a Necessary Dialogue between Human Development, Degrowth, and Buen Vivir." *Ecological Economics* 144: 304–313. doi: 10.1016/j.ecolecon.2017.08.025.

Belotti, Francesca. 2014. "Entre Bien Común y Buen Vivir. Afinidades a Distancia between Common Good and Buen Vivir: Long-Distance Affinities." *Íconos. Revista de Ciencias Sociales* 48: 41–54.

Blühdorn, Ingolfur. 2007. "Sustaining the Unsustainable: Symbolic Politics and the Politics of Simulation." *Environmental Politics* 16 (2): 251–275. doi: 10.1080/09644010701211759.

Brand, Ulrich. 2015. "Degrowth Und Post-Extraktivismus: Zwei Seiten Einer Medaille?" Working Paper der DFG KollegforscherInnengruppe Postwachstumsgesellschaften n° 5/2015, Jena.

Castoriadis, Cornelius. 1999. *L'institution imaginaire de la société.* Paris: Seuil.

Chakrabarty, Dipesh. 2000. *Provincializing Europe: Postcolonial Thought and Historical Difference.* Princeton University Press.

CMCD. 1996. *Notre diversité créatrice: rapport de la Commission Mondiale de la Culture et du Développement.* Editions UNESCO.

Cortez, David. 2011. "La Construcción Social Del 'Buen Vivir' (Sumak Kawsay) En Ecuador. Genealogía Del Diseño y Gestión Política de La Vida."

Cubillo-Guevara, Ana Patricia, Julien Vanhulst, Antonio Luis Hidalgo-Capitán, and Adrián E. Beling. 2018. "Entstehung, Institutionalisierung Und Veränderung Der Lateinamerikanischen Diskursen Zu Buen Vivir." *Peripherie* 4: 149.

Dryzek, John S. 2005. *The Politics of the Earth: Environmental Discourses.* Oxford University Press.

Eisenstadt, Shmuel Noah. 2000. "Multiple Modernities." *Daedalus* 129 (1): 1–29.

Escobar, Arturo. 1995. *Encountering Development: The Making and Unmaking of the Third World.* Princeton University Press.

———. 2011. "Sustainability: Design for the Pluriverse." *Development* 54 (2): 137–140. doi: 10.1057/dev.2011.28.

———. 2015. "Transiciones: A Space for Research and Design for Transitions to the Pluriverse." *Design Philosophy Papers* 13 (1): 13–23. doi: 10.1080/14487136.2015.1085690.

Espinosa, Cristina. 2015. "Interpretive Affinities: The Constitutionalization of Rights of Nature, Pacha Mama, in Ecuador." *Journal of Environmental Policy & Planning* 1–19. doi: 10.1080/1523908X.2015.1116379.

Gerlach, Joe. 2017. "Ecuador's Experiment in Living Well: Sumak Kawsay, Spinoza and the Inadequacy of Ideas." *Environment and Planning A: Economy and Space* 49 (10): 2241–2260. doi: 10.1177/0308518X17718548.

Gudynas, Eduardo. 2011. "Buen Vivir: Today's Tomorrow." *Development* 54 (4): 441–447.

―――. 2014. "El Postdesarrollo Como Crítica y El Buen Vivir Como Alternativa." In *Buena Vida, Buen Vivir: Imaginarios Alternativos Para El Bien Común de La Humanidad*, edited by Gian Carlo Delgado Ramos, 61–95. México: CEIICH – UNAM.

Hawkes, Jon. 2001. *The Fourth Pillar of Sustainability: Culture's Essential Role in Public Planning*. Common Ground.

Heyd, Thomas. 2004. "Themes in Latin American Environmental Ethics: Community, Resistance and Autonomy." *Environmental Values* 13 (2): 223–242. doi: 10.3197/0963271041159859.

―――. 2005. "Sustainability, Culture and Ethics: Models from Latin America." *Ethics, Place and Environment* 8 (2): 223–234.

Hidalgo-Capitán, Antonio Luis Hidalgo, and Ana Patricia Cubillo Cubillo-Guevara. 2014. "Seis Debates Abiertos Sobre El Sumak Kawsay." *Íconos: Revista de Ciencias Sociales* 48: 25–40.

Hopwood, Bill, Mary Mellor, and Geoff O'Brien. 2005. "Sustainable Development: Mapping Different Approaches." *Sustainable Development* 13 (1): 38–52. doi: 10.1002/sd.244.

Lalander, Rickard, and Javier Cuestas-Caza. 2017. "Sumak Kawsay y Buen-Vivir En Ecuador." In *Conocimientos Ancestrales y Procesos de Desarrollo*, edited by Ana Verdú Delgado, 30–64. Universidad Técnica Particular de Loja/Ecuador.

Latouche, Serge. 2009. *Farewell to Growth*. Cambridge/Malden: Polity Press.

Le Quang, Matthieu Le, and Tamia Vercoutère. 2013. *Ecosocialismo y buen vivir: diálogo entre dos alternativas al capitalismo*. Ed. IAEN.

Leff, Enrique. 2004. "Racionalidad ambiental y diálogo de saberes." *Polis. Revista Latinoamericana*, no. 7 (April).

Martinelli, Alberto. 2005. *Global Modernization: Rethinking the Project of Modernity*. SAGE.

McNeill, John Robert. 2001. *Something New Under the Sun: An Environmental History of the Twentieth-Century World (The Global Century Series)*. W. W. Norton & Company.

Nurse, Keith. 2006. "Culture as the Fourth Pillar of Sustainable Development." *Small States* 11: 28–40.

Porcedda, Aude, and Olivier Petit. 2013. "Culture et développement durable : vers quel ordre social ?" *Développement durable et territoires. Économie, géographie, politique, droit, sociologie* 2 (2, January). doi: 10.4000/developpementdurable.9030.

Rosa, Hartmut. 2010. *Alienation and Acceleration: Towards a Critical Theory of Late-Modern Temporality*. NSU Press.

Sachs, Wolfgang. 1997. "Sustainable Development." In *The International Handbook of Environmental Sociology*, edited by Michael R. Redclift and Graham Woodgate, 1st ed., 71–82. Edward Elgar Publishing.

Sousa Santos, Boaventura de. 2009. *Una epistemología del sur: la reinvención del conocimiento y la emancipación social*. Siglo XXI.

―――. 2010. *Refundación del estado en América Latina: perspectivas desde una epistemología del Sur*. Plural editores.

Steffen, Will, Wendy Broadgate, Lisa Deutsch, Owen Gaffney, and Cornelia Ludwig. 2015. "The Trajectory of the Anthropocene: The Great Acceleration." *The Anthropocene Review* 2 (1): 81–98. doi: 10.1177/2053019614564785.

Svampa, Maristella. 2016. "The Democratic Horizon of Emancipation." *Alternautas*, Dossier: The End of the Progressive Cycle, 3 (1): 142–154.

Therborn, Göran. 2003. "Entangled Modernities." *European Journal of Social Theory* 6 (3): 293–305. doi: 10.1177/13684310030063002.

United Nations Environment Programme. 2016. *Global Material Flows and Resource Productivity. An Assessment Study of the UNEP International Resource Panel*. Paris: UNEP.

Vanhulst, Julien. 2015. "El laberinto de los discursos del Buen vivir: entre Sumak Kawsay y Socialismo del siglo XXI." *Polis. Revista Latinoamericana* 14 (40): 233–261.

Vanhulst, Julien, and Adrián E. Beling. 2013. "Buen Vivir: La Irrupción de América Latina En El Campo Gravitacional Del Desarrollo Sostenible." *REVIBEC – Revista Iberoamericana de Economía Ecológica* 21: 15–28.

―――. 2014. "Buen Vivir: Emergent Discourse within or beyond Sustainable Development?" *Ecological Economics* 101: 54–63. doi: 10.1016/j.ecolecon.2014.02.017.

―――. 2017. "Esquisse Pour Une Généalogie Glocale Du Buen Vivir." *Synergies Chili*, 2017.

Vanhulst, Julien, and Edwin Zaccai. 2016. "Sustainability in Latin America: An Analysis of the Academic Discursive Field." *Environmental Development* 20 (November): 68–82. doi: 10.1016/j.envdev.2016.10.005.

Wagner, Peter. 2008. *Modernity as Experience and Interpretation: A New Sociology of Modernity*. Polity Press.

———. 2010. "Multiple Trajectories of Modernity: Why Social Theory Needs Historical Sociology." *Thesis Eleven* 100 (1): 53–60. doi: 10.1177/0725513609353705.

WCED. 1987. *Our Common Future*. Edited by Gro Harlem Brundtland. Oxford University Press.

Yencken, David, and Debra Wilkinson. 2000. *Resetting the Compass: Australia's Journey towards Sustainability*. CSIRO Publishing.

Zibechi, Raúl, and Susana Nuin. 2008. *Dibujando Fuera de Los Márgenes: ¿movimientos Sociales o Sociedad En Movimiento?: El Rol de Los Movimientos Sociales En La Transformación Socio-Política de América Latina*. 1. ed. Colección Huanacauri. Buenos Aires, Argentina: La Crujía.

# 10

# Responsibility

*Luigi Pellizzoni*

## Introduction

Responsibility literally means "ability to respond." Hence, it is a relational concept that connects an agent, provided with certain capacities, with a recipient (human or non-human; entity or process). Despite a record of debates that dates back to ancient Greece, the notion is far from settled. In common usage, it is loosely equated to "taking charge" – of what there is or will, or should, be; of what has happened or is going to, or may, happen.

The idea of taking charge – of the planet – is inbuilt in the very notion of sustainability. In this chapter, I address the evolution of the issue of responsibility in the context of current societal transformations. Specifically, I deal with the internal articulations of the notion, with an eye on the governance of social affairs and on how changes in the understanding of both the agent and the world acted upon affect the issue. Hence, my focus here is neither on philosophical ruminations on responsibility (which usually concentrate on moral responsibility, addressing cursorily other perspectives) nor on the issue of how different answers to the question "sustainability of what? and what for?" may impinge on responsibility ascriptions and assumptions (according, for example, to "weak" or "strong" conceptions of what it means to be sustainable, or to conservative or transformative political programs).

The chapter proceeds as follows. The next sections deal with the meaning of responsibility, how some features of modern society affect it and the problem of collective responsibility. Subsequently, I elaborate on the evolution of the problematic in the last decades. Changing views about reality, its knowability, and how both impinge on purposeful action affect the configuration of responsibility, as it was established in modernity. The conclusion notes that these trends do not rule out traditional forms of responsibility and that new social norms are arguably forming, which may be conducive to more sustainable ways of living on the planet.

## Two basic meanings

Genealogies of responsibility retrocede far back in history, up to ancient Greece and Rome.[1] However, to get closer to the current usages of the term, one needs to wait for the establishment of the modern understanding of individual autonomy and agency. In various European countries, the words "responsible" and "responsibility" gain salience in the 18th century, with reference either to the relationship between rulers and constituencies or to the obligation of repairing damage or suffering punishment (McKeon 1957).

Recent times have witnessed a growing interest in the issue. Authors have claimed that responsibility has become a "new cultural master frame of our time" (Strydom 1999: 76), though "slippery and confusing" (Miller 2007: 82), "ambiguous or multi-layered" (Giddens 1999: 8); a notion often addressed without mentioning it explicitly (Jamieson 2015). Responsibility concerning climate change, in particular, has witnessed a mushrooming of literature. This can be explained not only with the growing resonance of the issue but also because the latter highlights the conceptual and practical challenges that will be discussed in what follows.

Several taxonomies of responsibility have been proposed (among the latest: van de Poel 2011; Vincent 2011; Davies 2012; Arnaldi et al. 2018). Two basic meanings can be detected as underlying most accounts: *imputability* and *answerability*. For Ricoeur (1995), the grounds of the concept lie in the semantic field of the verb to impute, which means to attribute an action to someone as its actual author. If the agent is non-human, then natural or scientific causality applies (as in the sentence "draught is imputable to prolonged lack of rain"). If the agent is human, causality intersects questions of intentionality and agency. I can impute the event X to the (human) agent Y only if Y was able to represent, or to stand for, things, properties, and states of affairs in his or her mind, pointing accordingly to some goal, and if this mental representation corresponded to an actual possibility. Said differently, human imputability entails so-called "free causality," or free will (Hart 1968). Lack of or reduction in free will, due to mental conditions (age, involuntary impairment, etc.) or material circumstances (as when one invokes "force majeure," that is the aim to avoid worse outcomes), usually plays a role as extenuating factors, at moral, legal, and political levels. Of course, imputability may regard both acts and omissions, and recipients can be either generic ("humanity," "living beings," etc.) or identified with precision (Shue 2017).

Answerability connects directly, for its etymology, with the word responsibility. The latter comes from the Latin verb *respondere*, in its turn drawn from *spondere*, which means to pledge or vouch. Thus, *re-spondere* refers to a legally binding exchange of pledges (Benveniste 2016: 477 ff.). Authors stress either the retrospective orientation of the notion, as the duty of justifying a conduct (Schwartländer 1974), or the prospective one, as the guarantee concerning a future course of events (Villey 1977). In both cases, the focus shifts from causal connections to the reasons for an agent to behave in a certain way, and the presence of somebody to whom the agent is deemed accountable.

The conceptual distinction between imputability and answerability helps to grasp how the former is a necessary but often not sufficient condition for responsibility ascription (Jamieson 2015). For example, if my action produces damages, I can reject imputation showing how it is impossible for me to do otherwise and this impossibility does not depend on anything for which I can be blamed. This is a frequent justification of unsustainable practices. If one uses the car for commuting to work, it's because public transport is inefficient; and if one eats convenience food with high ecological impact (plastic packaging, ingredients coming from all over the world, heavy use of chemicals in farming and industrial processing), it's because one has not enough money or time to buy and cook non-industrial, locally produced food. Imputability and answerability typically mix factual and normative elements, claims about reality and how reality is to be conceived and addressed. This often brings to the fore social roles. An example comes from so-called "fault in supervising" (*culpa in vigilando*), that is failure to exercise due diligence in one's office (as a professional, parent, owner, user, etc.). Also, role distributions allow for distinguishing between first- and second-order responsibilities – for doing something and for checking that someone does something (as with the monitoring of industry compliance with emissions limits).

According to Schlenker et al. (1994), we talk of responsibility when a set of prescriptions applies to events, actors are perceived to be bound by these prescriptions because of their identity (features, beliefs, motivations, roles), and they seem connected to the events, with particular reference to their capacity of control of the latter. The stronger the links between each couple of elements, and the clearer the definition of each element (unambiguous prescriptions, absence of role conflict, etc.), the stronger the attribution of responsibility.

## Responsibility and modernity

Complications stem from some features of modern society, namely the rise of probability, the relevance of future and functional differentiation.

Since the 18th century, statistic regularities and probability calculations have played a growing role in social affairs, from the design and management of artefacts to welfare institutions and finance. Probability reduces "uncertainty to the same calculable status as that of certainty itself" (Hacking 1975: 237), making it possible to handle processes without controlling every element of them. Hence, responsibility can be based also on probabilistic action-event connections. Consider how one can be held liable for "moral hazard," for example by hiding a disease (or, today, a genetic disposition to develop such disease) or one's indulgence in dangerous hobbies, to get lower insurance rates. Consider also the idea of liability without fault ("strict liability"), as applied for example to industrial production. A firm is held liable for defective products without the need for the damaged user to demonstrate its negligence (a difficult task: it would require access to the firm's organisational sphere), given – and to the extent – that defects are amenable to probabilistic estimates, so that compensation for ensuing damages can be budgeted, usually by means of insurance. "To the extent" means that circumstances that cannot be anticipated work as an exonerating clause. Knowledge and predictability, thus, are crucial. I shall return later to this.

Till now we have mostly talked of responsibility in a retrospective sense, considering the causal chain that connects action and event and the factors "pushing" the agent to behave in a certain way. Yet, as we have seen, it is possible to talk of responsibility also in a prospective sense, considering the factors "pulling" the agent to behave in a certain way (goals, hopes, etc.). Borrowing from Alfred Schutz (1967), we can say that backward-looking responsibility is sensitive to the "because-motives"; forward-looking responsibility is sensitive to the "in-order-to motives." It is important, then, to consider that for modernity future is open and actionable. The interest in and the ability of impinging on future states of affairs, and the time span involved, have increased together with cognitive and technical capacities (but also life expectancy, organisational set-up, and other factors). Then, the more future-oriented society is, the more relevant prospective responsibility becomes. How future is anticipated is crucial. Probability calculation is just one approach, contingent on particular assumptions about human agency and the world. I shall return later also to this.

Modernity is also characterised by the pre-eminence of functional differentiation. The most elegant description of the latter and its implications arguably comes from Niklas Luhmann (1996). As he notes, functional differentiation means that specific codes have developed for describing, organising and responding to reality according to the task to be fulfilled. A firm, a research team, a court, a government, an ethical committee see – as such – the world in their own way, irreducible to others. This increases efficiency in doing their job – and just that. For the economic sector, the code is pay/not pay; for science, it is true/false; for law, right/wrong; for politics, powerful/powerless; for morals, good/evil. Of course, to apply the pay/not pay code, there have to be previous allocations of property rights, and Max Weber has shown how the legitimacy of state power stems from formal legality and factual efficacy. So, any logic of action presupposes the others. However, an act can be deemed morally good, yet its results legally wrong, politically ineffective, or economically nonsensical. Massive emissions in the atmosphere at no cost look fine to the entrepreneur, but sound crazy to the climatologist. The sportsman who eats lots of animal proteins to keep fit gets angry with the government for its inaction about global warming. In short, functional differentiation entails role conflicts, clash of institutional rationalities, and overall disunity of intents. Legal liability, political or financial accountability, scientific imputability, and moral answerability may, and frequently do, not match. This often works as a strong extenuating factor in the public discourse. Industries pollute – but they provide jobs. Intensive breeding is a major greenhouse gas (GHG)-emissions source – yet people want meat. Governments do not enforce stricter regulations – yet they are expected to defend their constituencies' lifestyles, whatever their ecological impact.

## Collective responsibility

We have considered so far responsibility as an attribute of individual human agents. The problem of collective responsibility has long been puzzling scholars. The issue boils down to whether, and in what sense, the same basic requirements for the responsibility of individual humans – intentionality and agency – apply to collective entities such as groups, clubs, corporations, or states. Philosophical hesitations come from the implicit or explicit reference to a Kantian account of the moral self, as presupposing rationality, freedom and self-consistency, or non-division; features that cannot be extended straightforwardly to fictional persons or aggregates. Yet, also methodological individualists usually admit that it may make sense to regard, for example, a crowd or a riot as a whole, in the sense of a strong interdependence of individual behaviours.[2]

Collective responsibility is obviously relevant to global sustainability. Once one accepts its conceivability (Smiley 2017), distributive issues immediately emerge concerning benefits and costs of, and remediation to, activities, at local and global levels. "Environmental Justice" mobilisations originated in the 1970s, focusing on local issues (pollution, land grabbing, speculation, etc.). They have subsequently been extended to the inequities associated with the differential impacts of global issues, from climate change to biodiversity loss, and, under the label of "Just Transition," to those associated with the social and geographical impacts of mitigation policies (Walker 2011; Snell 2018). Differential collective responsibilities have long been acknowledged in regard to climate change. For example, the 1992 United Nations Framework Convention on Climate Change talks of common but differentiated responsibilities among signatory countries (see Okereke this volume). The Kyoto Protocol (1997) includes differentiated commitments to GHG emissions reduction, according to country industrialisation and emissions levels.

The failure of Kyoto and the difficulties of subsequent attempts to allocate national burdens for climate policies are a clue to the distributive problems of shared responsibility. This applies also at individual level. For example, are American citizens to be held equally responsible (in whatever sense) for the huge ecological footprint of their country, with no distinction according to individual lifestyles? The question highlights a general problem: if an action produces consequences just by way of its unintentional aggregation with other independent actions, can the consequences be imputed to the agent? Implied in this issue is another: the capacity of anticipation and control over the consequences of action, and how this affects imputability and answerability. The evolution of the whole problematic of responsibility in the last decades is closely connected with such issue.

## Responsibility and risk

We have seen that responsibility requires imputability, and imputability requires establishing a connection between action and event. The way this can be done, however, is not self-evident. The anthropologist Mary Douglas observed that "in all places at all times the universe is moralized and politicized ... [and] for any misfortune there is a fixed repertoire of possible causes among which a plausible explanation is chosen" (Douglas 1992: 5). Hence, responsibility ascriptions are mediated by cultural views concerning the agent and its task environment. One thing is to conceive of reality as open-ended, hence the future actionable; another to see it as structured according to established patterns. One thing is to regard humans primarily as individuals: non-divisible, self-standing beings; another to regard personhood as premised on collective belonging. Modernity has chosen the first option in both cases (which, of course, are not independent of each other). Hence, in modern society misfortune or undesired events cannot be ascribed to deviations from a fixed order that sooner or later will re-establish itself, but to concatenations of events leading to specific, and often irreversible, states of the world. This gives salience to anticipation, connecting responsibility with risk and uncertainty.

Strictly speaking, risk refers to future events the probability of which can be calculated, with known degrees of error. This, as seen, is the rationale on which strict liability builds. Yet, it dictates also the latter's limits of application. If, for example, the dangers of a substance were unknown until a certain date, industries using this substance before such date cannot be held liable for resulting injuries. Countless trials around the world focus on the ascertainment of what an industry was supposed to know when damages were being produced (Jasanoff 1997). The implicit, and sometimes explicit (see e.g. European Commission 1999), assumption is that extending liability beyond predictability would discourage innovation, and that unpredictable adverse events have to be regarded as "development costs," which can be legitimately offloaded onto particular subjects because of the general interest in innovation, from which they also benefit. The problem escalates with the growth in organisational complexity and the pace of technical innovation, on which capitalism has become increasingly dependent. When Ulrich Beck (1992) famously talked of "organised irresponsibility" – that is, systematic exoneration from liability – as a key trait of the "risk society" engendered by modernity's own dynamics, he was referring exactly to this problem.

The question of organised irresponsibility, however, has two sides. The first concerns risk estimates. One can aim at reducing false positives (namely, the risk of detecting non-existing adverse phenomena) or false negatives (the risk of not detecting actually existing ones), but hardly both at once, as each choice entails different research designs, targeted to different purposes. The reduction of false positives, which often means designing mono-causal experiments, makes sense, for example, if the goal is keeping research on a sound track or avoiding redundant safety measures with related extra-costs. The reduction of false negatives, which often means designing multi-factorial experimental designs, typically makes sense for end users or exposed communities. Organised irresponsibility, thus, depends first and foremost on the inner logic of the marriage of science and industry, blessed by politics, in promoting technical innovation, since such marriage leads to privileging the reduction of false positives over false negatives in all the steps of the innovation process, from lab to market, which translates into offloading its costs onto specific groups, social categories, and locales.

## Responsibility and uncertainty

The second side of organised irresponsibility concerns the historical shift from a framework of risk to a framework of uncertainty. As distinguished from risk (which is that very particular type of uncertainty whose reach is calculable), uncertainty can mean different things, from unknown probabilities of known events to non-knowledge of events themselves, due to their "emergent" character (i.e. they result from the unpredictable interaction of elements) or sheer ignorance of what may happen and how to detect it. The limits of probabilistic prediction began to be conceptualised already in the 1920s. The economists John Maynard Keynes and Frank Knight, for example, reflected on situations where economic decisions, in the impossibility of probability calculations, have to rely on personal heuristics (intuition, foresight, experiential judgement, rules of thumb, etc.). Fast forward some decades and what Keynes and Knight regarded as the exception has become the rule. "An extensive and immensely influential managerial literature appearing since the early 1980s [...] celebrates uncertainty as the technique of entrepreneurial creativity, [...] the fluid art of the possible" (O'Malley 2004: 3), a condition that enhances danger and insecurity but is also "at the heart of what is positive and constructive" (O'Malley 2010: 502). From limiting purposive action, uncertainty has become an enabling condition (Pellizzoni 2016).

We'll try later to make sense of the implications of this sort of Copernican Revolution or paradigm change. For now, let's consider a crucial passage in its genealogy: the moment, which we can roughly locate in the 1970s, when in a number of fields indeterminacy and instability cease to be regarded as exceptional. In ecology, for example, equilibrium is replaced by competition, patchiness, fragmentation (Holling 1973). In chemistry and physics, attention focuses on "dissipative structures," that is, thermodynamically open systems characterised by the spontaneous formation of dissymmetry and bifurcations (Prigogine and Stengers 1979). In cybernetics, notions of homeostasis and selective openness/closure are supplanted by the idea of emergence (Hayles 1999). The management of radioactive waste engages science in out-of-the-lab, "trans-scientific" (Weinberg 1972) or "real-life" (Krohn and Weyer 1994) experiments, simultaneously physical and social. In subsequent years, a number of issues will pile up, from genetically modified (GM) crops to geoengineering and ecological restoration (Gross and Hoffman-Riem 2005). Faced with these changes, predictive knowledge, which needs circumscribed agential factors and regularity detections, loses grip. The theoretical and regulatory reply is the precautionary principle. Its conceptual foundations can be found in Hans Jonas's (1985) case for the "imperative of responsibility" based on a "heuristics of fear." The argument is that, faced with ever-more powerful technologies, one is ethically required to deal with worst case scenarios, envisaged yet incalculable long-term effects or suddenly deflagrating catastrophes. This idea is elaborated in the same years as an institutionalised reply to the growing salience of uncertainty in social affairs. The inclusion of the precautionary principle in declarations, codes of conduct, and legislations during the 1980s and 1990s indicates that Beck's "risk society" should more aptly be named the "society of uncertainty."

According to the original intentions, precaution expanded responsibility, asking that measures capable of avoiding or limiting major or irreversible damages be taken before conclusive evidence about a threat is achieved. Yet, the weakness of precaution in retrospective terms is evident. On which basis can someone be held politically or morally reproachable, let alone legally liable, for not having enacted sufficiently proportionate or timely measures against the actualisation of a threat, if by definition the latter's odds cannot be reliably assessed? This has been a strong argument for confining precaution to soft regulation (Pellizzoni 2009). Even worse, precautionary arguments have been often raised to weaken prospective responsibility. The lesser the predictability of the events, the easier the possibility of "manufacturing uncertainty" (Michaels 2006; Oreskes and Conway 2010), that is, of stressing the controversial import of data to make a precautionary case, not against the threat but against action against the threat. Well-known examples are climate change denial (Jacques et al. 2008, Brulle and Aronczyk, this volume) and the controversies over GM crops, electromagnetic fields, and a variety of chemicals.

Of course, in many cases responsibility is not only "theirs" (big interests, weak or complicit governments, etc.) but also "ours," as individual citizens and inhabitants of the planet. The question of shared responsibility and of the limits of control and anticipation comes back to the fore. According to Iris Marion Young, a liability type of responsibility, based on clear causal links, can't work with "structural injustices," that is social processes that put persons under a systematic threat of domination or deprivation, and which depend on extended and loose social connections. An example is global trade. Consumers can hardly be held liable for worker exploitation or hazardous emissions occurring somewhere in the world, along intricate production chains (Krogman, this volume). For Young, however, since they play their own role in the cooperative scheme enabling exploitation and hazard production, they should acknowledge and take up their part of responsibility. Yet, given the opacity of the causal chain, responsibility-taking cannot reside in the search for efficient, effective, and predictable courses of action, but rather in "carrying out activities in a morally appropriate way and aiming for certain outcomes" (Young 2006: 119).

In other words, if a detailed reconstruction of causal chains is impossible or extremely onerous, one has to base action on one's own commitment to a certain outcome. The deeper the uncertainty, the stronger the role of value orientations. As "the world's largest and most complex collective action problem" (Jamieson 2015: 30), climate change epitomises this problematic. Whether we buy organic food or scrap our old car replacing it with a less polluting one, we do it by committing ourselves to believe that this behaviour is conducive to "good" outcomes. In so doing we rely on sources, with which we often have no actual acquaintance, who tell us this is the right thing to do. In affluent countries an orientation of increasing appeal is "degrowth," that is, shrinking one's lifestyle, for example by turning to short production-consumption chains. Especially if they enable face-to-face interactions, short chains give renewed significance to causal connections – as they locally occur. As for their global social and ecological impact, however, we still have to rely on our normative assumptions, trusting that the overall outcome will be "good." And as for retrospective responsibility, it inevitably hollows out – whatever happens, we just did our best.

In sum, faced with uncertainty, decision has to build on an affective relation to the future (Cooper 2006). Uncertainty fosters prospective responsibility, yet not of a sort amenable to what Max Weber called the "ethic of responsibility," as decisions based on due consideration of their causal implications. Despite insistent calls for "sound science" and "evidence-based" policy-making, Weber's "ethic of ultimate ends," supposedly a legacy of premodern worldviews of lessening relevance, is gaining momentum (Pellizzoni 2018). Uncertainty, moreover, negatively affects retrospective responsibility, as long as the latter remains anchored to a causal basis for imputability.

## Responsibility in neoliberal times

This conclusion seems at odds with the earlier hinted overturning of the effects attributed to uncertainty, as enabling rather than constraining purposive action. The Kantian heritage leads to assume that growing agency entails growing responsibility. To make sense of why the opposite is increasingly the case, let's go back to the 1970s. The intellectual turn described earlier overlapped (hardly coincidentally) with the socio-ecological crisis engendered by the combination of energy scarcity and mounting environmental threats with stagflation and declining profits. Let's consider, then, the eventual outcome of this historical passage.

Theories of complexity and indeterminacy brought into question traditional scientific accounts of the world as based on stability and predictability, and with them the technical and social bases of Fordist industrialism and "command-and-control" regulation. In short, they implicitly, and sometimes explicitly, represented a libertarian critique of the post-war social order (Walker and Cooper 2011). Such critique, however, has become the basis of a new, "post-Fordist," regime of accumulation. Accounts of reality as open-ended, non-hierarchical, ever-changing have been translated into flexibility, networking, and permanent innovation (Boltanski and Chiapello 2005; Nelson 2015). The neoliberal turn of politics and regulation has been crucial to that. For neoliberalism[3] complexity makes any attempt at social planning bound to fail and lead to disaster. The market is the core institution because, as a blind mechanism, it ensures an optimal allocation of values, sanctioning ex-post the soundness of individual choices. Then, to make the market work properly, entrepreneurship, competition, endlessly growing aspirations, and preparedness to the unexpected have to become everyone's bread and butter (Dardot and Laval 2017). With adaptations to national contexts, this goal has been pursued all over the world via regulatory reforms, precarisation of work and massive interventions in the public discourse, academic organisation, and educational curricula (Lave et al. 2010).

If what precedes is well-known and widely discussed, a phenomenon far less understood in its implications is the emergence, at the turn of the century, of a new type of critique, not against stability and predictability but against the very grounds of the modern worldview: namely, the possibility to draw clear ontological distinctions in the texture of reality (language/things, mind/body, matter/cognition, living/non-living, nature/artefact, etc.). Think, for example, of how epigenetics has come to challenge the gene/environment and brain/body dichotomies (Papadopoulos 2011); how the inorganic realm is increasingly depicted as having vital connotations, and life as infused with dematerialised characterisations – textuality, information, codification (Keller 2011); how new prosthetics and brain-computer interfaces are questioning the divide between the organic and the inorganic (Rao 2013); how any aspect of existence, including the self, is increasingly portrayed as fluid, blurred, ever-changing (Coole and Frost 2010; Braidotti 2013).

It is hard to grasp in full the social and ecological implications of this *Weltanschauung*. Yet, one may at least take stock of some consequences already perceptible, for example concerning regulation in neoliberal times. Think of how biotech patents cover simultaneously genetic information and the organisms incorporating such information, blurring the distinction between knowledge and matter (Calvert 2007). Or how carbon trading is based on an entity, the "global warming potential" of GHGs, assumed at the same time as an abstract unity of measure and a concrete physical phenomenon (McKenzie 2009). Consider above all the notion of Anthropocene, the amazing success of which – well beyond academic circles – tells a lot about the scope of the new *Weltanschauung* (see also Mert, this volume). The idea that nature has been domesticated to the point that it can no longer be distinguished from technology may lead to either a strengthening or a watering down of responsibility. The modern self-standing agent is a figure that undoubtedly transpires from claims (coming from the very proponents of the Anthropocene notion) such as that "it's we who decide what nature is and what it will be" (Crutzen and Schwägerl 2011). Yet the issue is not so much contrasting this figure with post-humanist accounts of a "decentred," "nomadic" subjectivity, but the extent to which also such subjectivity retains a modern self-affirmative thrust, which the lack of a stable substance or identity liberates from any insuperable constraint. A decentred agent, then, may be more responsible, in the sense of an enhanced sensitivity and responsiveness to a world over which it cannot claim any supremacy; yet it may also be less (or non-)responsible, as long as the fluidity of the world is equated with its full pliancy to one's own will (Pellizzoni 2016). Sure, fluidity means also "vitality," unpredictability. Yet, the mantra of preparedness and alertness has already taken surprise on board, as a factor enhancing the opportunities of growth (Taleb 2012), simultaneously exonerating from charge for whatever may happen.

Cutting-edge social science scholarship tends to consider the overcoming of dualisms as conducive to a closer, caring, "affective" relationship with the world (Puig de la Bellacasa 2011). Yet, getting closer does not necessarily mean getting friendlier, or more respectful. This is especially the case, when myriads of techniques teach how to "affect" in order to "effect," and especially when, as with the controversial yet influential *Ecomodernist Manifesto* (Breakthrough Institute 2015), technologisation is equated with decoupling from nature, rather than growing intermingling with its processes.[4] Similarly, one should be careful before regarding "Responsible Research and Innovation" (RRI) as a strengthening of responsibility about technological change. The novelty of RRI as a policy approach, it is claimed, resides in its building on the "mutual responsiveness" of social actors and innovators, concerning not only the "how" but also the "why" of innovation (von Schomberg 2013). However, faced with ever-more powerful means of influence, the case for "responsiveness" becomes ambiguous. Even in the absence of manipulative intents, pursuing a mutual responsiveness while leaving unaddressed huge differentials in agency – for example between Big Pharma and patients or farmers – means offloading responsibilities from the more onto the less powerful actors. We "share" the choice, we'll "share" the outcomes.

Consider also what looks like a decline of the linear conception of time. Similarly to prediction, precaution builds on temporal linearity: past comes before, and affects future. If anything, linearity is strengthened by the irreversibility of the processes with which precaution typically deals. Yet, after a climax (around the year 2000), precaution has started to decline at both academic and political levels. Another type of anticipation has gained momentum: pre-emption. Pre-emptive war, told us G.W. Bush (2002), means "confronting the worst threats before they emerge," that is, without even the inconclusive but robust evidence required for precautionary measures. To confront merely guessed threats, one has to "incite" them, help them emerge: as the US "war on terror" in Iraq showed, fuelling rather than countering terrorism and war, pre-emption acts to create the reality which demonstrates that such very action was sound since the beginning. The process produces its own cause; truth becomes retroactive (Massumi 2007). Trends in "counter[ing] the unknowable before it is even realized" (Cooper 2006: 120), eliciting a constant alertness, can be observed in regard to conventional and bio-terrorism, and pandemics. Just military and security affairs? Not necessarily. Think of GM crops and suppose that sterility extended to non-GM ones, or super-resistant pests developed.[5] Or think of solar radiation management (SRM) and suppose that, after applications, devastating hurricanes or prolonged draughts occurred in areas where neither had previously come about. Even if causal connections were solidly established (hard job), liability allegations could be rebutted on the basis that what has happened, since it happened, is just the actualization of as yet unexpressed possibilities. Again, truth becomes retroactive. Moreover, it might be argued, even sterility and pests can be turned to advantage. And, to foster preparedness, better to elicit turbulence than wait for its unsolicited manifestations. Pre-emption, in short, is the ultimate expression of organised irresponsibility, whereby innovation has to be spurred precisely "to pre-empt its potential fall-out" (Cooper 2006: 121).

## Conclusion

This discussion aimed at diverting the issue of responsibility from philosophical speculations, addressing its evolution in the context of major societal transformations. The last sections highlighted trends of major relevance to sustainability. These are not unique to the latter, as they can be detected, for example, in political securitisation, labour precarisation, or economic financialisation. However, reclaiming responsibility for sustainability today entails tackling a cultural and policy backdrop that effectively promotes the figure of a "nomadic" yet self-affirming agent, faced with a task environment indeterminate and fluid even in its temporal structure. This requires appropriate theoretical and political approaches.

A positive note is that established types of responsibility are not ruled out. For example, the amazingly high contribution of a small number of large carbon producers to GHG emissions is increasingly documented (Heede 2014). Building on such evidence, a case for their duty of decarbonisation and rectification through fund disgorgement can be made (Grasso and Vladimirova, in press). This shows how backward-looking responsibility is anything but exhausted. Indeed, as happened with slavery, it may be conducive to social norms leading to a ban of carbon-intensive productions and consumptions, as morally unacceptable practices.

That new social norms are forming is suggested by the blossoming of mobilisations that replace protest with work at the level of the body and materiality, to build alternative forms of community organisation and material flows (community supported agriculture, community energy, open source seeds, participatory plant breeding, time banks, community-based credit systems, etc., see also Litfin, this volume). Their strengthening and diffusion may help reconnect action and outcome, circumventing deresponsibilising appeals to ultimate ends as the inevitable reply to uncertainty. Such norms, however, should steer clear from ambiguous equations between affect and effect, or naïve assumptions that the vanishing threshold between the human and the non-human, the technical and the natural, is by necessity conducive to non-dominative orientations. Only if getting closer never leads to assimilation, if vicinity does not cancel alterity, responsible relations – which first of all means respectful – are possible, and a sustainable human presence on the planet may be pursued with some chances of success.

## Acknowledgement

I wish to thank the editors and Robert Gianni for useful comments to earlier versions of the paper.

## Notes

1 For an overview see Gianni (2018).
2 This happens for example with accounts of collective action as a result of the balance of individual incentives and disincentives determined by the situation (Olson 1965).
3 Accounts of neoliberalism are notoriously controversial. A good working definition is the one proposed by David Hess: "public policies and economic thought that have guided a transition in many of the world's economies toward the liberalization of financial and other markets, the privatization of public enterprises, and the retrenchment of government commitments to social programs" (Hess 2013: 178).
4 Unsurprisingly, the notion of responsibility appears in this text just once, as physical causality.
5 This is seemingly happening already. GM crops are typically designed to stand massive use of pesticides, hence their possible greater responsibility for the development of super-resistant pests.

# References

Arnaldi, S., G. Gorgoni, and E. Pariotti. 2018. "Responsible Research and Innovation between 'New Governance' and Fundamental Rights." In *Responsible Research and Innovation: From Concepts to Practices*, edited by R. Gianni, J. Pearson, and B. Reber, 153–71. London: Routledge.

Beck, U. 1992. *Risk Society*. London: Sage.

Benveniste, E. 2016. *Dictionary of Indo-European Concepts and Society*. Chicago, IL: HAU Books.

Boltanski, L. and Chiapello, E. 1999. *The New Spirit of Capitalism*. London: Verso.

Braidotti, R. 2013. *The Posthuman*. London: Polity.

Breakthrough Institute. 2015. *An Ecomodernist Manifesto*. Oakland, CA: Breakthrough Institute.

Bush, G.W. 2002. "President Bush Delivers Graduation Speech at West Point, June 1." https://georgewbush-whitehouse.archives.gov/news/releases/2002/06/20020601-3.html.

Calvert, J. 2007. "Patenting Genomic Objects: Genes, Genomes, Function and Information." *Science as Culture* 16 (2): 207–23.

Coole, D. and S. Frost, eds. 2010. *New Materialisms*. Durham, NC: Duke University Press.

Cooper, M., 2006. "Pre-Empting Emergence." *Theory, Culture & Society* 23 (4): 113–35.

Crutzen, P. and C. Schwägerl. 2011. "Living in the Anthropocene: Towards a New Global Ethos." *Yale Environment* 360. https://e360.yale.edu/features/living_in_the_anthropocene_toward_a_new_global_ethos [accessed 24 August 2019].

Dardot, P. and C. Laval. 2017. *The New Way of the World: On Neoliberal Society*. London: Verso.

Davis, M. 2012. "'Ain't No One Here But Us Social Forces': Constructing the Professional Responsibility of Engineers." *Science and Engineering Ethics* 18 (1): 13–34.

Douglas, M. 1992. *Risk and Blame: Essays in Cultural Theory*. London: Routledge.

European Commission. 1999. *Liability for Defective Products*. Green Paper. COM(1999) 396.

Gianni, R. 2018. "The Discourse of Responsibility. A Social Perspective." In *Responsible Research and Innovation: From Concepts to Practices*, edited by R. Gianni, J. Pearson, and B. Reber, 11–34. London: Routledge.

Giddens, A. 1999. "Risk and Responsibility." *Modern Law Review* 62 (1): 1–10.

Grasso, M. and K. Vladimirova. in press. "A Moral Analysis of Carbon Majors' Role in Climate Change." *Environmental Values*. doi: 10.3197/096327119X15579936382626.

Gross, M. and H. Hoffmann-Riem. 2005. Ecological Restoration as a Real-World Experiment: Designing Robust Implementation Strategies in an Urban Environment. *Public Understanding of Science* 14 (3): 269–84.

Hacking, I. 1975. *The Emergence of Probability*. Cambridge: Cambridge University Press.

Hart, H.L.A. 1968. *Punishment and Responsibility: Essays in the Philosophy of Law*. New York: Oxford University Press.

Hayles, N.K. 1999. *How We Became Post-Human*. Chicago: University of Chicago Press.

Heede, R. 2014. "Tracing Anthropogenic Carbon Dioxide and Methane Emissions to Fossil Fuel and Cement Producers, 1854–2010." *Climatic Change* 122 (1–2): 229–41.

Hess, D. 2013. "Neoliberalism and the History of STS Theory: Toward a Reflexive Sociology." *Social Epistemology* 27(2): 177–93.

Holling, C.S. 1973. "Resilience and Stability of Ecological Systems." *Annual Review of Ecology and Systematic* 4: 1–23.

Jacques, P., R. Dunlap, and M. Freeman. 2008. "The Organisation of Denial: Conservative Think Tanks and Environmental Scepticism." *Environmental Politics* 17 (3): 349–85.

Jamieson, D. 2015. "Responsibility and Climate Change." *Global Justice* 8 (2): 23–42.

Jasanoff, S. 1997. *Science at the Bar*. Cambridge, MA: Harvard University Press.

Jonas, H. 1985. *The Imperative of Responsibility*. Chicago: University of Chicago Press.

Keith, D. 2013. *A Case for Climate Engineering*. Cambridge, MA: MIT Press.

Keller, E.F. 2011. "Towards a Science of Informed Matter." *Studies in History and Philosophy of Biological and Biomedical Sciences* 42 (2): 174–9.

Krohn, W. and J. Weyer. 1994. "Society as a Laboratory: The Social Risks of Experimental Research." *Science and Public Policy* 21 (3): 173–83.

Lave, R., P. Mirowski, and S. Randalls. 2010. "Introduction: STS and Neoliberal Science." *Social Studies of Science* 40 (5): 659–75.

Luhmann, N. 1996. *Social Systems*. Stanford, CA: Stanford University Press.

MacKenzie, D. 2009. "Making Things the Same: Gases, Emission Rights and the Politics of Carbon Markets." *Accounting, Organizations and Society* 34 (3–4): 440–55.

Massumi, B. 2007. "Potential Politics and the Primacy of Preemption." *Theory and Event* 10 (2). doi: 10.1353/tae.2007.0066.

McKeon, R. 1957. "The Development and Significance of the Concept of Responsibility." *Revue Internationale de Philosophie* 39(1): 3–32.

Michaels, D. 2006. "Manufactured Uncertainty: Protecting Public Health in the Age of Contested Science and Product Defense." *Annals of the New York Academy of Sciences* 1076 (1): 149–62.

Miller, D. 2007. *National Responsibility and Global Justice*. Oxford: Oxford University Press.

Nelson, S. 2015. "Beyond the Limits to Growth: Ecology and the Neoliberal Counterrevolution." *Antipode* 47 (2): 461–80.

Olson, M. 1965. *The Logic of Collective Action*. Cambridge, MA: Harvard University Press.

O'Malley, P. 2004. *Risk, Uncertainty and Governance*. London: Glasshouse.

O'Malley, P. 2010. "Resilient Subjects: Uncertainty, Warfare and Liberalism." *Economy and Society* 3 (4), 488–509.

Oreskes, N. and E.M. Conway. 2010. *Merchants of Doubt*. London: Bloomsbury.

Papadopoulos, D. 2011. "The Imaginary of Plasticity: Neural Embodiment, Epigenetics and Ectomorphs." *Sociological Review* 59 (3): 432–56.

Pellizzoni, L. 2009. "Revolution or Passing Fashion? Reassessing the Precautionary Principle." *International Journal of Risk Assessment and Management* 12 (1): 14–34.

Pellizzoni, L. 2016. *Ontological Politics in a Disposable World: The New Mastery of Nature*. London: Routledge.

Pellizzoni, L. 2018. "Responsibility and Ultimate Ends in the Age of the Unforeseeable: On the Current relevance of Max Weber's Political Ethics." *Journal of Classical Sociology* 18 (3): 197–214.

Prigogine, I. and I. Stengers. 1979. *La Nouvelle Alliance. Metamorphose de la Science*. Paris: Gallimard.

Puig de la Bellacasa, M. 2011. "Matters of Care in Technoscience: Assembling Neglected Things". *Social Studies of Science* 41 (1): 85–106.

Rao, R. 2013. *Brain-Computer Interface: An Introduction*. New York: Cambridge University Press.

Ricoeur, P. 1995. "The Concept of Responsibility: An Essay in Semantic Analysis." In *The Just*, 11–35. Chicago: University of Chicago Press.

Schlenker, B., T. Britt, J. Pennington, R. Murphy, and K. Doherty. 1994. "The Triangle Model of Responsibility." *Psychological Review* 101 (4): 632–52.

Schutz, A. 1967. *The Phenomenology of the Social World*. Evanston, IL: Northwestern University Press.

Schwartländer, J. 1974. "Verantwortung." In *Handbuch Philisophischer Grundbegriffe*, Band 6, edited by H. Krings, H.M. Baumgartner, and C. Wild, 1577–88. Munich: Kösel.

Shue, H. 2017. "Responsible for What? Carbon Producer $CO_2$ Contributions and the Energy Transition." *Climatic Change* 144: 591–96.

Snell, D. 2018. "'Just Transition'? Conceptual Challenges Meet Stark Reality in a 'Transitioning' Coal Region in Australia." *Globalizations* 15 (4): 550–64.

Strydom, P. 1999. "The Challenge of Responsibility for Sociology." *Current Sociology* 47 (3): 65–82.

Taleb, N.N. 2012. *Antifragile. Things that Gain from Disorder*. London: Penguin.

van de Poel, I. 2011. "The Relation between Forward-Looking and Backward-Looking Responsibility." In *Moral Responsibility: Beyond Free will and Determinism*, edited by N.A. Vincent, I. van de Poel, and I. van den Hove, 37–52. Dordrecht: Springer.

Villey, M. 1977. "Esquisse historique sur le mot responsable." *Archives de Philosophie du Droit* 22: 45–58.

Vincent, N.A. 2011. "A Structured Taxonomy of Responsibility Concepts." In *Moral Responsibility: Beyond Free will and Determinism*, edited by N.A. Vincent, I. van de Poel, and I. van den Hove, 15–35. Dordrecht: Springer.

Von Schomberg, R. 2013. "A Vision of Responsible Research and Innovation." In *Responsible Innovation*, edited by R. Owen, J. Bessant, and M. Heintz, 51–74. Chichester: Wiley.

Walker, G. 2011. *Environmental Justice*. London: Routledge.

Walker, J. and M. Cooper. 2011. "Genealogies of Resilience. From Systems Ecology to the Political Economy of Crisis Adaptation." *Security Dialogue* 4 (2): 143–60.

Weinberg, A. 1972. "Science and Trans-Science." *Minerva* 10 (2): 209–22.

Young, I.M. 2006. "Responsibility and Global Justice: A Social Connection Model." *Social Philosophy and Policy* 23 (1): 102–30.

# 11

# Religion

*Katharina Glaab*

## Introduction

In light of the multifaceted ecological crisis, ethical and normative questions have come to the forefront of public and political debates on the environment. Anthropogenic climate change not only has profound implications on environment, health, and economy, but it also poses a serious threat to human development and accelerates global inequalities. Debates around justice and moral responsibility are inevitable (Shue 1992), when one takes seriously that climate change leads to human suffering (Wapner 2014) and that particularly the poor and most vulnerable will be affected by climatic changes. Efforts to confront the multifaceted effects of climate change are high on national and international political agendas, but in order to address these challenges adequately, scholars increasingly argue that new approaches and more fundamental transformations in values and cultural practices are needed. Taking into account that changes in human-environment relations are constitutive for the ecological crisis, scholars point out that we need to go beyond a problem-solving approach and look at the ethical and moral foundations of these changes (Bergmann 2010). No wonder, the recent report to the Club of Rome calls for a "new enlightenment" that balances humans and nature, markets and the state, and short- versus long-term objectives (von Weizäcker and Wijnman 2017).

In this context, scholars and policymakers discuss the role of religion and its contribution to global sustainability governance. Scholars assume that religious values can lead to a change in consciousness and ultimately to more environmentally friendly behaviour (Johnston 2014). In contrast to economic, scientific, and technological solutions that are the usual point of discussion within national and international policy fora, religion provides values and normative frameworks that address the ethical dimensions of the current environmental crisis. Religion is not the only way to make a normative contribution to global sustainability governance, but religious values and practices are seen as one way to change views on the environment and human-nature relationship, nurture ethical perspectives, and motivate action. Considering that eight out of ten people worldwide identify themselves as religious (Pew Research Center 2012), scholars see a lot of potential for religious actors to drive global sustainability transitions by changing individual and community practices as well as influencing national and global environmental politics. More specifically, scholars argue that religions matter in fighting the impacts of the environmental crisis because of their close connection to adherents and their powerful position as a social actor (Haluza-DeLay 2014, 263). They would usually point to religions' ability to influence believers' worldviews, their broad reach based on their moral authority, their large institutional and economic resources, and their potential to connect through social capital (Haluza-DeLay 2014, 263; Gardner 2003). Accordingly, one should take seriously the claim that there is a religious answer to the global ecological crisis.

This chapter will address the role that religion plays in global sustainability governance and the contribution that religious actors can make to sustainable development. In order to understand the breadth that a consideration of such a diverse and multifaceted topic implies, this chapter will give an overview of the burgeoning literature on religion and sustainability, and indicate how religious actors contribute to global sustainability governance. The chapter will first explain the state of the scholarly debate on religion and sustainability. Second, recent developments in religious actors' involvement in sustainability transitions and the role that they play in environmental politics will be discussed. Third, it will focus on the normative contribution of religious actors to global sustainability discourses in relation to civil society actors. Finally, this chapter aims to discuss gaps in the literature and outline some new pathways in the study of religion and global sustainability governance that deserve our attention in future research.

## Religious perspectives on the environment

A discussion of the relationship between religion and sustainability governance requires a clarification of what the term religion encompasses. Religion, broadly defined, describes a system of cultural practices and beliefs that sets a relationship between humans and the spiritual/divine. More specifically, religion often refers to institutional religions like the world religions – including Islam, Christianity, Hinduism, or Buddhism – and their practices and traditions. The notion of religious actor takes into account that churches and world religions are not the only faith-motivated actors, and refers to organisations or individuals often in broader civil society that are motivated by religious traditions and values. Besides the world religions, religious actors include for instance faith-based development organisations or environmental groups.

The focus on institutionalised religion is often seen as too narrow to understand the breadth of beliefs and practices that religion can encompass in relation to the environment, as the distinction between substantive and functionalist conceptions of religion shows. While substantive conceptions of religion focus on religions as institutions with membership and the impact of their practices on the environment, functionalist conceptions understand religion as part of broader cultural practices of environmentalism (Berry 2013). Accordingly, people act for the environment because they believe and are motivated by religious convictions or spirituality (Lynch 2000). There is a leap of faith connected to environmental consciousness and sustainability itself refers to a broad range of values. It is concerned with questions of human purpose and "the good life" which often involves individual and collective sacrifices (Wapner 2010). This suggests that there is a religious element in environmentalism and some scholars have even conceptualised environmental movements as a "religious quest" or a "secular faith" itself (Dunlap 2006). This chapter will mostly discuss the role of substantive religions and other religious actors in global sustainability governance, but acknowledges that spiritual cosmologies or individual spirituality are important to understand motivations and actions of many environmental activists. Due to the dominant representation of Christian organisations in global governance institutions, which is historically developed and also related to religions' different degree of institutionalisation (Haynes 2014), this chapter will mostly focus on activities of Christian organisations with some reference to non-Christian organisations' activities.

The relationship between religion and the environment has particularly been addressed in theological literature. Other disciplines in the social sciences have just started to engage with that topic, and are exploring if and in what way religion can affect sustainability transitions. While there is increasing scholarly attention towards actual religious engagement and practices in the field of environmental politics, such as the global debate on sustainable development (Berry 2014; Glaab and Fuchs 2018), the role of religion in global climate politics at the United Nations Framework Convention on Climate Change (UNFCCC) (Glaab 2017), in local energy transitions (Köhrsen 2015), or in the context of political movements in the Global South and Global North (Veldman, Szasz, and Haluza-DeLay 2014a), scholars have pointed out that empirical studies on this phenomenon are still lacking (Haluza-DeLay 2014; Veldman, Szasz, and Haluza-DeLay 2014b).

Scholars of the ecology and religion debate in theology and religious studies who discuss how religions value nature, diverge over whether or not to ascribe religion a decisive transformative role in global sustainability politics. Those scholars that see a positive influence of religion on sustainable practices and politics point out that religions have entered their "ecological phase" (Tucker 2003) and that their foundational values define sustainable human-nature relationships. This argument was established as part of a conference series on "Religions of the World and Ecology" held at Harvard University that resulted in the founding of the Forum on Religion and Ecology and led to a series of publications on the specific contributions of respective faiths to ecological values. Although the different world religions are seen to provide diverse understandings of environmental ethics based on their respective faith traditions, religious ecologists argue that most religions share an attitude of care for the environment and support sustainable practices (Gottlieb 2006; Tucker and Grim 2001, 2–3). This enables them to play a supporting role in the fight against climate change and the environmental crisis (Tucker 2003).

Yet the assumption that religions are rapidly greening has been challenged. Some scholars point out that religions' relationship with nature is not necessarily benign, but can even be at odds with sustainability goals (Kalland 2005; White 1967). This perspective relates to an early study of Lynn White, published in *Science*, in which he prominently argued that the Judeo-Christian tradition nurtured an anthropogenic worldview that enabled environmentally destructive behaviour (White 1967). In more recent empirical studies, scholars have shown that there is no evidence that religious believers are more supportive of environmental causes than non-believers (Taylor 2015, 8; Taylor, van Wieren, and Zaleha 2016), or that religious actors make a distinctive contribution to sustainable development discourses (Berry 2014). Other research even shows that religious groups such as some Evangelical Christians within the United States tend to be more climate-sceptical (Carr et al. 2012; McCammack 2007), or, in the case of local energy transitions in Germany, that religious actors do not have an impact on sustainability transitions (Köhrsen 2015).

Environmental politics has been a particularly "secular" policy field as it is focused on scientific and technological solutions, often shows an implicit or overt tension between religion and science, and a bias towards "secular" solutions (Glaab 2018; Glaab and Fuchs 2018). This biased evaluation of the role of religion in global sustainability governance relates to how religion is defined and what role religious actors are presumed to take within society and the public sphere. Religion is usually seen as a personal and private matter. While proponents of secularisation theory would argue that religion has lost in importance in modern societies (Berger 1969), others have asked for a more nuanced debate and pointed out that while industrialised societies have become more secular, the world as a whole has become more religious (Norris and Inglehart 2004). In the field of global sustainability governance the assumed distinction between the religious and the secular matters, as it defines whether religious actors are seen to take a legitimate role in public and political debates on the environment and in finding solutions to the environmental crisis. In debates on the environment and climate change, secular environmentalists have often seen religious reasoning as irrational, uncomfortable, or unhelpful (Dunlap 2006; Wilson 2012). Notwithstanding, religion has become an increasingly important topic in sustainability politics.

## The role of religious actors in global sustainability governance

Despite these varied assessments of religion's potential influence on sustainability transformations, a growing number of scholars and international organisations argue that religions can and should play a bigger role in global sustainability politics. Renowned climatologist Mike Hulme, for instance, argued that religion has a part to play in our response to climate change (Hulme 2017). Scholars also point out that, empirically, increased religious activism for the environment can be witnessed (Tucker and Grim 2001; Veldman, Szasz, and Haluza-DeLay 2014a). In addition, secular organisations, civil society, scientists, and policymakers seem to be more interested in including religious actors and cooperating with them in projects and initiatives in global sustainability governance, especially in the field of climate change politics (Glaab 2017) or sustainable development (Johnston 2014). Indeed, religious actors are increasingly taking part in political debates on the environment: religious leaders such as Catholic Pope Francis I or Ecumenical Patriarch Bartholomew prominently argued for addressing climate change as a moral challenge to society. Actors of all faiths have started facilitating sustainability transitions in local communities and advocate change within global governance institutions. In addition, policymakers increasingly see religions as important change-makers. Former Secretary of the UNFCCC, Christiana Figueres (2014), encouraged religions to "set their moral compass" on climate change and former General Secretary of the United Nations, Ban Ki Moon (2009), sees religions as occupying a unique position in advocating sustainability transformations, indicating that international institutions are slowly moving away from a reductionist to a holistic approach to the environmental crisis.

In fact, religious actors have been present in debates on the environment since environmental concerns began to be more prominently articulated on a global level in the 1970s. For instance, religious actors worked on environmental issues in national offices, showed a strong presence at the historic Earth Summit on sustainable development in 1992, and inter-church organisations such as the World Council of Churches focused on environmental issues or concepts such as climate justice since the 1980s (Kerber 2014). In Bhutan, the Gross National Happiness approach to development, which aligns human development with ecological conservation and is rooted in its Buddhist traditions, has influenced the state's environmental politics and showed some success as two-thirds of its forest cover remain intact (Allison 2017).

More recently, the publication of the papal encyclical on the care for our common home confirmed that religious engagement against climate change has become an important part of global environmental politics (Pope Francis 2015). While the encyclical was praised by religious and secular groups, scientists, and politicians alike for its clear call for action on climate change, earlier Popes were equally attentive to environmental care. Already in 1986, Pope John Paul II facilitated the World Day of Prayer for Peace, which brought leaders from different faiths to Assisi, Italy, the birthplace of Francis of Assisi, patron saint of the environment and inspiration of Pope Francis papal name. The interfaith gathering helped to position ecology as a central concern in interfaith dialogues and gave rise to more cooperation between religious and secular organisations on the environment. Connected to that event, the World Wide Fund for Nature (WWF) launched the secular environmental organisation Alliance of Religion and Conservation (ARC) that provides support for faith-based environmental action worldwide.

Cooperation between different faith groups and with other civil society organisations is important in different fields of sustainability politics. The ARC facilitated for instance the Windsor Conference in 2009 that prepared an interreligious statement for the Copenhagen Climate Conference. Representatives from different religious traditions presented an interfaith statement at the Interfaith Summit on Climate Change (2014) to show their commitment to promote sustainable practices before the important Paris Climate Conference in 2015. In addition, a collaboration between civil society, Islamic scholars, academics, and UN representatives of the Islamic Declaration on Climate Change was published in the same year (Islamic Foundation for Ecology & Environment Sciences 2015). Religious organisations are involved in sustainability projects of the World Bank or the United Nations Development Programme (UNDP), but also work with the United Nations Educational, Scientific and Cultural Organisation (UNESCO), the United Nation Environment Programme (UNEP), or the UNFCCC. Those institutions often see religions as part of broader civil society and act on an understanding that solutions to environmental problems need to be culturally embedded. For instance, UNEP established a scientific and religious dialogue in *Interfaith Partnership for the Environment* (Bassett, Brinkman, and Pedersen 2000), while the UNFCCC has facilitated more communication and cooperation with religious groups in context of the global climate change negotiations (Glaab 2017).

Those instances of religious engagement on environmental issues show that religious actors operate across scales. Individuals, local faith communities, churches, as well as international and transnational faith-based development organisations and religious actors are part of a broad range of sustainability initiatives. This points to the unique role that religious actors can take in global sustainability governance by linking global politics to local communities and individual action. It reflects the transnational nature of religion, but also its embeddedness in local contexts. The cooperation with international organisations or with secular environmental organisations is for instance legitimised by arguing that religions can influence global sustainability agendas, but can at the same time implement sustainability projects such as energy or resource efficient initiatives in their communities. Based on the assumption that religions can influence individual consumption practices, activation and mobilisation through religious actors help to overcome the challenges in taking action that is known in sustainability studies as the knowledge-action gap (Johnston 2010; Wolf and Gjerris 2009).

## The contribution of religious perspectives to global sustainability governance

The attention that religion has received in scholarly discussions and its inclusion in political debates shows an increasing acknowledgement that religious actors and a religious perspective have something to contribute to address the global ecological crisis. However, as Haluza-DeLay (2014, 266) has pointed out, the answer to the question of how the world religions are responding to climate change cannot arrive at a simple answer as they are responding in many different ways. Hence, religious actors' specific normative contribution to global sustainability governance and the question of how religious perspectives on environmental ethics can be transformed and integrated into politics deserve attention.

Their contribution to sustainability transformations is often seen in material and discursive terms. Particularly religions' potential influence on sustainability politics is seen in material terms. In this context, scholars often refer to the number of adherents, large number of physical resources such as land and real estate ownership, their large share of financial assets which makes them important investors and shareholders, and their large media network (Gardner 2002). Their discursive contribution lies in the ability to influence political discourses which has the potential to bring about normative change. While religions have different environmental ethics, all religions value creation and share concerns for conservation and sustainable development. They see a moral responsibility to preserve nature and advocate stewardship for the earth. With this perspective, they already depart from the strong focus on scientific and technical solutions that have long been promoted in sustainable development and global climate politics. An analysis of sustainable development discourses by religious actors showed that religions advocate a more holistic understanding of sustainable development and emphasise the non-material dimension of development that has a religious or spiritual component (Glaab and Fuchs 2018).

Religious actors also tend to place greater emphasis on human suffering and the fate of the most poor and vulnerable. This perspective on human development is not unique to religious engagement on the environment but is embedded within religions' broader normative framework. Liberation theology, for instance, has long promoted the dignity of the poor (Lynch 2000) and it is also a key concern in other policy fields that religious actors are involved in. Taking climate change politics as a prominent example, religious actors use their unique role as mediators between the global and local level. They act as witnesses of detrimental climatic changes that are already happening in their communities in the Global South. These affect particularly the most vulnerable and religious actors communicate the impact of climatic changes in local communities within global policy fora. Religious actors also focus on principles of justice in their environmental activism and relate it to their concern with human development (Glaab, Fuchs, and Friederich 2018). Therefore, climate justice becomes a key concern. Religious conceptions of climate justice depart from state-centric understandings of justice, and focus on intergenerational, intra-societal, and increasingly inter-species aspects of justice instead (Glaab 2017).

In global sustainability governance, religious actors are often seen as part of broader civil society. Just as civil society is divided on the right course of action towards sustainability transformations, so are many religious actors involved in the debate. Civil society diverges for instance about the question of whether to ascribe to nature instrumental or intrinsic values. The division between those civil society proponents who advocate market-based approaches and prefer technical solutions, and those who advocate global justice approaches and prefer fundamental societal and politico-economic reforms is also visible among religious actors (Glaab, Fuchs, and Friederich 2018). While some advocate more radical shifts from global economic norms, others tend to promote reforms of existing instruments (Glaab, Fuchs, and Friederich 2018).

Religious actors do not just aim at changing global discourses on human-nature relations and sustainability politics. They also facilitate concrete changes in practices that do not just aim at changes in individual consumption practices but also at global efforts to contribute to structural changes. For instance, the global initiative to divest from fossil fuel has been broadly supported by religious organisations. Following the mandate of the papal encyclical, the Catholic Church has become a forerunner in the global divestment movement with more and more of its institutions pulling out investments from fossil fuel companies and diverting into renewable energies (Neslen 2017). Even before the release of the encyclical in 2015, other religious institutions and the World Council of Churches, which consists of over 300 churches that represent some 590 million people, joined the movement and decided to divest from coal, oil, and gas (350org 2014). Churches are large financial investors worldwide and their decision to take ethical considerations seriously and pull their assets out of fossil fuel investment is a strong signal that religions can transform their ethical principles into political action and impact.

## Further avenues for research

The study of religious dimensions of sustainability and the potential contribution of religious actors to sustainability governance is still in the beginning. While the theoretical discussion about religion and ecology is well known, empirical studies on religious activism in various contexts and the difference it makes are still needed. The increase in empirical work over the last decade shows that scholars pay attention not just to religious environmental ethics but also to their practical application in diverse settings and policy fields. Despite this upsurge in studies and their important contribution to our understanding of the practical impact of religion on sustainability governance, so far, research has mostly focused on more institutionalised religions such as the Christian faiths, reflecting a more general bias in global studies towards the study of faith traditions prevalent in Western countries (Smith et al. 2018). But with increased attention to environmental politics as a binding element of interreligious cooperation instead of separation, and an increasing acknowledgement by actors in the sustainability field that there is a religious answer to the environmental crisis, more efforts are being made to inquire into non-Christian groups and their environmental activism. First results from research on Islamic environmentalism indicate for instance that Muslim ways of sustainable practices help to activate action in countries with large Muslim populations (Amri 2014).

In addition, the focus on Western and institutionalised religions in discussions on religion and sustainability tends to exclude other forms of spirituality such as indigenous or contemporary spirituality like nature-oriented spirituality. However, these spiritualties may be able to make a specific contribution to sustainability goals, as "indigenous traditions may be more likely to be pro-environmental than other religious systems and that some nature-based cosmologies and value systems function similarly" (Taylor, van Wieren, and Zaleha 2016, 1000). Taylor (2009) showed that nature-spirituality is on the rise and has replaced traditional religion in many places. It brings forth the idea that nature is sacred and has intrinsic value and can be found among groups of radical environmentalists, lifestyle-focused bioregionalists, surfers, or new-agers (Taylor 2009). There are also indications that especially indigenous spirituality is finding its way into political discourses on the environment. Indigenous groups are represented in many political fora of environmental politics where they bring in a perspective of spirituality that is situated within indigenous cosmologies and knowledges (Smith et al. 2018). Indigenous groups have a spiritual relationship with their land which understands nature as sacred (Lightfoot 2016). Hence, scholars suggest that indigenous groups' discourses of justice are more broad and pluralistic and essentially represent their ability to practice their traditions in their relationship with nature and their land (Schlosberg and Carruthers 2010).

However, authors have cautioned against universalising indigenous cosmologies and environmental ethics which are often based on Western experiences. Scholars have for instance pointed to distinct African environmental ethics (Behrens 2014), and Konik (2018) argues that marginalised value systems such as the African ethic of *ubuntu*, which considers care for others and the environment as key to the development of personhood and esteem, may provide a powerful resource for social and environmental justice activism. In the same way, *buen vivir*, an indigenous expression in South America for living well not better has increasingly made its way into sustainable development debates (Villalba 2013; Vanhulst and Beling, this volume). Broadening our understanding of religion and further looking at the contribution of indigenous and other nature-based spiritualities will be an important avenue for further research contributing to the "worlding" of global sustainability governance by acknowledging different ontologies (Inoue, Ribeiro, and Resende, this volume).

Overall, the recent dialogues with natural sciences and politics about religion are one step to integrate humanities and social sciences to truly address the ecological crisis from an interdisciplinary perspective. Yet many practitioners in sustainability governance treat religion in an instrumental way and see it as just another cultural variable that can add a different perspective to the bigger picture of sustainable development. But for many scholars a mere addition of religious and ethical perspectives to studies of the ecological crisis is not enough; instead, they argue that religion can fundamentally revise them (Bergmann and Gerten 2010, 11). In order to fully understand the religious contribution to global sustainability governance, more work is needed and a better dialogue with other concepts, tools, and perspectives.

## Acknowledgements

I would like to thank the editors of this Handbook for their helpful feedback and suggestions in relation to this chapter.

## References

350org. 2014. "World Council of Churches Endorses Fossil Fuel Divestment." https://350.org/press-release/world-council-of-churches-endorses-fossil-fuel-divestment/ (Accessed 22 March 2018).

Allison, Elizabeth. 2017. "Spirits and Nature: The Intertwining of Sacred Cosmologies and Environmental Conservation in Bhutan." *Journal for the Study of Religion, Ethics and Culture* 11 (2): 197–226.

Amri, Ulil. 2014. "From Theology to a Praxis of 'Eco-Jihad': The Role of Religious Civil Society Organizations in Combating Climate Change in Indonesia." In *How the World's Religions are Responding to Climate Change*, edited by Robin G. Veldman, Andrew Szasz and Randolph Haluza-DeLay, 75–93. Milton Park: Routledge.

Ban Ki-Moon. 2009. *Many Heavens, One Earth: Faith Commitment for a Living Planet*. Windsor Conference, www.un.org/press/en/2009/sgsm12585.doc.htm (Accessed 4 June 2017).

Bassett, Libby, John T. Brinkman, and Kusumita P. Pedersen 2000. *Earth and Faith. A Book of Reflection for Action*. New York: UNEP.

Behrens, Kevin G. 2014. "An African Relational Environmentalism and Moral Considerability." *Environmental Ethics* 36 (1): 63–82.

Berger, Peter. 1969. *The Sacred Canopy*. Garden City, NY: Anchor Books.

Bergmann, Sigurd. 2010. "Dangerous Environmental Change and Religion." In *Religion and Dangerous Environmental Change. Transdisciplinary Perspectives on the Ethics of Climate and Sustainability*, edited by Sigurd Bergmann and Dieter Gerten, 13–37. Muenster: LIT Verlag.

Bergmann, Sigurd, and Dieter Gerten. 2010. "Religion in Climate and Environmental Change." In *Religion and Dangerous Environmental Change. Transdisciplinary Perspectives on the Ethics of Climate and Sustainability*, edited by Sigurd Bergmann and Dieter Gerten, 1–12. Muenster: LIT Verlag.

Berry, Evan. 2013. "Religious Environmentalism and Environmental Religion in America." *Religion Compass* 7 (10): 454–466.

Berry, Evan. 2014. "Religion and Sustainability in Global Civil Society. Some Basic Findings from Rio+20'." *Worldviews: Global Religions, Culture and Ecology* 18 (3): 269–288.

Carr, Wylie A., Michael Patterson, Laurie Yung, and Daniel Spencer. 2012. "The Faithful Skeptics: Evangelical Religious Beliefs and Perceptions of Climate Change." *Journal for the Study of Religion, Nature & Culture* 6 (3): 276–299.

Dunlap, Thomas R. 2006. "Environmentalism, a Secular Faith." *Environmental Values* 15 (3): 321–330.

Figueres, Christina. 2014. "Faith Leaders Need to Find Their Voice on Climate Change." *The Guardian* www.theguardian.com/environment/2014/may/07/faith-leaders-voiceclimate-change?CMP=twt_gu (Accessed 19 March 2018).

Gardner, Gary T. 2002. "Invoking the Spirit: Religion and Spirituality in the Quest for a Sustainable World." *Worldwatch Paper* 164: 1–62.

Gardner, Gary T. 2003. "Engaging Religion in the Quest for a Sustainable World." In *State of the World, 2003. A Worldwatch Institute Report on Progress Toward a Sustainable Society*, edited by Gary T. Gardner, Christ Bright, and Linda Starke, 152–175. New York: W. W. Norton & Company.

Glaab, Katharina. 2017. "A Climate for Justice? Faith-based Advocacy on Climate Change at the United Nations." *Globalizations* 14 (7): 1110–1124.

Glaab, Katharina. 2018. "Faithful Translation? Shifting the Boundaries of the Religious and the Secular in the Global Climate Change Debate." In *World Politics in Translation – Power, Relationality, and Difference in Global Cooperation*, edited by Tobias Berger and Alejandro Esguerra, 175–190. London: Routledge.

Glaab, Katharina, and Doris Fuchs. 2018. "Green Faith? The Role of Faith-based Actors in the Global Sustainable Development Discourse." *Environmental Values* 27 (3): 289–312.

Glaab, Katharina, Doris Fuchs, and Johannes Friederich. 2018. "Religious NGOs at the UNFCCC: A Specific Contribution to Global Climate Politics?" In *Religious NGOs at the UN: Polarizers or Mediators?*, edited by Claudia Baumgart-Ochse and Klaus-Dieter Wolf, 47–63. London: Routledge.

Gottlieb, Roger S. 2006. *A Greener Faith: Religious Environmentalism and Our Planet's Future*. Oxford: Oxford University Press.Haluza-DeLay, Randolph. 2014. "Religion and Climate Change: Varieties in Viewpoints and Practices." *WIREs Climate Change* 5 (2): 261–279.

Haynes, Jeffrey. 2014. *Faith-based Organizations at the United Nations*. New York: Palgrave Macmillan.

Hulme, Mike. 2017. "Climate Change and the Significance of Religion." *Economic and Political Weekly* 52 (28): 14–17.

Interfaith Summit of Climate Change. 2014. *Climate, Faith and Hope: Faith Traditions Together for a Common Future*. http://interfaithclimate.org/the-statement/ (Accessed 7 April 2017).

Islamic Foundation for Ecology & Environment Sciences. 2015. *Islamic Declaration on Global Climate Change* www.ifees.org.uk/declaration/ (Accessed 24 March 2018).

Johnston, Lucas F. 2014. *Religion and Sustainability: Social Movements and the Politics of the Environment*. New York: Routledge.

Kalland, Arne. 2005. "The Religious Environmentalist Paradigm." In *Encyclopedia of Religion and Nature*, edited by Bron Taylor, 1367–1371. London & New York: Continuum.

Kerber, Guillermo. 2014. "International Advocacy for Climate Justice." In *How the World's Religions are Responding to Climate Change*, edited by Robin G. Veldman, Andrew Szasz and Randolph Haluza-DeLay, 278–293. Milton Park: Routledge.

Köhrsen, Jens. 2015. "Does Religion Promote Environmental Sustainability? – Exploring the Role of Religion in Local Energy Transitions." *Social Compass* 62 (3): 296–310.

Konik, Inge. 2018. "*Ubuntu* and Ecofeminism: Value-building with African and Womanist Voices." *Environmental Values* 27 (3): 269–288.

Lightfoot, Sheryl. 2016. *Global Indigenous Politics: A Subtle Revolution*. London: Routledge.

Lynch, Cecelia. 2000. "Acting on Belief: Christian Perspectives on Suffering and Violence." *Ethics & International Affairs* 14 (1): 83–97.

McCammack, Brian. 2003. "Hot Damned America: Evangelicalism and the Climate Change Policy Debate." *American Quarterly* 59 (3): 645–668.

Neslen, Arthur. 2017. "Catholic Church to Make Record Divestment from Fossil Fuels." www.theguardian.com/environment/2017/oct/03/catholic-church-to-make-record-divestment-from-fossil-fuels (Accessed 22 March 2018).

Norris, Pippa, and Ronald Inglehart. 2004. *Sacred and Secular*. Cambridge: Cambridge University Press.

Pew Research Center. 2012. *The Global Religious Landscape: A Report on the Size and Distribution of the World's Major Religious Groups as of 2010*. Washington: Pew Research Center.

Pope Francis. 2015. "Encyclical Letter Laudato Si' of the Holy Father Francis on the Care for Our Common Home." http://w2.vatican.va/content/francesco/en/encyclicals/documents/papa-francesco_20150524_enciclica-laudato-si.html (Accessed 20 March 2018).

Schlosberg, David, and David Carruthers. 2010. "Indigenous Struggles, Environmental Justice, and Community Capabilities." *Global Environmental Politics* 10 (4): 12–35.

Shue, Henry. 1992. "The Unavoidability of Justice." In *The International Politics of the Environment. Actors, Interests, and Institutions*, edited by Andrew Hurrell and Benedict Kingsbury, 373–397. Oxford: Clarendon Press.

Smith, Peter J., Katharina Glaab, Claudia Baumgart-Ochse, and Elizabeth Smythe. 2018. "Conclusion." In *The Role of Religion in Struggles for Global Justice: Faith in Justice?*, edited by Peter J. Smith, Katharina Glaab, Claudia Baumgart-Ochse, and Elizabeth Smythe, 121–129. Milton Park: Routledge.

Taylor, Bron. 2009. *Dark Green Religion: Nature Spirituality and the Planetary Future*. Berkeley: University of California Press.

Taylor, Bron. 2015. "Religion to the Rescue (?) in an Age of Climate Disruption." *Journal for the Study of Religion, Nature and Culture* 9 (1): 7–18.

Taylor, Bron, Gretel van Wieren, and Bernard D. Zaleha. 2016. "Lynn White Jr. and the Greening-of-Religion Hypothesis." *Conservation Biology* 30 (5): 1000–1009.

Tucker, Mary E. 2003. *Worldly Wonder: Religions Enter Their Ecological Phase. The Second Master Hsüan Hua Memorial Lecture*. Chicago: Open Court.

Tucker, Mary E. and John A. Grim. 2001. "Introduction. The Emerging Alliance of World Religions and Ecology." *Daedalus, Journal of the American Academy of Arts and Science* 130 (4): 1–22.

Veldman, Robin G., Andrew Szasz, and Randolph Haluza-DeLay, eds. 2014a. *How the World's Religions are Responding to Climate Change: Social Scientific Investigations*. Milton Park: Routledge.

Veldman, Robin G., Andrew Szasz, and Randolph Haluza-DeLay. 2014b. "Social Science, Religions, and Climate Change." In *How the World's Religions are Responding to Climate Change*, edited by Robin G. Veldman, Andrew Szasz and Randolph Haluza-DeLay, 3–19. Milton Park: Routledge.

Villalba, Unai. 2013. "Buen Vivir vs Development: A Paradigm Shift in the Andes?" *Third World Quarterly* 34 (8): 1427–1442.

Von Weizäcker, Ernst U., and Anders Wijkman. 2017. *Come On! Capitalism, Short-termism, Population and the Destruction of the Planet – A Report to the Club of Rome*. New York: Springer.

Wapner, Paul. 2010. "Sacrifice in an Age of Comfort." In *The Politics of Sacrifice*, edited by Michael Maniates and John Meyer, 33–59. Cambridge: MIT Press.

Wapner, Paul. 2014. "Climate Suffering." *Global Environmental Politics* 14 (2): 1–6.

White, Lynn. 1967. "The Historical Roots of Our Ecologic Crisis." *Science* 155 (3767): 1203–1207.

Wilson, Erin K. 2012. "Religion and Climate Change: The Politics of Hope and Fear." *Local Global* 10: 20–29.

Wolf, Jakob and Mickey Gjerris. 2009. "A Religious Perspective on Climate Change." *Studia Theologica – Nordic Journal of Theology* 63 (2): 119–139.

# 12

# Sufficiency

*Anders Hayden*

## Introduction

The question of "how much is enough?" is increasingly urgent for sustainability governance. The idea of sufficiency – that "there can be enough and there can be too much" (Princen 2005) – has long been marginalised by dominant environmental approaches focused on "green growth" through improved eco-efficiency and technology. However, some critical approaches to sustainability, such as degrowth (e.g. Kallis 2018), doughnut economics (Raworth 2017), and strong sustainable consumption (Lorek and Fuchs 2013), incorporate the idea. So, too, do many chapters in this handbook, either explicitly or implicitly, including those that examine the 'good life' (Di Giulio and Defila) and *buen vivir* (Vanhulst and Beling); those analysing challenges related to a growth-centred development vision (Higgs), excessive reliance on technological solutions (Alexander), an extractivist/mining approach to natural resources (Princen), and the power of consumerist values (Krogman); and those proposing potentially transformative approaches such as consumption corridors (Fuchs), sustainable zero growth (Lange), beyond-GDP measurement (Philipsen), and work-time reduction (Larsson and Nässén).

This chapter examines sufficiency's meaning in the context of sustainability governance, and reasons why it persists – and indeed is increasingly important – in environmental debates despite significant obstacles to the idea. It emphasises the importance of seeing sufficiency as a political project. The chapter outlines some potential ways that sufficiency could be incorporated into policy and examples where the idea has made some inroads, and discusses opportunities and challenges in linking sufficiency to maintaining and even improving well-being in contemporary societies. It will argue that sufficiency deserves to be a core organising principle for society at a time of increasingly evident planetary limits, although other approaches are also needed to complement it.

## Sufficiency: what is it?

Many different interpretations of sufficiency – what it means to have enough – exist and debate continues over the concept's precise definition and boundaries. Although theorists

151

sometimes focus on only one side of the issue, "enough" involves two different thresholds: a minimum and a maximum (Spengler 2016). In the context of sustainability, sufficiency can be seen as "living well within limits" or having enough for a good life, but not consuming so much that it is ecologically excessive – that is, not consuming at a level that undermines possibilities for others, today and in the future, to also lead good lives (Di Guilio and Defila, this volume; Fuchs, this volume; O'Neill et al. 2018). For those living with very little, sufficiency may require more consumption – a point sometimes neglected in the sufficiency literature, which has largely focused on overconsumption among the globe's affluent consumers. For those individuals and the consumer societies they participate in, sufficiency involves limiting consumption and production volumes – with a goal of "one-planet living" or "fair share" levels that would be ecologically sustainable if replicated by all of the Earth's nearly eight billion people. Alexander (2015) sums up sufficiency in these demanding terms: "enough, for everyone, forever."

Sufficiency can be presented as an individual-level strategy to voluntarily reduce consumption (e.g. Alcott 2008). However, this chapter is based on the premise that sufficiency is a key principle for society as a whole (Princen 2005; Lorek and Fuchs 2013), requiring societal-level action and sufficiency-oriented policy and politics (Schneidewind and Zahrnt 2014; Darby and Fawcett 2018).

At the macro-economic level, a sufficiency lens leads to criticism of GDP growth as a dominant societal goal and to a quest for post-growth alternatives such as degrowth and a steady-state economy (e.g. Jackson 2017; Kallis 2018; Victor 2019; Lange, this volume). However, not all formulations of sufficiency challenge a growth-oriented economy (e.g. Association Négawatt 2018, 3). Sufficiency can also involve efforts to limit specific products, practices, or sectors considered excessive due to their social or ecological impacts such as meat and dairy consumption, air travel, and long-distance transport of food (Hayden 2014a, 2014b) – or other manifestations of the modern emphasis on "faster," "further," and "more" (Sachs 2001). Another variation is "sufficiency in performance levels" – for example, deliberately designing cars and trains with lower accelerating power and top speeds to limit energy use (Sachs 2001, 158).

Wolfgang Sachs (1993), a pioneering sufficiency theorist, explained that sufficiency can be about "less" in different ways: less speed, less distance, less clutter, and less market – i.e. more providing and making for oneself (see also Schneidewind and Zahrnt 2014). An alternative formulation is that sufficiency is about "slower, less, better, finer" (Hans Glauber quoted in Schneidewind and Zahrnt 2014). These words emphasise that sufficiency aims not for sacrifice and austerity, but for its own vision of a good life. Indeed, it has its own vision of more – "more of what matters," according to the Center for a New American Dream, or as a French degrowth movement slogan puts it, "*moins de biens, plus de liens*" (fewer goods, more connections) (Ariès 2010).

Sufficiency has an affinity with anti-capitalist critique; it can be linked not only to questioning the forces within capitalism that breed isolation and alienation, but above all to critique of the relentless pressures for economic expansion in the quest for capital accumulation. Some eco-socialists have incorporated variations on sufficiency themes, such as Kovel (2008, 9–10), who criticises the idea that capitalism's endemic overproduction can be "remedied by more consumption of more commodities – when in truth, the level of consumption already imposed by capital is the immediate instigator of ecological crisis…" However, proponents of sufficiency often sidestep an explicit critique of capitalism, while other political perspectives could also incorporate the idea. Muller and Huppenbauer (2016) argue that sufficiency – with its corresponding virtue of temperance – ought to be a core principle of liberal societies that evolve to respect planetary boundaries. Sufficiency could also take politically conservative forms if equity concerns were sidelined and consumption reduction concentrated among less-privileged people.

In terms of the IPAT equation – in which environmental impacts are a product of Population (P), Affluence (A, i.e. per-capita consumption), and technology (T, impact per unit of consumption) – sufficiency is about limiting the "A" variable, at least for high-consuming populations.[1] Alternatively, sufficiency can be considered one of three key components of a comprehensive ecological strategy – alongside efficiency and consistency, i.e. technologies and production methods consistent with natural processes such as renewable energy (Sachs and Santarius 2007, 158–165). Greater efficiency is about increasing the ratio of output to input, leaving no guarantee that it will result in lower total resource inputs. In contrast, sufficiency is about "living with absolute environmental limits" (Darby and Fawcett 2018, 4). It is the component of ecological strategy that mainstream "green growth" and ecological modernisation approaches neglect. A half-century ago, economist John Kenneth Galbraith commented on a similar neglect (1958, 92). Reflecting on the question "How Much Should a Country Consume?," he wrote:

> If we are concerned about our great appetite for materials, it is plausible to decrease waste, to make better use of stocks available, and to develop substitutes. But what about the appetite itself? Surely this is the ultimate source of the problem. … Yet in the literature of the resource problem this is the forbidden question.

## Why sufficiency?

There are powerful reasons why that question was long "forbidden." Businesses typically fiercely resist any suggestion of less output and consumer demand (at least for their own products, although not necessarily for their competitors). Politicians and even many environmentalists have been reluctant to challenge consumer aspirations for more, worrying about possible backlash. (Indeed, expectations for high consumption levels have spread beyond the global North to the growing middle class in many "developing" countries, where meeting such expectations is increasingly important for leaders to maintain political legitimacy.) Workers and labour unions fear that reigning in consumption will reduce employment. Governments have come to depend on high and rising consumption and production to meet core goals, including poverty reduction, limiting unemployment, and generating the tax revenues needed to satisfy voters and win re-election. Most generally, sufficiency has been seen as an idea that runs counter to the perceived imperative of economic growth (Hayden 2014c).

In light of the daunting obstacles to sufficiency, it might seem sensible to focus on less politically contentious approaches to ecological challenges such as "green growth" and ecological modernisation. Yet the idea of sufficiency and related post-growth perspectives have not faded away; in fact, they have gained renewed relevance in recent years in academic and social movement circles, if not yet to a large degree in mainstream politics. A number of factors have contributed to this persistence and resurgence (Hayden 2014c).

Most fundamental is the growing urgency of stronger action to address climate change and other ecological challenges. Recent research provides evidence that humanity has already crossed some planetary boundaries and is approaching others (Steffen et al. 2015); the global ecological footprint has overshot the limited biocapacity to provide resources and absorb emissions (GFN 2019), and only a narrow window of time remains for deep greenhouse-gas (GHG) reductions to avoid catastrophic climate outcomes (IPCC 2018). Given the need for rapid action, sufficiency-oriented policies to reduce demand have the potential advantage of being applied more quickly than developing and deploying new technologies (Schneidewind and Zahrnt 2014, 20–21). While most global energy scenario studies have ignored potential contributions from sufficiency in meeting climate goals (Samadi et al. 2017), some recent studies show the importance of substantially reducing energy demand, which would dramatically improve the possibility of meeting remaining demand through low-carbon sources and achieving a 1.5°C climate target without risky and unproven negative emissions technologies (Grubler et al. 2018; Wachsmuth and Duscha 2018). Such scenarios require a large role for efficiency and greener technologies, but also need sufficiency-based reductions in levels of energy-consuming activity. For example, analysis from France's Association Négawatt (2017, 27, 33; 2018, 3) estimates that energy sufficiency initiatives could reduce that country's energy consumption by 28% from 2015 to 2050, making a large contribution to 100% de-carbonisation.

Alongside the urgency of ecological action, limits to a "green growth" response are evident. A critique of that approach, and its excessive technological optimism, is outlined in detail by Alexander (this volume; see also Hayden 2014c, chapter 7). In brief, critics of "green growth" question the "myth of decoupling" (Paech 2012; Jackson 2017), noting that economic growth often outpaces relative decoupling (declining impacts per unit of output), resulting in no absolute decoupling (no decline in total impacts). Even apparent evidence of absolute decoupling may mask the transfer of impacts to other places and/ or fall short of the "sufficient absolute decoupling" needed (Raworth 2017). Other key limits to "green growth" include the lack of technological solutions for some key sources of ecological problems, the fact that technological solutions often have their own negative consequences, and the erosion of some – and sometimes all – of the benefits of efficiency due to rebound effects of various kinds (e.g. van den Bergh 2011; Santarius 2012). Such analysis does not mean that seeking improved eco-efficiency and technology is futile or unnecessary, but it is important to go beyond that approach and complement it with sufficiency (FOEE 2018).

Proponents of sufficiency and post-growth also point to the limits of income growth in improving well-being (e.g. Paech 2012; Alexander 2015; Gough 2017, 172) – and, more positively, highlight possibilities to maintain and even improve well-being with less consumption. In other words, while "green growth" seeks to decouple economic growth and environmental impacts, sufficiency aims to decouple well-being from income and consumption. It draws on studies showing relatively little, if any, connection between rising per-capita income and well-being or happiness beyond a threshold that high-income nations have already surpassed (e.g. Easterlin 1974; Frey and Stutzer 2002; Kasser 2003; Layard 2005; Easterlin et al. 2010). These issues remain contentious as some researchers maintain that an income-happiness link is still evident in the data (Deaton 2008; Stevenson and Wolfers 2013), or see a more complex relationship (Kahneman and Deaton 2010). However, even if one accepts that income growth can still boost subjective well-being in affluent nations, three points can be made. First, the benefits of more income diminish rapidly beyond a certain point (in other words, very large income increases become necessary to deliver small well-being gains). Second, some non-material factors – such as levels of trust in others and in institutions, close relationships, and sense of belonging to one's community – offer greater potential to deliver lasting well-being improvements without additional environmental burdens (Barrington-Leigh 2017). Third, the quest for ever-more income becomes counter-productive if it involves sacrificing time for more pleasurable or meaningful activities (Kasser 2003; Dolan 2019).

Global justice, including addressing North-South inequities (Orekere, this volume), is another key concern for proponents of sufficiency – a concept that is essential to "justice in an age of limits" (Sachs and Santarius 2007, 26). Sufficiency in the affluent global North involves leaving ecological space for poverty reduction in the South (e.g. Schor 2001), while acknowledging "ecological debts" from the North's historic and ongoing appropriation of resources from the South and indigenous lands (e.g. McLaren 2003; Salleh 2009). Sachs (2017, 9) writes: "More than anyone else, the poor are paying the price for the wealth of the Global North. In light of this, sufficiency could perhaps already be defined as mere refusal to live at the expense of others."

While one can frame sufficiency as a challenge to the perceived imperative of economic growth, in certain contexts it can bring economic benefits – even in conventional economic terms (Hayden 2014a). For example, sufficiency's contribution to reducing demand for energy and other resources can reduce the large monetary outflows from energy-importing nations or localities and improve their trade balance (Association Négawatt 2018, 10). A related benefit is improved energy security and reduced vulnerability to conflicts over resources that are likely to become increasingly scarce. Some sufficiency initiatives, such as promoting local food and repair of goods rather than disposal, can also boost local employment, adding to their political appeal.

Finally, it is possible that high-income nations have entered a protracted period of slow economic growth or "secular stagnation" (Gordon 2016; Summers 2016). Varied reasons have been put forward, including ageing populations, lingering effects of the financial crisis and overhang of public- and private-sector debt, growth of the service sector in which productivity gains are harder to achieve, and increasing inequalities that erode the foundations of mass consumption. Although long-term GDP stagnation is far from guaranteed, affluent societies may need to develop ways to manage without growth for non-ecological reasons as well (Büchs and Koch 2017, 52; Jackson 2019).

## Drivers of change and the need for policy

Having reviewed reasons why sufficiency-oriented thinking is increasingly important, we turn to factors that might drive a wider political role for it. The above-mentioned possibility of prolonged GDP stagnation would not only limit individual consumption; one political effect could be a new openness to ideas, projects, and policies capable of improving well-being and addressing problems such as unemployment, poverty, and inequality without more growth and consumption. Some observers look to bottom-up cultural change to drive a transition beyond consumerism, hoping that a "collective reframing of the idea of good life as less fixated on materialism … may be the engine of change" (Brown and Vergragt 2016, 311). Some evidence suggests that new consumer sensibilities are in fact emerging (Krogman, this volume), such as declining interest in car ownership among the young (Goodwin and Dender 2013; Cohen 2016) and declining meat consumption in some countries (Spiller and Nitzko 2015; Association Négawatt 2018, 8) – raising the possibility that "peak car" and "peak meat" may have arrived in some high-income societies. Some observers see the beginnings of a shift to a culture of access over ownership (e.g. Belk 2014), in which various forms of "sharing" replace the need to possess many goods, along with trends such as "minimalism" (Millburn and Nicodemus 2011). Change could also grow out of the many socially innovative, grassroots initiatives that are developing new forms of sharing and provisioning (Thompson and Schor 2014).

Taken together, these points suggest that evolution towards a society of sufficiency may already be occurring. Although such an evolution would be welcome, if it is happening, such change is occurring far too slowly given the need for rapid environmental action (e.g. IPCC 2018). A very different, but increasingly plausible scenario is that a consumerist, growth-oriented economy could end due to climate destabilisation and other forms of ecological and social breakdown (Bendell 2018). That outcome would be consistent with only one part a sufficiency vision – reduction of excess consumption – while eroding the possibilities for the other side of sufficiency, i.e. enough consumption for people to meet their basic needs and live good lives.

To accelerate any emerging trends towards norms of sufficient consumption, political and policy action is needed (Maniates, this volume). Indeed, a sufficiency lens can expand the policy options available to address environmental challenges. Opportunities exist to apply the principle of sufficiency in almost every policy area (Schneidewind and Zahrnt 2014). Among the many possibilities are energy policies that move beyond improving efficiency to demand reduction, such as revising building codes so that new buildings allow for thermal comfort without air conditioning (Darby and Fawcett 2018). Measures could be taken to moderate the size of new residential buildings by encouraging shared spaces (e.g. laundry and guest rooms) or inter-generational sharing of housing in ways that also reduce isolation of the elderly (Anderssen 2018; Association Négawatt 2018). Sufficiency-oriented transport policies include: car-free city centres to promote shifts to lower-carbon transport modes; reducing distances travelled by promoting teleworking or holidays closer to home; lowering speed limits; and curbing air travel by halting airport expansion, increasing its cost through carbon taxes, or a more egalitarian "green air miles allowance" with one tax-free flight per year and escalating taxes on frequent flyers (Schneidewind and Zahrnt 2014; Association Négawatt 2018; Darby and Fawcett 2018). Urban planning policies can limit suburban sprawl and reduce travel needs, supported by infrastructure investments to encourage cycling, walking, and public transit, while provision of non-commercialised public spaces such as parks and libraries can provide high-quality leisure opportunities with little need for income or pressure to consume (Schneidewind and Zahrnt 2014). Food policies can target food waste and, more controversially, encourage reductions in high-GHG meat and dairy consumption. Unnecessary purchases could be reduced by regulations that guarantee a "right to repair" and require products such as appliances to be more durable – as being debated in the European Union.

Of potentially greater importance than such sector-specific policies are policies that reflect and promote a more general reorientation of social priorities. One recurring idea in the sufficiency and post-growth literature is work-time reduction, which can create time affluence as an alternative to more material affluence – freeing up time for a less stressful, more convivial life – while also helping maintain employment and economic balance in a low-, non-, or de-growing economy (Lange, this volume; Larsson and Nässén, this volume; see also Hayden 1999; Schor 2001; Alexander 2016; Gough 2017; Kallis 2018). Moving beyond GDP as the dominant prosperity measure is a key part of a transition towards emphasising sustainable well-being rather than ever-expanding economic output (Philipsen, this volume; see also Alexander 2016; Hayden and Wilson 2017). Establishing caps that limit overall resource use and emissions to sustainable levels is another core idea in the sufficiency toolkit – one that could be particularly important in limiting rebound effects from efficiency gains (Alexander 2016; Büchs and Koch 2017; Alcott 2018). Distributive policies are also needed to ensure sufficiency both in terms of adequate minimum consumption and avoiding excess, with debate ongoing over the merits of options such as a basic income, minimum and maximum income, greater equity in ownership of economic assets, and "consumption corridors" (Fuchs, this volume; see also Büchs and Koch 2017; Gough 2017).

## Inroads

While the previous section showed that many sufficiency-oriented policy options are available, for the most part, political leaders and parties – even many Green Parties – shy away from talk of sufficiency, especially macro-level critique of a growth economy. That said, exceptions exist, such as Denmark's Alternative Party, whose leader has proclaimed that "Denmark is wealthy enough" (Elbæk 2016), and the UK Greens, who call for "a shift away from an obsession with economic growth" and policies such as a four-day workweek (Bartley and Berry 2018). As noted earlier, there is some embryonic cultural change to build on, as well as some evidence of widespread belief in the need for sufficiency. For example, surveys of Americans have found large majorities support the idea that the "country would be better off if we all consumed less" (Bowerman 2014).

Many jurisdictions have in fact implemented sufficiency-oriented policies of various kinds, including actions to reduce the volume of plastic bags and bottled water sales, policies to promote local food (i.e. reduce "food miles") and cut food waste (resulting in less food purchased), energy conservation policies that include both efficiency and sufficiency (and blur the lines between them), and policies to discourage purchases of large, gas-guzzling vehicles (which can also be framed in terms of greater efficiency), (Hayden 2014a, 2014c). Outdoor advertising has been severely restricted in cities as diverse as Thimphu, Bhutan; Grenoble, France; and Sao Paulo, Brazil. Sweden has sought to counter "throwaway consumer culture"– and create local jobs – by introducing tax breaks (lower value added tax, VAT) for repairs to items such as bicycles and clothes, and tax credits for repairs to bigger items such as fridges and washing machines (Starritt 2016). Meanwhile in Singapore, rising traffic congestion, lack of land to expand the road network, and a commitment to strengthen public transit led to a decision to stop increasing total vehicle numbers in 2018, meaning that for every new vehicle another must be de-registered (LTA 2017). The beyond-GDP movement has also made important steps forward in challenging the idea that more output necessarily equals greater well-being (Philipsen, this volume).

Despite such examples, sufficiency is still far from being an overarching organising principle for society, and it tends to have more supporters away from the centres of power than it does within those centres. One important task is continuing the academic work to show a viable economic path forward in a society emphasising sufficiency rather than endless expansion (Lange, this volume; Jackson 2017; Victor 2019) – along with developing a new social narrative around that vision. Winning broad public and political support will also depend, in part, on the degree to which sufficiency can be linked to high levels of well-being.

## Well-being and sufficiency: opportunities and challenges

As noted earlier, the predominant view among proponents of sufficiency is that well-being need not decline, and could even increase, in a post-growth economy with less emphasis on consumption. Possibilities exist to "live well and well within our means" (Princen 2010, 2), for "liberation from excess" (Paech 2012), and a new type of consumer freedom – the freedom not to buy (Lorek 2018), whether through individual choices not to bother keeping up with the Joneses or collective actions, such as provision of public infrastructure and sharing systems, that reduce the need for goods such as automobiles. Harris (2013, 173, 180) argues that since the "real sources of human happiness," i.e. social relations and quality leisure time, are undermined in consumer societies, a shift from consumerism to sufficiency does not require sacrifice; to the contrary, "what is needed is to encourage people to behave in ways that will enhance their interests."

A society less focused on getting, spending, and consuming would enable people to spend less time in paid work (Larsson and Nässén, this volume), freeing up time for family, friends, and freely chosen activities – or alternatively it would enable the choice of more meaningful and satisfying, but less remunerative work. Many sufficiency-oriented consumers and downshifters – or those pursuing what Juliet Schor (2010) calls plenitude or true wealth – are seeking precisely such changes, along with a stronger sense of belonging and social connection. Indeed, human conviviality need not require large quantities of material resources – offering opportunities for sustainable well-being. Meanwhile, reduced pressures to be constantly productive and keep up in the rat race could help address exhaustion and burnout. Less competitive striving could be supported by a commitment to greater social equality, which is a key element of most formulations of sufficiency and has been shown to have numerous well-being and societal benefits (Wilkinson, Pickett, and Reich 2010). Overall, the pleasures of a post-consumerist life – and reduced stress, congestion, and pollution – offer possibilities for an "alternative hedonism" (Soper 2008).

A "win–win" message of this kind is important politically to build support for a sufficiency vision (Hayden 2014c; Büchs and Koch 2017, 2). In a context where growing numbers of people globally have become accustomed to unsustainably high consumption levels, chances of success are higher to the extent that an appealing alternative can be offered. As Alexander (2019) points out, "calls for sufficiency will surely speak to a broader audience if people come to see that their own lives may actually be enriched by avoiding superfluous consumption...."

That said, it is also important to acknowledge the major challenges that exist in achieving such a vision within planetary limits (Büchs and Koch 2017). While high well-being levels need not require the consumption levels of high-income nations, let alone continued income and consumption growth in such societies, it is unclear whether they are compatible with deep reductions in consumption that – given existing technologies and provisioning systems – would be necessary to meet sustainability goals.

One challenge is supporting workers whose jobs are threatened by reduced consumption. A full discussion of that challenge is beyond the scope of this chapter, but one part of a just transition is policies that provide income support and training to enable workers to move from contracting sectors to others that can expand within environmental limits – such as renewable energy or caring occupations. Also important in this context are policies, mentioned earlier, to reduce work time (and thereby generate more employment from any given level of production) and to ensure more equitable distribution and greater economic security by establishing a minimum floor for income and consumption and more equitable asset ownership.

Another major challenge is that consumption levels that many people in high-income nations consider to be enough for a good life are, at present, beyond sustainable levels. One study found that if the entire UK population lived on a budget deemed sufficient for a "decent life," British GHG emissions would decline 37%. Although such reductions are significant, even this minimum consumption standard would generate emissions well above sustainable, fair-share levels, leaving a long way to go to the near 100% reductions needed by mid-century (Druckman and Jackson 2010; see also Gough 2017, 155). Meanwhile, "developing" countries above the poorest level also consume beyond their ecological means – raising questions about whether sustainable consumption can rise much above basic needs (Fritz and Koch 2016; Büchs and Koch 2017, 106). Some middle-income countries are much more efficient than high-income nations in turning resource consumption into high life satisfaction, notably Costa Rica, which tops the "Happy Planet Index" (Jeffrey, Wheatley, and Abdallah 2016). However, even this example, which offers some lessons in turning per-capita incomes one-fifth of US levels into high human development, comes with an ecological foot-print more than 50% above estimated sustainable levels (GFN 2019). Indeed, a recent study found that no country currently performs well in reaching the social thresholds necessary for a good life while simultaneously respecting the earth's biophysical limits – and countries with higher life satisfaction tend to cross more biophysical boundaries (O'Neill et al. 2018).

This evidence highlights the scale of the challenge in enabling good lives for a growing population (Coole, this volume) within planetary boundaries. To meet that challenge, a greater role for the idea of sufficiency is urgently needed, but so too, despite their limits, are greater efficiency and eco-technologies to enable greener, decarbonised production. As Gough (2017, 156) writes, "To make consumption sustainable within *existing* socio-technical structures would de-prive citizens of a vast range of goods and services they have agreed are necessary for effective participation in modern life."[2] O'Neill et al. (2018, 5–6) similarly conclude that two broad strat-egies are needed, the first being a "focus on achieving 'sufficiency' in resource consumption." In addition, improved provisioning systems are needed to enable good lives for all within planetary boundaries; this includes a role for greener technologies and resource efficiency, alongside social changes such as reducing inequality and greater reliance on non-material means to pursue qual-itative goals such as life satisfaction. As Alexander (this volume) points out, "Efficiency without sufficiency is lost." At the same time, sufficiency also needs help.

## Conclusion

Long marginalised by the dominant approaches to environmental reform, the idea of suf-ficiency has much to offer sustainability governance. This chapter began by examining sufficiency's meaning in the context of sustainability governance, before considering the substantial obstacles to the idea of enough within modern consumerist, capitalist societies built on the pursuit of more. Despite those obstacles, there are a number or reasons why suffi-ciency is increasingly relevant, starting with the urgency of much stronger action in response to ecological challenges. Sufficiency has the potential to address some key limits of "green growth" and its excessive technological optimism, notably the tendency of growing pro-duction and consumption volumes to undo the benefits of eco-efficiency improvements. By reducing consumption levels and resource demands, sufficiency can also substantially reduce the challenge involved in scaling up green(er) technologies to meet those demands. It is a key concept for achieving global justice within planetary limits, highlighting not only the need to reduce excess consumption, but also to ensure that all have enough to meet core needs. Sufficiency comes with its own, alternative visions of the good life and draws attention to the possibilities of achieving well-being in less materially intensive ways. Given its potential contribution to more effective governance of ecological problems, sufficiency deserves to be a core organising principle of contemporary societies.

In addition to reasons for an embrace of sufficiency, the chapter has considered social forces that might drive a greater role for sufficiency-oriented ideas in sustainability gover-nance. These include cultural changes and new consumer sensibilities – which have pro-duced initial signs that some forms of high-impact consumption may be peaking in affluent societies – along with sufficiency-oriented social innovation at the grassroots level. A possible long-term slowdown of economic growth in high-income nations would itself limit con-sumption levels and necessitate new approaches to meet needs and create well-being. To these potential drivers, one can add the efforts of environmental and social activists, as well as researchers, to raise awareness of sufficiency's importance and growing acknowledgement of the ecological imperative to find new ways to live within planetary boundaries.

One key argument of the chapter is that any evolution towards sufficiency, to the extent that it is happening, is occurring too slowly and policy measures are needed to accelerate it. Indeed, there are many options for applying the idea of sufficiency to policy, as well as examples of policy inroads to build on. That said, sufficiency still has a long way to go to being accepted as an overarching organising principle for society. Further advances for sufficiency in the political sphere will depend, in part, on the degree to which it can be linked to high levels of well-being, for which there are both opportunities and significant challenges.

The scale of ecological challenges is such that sufficiency alone is not sufficient – also needed are efforts to improve eco-efficiency and shift to greener technologies, among other changes to provisioning systems and economic structures. That said, one possible source of optimism in the face of these challenges is that the potential of sufficiency, a core component of a balanced ecological strategy, has still barely been tapped.

## Notes

1 For example, if the impact in question is GHG emissions that contribute to climate change, then IPAT could take the following form: GHG emissions = population x per-capita income (i.e. GDP/population) x GHG intensity (i.e. GHG emissions per unit of GDP). For more detail on the IPAT equation, see Alexander and Rutherford (this volume).
2 An alternative to taking into account what people say is sufficient for a good life would be to determine levels of allowable consumption based on scientific assessments of what is ecologically sustainable – possibly accompanied by a narrative that sacrifice is necessary in the short term for the greater long term good. Indeed, such sacrifices have been common in war-time and may again be called for in light of a growing sense of "climate emergency." The potential for sacrifice to achieve environmental goals may be greater than widely believed (Maniates and Meyer 2010), although the political challenges of gaining support for sacrifice are greater than cases where limiting consumption can be linked to maintaining or even improving levels of well-being.

## References

Alcott, Blake. 2008. 'The Sufficiency Strategy: Would Rich-World Frugality Lower Environmental Impact?' *Ecological Economics* 64 (4): 770–86.
———. 2018. 'Environmental Caps as a Solution to Rebound Effects'. In *Sufficiency: Moving beyond the Gospel of Eco-Efficiency*, edited by Leida Rijnhout and Riccardo Mastini, 9–13. Brussels: Friends of the Earth Europe.
Alexander, Samuel. 2015. *Sufficiency Economy: Enough, For Everyone, Forever*. Melbourne: Simplicity Institute.
———. 2016. 'Policies for a Post-Growth Economy'. MSSI Issues Paper No. 6. Melbourne: Melbourne Sustainable Society Institute.
———. 2019. 'What Would a Sufficiency Economy Look Like?' In *Just Enough*, edited by Matthew Ingleby and Samuel Randalls. 117–134. London: Palgrave Macmillan.
Anderssen, Erin. 2018. 'Seniors Have Too Much House. Millennials Have None. And a Business Model Is Born'. *The Globe and Mail*, 19 July 2018. www.theglobeandmail.com/canada/article-seniors-have-too-much-house-millennials-have-none-and-a-business/.
Ariès, Paul. 2010. *De la décroissance à la gratuité: Moins de biens, plus de liens*. Villeurbanne, France: Golias.
Association Négawatt. 2017. *Scénario NégaWatt 2017–2050 : Dossier de Synthèse*. Alixan, France: Association Négawatt.
———. 2018. *La Sobriété Énergétique: Pour Une Société plus Juste et plus Durable*. Alixan, France: Association Négawatt.
Barrington-Leigh, Christopher. 2017. 'Sustainability and Well-Being: A Happy Synergy'. *Great Transition Initiative*, April.
Bartley, Jonathan, and Siân Berry. 2018. 'Our Green Party Will Be Bold and Brave in Both Ideas and Actions'. *The Guardian*, 5 September. www.theguardian.com/commentisfree/2018/sep/05/green-party-bold-brave-ideas-britain.

Belk, Russell. 2014. 'You Are What You Can Access: Sharing and Collaborative Consumption On-line'. *Journal of Business Research* 67 (8): 1595–1600.

Bendell, Jem. 2018. 'Deep Adaptation: A Map for Navigating Climate Tragedy'. IFLAS Occasional Paper 2. Carlisle, UK: Institute of Leadership and Sustainability, University of Cumbria.

Bergh, Jeroen C. J. M. van den. 2011. 'Energy Conservation More Effective with Rebound Policy'. *Environmental and Resource Economics* 48 (1): 43–58.

Bowerman, Tom. 2014. 'How Much Is Too Much? A Public Opinion Research Perspective'. *Sustainability: Science, Practice & Policy* 10 (1): 14–28.

Brown, Halina Szejnwald, and Philip J. Vergragt. 2016. 'From Consumerism to Wellbeing: Toward a Cultural Transition?' *Journal of Cleaner Production* 132: 308–17.

Büchs, Milena, and Max Koch. 2017. *Postgrowth and Wellbeing: Challenges to Sustainable Welfare.* Basingstoke, UK: Palgrave Macmillan.

Cohen, Maurie J. 2016. *The Future of Consumer Society: Prospects for Sustainability in the New Economy.* New York: Oxford University Press.

Darby, Sarah, and Tina Fawcett. 2018. *Energy Sufficiency: An Introduction.* Oxford: Environmental Change Institute.

Deaton, Angus. 2008. 'Income, Health, and Well-Being around the World: Evidence from the Gallup World Poll'. *Journal of Economic Perspectives* 22 (2): 53–72.

Dolan, Paul. 2019. *Happy Ever After: Escaping the Myth of the Perfect Life.* London: Allen Lane.

Druckman, Angela, and Tim Jackson. 2010. 'The Bare Necessities: How Much Household Carbon Do We Really Need?' *Ecological Economics* 69 (9): 1794–1804.

Easterlin, Richard A. 1974. 'Does Economic Growth Improve the Human Lot? Some Empirical Evidence', In *Nations and Households in Economic Growth: Essays in Honor of Moses Abramovitz*, edited by Paul A. David and Melvin W. Reder, 89–125. New York: Academic Press.

Easterlin, Richard A., Laura Angelescu McVey, Malgorzata Switek, Onnicha Sawangfa, and Jacqueline Smith Zweig. 2010. 'The Happiness–Income Paradox Revisited'. *Proceedings of the National Academy of Sciences* 107 (52): 22463–68.

Elbæk, Uffe. 2016. 'Denmark Is Rich Enough'. *The Murmur*, 7 September. http://murmur.dk/denmark-is-rich-enough/.

FOEE. 2018. *Sufficiency: Moving beyond the Gospel of Eco-Efficiency.* Brussels: Friends of the Earth Europe.

Frey, Bruno S, and Alois Stutzer. 2002. 'What Can Economists Learn from Happiness Research?' *Journal of Economic Literature* 40 (2): 402–35.

Fritz, Martin, and Max Koch. 2016. 'Economic Development and Prosperity Patterns around the World: Structural Challenges for a Global Steady-State Economy'. *Global Environmental Change* 38 (May): 41–48.

Galbraith, John Kenneth. 1958. 'How Much Should a Country Consume?' In *Perspectives on Conservation: Essays on America's Natural Resources*, edited by Henry Jarrett, 89–99. Baltimore: Johns Hopkins University Press.

GFN. 2019. *National Footprint Accounts, 2019 Edition.* Oakland, CA: Global Footprint Network. http://data.footprintnetwork.org/#/.

Goodwin, Phil, and Kurt Van Dender. 2013. '"Peak Car"—Themes and Issues'. *Transport Reviews* 33 (3): 243–54.

Gordon, Robert J. 2016. *The Rise and Fall of American Growth: The U.S. Standard of Living since the Civil War.* Princeton, NJ: Princeton University Press.

Gough, Ian. 2017. *Heat, Greed and Human Need: Climate Change, Capitalism and Sustainable Wellbeing.* Cheltenham, UK: Edward Elgar.

Grubler, Arnulf, et al. 2018. 'A Low Energy Demand Scenario for Meeting the 1.5°C Target and Sustainable Development Goals without Negative Emission Technologies'. *Nature Energy* 3 (6): 515–27.

Harris, Paul G. 2013. *What's Wrong with Climate Politics and How to Fix It.* Cambridge, UK: Polity.

Hayden, Anders. 1999. *Sharing the Work, Sparing the Planet: Work Time, Consumption, & Ecology.* London: Zed Books.

———. 2014a. 'Enough of That Already: Sufficiency-Based Challenges to High-Carbon Consumption in Canada'. *Environmental Politics* 23 (1): 97–114.

———. 2014b. 'Stopping Heathrow Airport Expansion (For Now): Lessons from a Victory for the Politics of Sufficiency'. *Journal of Environmental Politics and Planning* 16 (4): 539–58.

———. 2014c. *When Green Growth Is Not Enough: Climate Change, Ecological Modernization, and Sufficiency.* Montreal: McGill-Queen's University Press.

Hayden, Anders, and Jeffrey Wilson. 2017. '"Beyond GDP" Indicators: Changing the Economic Narrative for a Post-Consumerist Society?' In *Social Change and the Coming of Post-Consumer Society: Theoretical Advances and Policy Implications*, edited by Maurie J. Cohen, Halina S. Brown, and Philip J. Vergragt, 170–191. New York: Routledge.

IPCC. 2018. *Global Warming of 1.5°C: Summary for Policymakers*. Geneva, Switzerland: Intergovernmental Panel on Climate Change.

Jackson, Tim. 2017. *Prosperity without Growth: Foundations for the Economy of Tomorrow*. 2nd ed. Abingdon, UK: Routledge.

———. 2019. 'The Post-Growth Challenge: Secular Stagnation, Inequality and the Limits to Growth.' *Ecological Economics* 156 (February): 236–46.

Jeffrey, Karen, Hanna Wheatley, and Saamah Abdallah. 2016. *The Happy Planet Index 2016*. London: New Economics Foundation.

Kahneman, Daniel, and Angus Deaton. 2010. 'High Income Improves Evaluation of Life but Not Emotional Well-Being'. *PNAS* 107 (38): 16489–93.

Kallis, Giorgos. 2018. *Degrowth*. Newcastle upon Tyne: Agenda Publishing.

Kasser, Tim. 2003. *The High Price of Materialism*. Cambridge, MA: MIT Press.

Kovel, Joel. 2008. 'Ecosocialism, Global Justice, and Climate Change'. *Capitalism, Nature, Socialism* 19 (2): 4–14.

Layard, Richard. 2005. *Happiness: Lessons from a New Science*. New York: The Penguin Press.

Lorek, Sylvia. 2018. 'Sufficiency in Sustainable Lifestyles'. In *Sufficiency: Moving beyond the Gospel of Eco-Efficiency*, edited by Leida Rijnhout and Riccardo Mastini, 26–29. Brussels: Friends of the Earth Europe.

Lorek, Sylvia, and Doris Fuchs. 2013. 'Strong Sustainable Consumption Governance – Precondition for a Degrowth Path?' *Journal of Cleaner Production* 38 (January): 36–43.

LTA. 2017. 'Certificate of Entitlement Quota for November 2017 to January 2018 and Vehicle Growth Rate from February 2018'. Singapore: Land Transport Authority. www.lta.gov.sg/apps/news/page.aspx?c=2&id=b010406e-6edf-4224-9cd1-928706cd6fe7.

Maniates, Michael, and John M. Meyer. 2010. *The Environmental Politics of Sacrifice*. Cambridge, MA: MIT Press.

McLaren, Duncan. 2003. 'Environmental Space, Equity and the Ecological Debt'. In *Just Sustainabilities: Development in an Unequal World*, edited by Julian Agyeman, Robert D. Bullard, and Bob Evans, 17–37. Cambridge, MA: MIT Press.

Milburn, Joshua Fields, and Ryan Nicodemus. 2011. *Minimalism: Live a Meaningful Life*. Asymmetrical Press.

Muller, Adrian, and Markus Huppenbauer. 2016. 'Sufficiency, Liberal Societies and Environmental Policy in the Face of Planetary Boundaries'. *GAIA* 25 (2): 105–9.

O'Neill, Daniel W., Andrew L. Fanning, William F. Lamb, and Julia K. Steinberger. 2018. 'A Good Life for All within Planetary Boundaries'. *Nature Sustainability* 1 (2): 88–95.

Paech, Niko. 2012. *Liberation from Excess: The Road to a Post-Growth Economy*. Munich: Oekom.

Princen, Thomas. 2005. *The Logic of Sufficiency*. Cambridge, MA: MIT Press.

———. 2010. *Treading Softly: Paths to Ecological Order*. Cambridge, MA: MIT Press.

Raworth, Kate. 2017. *Doughnut Economics: Seven Ways to Think Like a 21st-Century Economist*. London: Random House.

Sachs, Wolfgang. 1993. 'Die Vier E's: Merkposten Für Einen Maß -Vollen Wirtschaftsstil'. *Politische Ökologie* 11 (33): 69–72.

———. 2001. 'Development Patterns in the North and Their Implications for Climate Change'. *International Journal of Global Environmental Issues* 1 (2): 150–62.

———. 2017. 'The Sustainable Development Goals and Laudato Si': Varieties of Post-Development?' *Third World Quarterly* 38 (12): 2573–87.

Sachs, Wolfgang, and Tilman Santarius. 2007. *Fair Future: Resource Conflicts, Security and Global Justice*. London: Zed Books.

Salleh, Ariel, ed. 2009. *Eco-Sufficiency & Global Justice: Women Write Political Ecology*. London: Pluto Press.

Samadi, Sascha, Marie-Christine Gröne, Uwe Schneidewind, Hans-Jochen Luhmann, Johannes Venjakob, and Benjamin Best. 2017. 'Sufficiency in Energy Scenario Studies: Taking the Potential Benefits of Lifestyle Changes into Account'. *Technological Forecasting and Social Change* 124 (November): 126–34.

Santarius, Tilman. 2012. *Green Growth Unravelled: How Rebound Effects Baffle Sustainability Targets When the Economy Keeps Growing*. Berlin: Heinrich Böll Foundation.

Schneidewind, Uwe, and Angelika Zahrnt. 2014. *The Politics of Sufficiency : Making It Easier to Live the Good Life*. Munich: Oekom.

Schor, Juliet B. 2001. 'The Triple Imperative: Global Ecology, Poverty and Worktime Reduction'. *Berkeley Journal of Sociology* 45: 2–17.

———. 2010. *Plenitude: The New Economics of True Wealth*. New York: The Penguin Press.

Soper, Kate. 2008. 'Alternative Hedonism, Cultural Theory and the Role of Aesthetic Revisioning'. *Cultural Studies* 22 (5): 567–87.

Spengler, Laura. 2016. 'Two Types of "Enough": Sufficiency as Minimum and Maximum'. *Environmental Politics* 5 (5): 921–40.

Spiller, Achim, and Sina Nitzko. 2015. 'Peak Meat: The Role of Meat in Sustainable Consumption'. In *Handbook of Research on Sustainable Consumption*, edited by Lucia Reisch and John Thøgersen, 192–208. Cheltenham, UK: Edward Elgar.

Starritt, Alexander. 2016. 'Sweden Is Paying People to Fix Their Belongings Instead of Throwing Them Away'. *World Economic Forum*, 27 October 2016. www.weforum.org/agenda/2016/10/sweden-is-tackling-its-throwaway-culture-with-tax-breaks-on-repairs-will-it-work/.

Steffen, Will, Katherine Richardson, Johan Rockström, Sarah E. Cornell, Ingo Fetzer, Elena M. Bennett, Reinette Biggs, et al. 2015. 'Planetary Boundaries: Guiding Human Development on a Changing Planet'. *Science* 347 (6223): 1259855.

Stevenson, Betsey, and Justin Wolfers. 2013. 'Subjective Well-Being and Income: Is There Any Evidence of Satiation?' NBER Working Paper No. 18992. National Bureau of Economic Research. www.nber.org/papers/w18992.

Summers, Lawrence H. 2016. 'The Age of Secular Stagnation'. *Foreign Affairs*, March/April. www.foreignaffairs.com/articles/united-states/2016-02-15/age-secular-stagnation.

Thompson, Craig J., and Juliet B. Schor. 2014. 'Cooperative Networks, Participatory Markets, and Rhizomatic Resistance: Situating Plenitude within Contemporary Political Economy Debates'. In *Sustainable Lifestyles and the Quest for Plenitude: Case Studies of the New Economy*, edited by Juliet B. Schor and Craig J. Thompson, 233–49. New Haven: Yale University Press.

Victor, Peter A. 2019. *Managing Without Growth: Slower by Design, Not Disaster*. 2nd ed. Cheltenham, UK: Edward Elgar.

Wachsmuth, Jakob, and Vicki Duscha. 2018. 'Achievability of the Paris Targets in the EU—the Role of Demand-Side-Driven Mitigation in Different Types of Scenarios'. *Energy Efficiency*, May. doi: 10.1007/s12053-018-9670-4.

Wilkinson, Richard, Kate Pickett, and Robert B. Reich. 2010. *The Spirit Level: Why Greater Equality Makes Societies Stronger*. London: Penguin.

# Part 3

# Key challenges

# 13

# North-South inequity and global environmental governance

*Chukwumerije Okereke*

## Introduction

In this chapter, I review the critical challenges in the governance of global sustainability that are associated with North-South inequity. I argue that these are probably the most prominent and acute sets of problems that have faced the global governance of environmental change. The chapter demonstrates that from the 1972 United Nations Conference on the Human Environment (UNCHE) – when sustainability was first introduced as a focus for international governance – to the present day, every significant environmental summit, multilateral agreement, or global environmental institution has been severely challenged by issues related to North-South inequity and justice. The chapter will outline some of the core issues and describe the impact these have had on mainstreaming global environmental sustainability governance, with the primary focus on key global environmental summits and the multilateral agreements they have created.

In the end the chapter puts forward an argument for the urgent need to redress imbalances between the developed countries of the North and the developing countries of the South in relation to global environmental challenges. It surveys how governance in relation to these challenges has developed, including the ways in which that development has privileged the North's philosophies and priorities at the expense of those of developing countries. I argue that this structural imbalance at the roots of global environmental governance has now established radical inequalities between North and South which are perpetuating and exacerbating global environmental challenges. I go on to set out the ways in which these imbalances can be redressed to allow the South to bring its own approaches into the governance and management of the global environment at this crucial time. I make it clear that the North will need to cede ground and change the structures of governance if progress is to be made, both in equality with the South, and in solving our shared global environmental and development problems.

The focus on global environmental summits and agreements is not intended to suggest that other areas of global sustainability governance, such as the activities of international development agencies (e.g. the World Bank), transnational actors, and private enterprises (e.g. multinational corporations), are not crucial sites for North-South justice contestations. I focus on the international regime because this remains the central domain for global sustainability governance and the leading forum for the articulation of North-South equity issues/debates.

Section "Introduction" has offered an introduction and framework and set out my key arguments. In section "A brief history of global environmental governance," I provide a brief history of global environmental governance. Section "Defining core concepts" provides definitions of some of the key concepts. The bulk of the chapter is section "The four core issues," the discussion of core North-South inequity issues. This is followed in section "Summary and conclusion" by a short discussion and conclusion.

## A brief history of global environmental governance

The 1972 UNCHE, held in Stockholm, Sweden (the Stockholm Conference) is widely accepted as the birthdate of global environmental governance (cf. Elliott, 2004; Chasek, 2001). It was the first major global summit organised by the United Nations to promote international cooperation for addressing the world's environmental challenges. The Stockholm Conference produced a Declaration on the Human Environment containing 26 broad principles including an affirmation that every human has the right to enjoy a clean and healthy environment but also that humankind "bears a solemn responsibility to protect and improve the environment for present and future generations" (Principle 1). The Conference also produced a practical action plan containing 109 recommendations for achieving global sustainability, many of which covered proposals on how governments, intergovernmental agencies, and non-governmental organisations (NGOs) could work together to implement environmental protection strategies. The conference also created World Environment Day and resulted in the establishment of the United Nations Environment Programme (UNEP) which is headquartered in Nairobi, Kenya.

The principles and action plan agreed in the Stockholm Conference have inspired every subsequent international conference on the environment to date (Chasek, Downie, and Brown, 1999). It provided the momentum for many states to sign the Convention on International Trade in Endangered Species of Wild Fauna and Flora (CITES) in 1973 and to establish national ministries for the environment (Von Weizsäcker, 1994). Internationally the agenda led to the creation of the World Commission on Environment and Development (WCED) which subsequently popularised the concept of sustainable development through the Brundtland Report in 1987. Another landmark event was the United Nations Conference on Environment and Development (UNCED), otherwise known as the Earth Summit, which was held in Rio de Janeiro, Brazil on 4–10 June 1992. The Rio summit drew widely on the WCED report and process. The UNCED is notable because it was attended by an unprecedented number of heads of states and governments (172), NGOs, and media organisations. The UNCED produced three legally binding international environmental agreements, including the United Nations Framework Convention on Climate Change (UNFCCC), the United Nations Convention on Biological Diversity (UNCBD), and the United Nations Convention to Combat Desertification (UNCCD). Three other important documents were also produced, including the Rio Forest Principles and the Rio Declaration on Environment and Development and Agenda 21, an action plan for environmental management at national and local levels. The UNCED led to the establishment of a UN Commission on Sustainable Development (CSD) which functioned until 2013 when it was replaced by the High-level Political Forum on Sustainable Development.

Other major environmental summits are the World Summit on Sustainable Development (2002) and The United Nations Conference on Sustainable Development (Rio+20) which took place in Rio de Janeiro, Brazil on 20–22 June 2012. In addition to these summits and the resulting international agreements, global sustainability governance is implicated in a diverse number of other global initiatives. Notable examples include the Sustainable Development Goals and its precursor, the Millennium Development Goals, the World Trade Organization (WTO), the Food and Agricultural Organization (FAO), the United Nations Development Programme (UNDP), the World Health Organization (WHO), the International Labour Organization (ILO), and the International Strategy for Disaster Reduction (ISDR).

## Defining core concepts

Inequity in general terms refers to lack of fairness or justice in any given setting, including laws, systems, or processes (see also Kalfagianni et al. this volume). In the context of global environmental governance, North-South inequity refers to the absence or perceived lack of fairness felt by developed and developing nations over the cause of global environmental degradation as well as about the policies and processes involved in the governance of global environmental sustainability (Ikeme 2003; Okereke, 2010). The opposite of inequity would be a condition of environmental justice defined, in our case, as the fair involvement of developed and developing countries in the development and implementation of environmental institutions and policies, and the equitable distribution of global resources and of the benefits and responsibilities arising from interstate relations (Agyeman, Bullard and Evans, 2003; Okereke, 2007, 2010). In this chapter, I use the term justice and equity interchangeably.

The term "North-South divide" is used here as a broad classification that refers to developed and developing countries. Generally, the global South is made up of Africa, Latin America, and developing Asia, including the Middle East, while the North comprises the more affluent and industrialised countries. Nearly all of the developing nations belong to the Group of 77 (G77), a group formed in 1964 to increase their voice in trade negotiations. China participates in the G77 but does not consider itself to be a member. The North is home to all the members of the G8 and four of the five permanent members of the United Nations Security Council.

The economic success of some countries in the South means that the traditional divide between the North and the South corresponds less and less to reality and is increasingly challenged (Dimitrov, 2016). Important issues have emerged that threaten to fracture the Southern coalition (Falkner, 2016). However, the G77 and China continue to negotiate as a bloc, and despite notable internal differences, the general sense is often that what unites those in the South is more than what divides them (Okereke and Coventry, 2016).

It is helpful to note here that while the clamour for more equity in global environmental governance has emanated mostly from the South, with some Northern voices sympathetic to their cause, the demand for North-South equity is by no means the exclusive preserve of developing countries. It is instructive to note, for example, that the United States has made a perceived lack of equity a major argument in rejecting any binding treaty on climate change (Okereke, 2008). Similarly, inequity expressed by the absence of binding emission reduction targets for the rapidly developing countries was cited as the main reason why the United States refused to ratify the Kyoto Protocol. More recently the President of the United States, Donald Trump, has made repeated references to fairness as the reason for pulling the United States out of the Paris Climate Agreement (Tollefson, 2017).

## The four core issues

The multiple challenges posed to global environmental governance by North-South inequity can be organised around four central questions:

a    What is the problem to be solved?
b    Who has caused (and should therefore solve) the problem?

c    What tools and solutions are best utilised in addressing the challenge?
d    What is the process for making fair decisions?

It is around these core questions that the rest of this chapter will be organised. Here I examine them in turn.

## What is the problem to be solved?

One of the most pronounced challenges associated with North-South inequity when it comes to the governance of global environmental change is the wide divergence between the developed and developing countries in the framing or conceptualisation of the problem that needs to be solved (Mebratu, 1998). This divergence was prominently highlighted in the run-up to, as well as during, the 1972 UNCHE (the Stockholm Conference), which is widely accepted as the birthdate of global environmental governance.

In convening the Stockholm Conference, developed countries had framed the issue in terms of the imperative of international cooperation to address global environmental problems including transboundary air pollution, soil and water pollution, the risks associated with nuclear waste, and the increasing levels of global deforestation and loss of biological diversity. These were of course the types of environmental problems that were confronting the developed countries. The view was that the prevailing global environmental change facing the world had arisen chiefly from large-scale industrial development that needed to be controlled. Another essential subtext was the view that the rising global population (mostly in the developing countries) was a significant factor contributing to global environmental change, necessitating some form of state-sponsored global population control (see Coole this volume). The view that global population needed to be urgently controlled was central in the two leading publications that provided the intellectual foundation for the 1972 UN Conference. The first was a book entitled *The Population Bomb* by Stanford University Professor, Paul Ehrlich. The other was "The Limits to Growth" commissioned by the Club of Rome and published by Donella H. Meadows and her team from MIT.

In *The Population Bomb*, which sold more than two million copies in the first two years of publication, Ehrlich declared that population growth had already outstripped resources on the planet. He proclaimed that in the 1970s the world would undergo famines and that starvation will claim the lives of hundreds of millions of people regardless of any crash programmes that the world might decide to embark on. The only chance for saving the planet, he insisted, was in "determined and successful efforts at population control." He argued that "the battle to feed humanity was over" and advocated "the promotion of contraceptive technology, small-family incentives, and international guidance on population planning."

On their part, Meadows and her team used computer model simulations to predict the environmental impact of increasing population and economic growth globally. They found that populations were growing exponentially under the condition of finite resources and predicted that industrial and ecological collapse were imminent unless world governments intervened to design a "global equilibrium" which involved the careful management of industrial activity coupled with population control at the global level (cf. Conca, Alberty, and Dabelko, 1995, p. 20).

However, while some prominent voices within the developed countries conceived of global environmental change as something that required that both industrialisation and population be brought under control, developing countries, for their part, thought that the most pressing problems they were facing were those that could best be addressed with more, rather than less industrialisation. Furthermore, many developing countries were vehemently opposed to the idea of state-sponsored population control which they considered to be colonial and patronising. Developing countries were of the view that overconsumption and industrial-related pollution, for which the West was largely responsible, were by far the bigger problems and were the ones that required attention.

In the highly influential Founex Report (Founex Report, 1971, n.p.) released just before the commencement of the UN Conference by a group of the developing countries, the South provided a detailed analysis of how the differences in the levels of development and the everyday realities of the two groups of countries had resulted in a divergent conceptualisation of the problem. In no uncertain terms, they stressed that the core problems faced by the majority of the world population were poverty, unemployment, and lack of access to water, food, medical care, and decent housing. These challenges, they stressed, were best tackled through development. The impetus, they declared, should be towards more, not less, development.

Of course, developing countries recognised that some of the problems associated with industrialisation were already beginning to appear in the South and that these would require attention and careful management. They also noted the need to move the definition of development away from the narrow measurement of output in gross domestic product (GDP) and towards a broader understanding that covered social and environmental integrity. Yet, in unmistakable terms, they endorsed accelerated development as the solution to their problem. The disagreement that followed from this divergent conceptualisation of the problem was so acrimonious that the South eventually accused the North of contriving environmental sustainability as a means to impede or hold back the development of the South. The report says:

> It is evident that, in large measure, the kind of environmental problems that are of importance in developing countries are those that can be overcome by the process of development itself. In advanced countries, it is appropriate to view development as a cause of environmental problems. [...]. But, for the greater part, developing countries must view the relationship between development and environment in a different perspective. In their context, development becomes essentially a cure for their major environmental problems. For these reasons, concern for the environment must not and need not detract from the commitment of the world community -- developing and more industrialized nations alike -- to the overriding task of development of the developing regions of the world.
>
> *(Founex Report, 1971, n.p.)*

Continuing, the report went on to declare that, for the South, the quest for environmental sustainability:

> Underscores the need not only for a maximum commitment to the goals and targets of the Second Development Decade but also for their redefinition in order to attack that dire poverty which is the most important aspect of the problems which afflict the environment of the majority of mankind.
>
> *(Founex Report, 1971, n.p.)*

A look at the principles eventually formulated in Stockholm reveals several tensions caused by this divergence in the conception of the problem to be solved. For example, Paragraph 4 declared that the environmental problems in developing countries were the sort that could be addressed by more development, while at the same time urging developed countries to close the gap between themselves and the developing countries by scaling down their own industrial growth. Paragraph 5 suggested that population was a problem that needed to be addressed but quickly pivoted to say that population growth should be celebrated because humans are the most precious things on earth and that they have the power and creativity to transform their environment in positive ways:

> The natural growth of population continuously presents problems for the preservation of the environment, and adequate policies and measures should be adopted, as appropriate, to face these problems. Of all things in the world, people are the most precious. It is the people that propel social progress, create social wealth, develop science and technology and, through their hard work, continuously transform the human environment. Along with social progress and the advance of production, science, and technology, the capability of man to improve the environment increases with each passing day.
>
> (Stockholm Principles, 1972, n.p.)

A divergence in the understanding of how to define the problem to be solved under the rubric of global sustainability governance has remained a prominent feature of international cooperation for sustainability ever since. When the Brundtland Commission embarked on the process of drafting their report, they travelled the world's continents and took testimonies from thousands around the world. In almost every developing country they visited the message they were given was broadly the same: that people needed international governance arrangements to focus attention on everyday issues of poverty far more than they needed conservation, pollution control, and the other similar "exotic" environmental issues that tended to occupy the industrialised countries of the North. It was the force of these testimonies that led the Brundtland Commission to declare in their report that "poverty and inequality was one of the most prominent environmental issues" (WCED, 1987, 3). The report went on to declare that "it was futile to attempt to solve the global environmental problem without approaches that encompass addressing global poverty and inequality" (WCED, 1987, 3).

However, while the Brundtland Report highlighted the importance of framing the global environmental problem in ways that recognised the need for poverty eradication and a common future for both the South and North, it also questioned the sustainability of prevailing industrialised economic development practices suggesting that it "needed to change radically both in quality and quantity especially in developed countries" (WCED, 1987, 213). At the same time, some scholars (Lohman, 1990; Lélé, 1991; Redclift, 2002) have criticised the Brundtland Report for promoting wide-scale industrialisation as the solution to global environmental crisis and for not going far enough in their defence of community-based resource management by poor rural communities. We can certainly say that the report attempts to paint a homogenised picture of a common future for North and South that papers over important divides caused by longstanding issues of unfairness and inequality.

The tension in how to define the problem was also very palpable in the run-up to the Rio Earth Summit. Initially, the North wanted to keep the focus on issues of environmental degradation with links to development. However, the South insisted on making a stronger link between environmental protection and development. It was at the insistence of developing countries that the name of the conference was changed from the UNCHE II (Stockholm II) to the UNCED. For the South, this change in title was needed to underscore the fact that emphasis of the summit should be as much on development as it was on environmental protection. In his book *Earth Politics*, Von Weizsacker (1994) captured this divergence well.

He notes that while the North saw the Earth Summit as "devoted to the environment and its links to development," the dominant perspective of the South was that Summit was "devoted to global inequalities, development and their links to the environment" (p. 3).

During the Summit itself, a major source of controversy was the determination of the South to include statements that made development a right and the opposition of some developed countries (led by the United States) to such declarations. In the end the South prevailed, with Principle 1 of UNCED stating that "Human beings are at the center of concerns for sustainable development. They are entitled to a healthy and productive life in harmony with nature" and Principle 3 noting that "The right to development must be fulfilled so as to equitably meet developmental and environmental needs of present and future generations."

These sorts of differences in framing the problem are also abundantly manifest, and indeed a defining feature of present day global environmental treaties. In the case of climate change, for example, the North has tended to emphasise reduction of greenhouse gas emissions, while the poor South has been fighting for its own fair share of the atmospheric space to be able to pursue its development aspirations. There is also a focus on the adaptation considered necessary to limit the negative impact of climate change and protect their nascent economic development. Moreover, there have been some in the South who have seen climate change as an opportunity to achieve greater global equity, including a massive redistribution of wealth from industrialised counties to poor developing nations.

## Who has caused (and should therefore solve) the problem?

Another major flash point in the North-South relationship in the context of global environmental governance relates to the allocation of responsibility for the causes of global environmental change and, by implication, for solving the problem (see also Pelizzoni this volume). Developing countries assert that much of the global environmental challenges the world is facing have been caused by the industrialisation process in the North and by overconsumption of global resources by Northern countries. The developed countries are responsible for the emission of much of the chlorofluorocarbon (CFC) gases that burnt a massive hole in the ozone layer and a disproportionate amount of historic greenhouse gas emissions that are causing global climate change. Similarly, the North, which comprises only about 25% of the global population, accounts for the consumption of over 75% of global resources, including oil, precious metals, timber, paper, cement, and marine resources (Wackernagel and Rees, 1998). Developing countries also argue that much of the wealth acquired by the North has been achieved through decades of over-exploitation of commonly owned global resources and through dumping of wastes in global sinks. On this basis, developing countries argue that the industrialised nations who are the main cause of the problem should be fully responsible for solving the problem. They contend that it is grossly unfair for the North to impose an extra burden on the poor South to solve a problem for which the North is responsible and from which it has immensely benefited.

Developed countries do often acknowledge that they contributed more to, and benefited disproportionately from, the process that has caused global environmental change. They often acknowledge the need for equity considerations to guide international cooperation for global environmental governance. However, they also make some rebuttals to developing countries' arguments. First, they note that much of the natural resource exploitation and pollution by the North occurred when the harmful effects of such processes were not well established. Then, they argue that it is unfair to hold the present generation responsible for the ignorance of the past and the mistakes made by previous generations (for the subject of intergenerational justice, see Lawrence in this volume). Doing so, they argue, would be akin to punishing a child for the sins of the father. They go on to point out that developed countries never actively excluded the poor countries of the South from the use of global common resources. In other words, it is not the fault of the North that developing countries are, so to speak, latecomers to the party. They also point out that the process of industrialisation has resulted in scientific inventions and discoveries which have been of great benefit to developing countries, arguing that poor countries owe many of the improvements day-to-day life (water quality, reduction of child mortality rates, increase in life expectancy, access to electricity and medical care, etc.) to scientific innovations made on the back of industrialisation.

The North concludes that if they are not the only beneficiaries of industrialisation, nor should they be wholly responsible for tackling the problems associated with the process. Crucially, they argue that the economies of many countries in the South are growing rapidly, with the result that these countries are contributing significantly to global pollution. It would be unfair, the North argues, to focus exclusively on historical sources of the problem while ignoring current major causes of pollution and the consumption of global resources. They note especially that China is now the largest world emitter of greenhouse gases and other rapidly developing countries such as Brazil, Russia, India, and South Africa are major contributors to climate change. Finally, the North argues that the urgency and scale of the problem mean that the global community must move away from the blame game and adopt a concerted approach where all countries commit to doing whatever they can to address the problem. In stressing the need for collective action by all countries, some are quick to point out that it is poor countries that stand to suffer more from delayed action.

Years of dialogue and negotiations between the two groups of countries have resulted in several equity ideas and principles designed to overcome major differences. There is, for example, the acknowledgement that the global ocean and resources beyond the jurisdiction of states are the common heritage of mankind and that maintaining the integrity of global atmosphere and climate is a common concern of all states. There is the "polluter pays" principle, which recognises that the primary obligation for cleaning up pollution rests on those responsible for causing it. It is recognised in several international environmental treaties that environmental regulations and standards enforced in the North may not be applicable in developing countries where such measures might lead to unwanted social costs (Principle 23, UNHCE). The principle of leadership suggests that developed countries that have more responsibility for causing the problem and more capacity for dealing with it will take leadership in solving global environmental change. It is clear that the North will need to transfer "substantial quantities of financial and technological assistance as a supplement to the domestic effort of the developing countries" to address global environmental changes and tackle underdevelopment (Principle 9 UNHCE).

Scholars have often seen the Common but Differentiated Responsibility and Capability (CBDR+C) principle as the broad equity norm under which the earlier equity principle can be gathered. The idea recognises the importance of having all countries contribute their quota to solving the environmental problems, regardless of where they may be in their level of development. At the same time, it acknowledges that developed and developing countries have different levels of responsibility and obligations. However, while this clever principle has helped to enable both groups of countries to maintain cooperation on several issues, including climate change, forest and biological diversity conservation, and the management of ozone layer depletion, important differences remain over how to interpret and operationalise the CBDR in practice. Primarily, the North likes to emphasise the "capability" aspect of the principle, noting that they are willing to take the lead in environmental action and help the South because they have better technology and more financial resources. For their part, the South prefers to place weight on the "responsibility" aspect, arguing that financial and technological assistance from the North should be seen as obligatory rather than as a charitable act. There are also important debates about exactly how much the South needs, who should be responsible for determining the needs of the South, what proportion of the cost is "extra" to normal development spending, and who should decide how resources should be allocated and spent.

## What solutions are most suitable for addressing the challenge?

The disagreements between the North and the South over both the nature of the problem and who bears responsibility spill over to deep disagreements over the suitable means of addressing the challenge.

There are many dimensions to this, but the most important is the relative role of government and markets in solving global environmental problems. Developed and developing countries agree that both the government and the market have important roles to play in solving prevailing environmental challenges. However, developing countries are deeply suspicious of the tendency they see among developed countries to rely too much on the market and the private sector as the chief means of addressing global sustainability challenges.

For example one of the most contentious debates between developed and developing countries in the context of the UN Convention on the Law of the Sea was about how to handle the issue of mining in the deep seabed. Developing countries were of the strong view that an "Enterprise" jointly owned by the international community should be established and licensed to undertake the mining. Proceeds from the mining activity would be distributed among nations, with a significant portion devoted to tackling poverty in the South. Developed countries, led by the United States, favoured the idea of market competition and allowing private companies to apply to undertake mining activities and pay tax to the United Nations. When developing countries insisted on having an international Enterprise, the United States refused to sign the agreement and managed to keep it in limbo for over 20 years until the provision for the creation of the Enterprise was expunged from the treaty document.

In interstate climate negotiations, there is constant disagreement between the North and the South over how much market approaches should be relied on for climate mitigation and adaptation. Emission trading schemes, the Clean Development Mechanism (CDM) and voluntary carbon offsets markets, private sector-led climate insurance schemes, payment for forest conservation and other ecosystems services, and many more market-based solutions to environmental change have generally been promoted and pushed by industrialised countries. With differing degrees of passion, developed countries argue that the market should be deployed as far as possible to tackle global environmental challenges, because the market is an efficient means of value determination and the allocation and reallocation of resources. The United States, especially, like to point to their success in the use of emission trading to control NOx and SOx in the late 1970s and early 1980s and suggest that the world should follow their approach in placing a price on the global atmosphere, and on forests, water, and other vital environmental resources as a way to ensure their conservation.

But developing countries raise several objections to the sort of market approaches trumpeted by the North. Developing countries often point out that the market is rarely ever neutral but, rather, is often shaped by the interaction of social, economic, and political powers. Since the North is the dominant player in the global market place and has the advantage of economic, political, and technological powers, it is highly improbable that market approaches to environmental solutions will be favourable to developing countries. In the case of seabed mining, for example, the South would observe that it is Northern companies who possess the advanced technologies that are likely to win concessions for seabed mining and the concomitant. In the case of climate change, developing countries point out that the market cannot generate the vast amount of resources they need to cater for the additional cost of development caused by climate change. Furthermore, they argue that prevailing economic conditions in the South mean they may be forced to sell their resources at giveaway prices.

Developing countries pointed to the ugly events of the 1980s, when many developing countries accepted tons of hazardous wastes in return for minimal payment from companies based in developed countries. In one case, an Italian national working in Nigeria had obtained a product import licence and then substituted shipments of several thousand tons of highly toxic and radioactive wastes, including 150 tons of polychlorinated biphenyls, which are both carcinogenic and toxic (Vir, 1989). When this underhand was exposed and investigation conducted, it emerged that about 3,800 tons of these wastes had been dumped in a residential neighbourhood in Western Nigerian village called Koko (Atteh, 1993). In another case, in 1988, it was reported that Guinea-Bissau was offered a $600 million-dollar contract – four times its gross national product – to dispose of 15 million tons of toxic wastes over five years (Lipman, 1998). Although the contract was never enforced because of public concern, the perception remains that developing countries provide a disposal option at prices that are often a mere fraction of the equivalent cost in the state of origin. According to one study in the late 1980s, the average disposal costs for one ton of hazardous wastes in Africa were between US $2.50 and $50, while costs in industrialised countries ranged from US $100 to US $2,000 per ton (McKee, 1996). The lower costs generally reflect a lack of environmental standards, less stringent laws, and the absence of public opposition due to lack of information concerning the dangers involved (Liu, 1992).

More broadly, the South has always believed that international cooperation for sustainability should ideally offer the opportunity to address rather than exacerbate the wider issues of structural injustice that have characterised international relations. To this effect, they believe that it is the obligation of developed countries to transfer substantial amounts of finance and technology to the South to enable them not only to cope with the extra demands on their economic development posed by environmental protection but also to catch up with the North. According to this argument, the repeated attempt by the North to rely on market approaches represents at a minimum an abdication of responsibility and in some cases a form of environmental colonialism.

## What is the process for making fair decisions?

The fourth and last challenge posed by North-South inequity in global environmental governance is about how to ensure fair participation of developed and developing countries in global environmental decision-making arrangements and processes. Developing countries have repeatedly expressed concern that prevailing human and technical capacity constraints mean that they are not able to participate effectively in the many different arrangements by which critical decisions about global environmental governance are made. They argue that their limited participation leaves open the possibility for the North to agree policies that not only fail to address the problem but in fact make matters worse for developing countries.

This lack of quality participation raises the possibility that climate policies may be designed in ways that fail to address the interests of the poor countries and therefore exacerbate global inequalities.

Although their influence is not insignificant, it is more or less the case that developing countries are generally rule takers in international fora for environmental governance. There are many decisions and policies in international environmental treaties which developing countries are forced to live with because they do not have the power to effect the kinds of changes they would like to see. Developing countries fought for and lost on the proposal to establish an international Enterprise to undertake seabed mining. Instead, they were eventually forced to accept the United States' proposal for deregulation and private sector operation. Developing countries battled hard but were ultimately unsuccessful in getting a post-Kyoto agreement that preserved their exemption from making quantified emission reduction commitments. Instead, they were coerced to accept an agreement where all countries, rich and poor alike, are required to make nationally determined contributions to climate mitigation, albeit with targets that are non-binding. Developed countries have repeatedly rebuffed attempts by developing countries to demand compensation for damages caused by historical emissions for which developed countries are responsible.

Despite years of demands for adequate and predictable financial aid to help them cope with environmental degradation, developing countries have to live with the pittance offered by developed countries, which is in an ad hoc, incremental, and fragmentary manner. Furthermore, developing countries are made to complete large volumes of paperwork and effectively beg for the money, the allocation of which is decided by boards dominated and controlled by developed countries. Developing countries fought for decades before they could get adaptation to be elevated to almost the same level as mitigation in the climate regime. Developed countries have sometimes even opposed the inclusion of adaptation action in the nationally determined contribution to climate change. Developing countries would rather have forest protection schemes that draw on publicly funded money made available by developed counties who are responsible for consuming over 75% of global forest products. Instead, they have been made to accept a market-based approach to reducing both emissions from forests and land degradation, the funding of which comes from high-pollution sectors such as the aviation industries in the North. The list goes on. Clearly, developing countries are vastly outnumbered in the global conferences where important decisions are made and they are then made to live with countless policies that do not cater for their needs or address abiding global inequity. Lorraine Elliot (1997) was right when she noted that environmental degradation and ecological crisis are, for a vast majority of people, "symptomatic of a broader structural oppression and silencing."

## Summary and conclusion

Present day environmental cooperation is taking place under conditions of serious North-South injustice. Although it has now been acknowledged that environmental issues such as climate change include serious issues of inter- and intra-generational justice, global governance arrangements have not attended seriously to these issues of inequity. While ethics might not be a popular term in international affairs, it remains an inseparable aspect of every political process in so far as every choice is founded on different ideas of what is right or desirable. Distributional justice is not merely instrumental to, but a part of the package of environmental sustainability, forming an integral part of its socio-economic and political dimensions.

Achieving global sustainable development therefore requires more radical interrogations of the basic structure of international society and of patterns of social relations between the North and South. In short, questions of environmental justice must be seen as questions about the mode of wealth creation and appropriation itself rather than as an add-on optional extra. Given the equal and common dependence of humankind on one single natural system, the idea of global environmental or planetary citizenship should not be seen as a mere preachment but as something that deserves to be taken as a foundation upon which the institutions for international environmental governance ought to be built. To stand any chance of meeting the aspirations of majority of the global population, international governance approaches must strive harder to reflect responsible stewardship and the fact of our common inheritance and ownership of planetary resources.

## References

Agyeman, J., Bullard, R. D., & Evans, B. (Eds.). (2003). *Just sustainabilities: Development in an unequal world*. Cambridge: MIT Press.

Atteh, S. O. (1993). Political economy of environmental degradation: The dumping of toxic waste in Africa. *International Studies, 30*(3), 277–298.

Chasek, P. S. (2001). *Earth negotiations: Analyzing thirty years of environmental diplomacy*. Tokyo: United Nations University Press.

Chasek, P., Downie, D. L., & Welsh Brown, J. (2010). Global environmental politics. In Pamela Chasek, David Downie and Janet Welsh Brown, eds. *Global environmental politics*, 5th ed. Boulder, CO: Westview Press.

Conca, K., Alberty, M., & Dabelko, G. D. (ed.) (1995). *Green planet blues: Environmental politics from Stockholm to Rio*. Boulder, CO: Westview Press.

Dimitrov, R. S. (2016). The Paris agreement on climate change: Behind closed doors. *Global Environmental Politics, 16*(3), 1–11.

Elliott, L. (2004). *The global politics of the environment* (pp. 223–238). London: Palgrave.

Falkner, R. (2016). The Paris agreement and the new logic of international climate politics. *International Affairs, 92*(5), 1107–1125.

Ikeme, J. (2003). Equity, environmental justice and sustainability: Incomplete approaches in climate change politics. *Global Environmental Change, 13*(3), 195–206.

Lélé, S. M. (1991). Sustainable development: A critical review. *World Development, 19*(6), 607–621.

Lipman, Z. (1998). Trade in hazardous waste: Environmental justice versus economic growth. Environmental justice and legal process. In Proceedings of the *Conference on Environmental Justice*, Melbourne.

Liu, S. F. (1992). The Koko incident: Developing international norms for the transboundary movement of hazardous waste. *Journal of Environmental Science and Natural Resources, 8*, 121.

Lohmann, L. (1990). Whose common future?. *The Ecologist, 20*(3), 82–84.

McKee, D. L. (1996). Some reflections on the international waste trade and emerging nations. *International Journal of Social Economics, 23*(4/5/6), 235–244.

Mebratu, D. (1998). Sustainability and sustainable development: Historical and conceptual review. *Environmental Impact Assessment Review, 18*(6), 493–520.

Okereke, C. (2007). *Global justice and neoliberal environmental governance: Ethics, sustainable development and international co-operation*. London: Routledge.

Okereke, C. (2008). Equity norms in global environmental governance. *Global Environmental Politics, 8*(3), 25–50.

Okereke, C. (2010). Climate justice and the international regime. *Wiley Interdisciplinary Reviews: Climate Change, 1*(3), 462–474.

Okereke, C., & Coventry, P. (2016). Climate justice and the international regime: Before, during, and after Paris. *Wiley Interdisciplinary Reviews: Climate Change, 7*(6), 834–851.

Redclift, M. (2002). *Sustainable development: Exploring the contradictions*. Routledge.

The Founex Report. (1971). The Founex report on development and environment – 1971. www.unedforum.org/fileadmin/files/Earth%20Summit%202012new/Publications%20and%20Reports/founex_report_on_development_and_environment_1972.pdf.

The Stockholm Declaration. (1972). The Stockholm declaration on the human environment. In Report of the United Nations Conference on the Human Environment, UN Doc.A/CONF.48/14, at 2 and Corr.1.

Tollefson, J. (2017). Trump pulls United States out of Paris climate agreement. *Nature News, 546*(7657), 198.

Vir, A. (1989). Toxic trade with Africa. *Environmental Science & Technology, 23*(1), 23–25.

Von Weizsäcker, E. U. (1994). *Earth politics*. London: Zed Books.

Wackernagel, M., & Rees, W. (1998). *Our ecological footprint: reducing human impact on the earth* (Vol. 9). New Society Publishers.

WCED, U. (1987). *Our Common Future—The Brundtland Report*. Report of the World Commission on Environment and Development.

# Chapter 14

# Growth and development

*Kerryn Higgs*

---

[T]he bright and powerful vision that propelled economic growth—to provide the material basis for a better life for all—bears little resemblance to the current prospects of only accumulating the wealth of the richest while destroying the environment and livelihoods of future generations and the poorest and most vulnerable.

*(Matthias Schmelzer 2017)*

This chapter examines how, since the Second World War, economic growth defined by increasing gross domestic product (GDP)[1] has been advanced as the key process required to facilitate development. This strategy persists even though it has failed to distribute wealth equitably and has ignored encroaching problems of planetary limits, instead putting faith in "growing the pie" and "decoupling" economic expansion from environmental impacts.

An account of United Nations (UN) attempts at fostering "sustainable development" demonstrates how economic growth has functioned as the bedrock of international development practice. Moreover, although the 1987 World Commission on Environment and Development (WCED or Brundtland Commission) addressed distribution to some extent, market-driven GDP growth was always emphasised, and eventually became the primary strategy, still prominent in the Sustainable Development Goals (SDGs) of 2015.

Outcomes of this strategy in China and India illustrate the ecological risks of development through GDP growth and, in India, suggest that it does not guarantee development that reaches the mass of people. Development Alternatives (DA), a social enterprise functioning in central India, shows that a community-centred approach may be both sustainable and feasible.

## Economic growth and the development discourse

Since the end of the Second World War, Western-style economic development has been seen as the natural and inevitable model for all human development and the self-evident solution to poverty. These beliefs have foreclosed other models and made GDP growth the foundation of Western institutional development thinking – an index of development with no necessary reference to concrete benefits received by actual people (see also Fuchs, this volume).[2]

President Truman's (1949) inaugural speech marked a transition from imperial to postcolonial thinking in the developed world, and heralded the pursuit of development by expansion:

> More than half the people of the world are living in conditions approaching misery. Their food is inadequate, they are victims of disease. … [W]e should make available … the benefits of our store of technical knowledge in order to help them realise their aspirations for a better life. … *Greater production is the key to prosperity.*
>
> *(italics mine)*

By 1960, the new academic discipline of development economics had defined development in terms of economic growth and the exploitation of resources. Technology and capital accumulation were heralded as indispensable features of human progress. Walt Rostow (1960), for example, theorised five stages of growth – from traditional societies, through the centralised nation state and the introduction of modern technologies, to "take-off" where "growth becomes its normal condition"; following a period of consolidation, the once-traditional societies reach the zenith of economic maturity, "the age of high mass-consumption."

In reports from the International Monetary Fund (IMF), World Bank, Organisation for Economic Co-operation and Development (OECD), G20, UN bodies, and in politicians' speeches worldwide, economic growth is seen as imperative for human prosperity.

The scramble for economic growth in the post-war world was driven by several exigencies. For the North, full employment was regarded as an essential aim in the aftermath of the Great Depression, and economic expansion was thought to be the only acceptable means to achieve it; expansion also produced benefits that pacified large sections of unionised labour. In the South, where national liberation movements pressed for independence, growth could serve as an alternative to redistribution of land or resources. In her address to the Stanford India Conference, the IMF's Anne Krueger (2004) reflected this judgement as she spoke in glowing terms of India's economic liberalisation in the 1990s:

> Our job at the Fund is to support governments in their efforts to deliver the sustained and rapid growth needed to raise living standards and reduce poverty…. [T]he solution is more rapid growth—not a switch of emphasis towards more redistribution. Poverty reduction is best achieved through making the cake bigger, not by trying to cut it up in a different way.

"Making the cake bigger" has been the core development strategy of global institutions since 1950. In the initial period, state intervention and planning were considered suitable tools; from the 1980s, as neoliberal ideology began to capture the policy debate, market-driven "globalisation" took over. In both phases, growing economies were expected to provide "catch up" for "underdeveloped countries"; questions of distribution could be shelved. Harvard historian Charles Maier (1987, 130) described the situation succinctly: "the mission of planning became one of expanding aggregate economic performance and eliminating poverty by enriching everyone, not one of redressing the balance among economic classes." Under Christine Lagarde's leadership, the IMF questioned inequality and took climate crisis seriously, but continued to measure progress in terms of growth, now prefaced by "sustainable" or "equitable" (IMF 2017).

## Problems of growth solutions

The acceptance of economic growth as the centrepiece of development policy reflected its adoption as the fundamental ideology of the post-war world more generally. This view harboured several key assumptions, most of which proved problematic in practice. Economic growth was depicted as an inevitable stage of human civilisation, part of a natural and desirable progression from more "primitive" social forms to modernity; European history and practice were portrayed as a universal template. The benefits of GDP growth were assumed ultimately to pass through to whole populations and GDP growth "came to be presented as the common good, thus justifying the particular interests of those who benefited most from the expansion of market transactions as beneficial for all" (Schmelzer 2017). Critically, economic growth was assumed to be an indefinite process, unlimited in space or time.

This version of development did not take much account of the effects of the colonial legacy that had shaped the world and played a part in impoverishing countries now being designated as underdeveloped – or as stragglers in the race who could catch up if they tried. There is little evidence that the development path of the European nations was a realistic template for the postcolonial world. From Cortez to Cook, Europeans had sailed off into a world without significant defences and proceeded to expropriate the apparently boundless natural riches they found there. For centuries, Europeans relied on immense reserves of cheap (or stolen) resources, captive peoples who could be enslaved or fashioned into markets, and, in more temperate regions, convenient settlement opportunities for Europe's surplus population. The situation of the developing countries after the Second World War was very different. By 1950, there were few "empty" islands, let alone continents, even if the "stragglers" had commanded the necessary force to conquer them. For the developing world, there was no comparable platform of prior accumulation[3] or opportunity to create one.

### Problems of distribution

At least since 1980, distribution of the dividends of economic growth has favoured those who were already rich rather than those at the bottom of the wealth ladder, casting doubt on growth as a development strategy. On current trends, if we continue to pursue a bigger cake rather than adjust the distribution, many decades (perhaps centuries) of growth would be needed before reasonable material security reaches everyone. Oxfam (2018) found that 82 per cent of all wealth created globally in 2017 flowed to the top 1 per cent, with no increase whatsoever for the bottom 50 per cent.[4] The World Inequality Report (2018) observed that inequality has been rising since 1980. Apart from China,[5] developing countries also showed greater internal polarisation than Europe, Russia, or North America. Despite decades of extraordinary growth, prosperity is still concentrated among a privileged minority (Higgs 2014, 105–162; Hickel 2017).

### Problems of planetary limits

Distributional problems are not the only concern. *The Limits to Growth* (Meadows et al. 1972) was written by researchers at the Massachusetts Institute of Technology (MIT). They were not the first to consider the possibility that there are physical limits on the human economy's expansion, but they received substantial attention, especially in the 1970s, and brought the concept into mainstream thinking. The project was commissioned by the Club of Rome, which had gathered together a select group of individuals who wanted to address what they called the *problematique*, or "the predicament of mankind."

The MIT team identified five major aspects of this predicament: accelerating industrialisation, rapid population growth, extensive malnutrition, depletion of non-renewable resources, and environmental decline. These circumstances led them to ask this question: how could growing populations, locked into ever-expanding industrialisation, avoid immense environmental degradation, depletion of the resources on which everything depends, and the social chaos that would be likely to follow decline or collapse? To address it, they developed World3, a computer program that processed extensive data on interacting aspects of the economy and the environment in various combinations. The standard run (which assumed that industrial activity, resource extraction, and population growth would continue on the "business as usual" trajectory) resulted in collapse around the middle of the 21st century; immense technological advance merely postponed this outcome. The only scenarios that avoided collapse were those that stabilised population and reduced the scale and rate of material extraction and waste. Although economists ridiculed these ideas from the outset, researchers who analysed data for the years since 1972 found that real world outcomes correspond closely with the MIT team's "business as usual" scenario (Turner 2014).

The forceful opposition from mainstream economists reflects their assumption that the physical world is not essential to wealth-production: economic growth, they believe, results from two factors, capital and labour, with the economy depicted as a circular flow connecting producers and consumers.[6] For ecological economists and most physical scientists, on the other hand, the human economy is embedded in the physical world and requires its resources and sinks. Since the planet is finite, there is no option of unlimited growth in material flows; sustainable development is possible only if it is "development without growth" (Daly 1996). In this worldview, the growth model of development faces a dilemma: all human production must extract raw materials and dispose of wastes, but depletion of resources, including energy, and increasing pollution from wastes are inescapable consequences of its operation. These constraints gradually narrow the options for material expansion.[7]

No less troubling is the steady degradation of the natural world. The planetary boundaries team led by Will Steffen and Johan Rockström (Steffen et al. 2015) aims to identify the boundaries beyond which human activities fatally undermine the ecological integrity of the planet on which everything depends. In 2015 they argued that biodiversity loss and biogeochemical disruption (nitrogen and phosphorous cycles in particular) were in the danger zone (red light), while climate and land-use change were approaching it (amber). Subsequent research suggests that climate change may already be in the danger zone. Threats are interconnected, with land-use change, for example, affecting climate and biodiversity.

These limits (resource depletion, capacity of the environment to absorb production wastes, and potential for GDP growth to provide equitable prosperity) must all be considered in assessing realistic development possibilities.

## Problems of decoupling

It has become more widely acknowledged over the past decade that the ongoing expansion of the human economy is damaging the biosphere,[8] possibly irreparably over time spans that are meaningful for humans. However, it is not yet accepted that economic goals need to change. It is more often argued that economic growth is indispensable and technological advances will offer efficiencies that can prevent serious damage. Much of this optimism is advanced under the banner of "green growth," "dematerialisation," or "decoupling" (Alexander, this volume). Champions of ongoing economic expansion, including most of the global institutions involved in sustainability policy, argue that it is feasible to decouple GDP growth from environmental damage.[9]

While increased efficiencies, improved technologies, and circular patterns of production can mitigate environmental damage, and should therefore be pursued, actual permanent decoupling (where environmental impacts are stable, or reduced, while material flows increase) is improbable. Since 1990, there has been little improvement in global material efficiency – indeed, the global economy now requires more materials per unit of GDP than it did at the turn of the century (UNEP 2016a). Decoupling will not occur automatically and is particularly unlikely under the current market economy, where private interests conduct production and resist regulation.

The UNEP (2016a) report on material flows maintains optimism about decoupling, but concedes that efficiencies could cause a rebound effect and trigger accelerated economic growth, thus negating efforts to reduce gross material demand. UNEP (2016a, 29–30) also argues that "the level of well-being achieved in wealthy industrial countries cannot be generalized globally based on the same system of production and consumption," without encroaching further on environmental thresholds already under pressure. This system, however, remains in place.

In the case of non-substitutable resources such as land, water, raw materials, and energy, Ward et al. (2016) conclude that, "whilst many efficiency gains may be possible, such resources are ultimately governed by physical realities" and that permanent decoupling, whether absolute or relative, is impossible. Lenzen, Malik, and Foran (2016) also point out that the developed world's political class has a lamentable record on implementing measures aimed at decoupling such as establishing effective carbon pricing. If decoupling is to work, success at this kind of intervention is obligatory. Sustainable development, conventionally understood, depends on it.

## Towards sustainable development (SD)

The SDGs emerged in 2015 from the 50-year history of the UN's grappling with development. The United Nations Development Programme (UNDP) was set up in 1965 and continues as the UN's main development organisation. The 1970 Stockholm Conference and the WCED (1983–1987) were followed by the Rio "Earth Summit" (1992).[10]

### The Brundtland Commission

The terms "sustainability" and "sustainable development" have been prominent in development policy rhetoric since the Brundtland Commission released its report, *Our Common Future* (WCED 1987, 40–59), which envisaged "the integration of environment policies and development strategies." Its definition of SD is well known: "[It] seeks to meet the needs and aspirations of the present without compromising the ability to meet those of the future." The Commission saw economic growth as essential, but recognised that growth alone was not enough and developing countries needed to "reap large benefits." The needs of the poor warranted "overriding priority," while those of future generations were being compromised, especially by the excessive demands of the affluent, who were living beyond the "world's ecological means" and must reduce their per capita resource consumption.

The Commission (1987, 85) recognised the power of transnational corporations (TNCs)[11] and recommended international regulation and codes of conduct incorporating environmental principles for corporations, the World Bank, IMF, and General Agreement on Tariffs and Trade, (which became the World Trade Organization in 1995). It also insisted that the South must retain sovereignty over its own resources, without exception.

The WCED (1987, 5–6) contrasted the vast scale of economic growth generated in the 20th century[12] with the meagre improvements that this astonishing growth had brought to developing countries; issues of distribution, they argued, needed attention. However, as neoliberal economic orthodoxies gained ground, the Commission's modest redistributive proposals were discarded.

## The Rio "Earth Summit"

The United Nations Conference on Environment and Development (UNCED) or "Earth Summit" of 1992 was originally intended to put the Brundtland Commission's findings into practice, but in the brief period since the WCED reported, "free market" ideology had triumphed.[13] UNCED (1992) did not recommend reform of the international economic system and, although it acknowledged that debt, poor commodity prices, and poor terms of trade had deepened the poverty of developing countries, it did not tackle these obstacles directly, but made global free trade the centrepiece of its plans for SD. Whether addressing equity or environment, "free market" solutions were comprehensively adopted.[14]

Funding arrangements were minimal. Free trade was expected to "make economic growth and environmental protection mutually supportive." The sole concrete funding proposal was for developed countries to meet the long-established UN aid target of 0.7 per cent of GDP. Few had succeeded, but all agreed to reach it "as soon as possible." Seventeen years later, the average volume of aid from North to South was approximately 0.3 per cent of their GDP, a significant decline from an estimated 0.51 per cent in the late 1960s (Riddell 2009).

Imposing a free trade regime has been characterised as "kicking away the ladder," given that European and American prosperity was established in protected economies (Chang 2002; Higgs 2014, 113–115). Moreover, the promised aid did not eventuate, the UN Centre on Transnational Corporations was dissolved in 1993, and work on a corporate Code of Conduct was abandoned. Questions of landownership and the distribution of wealth were not addressed. The South had wanted UNCED to acknowledge sovereign rights of nations to exploit their own resources and the right of individuals to have freedom from hunger, poverty, and disease, and had called on the rich to accept the main burden of repairing the environmental damage they had caused. Only hints of these priorities survived the preparatory phases.

## Sustainable Development Goals, 2015

However ambiguous the notion of SD, it now underpins the UN's SDGs (UNDP 2015), which replaced the Millennium Development Goals (MDGs) of 2000. While the MDGs did not include environmental goals, other than the broad endorsement of "sustainable development," the SDGs incorporate three specific goals (13–15): urgent climate action; the conservation, sustainable use, and rehabilitation of land; and similarly of the oceans. A fourth goal (12), "sustainable consumption and production patterns," also implies environmental limits, though no reference is made to consumerism or discouraging the practice of fostering human desire in the interests of expanded consumption. While efficiency, the main strategy, could address issues such as the vast extent of food waste, it fails to engage with the ubiquitous incitement to consume.

The crucial weakness of the extensive set of new goals, targets, and indicators is the contradiction between the explicit environmental goals and many of the socio-economic goals. The word "sustainable" is attached to most of these, as if it can recuperate realities such as industrialisation, current agricultural practices, and GDP growth itself, simply by being mentioned.

## Trade-offs and contradictions

The internal contradictions within the SDGs are often referred to as trade-offs, a term consonant with market values. The reality of a trade-off, however, is that something must be sacrificed to achieve something else. Core contradictions exist between the goals that aim for economic growth or involve increasing material flows and the three central environmental goals. This raises the question of sustainable scale, defined by Costanza et al. (2016) as that which "cannot deplete natural capital or damage ecosystem services beyond a certain 'safe operating space,'" the term used by the planetary boundaries team.

For Costanza and colleagues (2016), the SDGs represent a major achievement for the world's people and our global institutions, in that the SDGs embody an agreement on shared goals for all humanity. This team recommends that an overarching goal should be formulated and adopted, to frame the SDGs and specify the overall direction – for example "sustainable, equitable and prosperous well being" for all. In this framework the true goal of development could be addressed, acknowledging perhaps that development is centred on and measured by outcomes for people and their well-being, rather than GDP growth.

Some of the specific goals illustrate these contradictions.

## Climate, hunger, land-use, and agriculture

SDG2, ending hunger and providing food security for all, raises difficult challenges for terrestrial health if methods conform to the growth paradigm, where increasing production and/or efficiency is emphasised. Expanding agricultural productivity through existing systems of industrial farming will almost certainly perpetuate forest clearing, wetland draining, and soil degradation, as well as adding greenhouse gases as energy demands rise and deployment of fertilisers, herbicides, and pesticides increases. Land-use emissions must decrease to zero by 2050 if warming is to be kept "well below 2°C" (Rockström et al. 2017). Modern farming affects all four of the planetary boundaries thought to be in dangerous territory (biodiversity loss, nitrogen and phosphorous pollution, land-use change, and climate). It is also geared to supply commodities to markets, including biofuels, luxury crops, and animal feed, rather than explicitly to feed the hungry.

This goal could, however, be more effectively aligned with climate and conservation objectives if improvements in productivity are made consistent with ecological sustainability. The world's smallholders make up a third of the world's population and half the world's poor, and produce about 70 per cent of its food on one quarter of its farmland (GRAIN 2014). Their representatives (Wolfenson 2013) have pointed out that the growth model imposed on them has deepened inequality and accelerated environmental decline, whereas the agro-ecology they practise has demonstrated its potential for genuine sustainability. Agro-ecology is based on the ecological processes that underlie all farming, rather than on technological fixes. Smallholders, however, who lack political influence and, often, secure land title,[15] are constantly losing ground to bigger farmers and agribusiness corporations.

Alongside agro-ecology, smallholders champion food sovereignty,

> a system that returns the land to its social function as the producer of food, puts the people who produce, distribute and consume food at the centre of decisions about food systems and policies, as opposed to the demands of markets and corporations.

Although Target 2.3 talks of supporting smallholders, it focuses on *doubling productivity per labour unit*, whereas productivity per unit of land and water would be more appropriate.[16]

Where labour is not the limiting factor, this target is mistaken and will encourage displacement of smallholders and consolidation of farmland under corporate control. The acreage of industrial crops (soy, canola, sugar cane, oil palm) quadrupled in the fifty years to 2011 and, during the current century, was facilitated by "land grabs" amounting to hundreds of millions of hectares (GRAIN 2014).[17] Increased productivity and the promise of employment are frequently offered to justify this trend, but there is little evidence of benefit for local smallholders who may get employment on miserable wages, while losing access to their communal land and water, which then supply food, fuel, and profits to foreign markets and investors (Daniel and Mittal 2009). More promising are the objectives in the "UN Declaration on the rights of peasants and other people working in rural areas," adopted by the UN General Assembly in December 2018 (Via Campesina 2018).

## Industry, energy, and climate

SDG9, which focuses on the expansion of infrastructure and industry, will also be difficult to align with the environmental goals. This kind of expansion has been associated historically with increases in energy demand and land-use change and has driven emissions growth for more than two centuries. Its continuation will lock in historical patterns and substantially shape future emissions (UNEP 2016b). Some 50 per cent of the urban infrastructure needed in the world in 2050 has not yet been built. If it relies on steel and concrete and is designed around private vehicles, rather than embracing low-carbon options, the Paris Agreement's mitigation objectives will be compromised. Some industrial sectors could be based on low-carbon energy sources however, and hopes for stabilising the climate are contingent on urgently embracing such potential.

## Growth, employment, and climate

Finally, and most difficult of all, SDG8 advocates "sustained, inclusive and sustainable economic growth" leading to jobs for all. While provision of many more jobs[18] is absolutely essential, little employment will arise through market-driven, capital intensive growth, where the goal of cost reduction favours mechanisation and job-shedding. Historically, economic growth has been strongly correlated with greenhouse gas emissions, resource depletion, and land-use change. Although "sustainable growth" is part of this goal's full title, sustainability is not actually mentioned in the text and appears in the targets as an injunction to *sustain economic growth*, which is hardly the meaning of "sustainability" – or, rather, reflects confusion about what must be sustained.

## Challenges of SDG contradictions

In line with UNEP's (2016a, 14) claim that "[d]ecoupling material use and environmental impacts from economic growth… will be essential for ensuring future human well-being based on much lower material throughput," Indicator 8.4 endorses that goal. However, if we accept that much lower material throughput is mandatory, and it also emerges that extensive decoupling is not realistic, as argued earlier, we may face insurmountable obstacles.

For Lenzen, Malik, and Foran (2016), the enthusiasm for decoupling is misguided: "it is illogical to expect technological progress to stretch to unprecedented limits in order to let populations enjoy unchecked growth in numbers and affluence." They propose, instead, that "affluence and population," not just technology, must be addressed (see also Coole this volume; Hayden this volume). Affluence and population are not considered in the SDGs, other than in advocating universal access to reproductive rights. But affluence in the North and its extension to élite minorities in the South, lies at the root of ecological unsustainability. Economist Paul Ekins (1991) calculated that "universal opulence" was simply unattainable; to provide even one fifth of Western affluence to the people of the South would require (at least) an immediate freeze on per capita consumption in the developed world. Rising inequality over the intervening decades has no doubt amplified that challenge.

The SDGs are compromised by their internal contradictions; resolving these depends on strategic and difficult choices, especially regarding agriculture, infrastructure, and industry. If we accept that some degree of physical economic growth is indispensable for the South, contraction in the developed world's consumption will also be necessary. Business as usual, relying on the current formula of setting the market free to do its work, will not suffice.

## China, development miracle

China's GDP growth accelerated from the late 1970s and millions of China's people have been lifted out of extreme poverty.[19] This has weighted the averages for the South as a whole, and given credibility to the notion that growth is the answer to poverty. At the same time, the social, environmental, and health costs have been cataclysmic. Land has been routinely confiscated from farmers, especially in the east, where vast new industrial cities have swallowed villages and fields. In 2005, China's deputy environment minister, Pan Yue, told *Der Spiegel* that

> desert areas are expanding... habitable and usable land has been halved over the past 50 years.... Acid rain is falling on one third of the Chinese territory, half of the water in our seven largest rivers is completely useless, while one fourth of our citizens does not have access to clean drinking water. One third of the urban population is breathing polluted air.

In Pan's view, this ruin indicates that China's "miracle" GDP growth has been inflated by the non-inclusion of its costs (Lorenz 2005).[20]

A decade and a half later, China's voracious appetite for minerals, energy, agricultural commodities, and timber involves it in "a dizzying variety of resource extraction, energy, agricultural, and infrastructure projects... that are wreaking unprecedented damage to ecosystems and biodiversity" across the world (Laurance 2017).[21] While the growth model has assisted many of the poorest people in China and China *is* investing in green technologies, inequality is growing and vast external resources are now imperative. Chinese activities should be assessed in the context of many centuries of extraction conducted by Europeans; China has, however, joined the developed world in exporting its environmental impact.[22]

## The "Gujarat Model" and Development Alternatives: India

The Gujarat model, as described at the website narendramodi.in (2014), echoes the language of the SDGs: "Gujarat's development journey is... inclusive and participative... consulting all the stakeholders." This development model, where the government establishes Special Economic Zones and rolls out tax breaks and subsidies, can certainly support GDP growth, whether conducted in Gujarat – considered an ideal example of successful globalisation – or elsewhere. However, its successes ignore the welfare of affected citizens; despite claims of inclusiveness, the proceeds of growth are little distributed to its people. Rather, businesses aim to compete on the global stage and typically adopt capital intensive technologies; little employment is generated per unit of investment capital or of output. Rapid industrial growth also involves destruction of natural resources and pollution of land, water and air, a most unfortunate situation for the large numbers of people who still depend on coast and forest for their livelihoods (Hirway and Shah 2011; Sood 2012).

Analysis of recent Reserve Bank of India data (Ghatak 2017) shows that, amongst India's 20 biggest states, Gujarat ranks in the top five on criteria related to growth in state domestic product, while it ranks at number ten or below on social indicators such as infant mortality, life expectancy, sex ratio, and poverty. Despite three decades of high growth, there has been little or no improvement in most social indicators – the "trickle-down effect" is absent. GDP growth in Gujarat's agricultural sector reflects a shift away from food to non-food and high value crops, with smallholder farms shrinking and the largest farmers increasing their acreage; from 2005, the state government facilitated transfer of village commons and wastelands to corporate use, both agricultural and urban (Sood 2012). People who relied on fishing and grazing in Kutch, for example, have been "stripped of their livelihoods, displaced, and the ecology of their villages and coast destroyed" (Perspectives 2012). The Gujarat model, an extreme version of market-based development, has secured neither environmental nor social improvement.

By contrast, the work of Development Alternatives (DA) shows that development can be accomplished outside the industrial growth model and, taking people's welfare as the benchmark, with outstanding success. Founded by Ashok Khosla (2015), who left his career in government and UN in 1982, DA is a social enterprise that has operated for more than 30 years, mainly in the central Indian region of Bundelkhand. For Khosla, the single most important priority is creating sustainable livelihoods on a large scale.

More than 70 per cent of India's people live in villages and small towns, as do most people throughout the South. For DA, genuine sustainable development requires two elements: it must maintain the natural world for current and future generations; and it must create sustainable livelihoods, enterprises that supply jobs that service the basic everyday needs of the majority of people. DA researches sustainable technologies and establishes or facilitates small to medium enterprises that produce items such as gasifiers for electricity, improved stoves and looms, affordable building materials such as mud blocks, handmade paper, and many others. Alongside these, DA regenerates land, water, and forest. Its "check dams," which slow down water flow and allow aquifers to recharge, have been widely adopted beyond Bundelkhand. DA also builds water and sanitation services and has designed a successful literacy program that teaches women to read and do basic maths in 56 days. This helps them become active citizens, participating in their villages and communities and running small enterprises, and encourages smaller families. Khosla's DA project has created millions of livelihoods at very modest cost (von Weizsäcker and Wijkman 2018, 108–113).

## Conclusion

As noted at the outset, GDP growth is still favoured as the path to development by governments, the corporate world, and global institutions. This preference has embraced "growing the pie" and "decoupling" economic expansion from environmental impacts while disregarding distribution and ignoring unwanted consequences. China's story shows that unconstrained growth pollutes severely and exhausts local resources, then moves outwards to extract them elsewhere (see Princen, this volume). Ultimately, if the growth model is continued, external sites for the rest of the world's resource needs will be few.

There is no evidence that growth will generate even a fraction of the employment needed to absorb the roughly three billion people in farming families throughout the world, whose land is being enclosed and who are expected to move to cities. Growth may continue to enrich élites in the global South, and contribute to the swelling middle class there, but growth as currently practised will not employ the rural masses, whether they are urbanised, as many predict, or not. The kind of development carried out by Khosla's DA, on the other hand, can provide people with jobs and security, assist them to retain what land they have access to, and moderate the rural exodus to burgeoning slums in the cities. Though smallholders may benefit from collaborative agricultural assistance, they first need support to retain their land and, along with the landless, get access to more. Although Khosla works with local businesses, DA is not primarily concerned with making a profit or competing with corporate giants in the world market. DA aims to empower people, preserve the natural world, and foster livelihoods.

To curb depletion, ecological decline, pollution, and inequality, development needs to be redefined according to human needs rather than greater production or GDP, and measured accordingly. GDP should urgently be discarded and an alternative indicator that measures human well-being adopted (Costanza et al. 2009; Philipsen, this volume). "Sustainable development" needs sharper definition – the word "ecologically" should preface the term, to clarify that SD involves preservation of the natural basis. "Sustaining economic growth" will be needed in some contexts, but should not be confused with ecological sustainability. While growth in material flows will be necessary to provide physical improvements such as clean water and sanitation, any such growth must meet actual human needs. Unlike the market paradigm which has dominated development thinking, this approach may yield genuine development.

## Notes

1 Growth and economic growth in this chapter mean GDP growth unless otherwise specified.
2 Critiques of and alternatives to GDP are widely available: Costanza et al. (2009), Higgs (2014, 158–160), Philipsen, this volume.
3 Term used by Adam Smith and Karl Marx ("primitive" accumulation) for the beginnings of accumulation, usually by enclosure and/or theft.
4 See also Chancel and Piketty (2017).
5 Inequality has, however, increased.
6 Robert Solow (1957) examined the empirical evidence that capital and labour could explain only a small fraction of observed economic growth in the United States. He estimated the "residual" to be 87.5 per cent and hypothesised that "technical change" could account for it. Later, Robert Ayres and Benjamin Warr suggested that the missing factor is actually energy or, more precisely, "exergy," the increasing efficiency with which energy and resources can be made into useful work. Thus, technical innovation has a role, but "labour and capital extract energy; they don't make it" (Ayres and Ayres 2009, 9–18). Energy is a prerequisite for, not a product of, economic activity.
7 See Lange this volume for exploration of sustainable zero growth options within the current market system.
8 UNEP (2011), OECD (2011), Spence (2012), Hayden (2014), Schmelzer (2017).
9 OECD (2011), World Bank (2012), Hatfield-Dodds (2015), UNEP (2016a).
10 Higgs (2014, 38–39, 125–133).
11 TNCs controlled trade in tea, coffee, cocoa, cotton, timber, tobacco, jute, copper, iron ore, and bauxite.
12 Industrial production was multiplied 50 times, 80 per cent of that since 1950.
13 See Higgs (2014, 239–254), for some processes that achieved this outcome.
14 Deregulation, privatisation, cutting taxes and government programs, and trade liberalisation.
15 Where title is granted, customary title is preferable to individual title, which can facilitate sale.

16  See Higgs (2014, 146), on defining agricultural productivity.
17  See Magdoff (2013) for an overview.
18  Or a guaranteed basic income.
19  Perhaps assisted by China's high levels of equality beforehand.
20  Also Economy (2007), Higgs (2014, 135–162).
21  Also Simons (2014), Economy and Levi (2014).
22  See Lenzen et al. (2012).

# References

Ayres, R. and E. Ayres. 2009. *Crossing the energy divide.* Upper Saddle River, NJ: Pearson Education.

Chancel, L. and T. Piketty. 2017. Indian income inequality, 1922–2014: From British Raj to Billionaire Raj? WID.world. Working Paper Series 2017/11. http://wid.world/document/chancel piketty2017widworld/.

Chang, H.-J. 2002. *Kicking away the ladder.* London: Anthem Press.

Costanza, R., L. Daly, L. Fioramonti, et al. 2016. Modelling and measuring sustainable wellbeing in connection with the UN Sustainable Development Goals. *Ecological Economics* 130 (1): 350–355.

Costanza, R., M. Hart, S. Posner and J. Talberth. 2009. Beyond GDP: The need for new measures of progress. *Pardee Papers 4.* Boston: Boston University.

Daly, H. 1996. *Beyond growth: The economics of sustainable development.* Boston, MA: Beacon Press.

Daniel, S. and A. Mittal. 2009. *The great land grab: Rush for world's farmland threatens food security for the poor.* Oakland, CA: Oakland Institute. www.oaklandinstitute.org/sites/oaklandinstitute.org/files/LandGrab_final_web.pdf.

Economy, E. 2007. The great leap backward? *Foreign Affairs,* Sep/Oct. www.foreignaffairs.com/articles/asia/2007-09-01/great-leap-backward.

Economy, E. and M. Levi. 2014. *By all means necessary: How China's resource quest is changing the world.* Oxford: Oxford University Press.

Ekins, P. 1991. The sustainable consumer society: A contradiction in terms? *International Environmental Affairs* 3 (4): 243–258.

Ghatak, M. 2017. Gujarat model: The gleam of Gujarat's high growth numbers hides dark reality of poverty, inequality. *Scroll.in.* 25 October. https://scroll.in/article/855027/gujarat-model-the-gleam-of-states-high-growth-numbers-hides-dark-reality-of-poverty-inequality.

GRAIN. 2014. Hungry for Land: Small farmers feed the world with less than a quarter of all farmland. www.grain.org/article/entries/4929.

Hatfield-Dodds, S., H. Schandl, P. Adams, et al. 2015. Australia is 'free to choose' economic growth and falling environmental pressures. *Nature* 527 (7576): 49–53.

Hayden, A. 2014. *When Green growth is not enough.* Montreal: McGill-Queens University Press.

Hickel, J. 2017. *The divide: A brief guide to global inequality and its solutions.* London: William Heinemann.

Higgs, K. 2014. *Collision course: Endless growth on a finite planet.* Cambridge: MIT Press.

Hirway, I. and N. Shah. 2011. Labour and employment under globalisation: The case of Gujarat. *Economic and Political Weekly* 46 (22): 57–65.

IMF. 2017. World Economic Outlook. Seeking sustainable growth. www.imf.org/en/Publications/WEO/Issues/2017/09/19/world-economic-outlook-october-2017.

Khosla, A. 2015. *To choose our future.* New Delhi: Academic Foundation.

Krueger, A. 2004. Letting the future in: India's continuing reform agenda. 4 Jun. www.imf.org/en/News/Articles/2015/09/28/04/53/sp060404.

Laurance, William. 2017. The dark legacy of China's drive for global resources. *Yale Environment 360,* 28 Mar. http://e360.yale.edu/features/the-dark-legacy-of-chinas-drive-for-global-resources.

Lenzen, M., A. Malik and B. Foran. 2016. Letter. *Journal of Cleaner Production* 139: 796–798.

Lenzen, M., D. Moran, K. Kanemoto, B. Foran, L. Lobefaro and A. Geschke. 2012. International trade drives biodiversity threats in developing nations. *Nature* 486 (7401): 109–112.

Lorenz, A. 2005. The Chinese miracle will end soon: *Spiegel* interview with China's deputy minister of the environment. *Der Spiegel,* 7 Mar. www.spiegel.de/international/spiegel/spiegel-interview-with-china-s-deputy-minister-of-the-environment-the-chinese-miracle-will-end-soon-a-345694.html.

Magdoff, F. 2013. Twenty-first-century land grabs: Accumulation by agricultural dispossession. *Monthly Review* 65 (6): 1–18.

Maier, C. 1987. *In search of stability*. Cambridge: Cambridge University Press.

Meadows, D., D. Meadows, J. Randers, and W. Behrens III. 1972. *The limits to growth: A report for the Club of Rome's project on the predicament of mankind*. London: Pan Books.

Modi, N. 2014. The Gujarat model. www.narendramodi.in/the-gujarat-model-3156.

OECD. 2011. Towards green growth. www.oecd.org/greengrowth/48012345.pdf.

Oxfam International. 2018. Reward work, not wealth. www.oxfam.org/en/research/reward-work-not-wealth.

Perspectives. 2012. Swimming against the tide: Coastal communities and corporate plunder in Kutch. *Economic and Political Weekly* 47 (29): 12–17.

Riddell, R. 2009. Is aid working? *Open Democracy*, 20 November. www.opendemocracy.net/en/is-aid-working-is-this-right-question-to-be-asking/.

Rockström, J., O. Gaffney, J. Rogelj, M. Meinshausen, N. Nakicenovic and H. Schellnhuber. 2017. A roadmap for rapid decarbonization. *Science* 355 (6331): 1269–1272.

Rostow, W. 1960. *The stages of economic growth: A non-communist manifesto*. Cambridge: Cambridge University Press.

Schmelzer, M. 2017. *The hegemony of growth: The OECD and the making of the economic growth paradigm*. Cambridge: Cambridge University Press.

Simons, C. 2014. *The devouring dragon*. New York: St. Martin's Griffin.

Solow, R. 1957. Technical change and the aggregate production function. *Review of Economics and Statistics* 39 (3): 312–320.

Sood, A. 2012. Rousing growth amidst raging disparities. In *Poverty amidst prosperity*, ed. Atul Sood, 1–38. New Delhi: Aakar Books.

Spence, M. 2012. The sustainability mindset. *Project Syndicate*, 17 February. www.project-syndicate.org/commentary/the-sustainability-mindset.

Steffen, W., K. Richardson, and J. Rockström, et al. 2015. Planetary boundaries: Guiding human development on a changing planet. *Science* 347 (6223): 736–746.

Truman, H. 1949. Inaugural address, January 20, 1949. Harry S. Truman Library and Museum. www.cbsnews.com/news/harry-truman-inaugural-address-1949/.

Turner, G. 2014. Is global collapse imminent? MSSI Research Paper 4, Melbourne Sustainable Society Institute, University of Melbourne. https://sustainable.unimelb.edu.au/publications/research-papers/is-global-collapse-imminent.

UN Conference on Environment and Development. 1992. *Agenda 21*. https://sustainabledevelopment.un.org/content/documents/Agenda21.pdf.

UNDP. 2015. Sustainable Development Goals. www.undp.org/content/undp/en/home/sustainable-development-goals.html.

UNEP. 2011. Decoupling natural resource use and environmental impacts from economic growth, a report of the working group on decoupling to the International Resource Panel. www.resourcepanel.org/reports/decoupling-natural-resource-use-and-environmental-impacts-economic-growth.

UNEP. 2016a. Global material flows and resource productivity. Summary report. www.resourcepanel.org/reports/global-material-flows-and-resource-productivity-database-link.

UNEP. 2016b. Emissions gap report. www.unenvironment.org/resources/emissions-gap-report-2016.

Via Campesina. 2018. https://viacampesina.org/en/finally-un-general-assembly-adopts-peasant-rights-declaration-now-focus-is-on-its-implementation/.

von Weizsäcker, E. and Wijkman A. 2018. *Come on! capitalism, short-termism, population and the destruction of the planet: A report to the Club of Rome*. New York: Springer.

Ward, J., P. Sutton, A. Werner, R. Costanza, S. Mohr and C. Simmons. 2016. Is decoupling GDP growth from environmental impact possible? *PLoS ONE* 11 (10): e0164733.

Wolfenson, K. 2013. Coping with the food and agriculture challenge: Smallholders' agenda. Preparations and outcomes of the 2012 United Nations Conference on Sustainable Development (Rio+20). www.fao.org/3/a-ar363e.pdf.

World Bank. 2012. Inclusive green growth—the pathway to sustainable development. https://openknowledge.worldbank.org/handle/10986/6058.

World Commission on Environment and Development. 1987. *Our common future*. Oxford: Oxford University Press.

World Inequality Lab. 2018. World inequality report. http://wir2018.wid.world/.

# 15

# The mining dilemma

*Thomas Princen*

Until modern industrial times, the goal of sustainability was irrelevant. The game was to subsist or create wealth or build an empire. Or it was to defend oneselves from the predations of others and the vagaries of nature. Sustainability arose as a goal when pollution clean-up was not enough. Through its many iterations sustainability has been constructed as an end state, a target, a set of variables that, if optimised, would "sustain" society.

Here I wish to take a different tack, one grounded in that which must be sustained before all else, namely, *natural resources*, as the precondition for sustaining peoples and their institutions. What's more, I wish to emphasise *practices*, dividing economies by the practices that contribute to "sustaining economies" and those that do not, what I will call "mining economies." The significance of this dichotomy of practices – sustaining and mining – is that it steers attention to that which fundamentally grounds an economy, to its material provisioning, its substantive foundations, its natural and necessary infrastructure as opposed to its constructed and ephemeral superstructure. It pays primary attention to soil and water, timber and food rather than gadgets and offices, sales and money. It prioritises relations – especially human-material relations – over wealth and impacts.

The general argument then is this. While all societies mine – that is, use resources in ways that permanently use up those resources – only societies that block mining practices from being applied to renewable resources can be truly sustainable. It is a dilemma that rarely plagued earlier societies yet is now pervasive in modern, industrial, growth-dependent, efficiency- and speed-obsessed, consumerist, commercialist, fossil fuelled societies: practices that make perfectly good sense applied to minerals and petroleum and fossil water – i.e., mining practices – jeopardise the generativity of all other resources, including those resources that make mining possible.

The mining dilemma arises when a collectivity intends to sustain itself for a long time, what, to my read, all societies do, subsistence and industrial, constitutional and authoritarian, and yet modern industrial consumerist societies have organised their resource use to make it hard to distinguish mining and sustaining: it's all growth, all wealth formation, all progress. One result is that, in its own terms, in its language of endless expansion, it sees no reason to block collateral mining, that is, the mining that spills over from the mining of non-renewables (the only thing one can do with such resources) to the mining of renewables (which otherwise can be sustained).

Resolution of the dilemma cannot be found in the practices and institutions of mining, nor in the language of mining – e.g. total use, growth, efficiency, consumer sovereignty, domination (Daly 1996; Jackson 2009; Worster 1985; see also Higgs, this volume). Rather, a wholly different language is needed, a language that is not expansionist, exploitative, and dominating, objectified and mechanistic, commercial and consumerist. What is needed is a language of grounding and sufficiency (Hayden, this volume; Princen 2005; Princen, Maniates and Conca 2002).

In a nutshell, modern economies are, from a natural resource perspective, indeed an eco-logical perspective, *mining economies*. Literal mining and practices that resemble literal mining dominate material exchange, physical and monetary. Because mining is, fundamentally, the permanent and irreversible removal of a resource it cannot be sustained, not over the long term. Moreover, domination means – again, from a natural resource perspective – that mining practices supersede and crowd out sustaining practices in the short term. A mining economy thus consumes its resource base, both those resources which can only be mined (minerals, pe-troleum, coal, fossil water) and that which can, with non-mining practices, be sustained. Value and wealth accrue to extraction in the first instance, what I'll call "easy wealth," production the second (which requires "real work," not symbolic manipulation), and exchange only the third, possibly a distant third in the long term (money ultimately does not reflect long-term, ecological underpinnings), claims of financialisation notwithstanding (Clapp 2014).

As a result, I argue, modern economies

i   chronically experience booms and busts (as all mining operations do) at a variety of spa-tial and temporal scales;
ii   are logically unsustainable (that which is mined cannot regenerate); and
iii   are inherently unjust (domination begins with extraction, which pushes out subsistence).

It follows, then, that the modern economy will, like all mining operations, end. The two-century boom will go bust; for biophysical, ecological reasons it cannot be sustained; inequities mount until either the oppressed rebel or critical resources deplete or both. The transition, from extractive, easy wealth to productive, regenerative wealth could take a cen-tury or two, itself punctuated no doubt by unpredictable emergencies and crises, "natu-ral" (e.g. earthquakes), human-made (e.g. financial collapse), and, increasingly, hybrid (e.g. "wild" fires, floods, heat waves). The material task then shifts from "grow the economy to solve society's problems (including the problems of excess growth)" to "find sufficient wealth in both mining and sustaining."

The *politics of this transition* will be, in the first instance, one of *acceptance*, coming to accept the reality and inevitability of fundamental change, in this case the end of the mining boom. In the second, it will be developing institutional mechanisms that select out the destructive, extractive practices before circumstances force them out. In the third, it will be creating a regenerative economy. Overarching these three components of a politics of transition is a nor-mative shift – from an expectation of boom times forever (endless growth) to one of ineluc-table decline in overall material availability (because the easy-to-get stuff is already extracted and technologies do not create material and energy). The normative shift will also be from

a   value as exchange where invented exchanges such as financial derivatives and virtual currencies purport to create value out of nothing; to
b   value as use where among the highest uses is that which is grounded in the land and sustains all members of a society.

## The mining dilemma

To mine is to use a resource and permanently and irreversibly use it up, all to gain temporary value. This statement of the essence of mining entails no judgement of good or bad, right or wrong; it is only a description of a universal mode of resource use (Bridge 2004; Gadgil and Guha 1992). What is more, I use extraction not as a synonym for mining but as "taking." The resource may be a deposit of gold, a seam of coal, a pool of oil; or it may be a layer of soil, an aquifer, a grassland, or forest. Once the resource, say the pool of oil or the forest, is used and used up, that is, the oil is pumped out or the forest is cleared (and only scrub can succeed the trees because the soil washes away) its extraction is permanent; it can't be reversed. The extracted value is only temporary, that is, until the resource is used up and that value works its way through the so-called value chain. To gain more such value one must go elsewhere and mine some more, or find a substitute. *Extracting irreversibly* and *moving on* are the essential practices of mining. The essential practices of sustaining, I argue elsewhere, are restrained extraction and buffering (Princen 2005).

Superficially, mining is no different from killing the cow to eat it. Ecologically, though, after slaughter, the grass continues to grow, cows continue to graze, and calves continue to be born. A cow, even a herd, cannot be mined. A breed of cow or the cow as a species can be mined, that is, permanently and irreversibly used up. Killing the cow does not, by itself, use up the breed or species. Cows do regenerate. But when the genetic stock of the breed or species is lost, the breed or species is lost, wholly and forever. This, then, is mining. It could be highly beneficial to eliminate a breed or the species (to alleviate hunger among the users, say, or exterminate a pathogen). Once again, there is no value judgement in describing mining with terms like "permanent" and "irreversible," "lost" and "forever" and "used up"; mining simply yields value, albeit temporary.

To mine the breed or species is analogous to the mining of minerals. In fact, the only difference between mining minerals and mining living systems (breeds, species, ecosystems) is that minerals have no capacity to regenerate (in human relevant time scales) and living systems do. Most significantly, though, from a long-term sustainability perspective, both can be mined.

The mining dilemma arises when the two look alike, when mining living systems and mining minerals are indistinguishable. I clear cut the forest, pump out the aquifer, assuming the two resources will come back. Alternatively, it is when a society organises itself *as if* the two are indistinguishable, as if it makes no difference to mine or sustain (over the long term). *Operationally*, both are extractive, both destroy to gain value, both reward effort: a farmer destroys native prairie, a gold miner fish habitat. In fact, both are routinely called "production": a farmer produces bushels of wheat; a wildcatter produces barrels of oil. Behaviourally, both engender a sense of just desert, of entitlement: I cleared the land, seeded the pasture, built up the herd, risked my savings, organised a farmers cooperative, found markets; I deserve my rewards. I found the deposit, built the mine, hired the workers, risked my capital, not to mention life and limb; I deserve a return.

*Thermodynamically*, in both, "production" entails the degradation of ordered, low entropy material (think cutting the tree, cracking the oil, processing the corn) for service value (house, heat, food) resulting in relatively disordered, high entropy material (boards and sawdust, gasoline and $CO_2$, cellular energy, water and $CO_2$). The process, whether applied to a renewable (tree or corn) or non-renewable (oil), appears the same. In fact, mainstream economics and legal doctrine often treat the two the same. Depletion is an acceptable option if it can be shown to be the "best use" compared to all other options, including investing monetary proceeds from extraction in a bank and earning a "return."

From a neoclassical economic perspective and a rational legal perspective no distinction between mining and sustaining is necessary (Dryzek 1987). *Economistically*, resources are degraded to create service value. Whether it is irreversible has no bearing on the generation of such value. The tree, the corn plant, the oil is necessarily degraded. Indeed from the production angle what matters is precisely production. As oil people say, after that it's just "disposal" – that is, the true business of oil is in exploration and extraction – i.e. "production;" those who distribute and sell are just getting rid of the stuff so more can be produced. With groundwater, western water law in the United States is constructed to promote frontier expansion. A Utah water lawyer, Clesson Kinney, wrote a seminal book in 1894 that took as given that water underground, unlike surface water, was inexhaustible, flowing underground in watercourses that could not and need not be regulated. "Unfortunately," writes contemporary water law professor, Robert Glennon, "American groundwater law has never recovered from the contributions of Clesson Kinney. ... Groundwater rules [in] most states routinely allow the drilling of new wells," which is to say more wells, more pumping, more depletion. The result, Glennon writes, is that "throughout the country water tables are dropping as pumping exceeds recharge. Overdrafting or 'mining' groundwater creates serious problems" (2002, 30, 32).

*Ecologically*, mining is rarely just the removal of a mineral or volume of water or stand of trees. Extraction is linked, often invisibly or with significant time lags, to other resources, including potentially renewable resources. Hard rock mining consumes vast quantities of water, for instance. Pumping out a fossil aquifer eliminates dependent plant and animal life. I call this "collateral mining."

Mining of a non-renewable (the only way a mineral or, say, fossil groundwater can be used) reverberates through ecosystems, risking the permanent and irreversible elimination of potentially renewable resources. Perhaps the biggest risk is to the source of a renewable resource – the soil that nurtures the tree and the corn plant, the water supply that serves a city.

In all this, the mining dilemma arises when, for key actors in the society, everything comes to look like a non-renewable (Figures 15.1 and 15.2). Then, when everything is mining, degradation, depletion, extermination are normal; they are integral to resource use, the "cost of doing business," the opportunity cost of extraction. Indeed, a cursory sampling of journal article titles reveals a ready conflation of sustaining and mining. Or, as Gavin Bridge puts it,

> The reduction of sustainability to a set of management tools that obscure underlying resource and environmental politics has led some observers to argue that sustainable development—initially a call for a new set of development goals—has become the means du jour to the conventional ends of resource access and extraction.
>
> *(2004, 234)*

But even if mining minerals – metals, petroleum, coal – makes perfectly good sense, even ecological sense, which is to say there is no collateral mining, mining renewables makes no ecological sense, not if the integrity of the existing ecosystem (the forest, the grassland, the hydrologic system), a necessary condition for a sustainable economy, is a primary concern. Rationalisations for mining renewables – directly or collaterally – are, of course, ubiquitous: it creates jobs and grows the economy; it is the most efficient use of the resource; substitutes will be found; new technologies will fill in; other populations did it and got wealthy; hungry people must be fed.

*Historically*, to say that mining makes sense for renewables and non-renewables alike may have been a reasonable position when, for instance, minerals and fossil fuels were of minor significance in pre-industrial economies, especially in relation to resources such as forage, wood, and grain. It is hard to imagine, e.g., that with a global human population of 5 million in 5000 BCE, any amount of mining, let alone any method of mining (what we would now call artisanal), would degrade ecosystems. This of course changed with industrialisation and the onset of the fossil fuel era (when fossil fuels eclipsed wood and other energy sources).

The mining dilemma thus has a temporal dimension along with the conflation of mining and sustaining practices. What made sense in the past, indeed for some 99% plus of all human history, suddenly (historically speaking) does not. Only a century or so into the fossil fuel era and humans have demonstrated beyond doubt their inability to handle these substances (Princen, Manno, Martin 2015a; VanDeveer 2013). What's more, some practices can appear to be sustainable in the short term, even across generations, but not actually sustainable over the very long term. "Depending on the erosion rate, thick soil can be mined for centuries before running out; thin soils can disappear far more rapidly," writes geologist David Montgomery (2012, 23). So what was once adaptive, or at least not maladaptive, now threatens survival due to massive increases in population, consumption, and technological capacity.

In sum, operational, thermodynamic, economistic, ecological, and historical perspectives on mining all shift attention away from the immediate extractive and commercial value (a gold mine or alluvial farm is "worth" its net monetary return until it is mined out) to the long term, to value that is linked to all that is extracted – the target mineral, say, plus other resources and their potential uses over the long term, that is, that can be sustained. The mining dilemma arises when the two – mining and sustaining – conflate. This is more likely, I argue next, when the entire economy is organised as a mining operation.

## A mining economy

The preceding comes at mining via *practices*. Together, over time and across institutions such practices form a mining economy, an economy dominated by such practices and their biophysical and social consequences, positive and negative. Whether at the scale of a tiny mountain mining town or an industrialised country or the global economy (what defines the fossil fuelled industrial era of the past century or so), a mining economy's wealth is overwhelmingly dependent on extraction, that is, permanent and irreversible and, consequently, not sustainable. Once again, the mining in question here may be of minerals and other non-renewables and/or of living systems and other renewables; it is an *ecological definition* of mining I offer here, *not an industrial definition*.

The mining dilemma at this aggregate, collective scale manifests as a *path dependency* with some five features.

1   *Benefits are immediate, visible, and powerful.* Mining is at a scale and intensity that its benefits overshadow the benefits from extracting food and fibre from plants and animals and generating energy from wood, wind, and sun, that is, from regenerative practices. Moreover, mining benefits create a positive, self-reinforcing feedback loop: the benefits of extraction enable more extraction. Unlike with renewables, negative, checking feedback is mostly absent, especially in the short term. And the *short term* becomes the dominant time scale; the long term is presumed to take care of itself with yet more mining.

2   *Depletion is integral to mining practices and a mining economy.* Irrespective of the resource, it is normal to use up the resource and then find more, or substitute other resources, or employ new technologies to avoid problems. Each new source creates a frenzy of extraction and wealth accumulation followed ineluctably by decline. "Local" may be a mine site, a watershed, or, in the case of the resource curse, a country (Karl 1997; Ross 2012; VanDeveer 2013).

3   *Expansion follows depletion.* The unchecked, self-reinforcing positive feedbacks of mining determine the response to depletion: find more – expand exploration geographically, expand digging and drilling technologically. Combining expansion and depletion, local boom and bust cycles are thus endemic, along with human migration. Again, "local" can range in scale but now it can be global. That is, the world's national economies, each mostly a mining operation (Figure 15.1), aggregate, or integrate, into a *global mining operation.*

4   *Expansion of mining operations crowds out sustaining operations.* In part, this is because the benefits of mining are immediate, visible, and powerful whereas the benefits of sustaining operations are deferred (consumption restraint ensures future consumption), less tangible (the dollar value or power increment of a long-term orientation is hard to assess), and more diffuse as a source of power (it is less countable, less taxable) (see Figure 15.2).

5   *Mining's short-term gains eventually give way to its own accumulated losses, as well to the losses of regenerative capacities that mining depends on.* Mining thus has a *dual dependency* – on new sources and on renewable sources, both natural resources (think water, soil) and waste sinks (think atmosphere, oceans). Mining, in short, cannot be sustained. The global mining boom will go bust.

For a general proposition, as mining practices aggregate into and dominate an economy, a threshold is eventually passed in the prevalence and dependency of that society on such mining to create wealth and maintain itself. Mining becomes normalised for both non-renewables and for renewables, the two being indistinguishable at the point of "production" – it's all "resources" – and via production metrics – it's all wealth, all money. Social practices such as exploration, new sources, and migration are likewise normalised, along with organising principles such as total use, efficiency, consumer sovereignty, and, above all, growth (Princen 2014).

Materials flow studies provide some of the best empirical evidence that modern, industrial, consumerist economies are indeed mining economies.[1] For instance, one study found that the linear, one-way flow of nonwater materials – minerals, fossil fuels, metal ores, and biomass – comprises some 90.9% of the global economy's material flow by weight, materials that "are being used by society as short-lived products reaching their end-of-use typically within a year" (Figure 15.1).

A 2017 US Geological Survey study found that, for over a century, the United States has become "increasingly dependent on non-renewable materials to sustain its standard of living" (Figure 15.2). In 2014 that dependence was 96% of non-food and non-fuel raw materials. Curiously, after charting the century-long economic shift to mining and asserting that this trend is "essential for the economic well-being of citizens and national security," the Survey seems to conclude that it is not mining:

> advances in technology for the development of resource substitutes, improvements in technology for the discovery and extraction of mineral resources, and improved recycling efficiencies may reduce the possibility of exhausting the supply of nonrenewable resources and secure a sustainable supply of materials for society.

In other words, if we find substitutes for water and soil, and mine better, we'll be sustainable.

Another group of materials scientists calculated a "material footprint" to examine the biophysical basis of trade. It found that what appears to be productivity improvements in

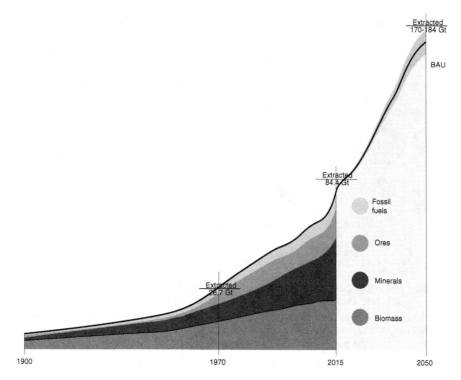

*Figure 15.1*    Mining as linear flow of extracted materials. Global material extraction of fossil fuels, ores, minerals, and biomass (not water) in gigatons (Gt) between 1900 and 2015

*Source*: Marc de Wit et al. (2018). The Circularity Gap Report: An Analysis of the Circular State of the Global Economy, January 2018. Circular Economy. www.circle-economy.com/the-circularity-gap-report-our- world-is-only-9-circular/.

wealthy countries is actually the offshoring of raw material extraction and concomitant "environmental impacts, such as water resource depletion; soil erosion; biodiversity loss; or pollution through agrochemicals, mine tailings, or oil spillages" (Wiedmann et al. 2015, 6275). In other words, the global mining economy only works because there is a flow of resources with concomitant depletion from some countries and claims of earned wealth from others. Because that flow is largely one-way, not all countries can "develop"; many can only extract, export, and deplete. What's more, because such dependency appears to be largely about mined resources (as mentioned earlier), for biophysical reasons it cannot continue forever: it cannot be sustained.

Similarly, a 2019 materials extraction report by the United Nations Environment Programme concludes that

> our consume and throwaway models of consumption have had devastating impacts on our planet. ... [Some] 90 per cent of biodiversity loss and water stress are caused by resource extraction and processing. These same activities contribute to about half of global greenhouse gas emissions.

> *(Oberle et al. 2019, 7)*

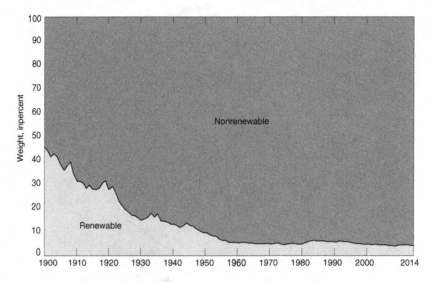

*Figure 15.2* Mining dependency. Percentage of US annual material usage by weight of renewable and non-renewable materials, excluding food and fuel, from 1900 to 2014

*Source*: US Geologic Survey and US Department of the Interior. 2017. "Use of Raw Materials in the United States from 1900 through 2014." Fact Sheet 2017-3062, August 2017. https://pubs.usgs.gov/fs/2017/3062/fs20173062.pdf.

## Just mining?

Is mining inherently unjust? A full comparative study, across time and cultures, would be needed to answer the question. But three features of mining as defined here – i.e., defined ecologically – suggest that it is.

First, many of the resources that lend themselves to mining – minerals, sand and gravel, oil, coal, old growth timber, aquifers – are concentrated in place. This is unlike the resources of hunter-gatherers or pastoralists that are dispersed. Concentrated resources lead to concentrated wealth, concentrated economic power, and concentrated political power, all of which increases inequality. The most vivid example in the late 19th, 20th, and early 21st centuries is oil (Coll 2012; Yergin 1991).

Second, no social structure better epitomises concentration than the state. In its earliest form it concentrated people (cities), food production (especially monocrop grains), livestock (for food and muscle power), and decision-making (administration) (Scott 2017). Modern versions are more complex and ever more concentrated, especially with machinery and fossil fuels. All this concentrates wealth and power which is used, first and foremost, to write the rules of the game, that is the game of maximum extraction irrespective of the kind of resource (material or human, renewable or non-renewable), and further concentrates wealth and power. Inequality intensifies.

What's more, because mining operations are visible and their products countable and hard to hide, the state, always on the lookout for taxable wealth, favours mining over sustaining. In sustaining operations, surpluses are minor and hard to count.

Third, if concentrated, state-sponsored mining is pitted against, say, tribal hunting and gathering, nomadism, shifting agriculture, or peasant farming, there is little question which prevails. The very organisation of mining is a major reason. Mining operations tend to be singularly focused on the mined resource, neglecting not only other resources but also other peoples. Mining success goes to the aggressive, risk-taking, competitive, even ruthless, qualities that overwhelm small-scale settled or dispersed producers.

The politics of mining thus aligns with an insatiable state. Concentration, state favouritism, competitiveness discriminate against sustaining. Inequity is inherent in the state-mining complex. Resistance politics is expected, especially when combined with cycles of boom and bust and a state's propensity to rise and fall. Until recently, though, the essential biophysical condition has been an ecologically empty world. Beginning roughly mid-20th century, resource frontiers are no more. The state can push the support population (Sale 1980; Tainter 1988), technologists can concoct virtual resources, financiers can fabricate currencies, but soon the mining state undermines itself. The 21st-century transition therefore is not just from mining to sustaining as if the bad resources are swapped out and the good ones swapped in. This transition, unlike all previous ones, is one from state as enabler of, and dependent on, mining, to the state as guarantor, even, dare I say, steward of sustaining. For that, the relative power of the support population vis-à-vis the extractive population will increase, a process aided by the eventual decline in mining's relative importance.

In sum, mining is never "just" mining, in two senses. Materially, as argued, it almost always entails collateral mining, especially in economies that follow the ethic of total use, efficiency, and growth (Princen, Manno and Martin 2015a; Worster 1985). Ethically, it almost always concentrates power in the hands of a few who tend to further concentrate their power opening up chasms of inequality. Mining *as practice*, mining as a defining characteristic of an economy, not just mining as wealth and impacts, is almost unavoidably unjust.

## Whither a sustaining economy?

A sustaining economy, by contrast, supports its members by extracting primarily that which renews, whether through self-generation (the "wild" resources hunters and gatherers depend on) or through domestication (the crops and livestock agriculturalists depend on). Mining of non-renewables still occurs but at a scale and rate that does not jeopardise renewables, i.e. there is no collateral mining. Excess – exceeding regenerative capacities – is avoided inadvertently (e.g. through high infant mortality and limited technology) or deliberately (via population control measures, practices that limit expansion or entrants) (Ostrom 1990). A sustaining society thus supports itself by, in the first instance, sustaining its resource base, wittingly or not. It *grounds* itself in the biophysical, the ecological, the cycles, and regenerative processes responsible for life on this planet.

Mining and sustaining are both adaptive strategies and are both risky (Kerschner et al. 2013; Turner et al. 2003). Mining economies can fail to find new sources; sustaining economies are vulnerable to exogenous shocks – extreme weather, for instance, or invading hordes. But because the mining economy depends on endless extraction and new sources or substitutes and the sustaining on limited but regenerative resources, mining over the long term appears more vulnerable.

## Normative implications: sufficient mining

If mining has been largely innocuous for all but the last couple centuries then there is probably a scale and rate of mining that is compatible with sustainability, albeit far smaller and far slower than that of the past couple centuries. Put differently, just because mining is the opposite of sustaining does not mean that a sustainable society must banish mining. The foregoing argument does lead to the conclusion, however, that, to be sustainable over the very long term, the ecologically and geologically long-term, collateral mining practices must be blocked. Blocking would then slow the mining of non-renewables.

What's more, because sustaining practices build in restraint, because they enact sufficiency – using enough but not too much of a resource – allowable mining in a resource regime where sustaining is hegemonic must itself build in restraint (Princen 1997). So the question becomes, what might be "sufficient mining"? (Sufficient mining here is not the same as so-called "sustainable mining," the industry's attempt to burnish its image by appropriating the term sustainability to extract with less environmental and social impact.) Here I entertain three possibilities.

First, like lobster harvesting with inefficient wire traps, gold mining could be restricted to independent operators (with, say, a couple hired hands), using pick axes and pans, no mercury or cyanide, and all operations completely transparent (a person can dream). Sure, big operators would scream, but no more than the transnational companies now banned entirely from mining gold in El Salvador, soon possibly in the Philippines (Broad 2017). Panning would severely limit overall extraction. It might even democratise gold mining in the sense that nearly anyone could do it. Consequently, gold would be extremely rare, and expensive. And, I surmise, the world would do just fine without it, or with very little of it. Certainly the child labourers in Africa and Asia would do better.

Second, not all mining is necessary or desirable. Just because it can be done, just because there is a market, just because some individuals gain great wealth and some governments generate major tax revenues, doesn't mean it is right. In fact, much about mining, from extraction and manufacture to distribution and adornment, is *excessive*: it exceeds regenerative capacities ecologically; it exceeds management capacities organisationally; it exceeds standards of decency ethically.

That is, *there are some things humans simply cannot handle*. The international community of nations has come to this conclusion with respect to trade in endangered species (Appendix A of CITES), ozone depleting substances (Montreal Protocol), mile-long drift nets, resource extraction generally in Antarctica, land mines, and some day, hopefully, nuclear weapons. For so much of what humans have conjured up, especially technologically, but also institutionally, the zero option is the best option (Princen 1996). Sufficient gold use then would be enough to display, to see and feel its special qualities but not so much as to spur a gold rush, claim superiority, or initiate collateral mining.

Third, fossil fuels are highly distanced, commoditised, placeless global substances. They are ubiquitous and cheap, available with the flip of a switch or the click of a nozzle. All uses are deemed legitimate – growing rice in a desert, heating driveways in mountain resorts – as long as the user can pay for it. Interruptions are, like interruptions to food supplies, an offense no politician wishes to encounter.

So how might fossil fuels be different, something so precious that we would use very little? A first step might be a campaign of delegitimisation, not of fossil fuels per se (or other non-renewables) (they are, after all, perfectly "natural"), nor of mining companies (consumers would logically have to be included, hence everyone) but of *uses*. A second step would be to treat fossil fuels like fine jewellery or a dangerous weapon, that is, use them only on special occasions for special purposes and otherwise keep them under lock and key at all other times. To lock away fossil fuels is to say that we will use them only when we must, only when no alternative exists – including the alternative of non-use – only when we are sure, or as sure as we can be, that such limited use will not cause cascades of depredations across landscapes and through the body politic as they are now doing. A third step would be to resurrect the magic in fossil fuels, the early magic, the magic that made such substances indeed special. Updated for the 21st century, that magic would necessarily derive from highly limited use. The magic of fossil fuels would then derive not from their scarcity (physically they are not at all scarce) but from their abundance, an abundance experienced as having plenty because their uses are so special (Princen 2015).

Making fossil fuels special would be a cultural act to complement the materials management acts of conventional policymaking. Thus, at the same time a politics of ending the fossil fuel era begins by *delegitimising* fossil fuels, that era ends by *celebrating* fossil fuels too. The delegitimisation is about excess rates, the celebration about the remaining modest rates. The result would, of course, be a profound cultural shift, precisely what is needed, I submit, in a time of great urgency and policy deadlock (Princen, Manno and Martin, 2015b).

In the end, then, sufficiency in the use of fossil fuels and other non-renewables would not be sacrifice or "doing without." It would be doing well now and into the future by using less fossil fuels and minerals, much less, than the most possible now. The energetic basis of a sustainable economy, then, would be relying mostly on regenerative sources, keeping within their regenerative capacities, and, when the occasion demands, a bit on non-renewables.

## Conclusion

"Ecological mining" seems oxymoronic. "Sustainable mining" certainly is. In this chapter I've argued that an ecological view of mining is not only possible; it creates a potentially useful lens on the role of mining in envisioning a sustainable society. One reason ecological mining makes sense is that humans are just one creature interacting with its environment. It so happens that a part of that interaction entails irreversibilities over a long time (itself a human perspective). A second reason is that minerals are themselves perfectly "natural," as is physical and chemical transformation. To mine is not so much to degrade (a human judgement) as to transform, what the earth and its inhabitants do constantly.

But mining can be ecological in a more practical sense. By practical I mean as it relates to human decision-making, human time scales, and human values such as perpetuating the species, leaving one's descendants a legacy, and honouring nature's (God's/the cosmos's) creation. Activities that undermine ecosystems and disrupt critical cycles (e.g. nutrient, hydrological, carbon, nitrogen, phosphorous) eventually undermine human cycles (e.g. reproductive, food producing, governing). They diminish humans' quality of life; they even take lives. They diminish one's place in a larger scheme. When they entail practices of domination (e.g. dam and divert a river until it no longer reaches the sea) that domination too easily spills over to human domination. Mining, as argued, is inherently unjust and so too are irreversibilities. Such societies cannot endure; they are not sustainable.

This proposition derives from ultimate causes that are biophysical, namely, declining availability of key natural resources, increasing defensive expenditures from past and ongoing environmental abuses, and "surprises" from a panoply of biophysical and social developments. Proximate causes are likely to be economic (e.g. more speculation, more bubbles, more bursting bubbles) and political (e.g. retreat into enclaves, national to local, with attendant demagoguery and exclusion).

Ecological mining is possible when

1    collateral mining is blocked;
2    structures of domination – e.g. concentrated ownership and decision-making, heavy capital investment, coercive forces (police, army) are not the norm in resource extraction;
3    sustaining practices integrated into a sustaining and just economy are hegemonic.

Put differently, ecological mining is, in the first instance, about ecological relations, human and non-human, and only in the second, a distant second as I see it, actual extraction. In a sustainable world mining does not threaten that which can be sustained; mining practices are subordinate to sustaining practices. That which is mined *supports* that which is sustained.

## Note

1 I thank Tom Makled for research assistance in collecting these data and assembling.

## References

Bridge, G. 2004. Contested Terrain: Mining and the Environment. *Annual Review of Environment and Resources* 29: 205–259.

Broad, R. 2017. *Email Communication*, January 4.

Clapp, J. 2014. Financialization, Distance and Global Food Politics. *The Journal of Peasant Studies* 41 (5): 797–814.

Coll, S. 2012. *Private Empire: ExxonMobil and American Power.* New York: Penguin Press.

Daly, H. E. 1996. *Beyond Growth: The Economics of Sustainable Development.* Boston, MA: Beacon Press.

Dryzek, J. S. 1987. *Rational Ecology: Environment and Political Economy.* New York: Basil Blackwell.

Gadgil, M. and R. Guha 1992. *This Fissured Land: An Ecological History of India.* Berkeley: University of California Press.

Glennon, R. 2002. *Water Follies: Groundwater Pumping and the Fate of America's Fresh Waters.* Washington: Island Press.

Jackson, T. 2009. *Prosperity without Growth? The Transition to a Sustainable Economy.* London: Sustainable Development Commission.

Karl, T.L. 1997. *The Paradox of Plenty: Oil Booms and Petro-States.* Berkeley: University of California Press.

Kerschner, C., C. Prell, K. Feng, and K. Hubacek. 2013. "Economic Vulnerability to Peak Oil." *Global Environmental Change* 23 (6): 1424–1433.

Montgomery, D. R. 2012. *Dirt: The Erosion of Civilizations.* Berkeley: University of California Press.

Oberle, B., S. Bringezu, S. Hatfeld-Dodds, et al. 2019. *Global Resources Outlook 2019: Natural Resources for the Future We Want. A Report of the International Resource Panel. Summary for Policymakers.* Nairobi: United Nations Environment Programme.

Ostrom, E. 1990. *Governing the Commons: The Evolution of Institutions for Collective Action.* New York: Cambridge University Press.

Princen, T. 1996. The Zero Option and Ecological Rationality in International Environmental Politics. *International Environmental Affairs* 8 (2): 27–71.

Princen, T. 1997. Toward a Theory of Restraint. *Population and Environment* 18 (3): 233–254.

Princen, T. 2005. *The Logic of Sufficiency.* Cambridge: MIT Press.

Princen, T. 2014. Sustainability: Foundational Principles (189–204) in P.G. Harris (Ed.) *Routledge Handbook of Global Environmental Politics.* London: Routledge.

Princen, T. 2015. The Cultural: The Magic, the Vision, the Power (53–95), in T. Princen, J. P. Manno and P. L. Martin (Eds.), *Ending the Fossil Fuel Era.* Cambridge: The MIT Press.

Princen, T., J. P. Manno and P. L. Martin (Eds.). 2015a. *Ending the Fossil Fuel Era.* Cambridge: MIT Press.

Princen, T., J. P. Manno and P. L. Martin. 2015b. The Problem. In T. Princen, J. P. Manno and P. L. Martin (Eds.), *Ending the Fossil Fuel Era.* Cambridge: The MIT Press.

Princen, T., M. Maniates and K. Conca (Eds.). 2002. *Confronting Consumption.* Cambridge: MIT Press.

Ross, M. L. 2012. *The Oil Curse: How Petroleum Wealth Shapes the Development of Nations.* Princeton: Princeton University Press.

Sale, K. 1980. *The Decentralist Tradition.* New York: Perigee Books.

Scott, J. C., 2017. *Against the Grain: A Deep History of the Earliest States.* New Haven: Yale University Press.

Tainter, J. A. 1988. *The Collapse of Complex Societies,* Cambridge: Cambridge University Press.

Turner, B. L., R. E. Kasperson, P. A. Matson, J. J. McCarthy, R. W. Corell, L. Christensen, N. Eckley, J. X. Kasperson, A. Luers, M. L. Martello, C. Polsky, A. Pulsipher and A. Schiller. 2003. A

Framework for Vulnerability Analysis in Sustainability Science. *Proceedings of the National Academy of Sciences of the United States of America* 100 (14): 8074–8079.

VanDeveer, S. D. 2013. *Still Digging: Extractive Industries, Resource Curses, and Transnational Governance in the Anthropocene.* Washington, DC: Transatlantic Academy.

Wiedmann, T.O., H. Schandl, M. Lenzen, D. Moran, S. Suh, J. West and K. Kanemoto. 2015. The Material Footprint of Nations. *PNAS* 112 (20): 6271–6276.

Worster, D. 1985. *Rivers of Empire: Water, Aridity, and the Growth of the American West.* New York: Oxford University Press.

Yergin, D. 1991. *The Prize: The Epic Quest for Oil, Money, and Power.* New York: Simon and Schuster.

# 16

# Financialising nature

*Jennifer Clapp and Phoebe Stephens*

## Introduction

The growing importance of financial markets, motives, actors, and institutions in the world economy – often referred to as "financialisation" – has significant implications for the environment and sustainability. There has been a burgeoning literature on the phenomenon of financialisation in the global economy in recent decades (e.g. Epstein 2005; Krippner 2011; van der Zwan 2014). Financialisation has been felt across a range of sectors, from manufacturing, to food and agriculture, to mining and energy. Certainly, the interest of financial capital in nature and natural resources is nothing new. Financial investors have long provided capital backing and received profit from the extraction of nature through economic activities such as mining, forestry, oil and gas production, water taking, and agriculture (see also Princen, this volume). In recent decades, however, we have seen the rise of new kinds of financial instruments that enable investors to profit from these sectors in novel ways – a key characteristic of modern-day financialisation – which, in turn, has important implications for environmental sustainability.

In this chapter, we examine the implications of financialisation for environmental sustainability. We outline the long attraction between financial investment and nature, and make the case that this relationship has seen important changes in recent decades as the process of financialisation has unfolded. In particular, financialisation has encouraged the rise of new kinds of financial instruments that are tied to natural resources and environmental change. We show that these new financial instruments have relied on an abstraction of nature from its material form, and have transformed elements of the natural world into purely financial assets. These kinds of new financial tools are often based on indexes or pooled funds that track the performance of real things, such as natural resources, land, carbon, or the weather. But the fact that nature ultimately forms the underlying base for this financial investment means that this financial activity can, and often does, have real world consequences. These effects, however, are often distanced from their financial origins, and are not always accounted for in sustainability policy and governance.

## From finance and the environment to financialisation of the environment

Finance has long underpinned activities that have mattered for environmental outcomes. Dating back centuries, financial investors have provided funding for activities ranging from oil exploration, forestry operations, mining, and agricultural commodity trading – each of which has direct implications for environmental outcomes, including climate change, deforestation, toxic pollution, and land degradation. Financial investors have also provided project financing for infrastructure projects ranging from large-scale dams, to power plants, to road construction across the world, activities that have also historically been associated with environmental harm due to deforestation, biodiversity loss, and carbon emissions (Clapp and Dauvergne 2011). Because these kinds of activities were widely assumed to be important drivers of economic growth, the environmental costs associated with them were, until relatively recently, often considered to be simply the cost of doing business. In effect, for hundreds of years the environmental externalities associated with investment financing were neither priced nor accounted for.

It is only in the last 50 years that financial investment in these sorts of economic activities has been subject to closer scrutiny by policy makers for its environmental impacts. This concern rose to a head in the 1980s and 1990s, when the concept of "sustainable development" began to gain traction in environmental policy and governance. Activists began to target global financial institutions such as the World Bank over concerns about the environmental implications of its development lending (Rich 1994). Alongside efforts to "green" project lending by public international financial institutions, there was a push in the 1990s to encourage more environmentally friendly activities by private banks through initiatives such as the United Nations Environment Programme's Finance Initiative (UNEP FI 2012). Similarly, the Equator Principles, adopted in 2003, call upon the private banking industry to pledge to ensure its project lending takes environmental considerations into account (Wright and Rwabizambuga 2006).

Beyond a focus on lending institutions, financial investors themselves – individuals as well as asset management firms, pension funds, and other institutional investors – are now actively and routinely encouraged to invest in socially and environmentally responsible ways. As awareness has increased, there has been a recognition of the responsibility of investors to minimise any potential negative environmental effects from the activities their financing supports. Responsible investment initiatives emerged in the mid-2000s, including the UN Principles for Responsible Investment (UN PRI), which encourage investors to take environmental, social, and governance considerations under advisement before making investments in major firms. Such initiatives fall under the broad heading of socially responsible investing (SRI), which aims to reduce socially and environmentally harmful investments and is now well established amongst mainstream institutional investors (O'Donohoe et al. 2010; Geobey 2014). More recently, interest has grown in not merely avoiding the negative impacts of financial investments, but also in generating positive ones. This movement falls under the wide umbrella of "social finance" and includes a variety of tools and approaches such as impact investing, venture philanthropy, alternative currencies, ethical banking, and social impact bonds (Rizzi et al. 2018). The guiding rationale behind these approaches is that private sector capital must be mobilised in order to tackle the complex socio-ecological problems of our times.

At the same time as calls have grown for more sustainable investment finance, however, there has been an increased "financialization" of the economy, that is a growing importance of financial actors, institutions, and motives in guiding key decisions within firms and in the economy more broadly (Epstein 2005; Krippner 2011). This growing role for finance in the economy is widely seen to be the product of neoliberal economic policies that became vogue across most governments in the 1980s, and which saw a significant scaling back of government regulations, including in the financial sector (Helleiner 1994). Financialisation has generated new investment tools that enable investors to accumulate profit in new arenas and sectors through complex derivatives and other financial products. Derivatives are financial instruments whose value is derived from the price of an underlying asset. For example, a futures contract, which arranges for the payment for an asset at a future date, is a common type of derivative. Although a derivative instrument is based on an underlying real asset, it is abstracted from those assets and is a purely financial instrument (Krippner 2011; van der Zwan 2014).

The development of new financial instruments that are designed to draw new revenue streams from natural resources and the environment creates a new arena for financial profit by investors. These nature-based financial products enable investors to speculate on the changing prices of natural resources and the costs of environmental change, and are based on what Loftus and March (2015) call "proxy commodities" that are purely financial in nature. Some of these new investment tools seek to profit from the development and use of traditional natural resource commodities such as energy, minerals, agriculture, and land. Others have sought to generate profit from efforts to address environmental problems such as water scarcity, weather risks, and climate change (Ouma et al. 2018). For these non-traditional commodities, markets did not previously exist and had to be created before the natural phenomenon could be financialised. In other words, water, weather, and climate had to be commodified, that is, given a price so that they could be traded on a market. As Sullivan (2012, 8–9) notes, the commodification of nature involves developing the means to measure it numerically, the attachment of monetary value to the units of measurement, and then the creation of markets for it, typically by governments. The creation of markets for environmental costs and services then paves the way for the financialisation of those markets by the private financial sector. Importantly, the rapid expansion of financialisation via novel investment tools has been associated with heightened levels of resource extraction and environmental degradation in a number of cases, even though these tools are marketed as purely financial and not as directly trading actual "things" like commodities, units of carbon, or water.

Large-scale institutional investors with long-term outlooks – including insurance companies, pension funds, mutual funds, hedge funds, sovereign wealth funds, commodity trading firms, and university and foundation endowments – have been especially interested in these new types of financial investment vehicles. These types of investors typically invest with what are termed "passive" investment strategies. That is, they prefer to invest in assets that have returns over the long term, but that do not require active management based on current market conditions. For this reason, these investors have been especially drawn to investment vehicles that pool funds managed by others, or that follow an established index that tracks various sets of underlying assets such as commodities, firm shares in specific sectors, or real estate. In these types of investment vehicles, investors are tracking the performance of an underlying real asset, without actually owning those real assets directly.

Below, we provide an overview of some of the new investment vehicles of this type that are in effect financialising nature by abstracting natural resources and the environment for the purpose of financial investment. We explain how these investment vehicles work, the forces behind their emergence, the size of these investments, and the key investors, as well as their implications for the environment and sustainability. Because these financial products are several steps removed from actual commodities, the linkages between the financialisation of nature and its sustainability impacts are often obscured. Indeed, this distancing between finance and outcomes on the ground makes it challenging to draw a direct connection between a single financial investment and a particular outcome. Still, it is possible to identify the mechanisms by which the various nature-based investment tools affect economic activities, which, in turn, can have an environmental impact, and we try to do so for each type of instrument.

## Commodity index funds

Futures contracts for commodities – such as agricultural products, minerals, and energy – have been traded by investors for hundreds of years on commodity exchanges around the world. Concerned about the potential for financial speculation to distort the markets for these commodities, regulators in the United States put rules in place that limited the number of commodity futures contracts that purely financial investors could hold, and required transparency in the form of reporting on such trades. These rules remained in place for over 60 years. But with the rise of neoliberal economic policies in the 1980s–1990s, there was growing pressure to deregulate the activities of financial actors engaged in commodity markets. A series of deregulatory moves in the 1980s and 1990s was codified in the 2000 Commodity Futures Modernization Act in the United States, home to the largest commodity markets in the world (Ghosh 2010). With this new legislation, over-the-counter (OTC) derivatives were effectively exempted from regulatory oversight (Russi 2013). This move brought the US commodity futures markets more in line with European commodity markets, which had only light regulations (van Tilburg and Vander Stichele 2011).

Financial institutions pushed for deregulation in large part because they were keen to sell new kinds of commodity investment vehicles to investors, including what are known as commodity index funds (CIFs) (Clapp and Isakson 2018). CIFs track the performance of an index that includes the prices of futures contracts for a range of commodities, typically including petroleum products, minerals and metals, livestock, and agricultural commodities. Standard and Poor's Goldman Sachs Commodity Index (GSCI) and the Bloomberg Commodity Index (BCOM) are the most popular indexes on which this kind of investment product is based, and products based on these indexes are sold by financial actors such as asset management companies and investment banks (Meyer 2015). Some indexes focus on only agricultural commodities; some on only minerals; and some just on energy. Some commodity index funds are sold "over-the-counter" (OTC), that is, directly to investors, while others are traded on exchanges. The latter are known as exchange traded funds (ETFs) (Russi 2013). The novel feature of CIFs is that they create new arenas for financial accumulation that are open to a wider range of investors than was previously the case. Greater access to these investments has been facilitated by the fact that with these new investment vehicles, investors no longer need to own the commodity, or even any commodity futures contracts, to profit. Rather, they simply buy a share of a financial product through a financial intermediary.

Financial investors became especially interested in CIFs as a vehicle to diversify their investment portfolios after 2000 (Meyer 2010). Institutional investors, who typically have long-term outlooks, find CIFs attractive because they are low-maintenance investments that provide a hedge against inflation. As food and energy prices began to rise in the 2005–2012 period, investors moved into the sector in large numbers. Total financial assets under management in the commodities sector climbed from around US$10 billion in 2000 to US$150 billion just before the 2008 financial crisis. By 2011, it was worth over

US$450 billion (Meyer 2015; UNCTAD 2015, 21). Financial investment in commodity ETFs shot up markedly, from under US$10 billion in 2006 to over US$200 billion in 2012 (UNCTAD 2015). A decline in commodity prices after 2013 led to lower interest among investors, and the amount invested in commodities fell to US$161 billion by the end of 2015. However, financial investment picked up again in 2016 to reach US$235 billion in commodity assets (Hume and Sanderson 2016).

Although CIFs are purely financial products in that the investors do not own physical commodities or even commodity futures contracts, this does not mean that investment in them is neutral with respect to sustainability. On the contrary, investment in CIFs can bid up prices and encouraged new forms of physical extraction of those commodities. Indeed, over the 2007–2013 period, prices for fossil fuels, agricultural commodities, and minerals and metals all shot up markedly, sparking increased interest in oil extraction, biofuel production, mining, and agricultural commodity production, each of which have enormous environmental implications including carbon emissions, soil degradation, and biodiversity loss. Shale and tar sands oil production, for example, became economically more attractive as prices for fossil fuels rose, leading to widespread concerns about the ecological consequences not only of the extraction of these resources, which includes water use and toxin release, but also over the carbon released when burned (Willow and Wylie 2014). Biofuel production also increased, leading to concerns about deforestation from land clearing to increase the area available for the production of biofuel crops (Neville 2015). Because they are typically bundled together in CIF financial products, prices for these commodities are increasingly moving together (UNCTAD 2011). Thus, if one commodity sees a price increase, the others do as well. As prices for these commodities become increasingly entangled with one another, the environmental and sustainability implications can become amplified, as market dynamics for any one of the commodities in the index can drive heightened extraction, and, in turn, the environmental impact, of all of them.

## Real estate investment trusts (REITs)

A REIT is a company that owns and usually operates income-generating properties whose shares are publicly traded on stock exchanges (Chiang et al. 2017, 2). REITs have transformed the investment landscape by rendering a traditionally illiquid and privately owned industry – the purchase and sale of physical land – into an abstract, liquid, and publicly owned one through the merging of several financial assets together into one instrument that is packaged and resold to investors (Chiang et al. 2017, 2). This financial instrument allows individual investors of more modest means to participate in landownership without having to own and manage the properties directly.

As Fairbairn explains, timberland REITs emerged in the 1980s as a result of "economic transformations that began in the 1970s – the increasing size and power of institutional investors and the corporate takeover movement" (Fairbairn 2014, 787). Institutional timberland investors include pension funds, insurance companies, foundations, and church institutions who "value the sustainability of this asset class but also the composition of the returns and the risk profile" (Aquila Group 2015). Moreover, because of the wide range of share values offered, individuals also readily participate in these investment vehicles.

Farmland REITs developed more recently and have attracted investors looking for new ways to capitalise on (narratives of) food and land scarcity. In the shadow of the 2007/2008 food and financial crises, there was a substantial increase in investor interest in agricultural land, as investors sought a secure place to invest capital (Visser 2017). Farmland fit the bill as it is perceived as a low risk investment that helps to diversify and balance portfolios (Clapp and Isakson 2018). Hedge funds and private equity are also involved in REITs, particularly in the case of farmland. Farmland REITs are still small in number but hold staggering amounts of farmland. Farmland Partners REIT possesses approximately 150,000 acres of farmland across 16 US states, and the Gladstone REIT holds 54,000 acres in seven states (Clapp and Isakson 2018, 93).

Timberland REITs are often positioned as sustainable investments because of the vital role of forests in sequestering carbon from the atmosphere. As such, institutions may invest in a timberland REITs to satisfy environmental targets. The degree to which this is true varies across jurisdictions. In the United States, for example, forestry is regulated federally and by the state forestry commission, resulting in relatively tight guidelines. As a result, all of the timberland REITs in the United States have their land certified by either the Sustainable Forest Initiative or the Forest Stewardship Council (Lerner 2015). However, in the case of palm oil REITs in Malaysia, the sustainability track record is less positive due to the environmental challenges associated with palm oil production such as deforestation and biodiversity loss.

In the case of farmland REITs, these investments are still nascent and it is unclear exactly what their environmental impact will be. However, if the short-termism typical of speculative investment infiltrates the space, it could lead to "careless environmental governance" (Fairbairn 2015, 244). Moreover, the lack of farming expertise of many investors has meant that in some cases, industrial farming technologies have been applied in an irresponsible way, adding stress to the environment while contributing further to greenhouse gas emissions (Clapp and Isakson 2018, 101). Finally, the push for high returns may place pressure on farmers to adopt high-yield production methods and chemical use to control pests, rather than supporting slower growth but more sustainable production methods (Knuth 2015, 165).

## Weather derivatives

Weather derivatives are a relatively new financial product that packages the weather so that it can be traded for profit. Because the weather *itself* cannot be traded, weather derivative contracts – assets like futures or options – rely on underlying values such as meteorological indices to create a market. Weather derivatives emerged in response to increased weather variability due to climate change, which presents new risks for firms in a variety of sectors and also creates a demand from investors to "climate proof" their investment portfolios (Isakson 2015, 575). The evolution of meteorological science that allows for more accurate measurement and forecasting enabled the commodification of information about the weather (Pollard et al. 2008, 619). Deregulation in the US energy industry in the mid-1990s, which exposed companies to greater weather-related risk (Pollard et al. 2008; Isakson 2015), prompted a number of energy utilities to develop the first privately negotiated, OTC weather derivative in 1997 by a US power company, Aquila Energy (Climetrix 2010, Clark 2010). Two years later, the Chicago Mercantile Exchange (CME) listed weather derivative contracts and is now the world's largest weather derivatives exchange, listing more than 60 contracts (Till 2014; Carabello 2018). Though OTC contracts still dominate the market, weather derivatives are now listed on a number of commodity exchanges.

The most commonly used weather index is based on average temperature but some derivatives track wind-speed, rainfall, and even humidity. According to Till, weather derivative contracts are priced "using actuarial analysis of historical payouts, factoring in recent weather trends and climatic trends" (Till 2014). These contracts differ from traditional insurance products in that weather insurance is geared towards high-risk, low probability events such as hurricanes and floods, whereas weather derivatives focus on low-risk high probability events such as dry summers or warm winters (Pollard et al. 2008, 618). Firms use weather derivatives to hedge against weather-related losses but speculators also participate in the weather derivatives market. This strategy is considered more nimble and cost effective than traditional insurance products tied to disaster relief (Mandel et al. 2010).

The weather derivatives market is now valued at an estimated US$8 billion and is growing quickly (Carabello 2018). The market has expanded beyond the energy sector and now includes municipal governments, agrifood businesses, and event management companies (Seth 2018). Insurance companies, investment banks, commercial banks, commodity traders, and hedge funds are all actively involved in the weather derivatives market (Carabello 2018). As renewable energy generation becomes increasingly popular, new weather derivatives contracts for wind have emerged, though these are still in the early stages of development (Gandel 2017).

Index-based agricultural insurance (IBAI), which is a form of weather derivative, clearly illustrates the positive and negative ways in which these financial products interface with the environment. IBAI was developed in the late 1990s, with support from the World Bank, as a way to allow developing country farmers to reduce the impact of climatic variability on their crops (Clapp and Isakson 2018, 68). It is positioned as a benefit to poor farmers who may be ineligible for traditional forms of insurance as IBAIs typically do not require proof of assets, which reduces surveying costs and ultimately increases the affordability of insuring small plots of land (Isakson 2015, 270). At first glance this approach may appear to be a relatively benign and even efficient way of hedging risk. However, Isakson (2015) explains how IBAIs affect land-use patterns in troubling ways because of their interconnection to agricultural modernisation. In addition, this type of insurance is tied to the purchase of modern inputs – locking farmers into an industrial model of agriculture that is often insensitive to the ecological surroundings (Isakson 2015). In other words, the proliferation of weather derivatives is tied to agricultural modernisation, which tends to reduce the biodiversity of agricultural landscapes and their resilience to extreme weather events typical of climate change (Isakson 2015).

## Water index funds

The privatisation of water utilities occurred globally during the 1990s, and was an enabling condition for financial investment in the water sector (Bayliss 2014). Today, the focus has shifted away from discourses of efficiency, towards water security, as awareness grows around the issue of water scarcity (Ahlers and Merme 2016, 769). According to some estimates, by 2035 nearly 3 billion people will be water stressed as a result of a changing and warming climate, growing demand from a widening variety of industries, and ballooning urban populations (Kaufman 2012, 470). This worrisome scenario has led to calls for large infrastructure projects to ensure a safe and consistent supply of water. These projects require enormous injections of capital, which is where financial investment comes into play. In this context, four water ETFs were launched in 2005 (Rompotis 2016, 104). These funds typically track the value of stocks of water-related businesses (Kaufman 2012, 470) such as those involved in water conservation, purification, and treatment (Rompotis 2016, 103). In the case of water index funds, actual units of water are not being traded. However, the recent launch of a water futures exchange in Australia suggests that such markets could become more mainstream over the coming decades (Curran 2014).

According to Bayliss (2014, 298), the four original water ETFs that were launched in 2005 had US$1.4 billion in assets under management by 2014. Today, more than 100 indices are involved in tracking and measuring water-related stocks (Kaufman 2012, 470). This mushrooming market is attracting investors beyond the traditional public agencies and private water industry. Lenders, institutional investors, sovereign wealth funds, private equity investors, water funds, and new multilateral banks are also participating in the nascent market (Ahlers and Merme 2016, 767).

Since water is not being physically traded by ETFs (and more importantly, ETFs are tracking firms investing in water infrastructure, not water futures), these instruments have not directly affected the price of water (Bayliss 2014, 301). However, the connection between financial investment and large infrastructure projects such as dams has significant adverse impacts on the surrounding physical environment. The development of a global water futures market would likely lead not only to the price volatility seen in other futures markets (such as agriculture), but could also create a scenario where the highest bidder wins. As Kaufman (2012, 471) points out, "if the natural-gas industry can pay more for water than soy farmers, then the gas drillers will get the water and the soya will not." Such a market would not inherently produce water-conserving efforts. However, Australia has worked to ensure that its nascent water market is linked to its conservation efforts. It has done so by requiring that water rights be registered with Australian states that are in charge of managing water supplies. This serves to limit the aggregate drawdown of water, positioning the government initiative as a form of a cap-and-trade program (Curran 2014). Because the water futures market is still new, its sustainability impacts are merely speculative at this point in time.

## Carbon derivatives

The rise of carbon markets in the past few decades as a state-sanctioned means to address climate change has been accompanied by the development of complex financial derivatives associated with carbon trading (Layfield 2013). Carbon markets operate based on a fixed number of permits to emit carbon (allowances), which require firms to either reduce their emissions or purchase carbon emission rights or offsets (carbon credits that are verified units of avoided carbon emissions) from other firms if they are unable to meet their targets. The trading of carbon permits effectively sets the price of carbon, based on the supply and demand of allowances and credits. While the base of the market is the trade in carbon products between firms, there is also a demand for carbon-based financial derivatives that can help firms to manage risk associated with changing carbon prices (Bryant 2018, 610). This market in carbon derivatives – which includes financial products that bundle various carbon-based assets that, in turn, are marketed not just to firms but also to large institutional investors such as pension funds – is the fastest growing part of the carbon market (Layfield 2013, 908).

While some analysts have raised concerns about the emergence of financial derivatives based on carbon markets as an impediment to addressing climate change (Lohmann 2010), others have seen some promise in harnessing financial markets for the benefit of carbon reduction (Newell and Paterson 2010). In both cases, there was an expectation that the carbon markets would inevitably grow to be massive in size after they were endorsed as part of the Kyoto Protocol in 1997. However, after initial growth in the first years after the European Union (EU) Emissions Trading Scheme (ETS) was established in 2005, and the carbon credit provisions associated with the Clean Development Mechanism (CDM) of the Kyoto Protocol came into force in 2008, the markets shrank significantly and have remained stagnant in recent years despite various other carbon trading schemes being established in a number of countries around the world. Having reached a value of US$176 billion in 2011, the size of carbon markets in 2018 was approximately US$50 billion (Bryant 2018). While larger than weather derivatives, this amount is significantly lower, for example, than standard commodity derivatives markets. According to Layfield (2013, 908), however, the value of financial and technical trades associated with the carbon markets, at least up until 2013, surpassed the value of actual trades in carbon permits for compliance purposes. Nonetheless, overall, carbon has not turned into the accumulation opportunity many had expected (Bryant 2018, 612).

The idea behind carbon markets, and ultimately the financial derivatives associated with them, is that this trade will support major reductions in carbon emissions by making it costly to obtain carbon emission permits, and attractive to reduce carbon emissions in developing countries for CDM credits, even as it opened up opportunities for financial actors to profit. But because carbon credits as a commodity are unique in that they are highly abstract – i.e. they represent something that is *not* emitted – it is extremely difficult not only to create a functioning market for carbon but also to measure and verify that it results in reduced emissions (Layfield 2013). A recent EU study (Cames et al. 2016) concluded that the carbon crediting procedures under the CDM have fundamental flaws, raising questions about the extent to which these credits really represent reduced emissions. And with the markets now stagnant, it is hard to see how they can make a significant dent in mitigating greenhouse gas emissions, at least in the near future.

## Implications for policy and governance

As the earlier examples show, the current era of heightened financialisation in the global economy has spawned new financial instruments that are tethered to natural resources and the environment in novel ways. Some of these new financial instruments have attracted more capital than others, with more traditional commodities – agricultural crops, energy, minerals, timber, and land – garnering a greater share of investment than assets associated with more recently constructed markets such as carbon, water, and weather. But despite the unevenness, it is clear that nature and natural resources are increasingly viewed as arenas in which to build new regimes for financial profit-making. While these new investment products are purely financial in nature and attract financial investors who are far removed from the physical assets on which their investments are ultimately based, they nonetheless have the potential to generate ecological side-effects that are problematic for sustainability. As Bracking (2015) notes, the virtual nature of financial investments is not without material impact, as it enables powerful actors more access to and control over natural resources and energy systems. This privileged access, in turn, can result in increased extraction that is not always mindful of sustainability considerations.

Although analysts have pointed to the potential for negative environmental outcomes, efforts to put policy and governance into place to address those concerns have been extremely weak, and in some cases nonexistent. There are a number of challenges to addressing these policy weaknesses. First, the environmental impacts of these natural resources and nature-based financial instruments occur through a complex set of economic processes that are not always easy or straightforward to identify, making the connection between cause and effect fuzzy. The effects often occur through price signals and other economic incentives built into the instruments, which can influence rates of exploitation of the natural resource or the technologies utilised, as in the case of IBAI outlined previously. Moreover, the highly abstract nature of derivatives and other purely financial investment tools tends to increase "distance" between investors and outcomes, making it difficult to assign responsibility to any specific financial actor for any environmental degradation that occurs as a result of the investment (Layfield 2013; Clapp 2017, see also Pellizzioni, this volume, on the challenges associated with assigning responsibility in complex economic structures). And, in the case of new financial instruments linked to markets for weather risk, carbon, and water, powerful actors portray these tools as key in solving environmental problems, without recognition of their potential negative impacts (Ouma et al. 2018, see also Brulle and Aronczyk, this volume).

Second, even if the linkages between financial investment tools associated with nature and natural resources and environmental outcomes were clear to regulators, it would be extremely difficult to put new financial rules in place to address them. As the battles over financial regulations in the United States and EU have shown following the 2007/2008 financial crisis, powerful financial lobbies fight hard to water down any legislation that may rein in their activities (Helleiner 2014). In this context, voluntary responsible investment initiatives have emerged as the primary response to the potential negative impacts of new forms of financial investment. Consequently, voluntary initiatives of this type have cropped up in some areas such as farmland (e.g. the PRI's Farmland Principles) and agricultural investment generally (e.g. the Principles for Responsible Agricultural Investment promoted by the World Bank, and the Food and Agriculture Organization of the United Nations (FAO) led Principles for Responsible Investment in Agriculture and Food Systems). But these initiatives do not specifically target financial derivatives, and more generally suffer from a number of weaknesses that limit their ability to ensure that investment protects the environment (for a review, see Clapp 2017).

Finally, advocating for stronger policies and governance initiatives to address the environmental side-effects of new financial instruments is harder than in the past because the number of participants in these types of investment is enormous. In the 1980s, activists were able to successfully target a very visible public lender, the World Bank, as it was concerned about its public image and consequently had to respond to calls for greater accountability. In the 1990s and early 2000s, the focus expanded to commercial banks, and while they are numerous, it was possible to get a significant number to sign on to a statement to ensure that their lending takes environmental concerns into account. With highly financialised investment in the current era, there are numerous players involved, many of which remain opaque to the public, including large institutional investors such as pension funds, hedge funds, sovereign wealth funds, and asset management companies, not to mention the millions of individuals whose money is also invested in pension funds and private savings. Because their investments are financial in nature, tracking an index rather than owning "real things," these investors typically do not see themselves as a problem. They are also often unaware of not only the impact of their investments but also the nature of their investments, because in many cases they have turned over their portfolios to professional asset managers.

In sum, given the challenges of enacting stronger policy and governance initiatives on this issue, we are faced with a continuation of a troubling dynamic in which the short-term demands of the financial sector for immediate returns tend to override the longer-term needs of ecological cycles and processes (Knox-Hayes 2013). In such a context, it is imperative that researchers continue to examine these dynamics, and bring them forward in ways that can make meaningful contributions to policy debates regarding the sustainability implications of the financialisation of nature and natural resources.

## References

Ahlers, Rhodante, and Vincent Merme. 2016. "Financialization, Water Governance, and Uneven Development." *Wiley Interdisciplinary Reviews: Water* 3 (6): 766–74.

Aquila Group. 2015. *Real Assets – Investments in Timberland.* Hamburg, Germany. www.aquila-capital. de/sites/51bb1bb94603d65165000005/content_entry51dc08dc4603d6340000007d/56950a38b-58f6b57f50018f4/files/2015-11_AC_White_Paper_Investments_in_Timberland.pdf.

Bayliss, Kate. 2014. "The Financialization of Water." *Review of Radical Political Economics* 46 (3): 292–307.

Bracking, Sarah. 2015. "Performativity in the Green Economy: How Far Does Climate Finance Create a Fictive Economy?" *Third World Quarterly* 36 (12): 2337–57.

Bryant, Gareth. 2018. "Nature as Accumulation Strategy? Finance, Nature, and Value in Carbon Markets." *Annals of the American Association of Geographers* 108 (3): 605–19.

Cames, Martin, Ralph O. Harthan, Jürg Füssler, Michael Lazarus, Carrie M. Lee, Pete Erickson, and Randall Spalding-Fecher. 2016. "How Additional Is the Clean Development Mechanism?" Berlin: Öko-Institut e.V. www.atmosfair.de/wp-content/uploads/clean_dev_mechanism_en.pdf.

Carabello, Felix. 2018. "Market Futures: Introduction to Weather Derivatives." Investopedia, March 6. Accessed April 17, 2018. www.investopedia.com/trading/market-futures-introduction-to-weather-derivatives/.

Chiang, Kevin C. H., Gregory J. Wachtel, and Xiyu Zhou. 2017. "Corporate Social Responsibility and Growth Opportunity: The Case of Real Estate Investment Trusts." *Journal of Business Ethics*. Accessed April 12, 2017. doi:10.1007/s10551-017-3535-1.

Clapp, Jennifer. 2017. "Responsibility to the Rescue? Governing Private Financial Investment in Global Agriculture." *Agriculture and Human Values* 34 (1): 223–35.

Clapp, Jennifer, and Peter Dauvergne. 2011. *Paths to a Green World: The Political Economy of the Global Environment*, 2nd ed. Cambridge: MIT Press.

Clapp, Jennifer, and S. Ryan Isakson. 2018. *Speculative Harvests: Financialization, Food, and Agriculture.* Halifax: Fernwood Press.

Clark, A. 2010. "Chicago Mercantile Exchange Starts Offering Rainfall Futures and Options." *The Guardian*, November 7. www.theguardian.com/business/2010/nov/07/chicago-exchange-rainfallfutures-options.

Climetrix. 2010. "Weather Market Overview." Accessed April 17, 2018. www.climetrix.com/WeatherMarket/MarketOverview/.

Curran, Rob. 2014. "How to Bet on the Price of Water." *Fortune*, June 25. http://fortune.com/2014/06/25/water-futures-markets/.

Epstein, Gerald A. 2005. "Introduction: Financialization and the World Economy." In *Financialization and the World Economy*, edited by Gerald A. Epstein, 3–16. Cheltenham: Edward Elgar Publishing.

Fairbairn, Madeleine. 2014. "'Like Gold with Yield': Evolving Intersections between Farmland and Finance." *Journal of Peasant Studies* 41 (5): 777–95.

Fairbairn, Madeleine. 2015. "Finance in the Food System." In *Handbook of the International Political Economy of Agriculture and Food*, edited by Alessandro Bonanno, 232–49. Cheltenham: Edward Elgar Publishing.

Gandel, Stephen. 2017. "Wall Street Still Can't Figure Out How to Make Money off Snow." *Fortune*, February 9. http://fortune.com/2017/02/09/wall-street-snow-derivatives/.

Geobey, Sean. 2014. "Measurement, Decision-Making and the Pursuit of Social Innovation in Canadian Social Finance." PhD diss., University of Waterloo.

Ghosh, Jayati. 2010. "The Unnatural Coupling: Food and Global Finance." *Journal of Agrarian Change* 10 (1): 72–86.

Helleiner, Eric. 1994. *States and the Reemergence of Global Finance: From Bretton Woods to the 1990s.* Ithaca, NY: Cornell University Press.

Helleiner, Eric. 2014. *The Status Quo Crisis: Global Financial Governance after the 2008 Meltdown.* Oxford: Oxford University Press.

Hume, Neil, and Henry Sanderson. 2016. "Commodities Attract Biggest Bets Since 2009." *Financial Times*, August 5.

Isakson, S. Ryan. 2015. "Derivatives for Development? Small-Farmer Vulnerability and the Financialization of Climate Risk Management." *Journal of Agrarian Change* 15 (4): 569–80.

Kaufman, Frederick. 2012. "Futures Market: Wall Street's Thirst for Water." *Nature* 490: 469–71.

Knox-Hayes, Janelle. 2013. "The Spatial and Temporal Dynamics of Value in Financialization: Analysis of the Infrastructure of Carbon Markets." *Geoforum* 50 (December): 117–28.

Knuth, Sarah Elizabeth. 2015. "Global Finance and the Land Grab: Mapping Twenty-first Century Strategies." *Canadian Journal of Development Studies* 36 (2): 163–78.

Krippner, Greta. 2011. *Capitalizing on Crisis: The Political Origins of the Rise of Finance.* Cambridge: Harvard University Press.

Layfield, David. 2013. "Turning Carbon into Gold: The Financialisation of International Climate Policy." *Environmental Politics* 22 (6): 901–17.

Lerner, Michele. 2015. "Timberland REITs Standing Tall." *Nareit*, July 20, 2015. www.reit.com/timberland-reits-standing-tall.

Loftus, Alex, and Hug March. 2015. "Financialising Nature?" *Geoforum* 60 (March): 172–75.

Lohmann, Larry. 2010. "Uncertainty Markets and Carbon Markets: Variations on Polanyian Themes." *New Political Economy* 15 (2): 225–54.

Mandel, James T., C. Josh Donlan, and Jonathan Armstrong. 2010. "A Derivative Approach to Endangered Species Conservation." *Frontiers in Ecology and the Environment* 8 (1): 44–49.

Meyer, Gregory. 2010. "Gurus Who Sparked Commodities Rush Are Betting on the Long Term." *Financial Times*, February 10.

Meyer, Gregory. 2015. "Commodity Indexing Embraces New Methods." *Financial Times*, August 19.

Neville, Kate J. 2015. "The Contentious Political Economy of Biofuels." *Global Environmental Politics* 15 (1): 21–40.

Newell, Peter, and Matthew Paterson. 2010. *Climate Capitalism: Global Warming and the Transformation of the Global Economy.* Cambridge: Cambridge University Press.

O'Donohoe, Nick, Christina Leijonhufvud, and Yasemin Saltuk. 2010. "Impact Investments: An Emerging Asset Class." In *The Rockefeller Foundation, and Global Impact Investing Network,* edited by J. P. Morgan. November 28. https://assets.rockefellerfoundation.org/app/uploads/20101129131310/Impact-Investments-An-Emerging-Asset-Class.pdf.

Ouma, Stefan, Leigh Johnson, and Patrick Bigger. 2018. "Rethinking the Financialization of 'Nature.'" *Environment and Planning A: Economy and Space,* Accessed January 30, 2018. doi;10.1177/0308518X18755748.

Pollard, Jane S., Jonathan Oldfield, Samuel Randalls, and John E. Thornes. 2008. "Firm Finances, Weather Derivatives and Geography." *Geoforum* 39 (2): 616–24.

Rich, Bruce. 1994. *Mortgaging the Earth: The World Bank, Environmental Impoverishment, and the Crisis of Development.* London: Earthscan.

Rizzi, Francesco, Chiara Pellegrini, and Massimo Battaglia. 2018. "The Structuring of Social Finance: Emerging Approaches for Supporting Environmentally and Socially Impactful Projects." *Journal of Cleaner Production* 170: 805–17.

Rompotis, Gerasimos G. 2016. "Evaluating a New Hot Trend: The Case of Water Exchange-Traded Funds." *The Journal of Index Investing* 6 (4): 103–28.

Russi, Luigi. 2013. *Hungry Capital: The Financialization of Food.* Hants: Zero Books, John Hunt Publishing.

Seth, Shobhit. 2018. "How Do You Trade the Weather?" *Investopedia,* January 31. www.investopedia.com/articles/active-trading/111014/it-possible-trade-weather.asp.

Sullivan, Sian. 2012. *Financialisation, Biodiversity Conservation and Equity: Some Currents and Concerns.* Penang: Third World Network. https://siansullivan.files.wordpress.com/2010/02/sullivan-financialisation-biodiversity-2012-twn-end16.pdf.

Till, Hilary. 2014. *Why Haven't Weather Derivatives Been More Successful as Futures Contracts? A Case Study.* Nice: EDHEC-Risk Institute. www.edhec.edu/sites/www.edhec-portail.pprod.net/files/publications/pdf/edhec-working-paper-why-haven-t-weather-derivatives_1436278088665-pdfjpg.

UNEP FI. 2012. "UNEP FI Position Paper on the United Nations Conference on Sustainable Development (Rio+20): A Financial Sector Perspective." Accessed April 17, 2018. www.unepfi.org/fileadmin/documents/UNEP_FI_Position_Paper_Rio20.pdf.

United Nations Conference on Trade and Development (UNCTAD). 2011. *Price Formation in Financialized Commodity Markets: The Role of Information.* New York and Geneva: United Nations. https://unctad.org/en/docs/gds20111_en.pdf

United Nations Conference on Trade and Development (UNCTAD). 2015. *Trade and Development Report, 2015: Making the International Financial Architecture work for Development.* New York and Geneva: United Nations. unctad.org/en/PublicationsLibrary/tdr2015_en.pdf.

van der Zwan, Natascha. 2014. "Making Sense of Financialization." *Socio-Economic Review* 12 (1): 99–129.

van Tilburg, Rens, and Myriam Vander Stichele. 2011. *Feeding the Financial Hype: How Excessive Financial Investments Impact Agricultural Derivatives Markets.* Amsterdam: SOMO. papers.ssrn.com/sol3/papers2.cfm?abstract_id=1974405.

Visser, Oane. 2017. "Running Out of Farmland? Investment Discourses, Unstable Land Values and the Sluggishness of Asset Making." *Agriculture and Human Values* 34 (1): 185–98.

Willow, Anna J., and Sara Wylie. 2014. "Politics, Ecology, and the New Anthropology of Energy: Exploring the Emerging Frontiers of Hydraulic Fracking." *Journal of Political Ecology* 21 (1): 222–36.

Wright, Christopher, and Alexis Rwabizambuga. 2006. "Institutional Pressures, Corporate Reputation, and Voluntary Codes of Conduct: An Examination of the Equator Principles." *Business and Society Review* 111 (1): 89–117.

# 17

# Environmental countermovements

## Organised opposition to climate change action in the United States

*Robert Brulle and Melissa Aronczyk*

In June 1988, NASA scientist Dr James Hansen testified before Congress that "the greenhouse effect has been detected, and it is changing our climate now" (Hansen 1988: 40). Hansen's testimony forced the issue of climate change into American public consciousness. It also marked the beginning of a long political struggle to deal with carbon emissions in the United States. Twenty-nine years later, in a surprisingly festive ceremony in the Rose Garden on 1 June 2017, US president Donald Trump announced that the United States was withdrawing from the Paris Climate Accord, an international agreement signed a year earlier by 195 countries to mitigate greenhouse gas emissions.

These two events bookend the ongoing political contention over US responses to climate change. During this time period, atmospheric scientists have provided increasingly clear and dire evidence of alterations in the earth's climate system, including five United Nations Intergovernmental Panel on Climate Change (IPCC) Assessment Reports, and four US National Climate Assessments. There have been over 600 Congressional hearings on the topic, and hundreds of bills introduced to address this issue. Yet neither scientific evidence nor political mobilisation has resulted in meaningful action in the United States to curb the effects of climate change.

Instead, efforts to take action on climate change have encountered substantial social inertia as well as cultural, institutional, and individual resistance. There are multiple explanations for this failure to mobilise. One dominant explanation centres on institutionalised efforts to oppose action on climate change (Dunlap and McCright 2015). In this chapter, we provide an overview of research on institutional strategies to oppose limits on carbon emissions (for a discussion of institutional barriers to decarbonisation, see Lane in this volume). This research reveals that organised opposition to climate change action is a sophisticated system that operates at multiple institutional levels and within multiple time frames. It uses a wide variety of tactics to influence cultural perceptions of, and political action towards, climate change. Media institutions also play a role through the promulgation of climate misinformation, which has had a significant impact on public opinion and communication.

The US case is relevant to global sustainability efforts for a number of reasons. As one of the regions of the world most responsible for contributing to climate change, the United States is broadly deemed responsible for leadership in responding to its effects. Organised opposition to climate change broadcasts American exceptionalism globally and threatens to weaken transnational agreements and organisations. Second, to the extent that the basis of US opposition to climate change is located in the promotion of scientific uncertainty (Freudenberg, Gramling, and Davidson 2008; Oreskes and Conway 2011) some have identified a psychological "seepage" of this uncertainty into the global arena, affecting international scientific and popular discourse on climate change mitigation that reinforces "human tendencies towards preference for preservation of the status quo" (Lewandowsky et al. 2015).

A third reason to consider how US opposition to climate change impacts global sustainability initiatives is the political influence of powerful business interests worldwide (e.g. Fuchs and Feldhoff 2016; Wright and Nyberg 2017). On the one hand, multinational corporate lobbies intervene in public policy and foreclose upon legislative initiatives to control industrial output, further restricting climate change action. On the other, the "new corporate environmentalism," steeped in market-friendly and technological approaches to sustainability, regularly undervalue or reframe environmental concerns to fit shareholder obligations.

We start with a historical overview of social trends that have informed oppositional behaviours towards environmental sustainability. We then discuss how these elements have coalesced into institutional forms and practices that adopt short-term, medium-term, and long-term horizons to counter scientific evidence of global warming and undermine political support to address it. We conclude by highlighting some of the implications of this organised countermovement for ongoing efforts at political and cultural mobilisation to provoke meaningful action on climate change.

## Historical development of climate change opposition

A number of organisations make up the organised opposition to environmental sustainability efforts in general and climate change action in particular, including corporations, trade associations, conservative think tanks, philanthropic foundations, advocacy groups, lobby groups, and public relations (PR) firms, promulgated by a network of blogs and media outlets (Dunlap and McCright 2015). These various organisations act in different political and cultural arenas and employ different time horizons to achieve a range of objectives. For these reasons, we cannot refer to the organised efforts to block or delay climate action in monolithic terms. Rather, these efforts form an amalgam of loosely coordinated groups that can be understood as a countermovement. Countermovements are

> networks of individuals and organizations that share many of the same objects of concern as the social movements that they oppose. They make competing claims on the state on matters of policy and politics and vie for attention from the mass media and the broader public.
>
> *(Meyer and Staggenborg 1996: 1632)*

As Gale (1986) notes, countermovements "typically represent economic interests directly challenged by the emergent social movement." Countermovement organisations are often elite-driven efforts to mobilise economically impacted populations or populations that share similar interests or ideologies (207). Indeed, the climate countermovement is not simply made up of industries attempting to preserve their market position. There is also a strong component of ideologically motivated action that reflects its historical development. There are three main historical threads woven into the current climate countermovement.

## *The wise use movement and opposition to government regulation*

An important set of efforts to oppose action on climate change originated in the western United States in response to government regulation of public lands. Stemming from the 19th-century belief in Manifest Destiny (Brulle 2000), territorial expansion and stewardship were seen as "man's" divine right; any threat of control over the unbridled use of natural resources for economic development was an affront to this central tenet. As noted by Lo (1982: 119), countermovements generally arise to defend established societal myths. The myth of Manifest Destiny extends throughout a long history of western states' opposition to any form of government regulation of public property.

Laws passed in the 1960s and 1970s were designed to restrict the use of federal lands. The Wilderness Act (1964) was aimed at preservation of the remaining natural areas in the national forests. The Endangered Species Act (1973) placed limits on land use, even by private property owners (Cawley 1993: 161). In response, a large-scale social protest emerged between 1979 and 1983, initially known as the "Sagebrush Rebellion" (Shabecoff 1993: 164). This movement, representing western economic interests (primarily ranchers and miners), attempted to return control of federal lands to local economic groups. The election of Ronald Reagan gave this movement access to power. Reagan appointed James Watt, a leader in the Sagebrush Rebellion, as Secretary of the Interior (he served between 1981 and 1983). This controversial Cabinet appointment signalled the Reagan administration's intention to dismantle environmental regulations. Even in the face of several reports on the progression of environmental damage, including two major studies on climate change in 1983 (NRC 1983; Seidel and Keyes 1983), no federal action was taken.

By the late 1980s, the original Sagebrush Rebellion had been reorganised into a more comprehensive movement known as the Wise Use movement (Brulle 2000: 115–131). Helvarg (1994: 9) sees the Wise Use movement as a "counterrevolutionary movement, defining itself in response to the environmental revolution of the past thirty years." The intellectual roots of this movement combine the idea of Manifest Destiny, states' rights, and property rights to call for the reversal of environmental restrictions. Expanding on Milton Friedman's (1962) injunction that the free market is best positioned to deal with environmental problems, the Wise Use movement argued that market mechanisms ought to dictate natural resource management. It is estimated that the movement comprised 200 member-groups in 1988, expanding to over 1,500 groups nationwide in 1995 (Brulle 2000). The Wise Use movement became a powerful lobby. It also developed a popular strategy known as Strategic Lawsuits Against Public Participation (SLAPPs). These lawsuits sought to limit the ability of environmental activists to voice their concerns by burdening them with the costs of legal defence (Canan and Pring 1988).

Many of the original Wise Use groups went on to oppose climate change action, and to form coalitions to increase their lobbying power. According to a recent study documenting opposition to the Kyoto Protocol (a 1997 international agreement to set binding emission reduction targets), several national Wise Use organisations joined forces with industry-sponsored coalitions to oppose this treaty, including the American Land Rights Association and the American Policy Center (Savitt et al. 1997).

In recent years, the Wise Use movement has become embedded in multipurpose American organisations such as the anti-government, populist Tea Party movement and the Koch-funded, right-wing political advocacy group Americans for Prosperity. While its distinct identity has become diffused as a result, this movement generated a nucleus of activists and organisations opposed to climate action who now staff a wide range of conservative organisations.

## Neoliberalism and the rise of the conservative movement

The second historical development contributing to the growth of organised efforts in opposition to climate action, especially in the United States, was the rise of the conservative movement. While conservative organisations had campaigned in opposition to the New Deal in the 1930s and 1940s, it was the consolidation of a well-financed and coordinated conservative social movement starting in the 1950s that would become one of the core components of institutionalised efforts against climate action.

Founded by the Austrian economist Friedrich von Hayek, the Mont Pelerin Society first met in 1947. It was an intellectual society devoted to advancing a philosophy of market deregulation and low state intervention, partly in opposition to socialism. The society developed a robust neoliberal perspective that gained powers of expression in the highly regarded economics department of the University of Chicago, which advocated the application of free-market economics to public policy issues and the rollback of the welfare state. It was a philosophy that would come to define the modern conservative movement's approach. The neoliberal position equates democracy with "economic freedom" or "free enterprise" – property rights, contracts, and consumer choice. It rejects the notion of public goods and opposes regulation, taxation, and other state-led intrusions into the market.

This movement gained momentum when the corporate lawyer Lewis Powell wrote a now-infamous memo warning about the social and political threat to the American free enterprise system, arguing that corporations and industry must wrest control of the economy from leftist inclinations (Powell 1971).[1] "Conservatives must capture public opinion by exerting influence over the institutions that shape it: academia, media, church, courts," he wrote (Powell 1971; Mayer 2016: 75). Powell would go on to become a Supreme Court Justice with considerable influence over such institutions. Meanwhile, neoconservative and religious organisations joined business interests to campaign against social liberalism and the welfare state and promote free-market ideology. The emergence of a network of conservative think tanks and foundations lent material and symbolic support to calls for deregulation, privatisation, welfare reduction, and decreased taxation to revive corporate profits and economic growth (Stefancic and Delgado 1996). This network consolidated its access to political power via the election of Ronald Reagan. In subsequent decades, this neoliberal network in the United States sought to weaken the environmental regulations and oversight agencies created in the 1970s and blunt the environmental movement's ongoing efforts to extend this discipline.

Perceived threats posed by climate change discourse intensified this opposition, mobilising energy companies and related industries as well as broader free-market forces. The discrediting of global climate change science began in earnest in 1989, when the George C. Marshall Institute issued the first report in its "Climate Change Policy" program, a program designed to promote uncertainty over mainstream scientific consensus (Oreskes and Conway 2011: 186). Other conservative think tanks followed the Marshall Institute's lead, aiming to delegitimise climate science and limit public and political action. As noted by Jacques, Dunlap, and Freeman (2008), conservative think tanks were at least partly responsible for 92% of all books published between 1972 and 2005 that expressed scepticism towards climate change science.

As the US conservative movement expanded, opposition to climate change action became a critical component of its political program. What emerged was a well-developed effort composed of a number of conservative foundations connected to nearly 100 conservative think tanks (Brulle 2014) that took on opposition to climate change action as part of their mission. For many conservatives, climate change is an issue that provides license for wholesale government intervention into the economy and is thus a major threat to economic liberty. Coordinated by peak meetings of funders, such as the annual Koch brothers' summits or meetings at the Philanthropy Roundtable, the conservative network of institutions has become one of the core components of institutionalised opposition to climate action. Opposition to climate change action has now become the countermovement's pivotal issue in battles against environmental regulations.

## Expansion of public relations activities into the political sphere

The final component that developed into an institutionalised component of opposition to climate change action was the refinement of promotional campaigns by industrial interests in an effort to influence public opinion and thereby combat political control. The development of this phenomenon was theorised by Habermas (1962) through his discussion of the "refeudalization of the public sphere." If the 17th and 18th centuries saw the expansion of civil society and a new notion of "publicness" as private persons came together in public spaces to debate critical issues of the day, this bourgeoning public sphere was circumscribed in the early 20th century by the twin rise of the market and the state as intervening entities. As the state expanded in the latter half of the 19th century and first half of the 20th century, it took on responsibilities to manage national economies, linking market organisation and outcomes to the vagaries of the state. This created an enormous incentive for private interests to attempt to mould public policy. Rather than the "rational-critical" discussion germane to democratic exchange in an ideal-typical setting, public discussions entered into compromise on the basis of negotiations among competing "interests" (Calhoun 1992: 22). And rather than enter into political debates to ascertain the common public good, institutions could use publicity techniques and intervene in civil society to secure a political and cultural advantage (e.g. Magan 2006: 32). The goal of these activities is to generate goodwill and prestige for a certain position, strengthening public support for a given position without ever making the position a matter of formal public inquiry. The consensus that results is elicited through persuasive appeals through the application of promotional publicity techniques (Sievers 2010: 136; Walker 2014). In this "promotional" public sphere, organisations with sufficient economic, political, or organisational capacities to generate publicity campaigns maintain a distinct advantage. By generating support for their positions, these organisations set the terms of the debate and disadvantage participation by community organisations or those with dissenting views (Greenberg et al. 2011: 69).

Habermas (1989: 193–194) traces the origins of PR and public opinion management to the application of publicity techniques by the press agent Ivy Lee before the First World War. Lee's most infamous work was for John D. Rockefeller's Colorado Fuel and Iron Company, the largest coalmine operator in the region. Lee was called in to manage public perception of the company and industry following the "Ludlow Massacre," a labour strike in 1913 that led to intense violence between company guards, the National Guard, and the mine workers. Lee saw in the tragedy an opportunity to promote cooperation. He created a "Colorado Fuel and Iron Company Bulletin," a newsletter publication advertising the multiple benefits labourers enjoyed as employees of the company. When, a year later, Rockefeller established a labour union for his workers, Lee promoted this as a natural extension of existing employee welfare programs. Lee then brought these innovations to Standard Oil and Socony-Vacuum Oil (also Rockefeller properties) as well as the steel industry, working with Bethlehem Steel to develop an "employee representation plan" for the company. Following the war, Ivy Lee worked for a variety of clients to improve their public image (Cutlip 1994; Miller and Dinan 2008).[2]

Edward Bernays, the nephew of Sigmund Freud, further developed the use of PR to burnish the image of corporate entities, including General Electric and the American Tobacco Company. He also worked to drum up political support for political candidates and the National Association for the Advancement of Colored People (NAACP). The first professional political consulting firm was founded in 1933. Called Campaigns, Inc., the firm pioneered the use of indirect lobbying (also known as grassroots lobbying) as a political campaign tactic (Lepore 2012; Walker 2014: 53–55).

The large-scale application of corporate PR to combat environmental concerns began earlier than most studies allow. The monopoly companies of the early 20th century in environmentally compromising industries like rail, steel, and coal faced considerable anxiety among Americans over their size and power. Public opinion polling informed corporate leaders that campaigns to engage with communities and political influencers would help stem growing public outcry over the lack of regulation and government control over these industries. These efforts became more sophisticated after the Second World War, as many PR and ad men who had helped develop large-scale propaganda campaigns for the war effort returned to public life. The American Petroleum Institute (API) instituted a two-million-dollar PR campaign beginning in 1946 to promote goodwill among local residents. The campaign included the development of "field operations" (opening and staffing 13 units across the country), extensive magazine advertising, and a weeklong "Oil Progress Week" to educate communities about the benefits of oil to their lives (Potter 1990: chapter 6).

The development of a specialised industry of environmental PR would emerge in the 1970s, ten years after the public outcry following the publication of journalist Rachel Carson's research on the toxic legacy of chemical pesticides. Serialised in the *New Yorker* magazine in June 1962, Carson's book, *Silent Spring,* shocked and galvanised a public formerly unaware of the devastating impact of dichlorodiphenyltrichloroethane (DDT) and other pesticides on animal, plant, and human life. The chemical industry and its trade association, the Chemical Manufacturers Association (CMA), launched a massive PR campaign to sow doubt about Carson's methods and findings (Murphy 2005). But the book's findings were too well researched, and perhaps too damning, to be overturned. Attention to the issue by President John F. Kennedy as well as multiple scientific and political bodies grew. DDT was eventually banned; and a host of environmental regulations passed in the wake of focused attention on industrial overreach and its impacts on human and natural environments.

Though ultimately unsuccessful in stemming the tide of change in public and political attitudes evoked by *Silent Spring,* this PR campaign taught industrialists that managing public perception and political decision-making around environmental issues was not only a worthwhile but also a necessary investment. In 1973, a "boutique" environmental PR firm, E. Bruce Harrison Co., opened its doors in Washington, DC. Harrison, who had helped to manage the CMA's response to *Silent Spring,* founded his eponymous firm by assembling an industry-labour coalition he called the National Environmental Development Association (NEDA), a group of contractors, corporations, labour unions, and other interests that was collectively "opposed to some types of environmental control" (*Publicist* 1982: 1). Harrison drafted a "declaration of principles" (*Publicist* 1982: 2) which helped to recruit additional members, and hosted regional and national conferences "as part of a continuous research and education program" (NEDA 1979) to diffuse these principles. Over the next 30 years NEDA became the umbrella structure for a series of issue coalitions organised around "softening" specific federal legislative initiatives. By the late 1980s, the firm was highly successful, consistently ranked by *O'Dwyers* (the leading trade publication) among the top ten environmental

PR firms in terms of billings. In 1992 the firm claimed to represent, "through coalitions and direct service...more than eighty of the Fortune 500" (Harrison 1992).

Over time, the US corporate community has integrated PR and lobbying into its business strategy. Between the 1970s and the 1990s, corporate PR agents and firms built advocacy structures to anticipate and manage public policy issues. These advocacy structures are more sophisticated, and more socially embedded, than classic movement structures such as constituencies, coalitions, and networks. They include public–private sector partnerships, events and sponsorships; industry benchmarking and reporting; awards/certification programs; media training seminars; and international technology transfer systems (Aronczyk 2018). These advocacy structures work in an integrative manner to actively shape public policy decisions in ways favourable to overall corporate interests (Barley 2010).

One key tactic developed by corporate leaders during this time was to emulate the grassroots advocacy that led to the citizens' and environmental movement's successes in the 1960s. Helped by corporate communicators such as PR counsel, companies developed an approach of "corporate political activism," adopting not only the title but also the techniques of public interest groups. Corporations and trade associations developed the capacities to simulate and mimic grassroots mobilisation activities through the creation of front groups.

Corporations also engaged law firms to facilitate legal activism (such as the SLAPP libel lawsuits discussed earlier) and created corporate foundations, which, in turn, funded ideologically motivated think tanks. Trade associations were strengthened and enabled as major political actors, and corporate political activities were facilitated through the rise of political action committees. This effort has evolved into a complex organisational field that focuses on the development and promulgation of a uniform ideological message. This organisational strategy simulates a unified front, as there appear to be multiple diverse voices simultaneously advocating for a uniform position. This perception is enforced through the use of different communication channels, including academic journals, policy papers, briefings, media coverage, and advertising to reach targeted audiences. Manheim (2011) details how these campaign efforts have developed into a common practice that corporations, government, and advocacy organisations regularly employ to realise their political objectives. Manheim calls these Information and Influence Campaigns (IICs): "systemic, sequential and multifaceted effort[s]" (2011: 18) to promote information that orients the political decision-making process towards desired outcomes either through direct persuasion, or by persuading other parties to bring pressure on decision makers. IICs involve comprehensive and coordinated actions to identify and segment target audiences, incorporate diverse messaging schemes, and vary timing and distribution channels to achieve their political goals.

## The institutional structure of climate politics in the United States

The historical origins of opposition to environmental action have yielded a complex network of institutional actors. These organisations' efforts span multiple time frames, with different objectives, strategies, and institutions involved at each level. Since climate change activities are an integral focus of the conservative movement, it is not helpful to try to separate climate change from other topics and themes relevant to the conservative movement. Opposition to environmental action should be seen as a large scale, multi-decade institutionalised effort that spans the entire range of cultural and political institutions over multiple time scales.

Driven primarily by strategic investments by conservative foundations, the conservative movement was animated by a focus on three distinct time periods: long term, intermediate term, and short term (Covington 1997). This strategy has a long genesis, starting with the writings of Frederick Hayek in 1949, and has developed incrementally over the last 70 years. It is a directed and coherent approach that informs the funding activities of conservative philanthropies.

The conservative movement's institutional goal was to build an intellectual and ideological infrastructure over the long term that can facilitate political action in the short term and inform policy proposals in the medium term. This philosophy was stated succinctly in 1996 by Richard H. Fink, onetime president of the Charles Koch Foundation and former executive vice-president of Koch Industries, one of the largest privately held companies in the United States. In a 1996 article in *Philanthropy Magazine* titled "From Ideas to Action," Fink elaborated the strategic underpinnings of the foundation's philanthropy, a model he called the "Structure of Social Change" (Fink 1996). Inspired by Hayek's model of production, which articulated three stages of production to transform raw materials into value-added products for consumers, Fink saw a metaphorical link with the transformation of ideas into social and political action:

> When we apply [Hayek's] model to the realm of ideas and social change, at the higher stages we have the investment in the intellectual raw materials, that is, the exploration and production of abstract concepts and theories. In the public policy arena, these still come primarily (though not exclusively) from the research done by scholars at our universities. At the higher stages in the Structure of Social Change model, ideas are often unintelligible to the layperson and seemingly unrelated to real-world problems. To have consequences, ideas need to be transformed into a more practical or useable form.
>
> In the middle stages, ideas are applied to a relevant context and molded into needed solutions for real-world problems. This is the work of the think tanks and policy institutions. Without these organizations, theory or abstract thought would have less value and less impact on our society.
>
> But while the think tanks excel at developing new policy and articulating its benefits, they are less able to implement change. Citizen activist or implementation groups are needed in the final stage to take the policy ideas from the think tanks and translate them into proposals that citizens can understand and act upon. These groups are also able to build diverse coalitions of individual citizens and special interest groups needed to press for the implementation of policy change.[3]

A number of authors (Covington 1997; Barley 2010; Dunlap and McCright 2011, 2015, Farrell 2016) have examined the organisational structure of the conservative movement. Despite different intellectual perspectives, there is a strong consensus regarding the focus and nature of the organisations involved in the climate change countermovement.

## Long-term activities: actions and institutions

The first set of institutions focuses on long-term efforts that range from five years to decades in duration. The goal is to build and maintain a cultural and intellectual infrastructure of organisations that supports the development of ideas and policies favourable to conservative or industry viewpoints. This includes efforts to create and maintain academic programs in institutions of higher education, endow academic chairs, and provide educational support for students in these programs (Mayer 2016). We can see the outcome of some of these efforts in the proliferation of programs in economics and law that advocate Chicago School styles of neoliberal economics (Teles 2008). Efforts to promulgate conservative ideas have also been documented in elementary and secondary schools (Goozner and Gable 2008; Washburn 2010). A second, longstanding set of efforts includes the corporate manipulation of scientific research to develop and promote results favourable to industrial interests. In the 1950s, for example, the American Petroleum Institute's Smoke and Fumes Committee sponsored ongoing research into air pollution in the wake of growing public alarm. At that time smog in the city of Los Angeles had reached dramatic levels; pending regulation or legislation by the state of California would be costly for the petroleum industry in terms of both production and reputation. The API drew on industry-friendly findings by the Stanford Research Institute that questioned the link between oil refineries and air pollution, calling for

> the need for thorough investigation and development of complete and accurate information on a given pollution situation before legislation designed to cover it is enacted...a law designed to improve air pollution might have a reverse effect if not properly and wisely drawn.
>
> *(Jenkins 1954)*

According to Conley (2006), by the mid-1950s "the Smoke and Fumes Committee was sponsoring ten ongoing projects" to study industrial pollutants in order to cast doubt on the connection between air pollution and oil products (23). Such practices to "institutionaliz[e] delay" (Brulle 2014) – selectively interpreting research or commissioning studies to develop more industry-positive findings – are an important part of the cultural arsenal of the conservative anti-environmental movement.

Another set of activities that allowed industry to make major inroads into popular legitimacy was their sponsorship of cultural events and forms. One the best known examples is Mobil Oil's decades-long sponsorship of Masterpiece Theatre, the dramatic television series on Public Broadcasting Service. From 1970 to 2004, Mobil funded Masterpiece Theatre (along with Upstairs, Downstairs, another popular drama) (Kerr 2005). Although Mobil was technically not allowed to advertise directly on PBS, the station did offer Mobil airtime through its infamous announcer-read tagline: "Made possible by a grant from Mobil Corporation, which invites you to join with them in supporting public television." But Mobil saw its relationship with PBS as fruitful in other ways. Mobil had sway over the choice of host, Alastair Cooke, and all advertising of the show (in TV guides, on posters, etc.). Mobil announced its sponsorship of Masterpiece Theatre and Upstairs, Downstairs in press releases and ad campaigns (referring to the show as "Mobil Masterpiece Theatre"). Mobil also ran publicity stunts like inviting (and paying for) TV critics to come to New York to preview the series (Ledbetter 1997).[4] David H. Koch has also been a supporter of public television (e.g. the science show, Nova). This approach was characterised as "affinity of purpose" advertising, which seeks to improve the corporate public image by association with scientific and cultural achievements (Schmertz 1986).

## Medium-term activities: actions and institutions

The second set of activities focuses on the intermediate time horizon of three to five years. This stage involves the translation and promulgation of academic ideas into concrete policy proposals. A wide range of distribution channels are employed, from mass media to published books and testimony provided in congressional hearings. The major institutional actors in this time frame are think tanks, advocacy organisations and PR firms, which serve as credible third-party spokespersons to boost the legitimacy of these policy arguments (on the importance of legitimacy in reinforcing policy, see Bexell in this volume). PR firms play a further role during this stage by developing and promulgating materials that support policy objectives and by securing media contacts.

## Short-term activities: actions and institutions

The third component focuses on short-term political outcomes such as elections or pending issue legislation. Considerable effort is put into influencing public opinion around climate change. One style of public opinion management is to promote positive perceptions of fossil fuel corporations through the extensive use of advertising campaigns. A second tactic, as mentioned earlier, involves citizen mobilisation and/or the creation of front groups to demonstrate popular support for a political position. A third approach involves lobbying activities, either directly by corporations or trade associations, or indirectly through the employ of public affairs firms to influence legislative outcomes.

## Conclusion

Opposition to climate action in particular, and environmental protection in general, is maintained by a comprehensive set of institutional mechanisms that work integrally to develop, promulgate, and advocate for a series of conservative policies across political and cultural arenas. Clearly, the work to counter climate misinformation campaigns requires a new strategy. It has been nearly three decades since Dr Hansen's dramatic 1988 testimony placed the issue of climate change squarely in the public arena. Despite nearly three decades of activism, however, meaningful action on climate change remains elusive.

There are a few bright spots in both the US context and internationally. Some have pointed to strong urban governance (e.g. at the state and municipal level) as a means to address United Nations Sustainable Development Goals, among other ambitions, which can offer at least a partial solution to federal opposition (Aust and du Plessis 2018). The Global Covenant of Mayors for Climate and Energy is an international coalition of urban mayors, including several from US cities, that support climate change action and have committed to promoting sustainability initiatives (globalcovenantofmayors.org).

Still, these initiatives cannot match the extent of institutionalised activities and financial resources that have been employed to either advance or delay US government actions to address climate change. While considerable sums have been spent by all actors in this political arena, the comparative spending priorities and levels remain poorly understood. This has important consequences for different approaches required to address climate change. Responding to climate change is an inherently political process. To be successful, actors must have a firm and well-grounded understanding of the different organised interests engaged in this process. This understanding can form the basis for the formulation of efficacious strategies and approaches that can inform both funding priorities and the actions of engaged organisations.

While there are many avenues for climate action in the United States, the political and cultural arena is a critical component. Without success in this area, other actions will be partial and limited. After decades of action, it is clear that the political barriers to climate action have not been overcome. Further targeted research as discussed earlier can help provide the basis for the discussion and development of more efficacious strategies and approaches that can inform both funding priorities by foundations and the actions of engaged organisations.

## Notes

1 Powell was a board member of several major American corporations, including Phillip Morris.
2 It is rumoured that Lee crafted PR campaigns for the API in its early years (1910s). However, the fledgling API did not develop a dedicated PR program until the Second World War. More likely is the version put forward by Potter (1990: 4), which describes Lee's status on retainer for API in its first decade. Because of one of the association founder's general distrust of PR, however, it appears Lee's services were not put to use at API.
3 See page 4 of online version of Structure of Social Action, available at https://archive.org/stream/TheStructureOfSocialChangeLibertyGuideRichardFinkKoch/The+Structure+of+Social+Change+_+Liberty+Guide+_+Richard+Fink+_+Koch_djvu.txt.
4 The positive image associated with PBS benefited Mobil in other ways. During the 1991 Gulf crisis, when Mobil was being attacked for gouging, the company leveraged its relationship with PBS to soften the attacks.

## References

Aronczyk, Melissa. 2018. "Public Relations, Issue Management, and the Transformation of American Environmentalism, 1948–1992." *Enterprise & Society* 19(4): 836–863.

Aust, Helmut Philipp, and Anél du Plessis, 2018. "Good Urban Governance as a Global Aspiration: On the Potential and Limits of SDG 11." In Duncan French and Louis J. Kotzé, eds. *Sustainable Development Goals: Law, Theory & Implementation,* 201–221), Cheltenham: Edward Elgar Publishing.

Barley, Stephen. 2010. "Building an Institutional Field to Corral a Government: A Case to Set an Agenda for Organization Studies." *Organization Studies* 31(6): 777–805.

Brulle, Robert J. 2000. *Agency, Democracy, and Nature: U.S. Environmental Movements from a Critical Theory Perspective.* Cambridge: MIT Press.

Brulle, Robert J. 2014. "Institutionalizing Delay: Foundation Funding and the Creation of US Climate Change Counter-movement Organizations." *Climatic Change* 122: 681–694.

Calhoun, Craig. 1992. "Introduction: Habermas and the Public Sphere." In *Habermas and the Public Sphere,* edited by Craig Calhoun, 1–50. Cambridge: MIT Press.

Canan, Penelope, and George W. Pring. 1988. "Strategic Lawsuits against Public Participation." *Social Problems* 35(5): 506–519.

Cawley, R. McGreggor. 1993. *Federal Land, Western Anger: The Sagebrush Rebellion and Environmental Politics.* Lawrence: University Press of Kansas.

Conley, Joe Greene. 2006. *Environmentalism Contained: A History of Corporate Responses to the New Environmentalism.* PhD Dissertation, Princeton University, New Jersey.

Covington, Sally. 1997. *Moving a Public Policy Agenda: The Strategic Philanthropy of Conservative Foundations.* National Committee for Responsive Philanthropy (July). 55 pages.

Cutlip, Scott. 1994. *The Unseen Power: Public Relations, a History.* Hillsdale: Lawrence Erlbaum Associates.

Dunlap, Riley, and Aaron McCright. 2011. "Organized Climate Change Denial." In *The Oxford Handbook of Climate Change and Society,* edited by John Dryzek, Richard Norgaard, and David Schlosberg, 144–174. New York: Oxford University Press.

Dunlap, Riley, and Aaron McCright. 2015. "Challenging Climate Change: The Denial Counter-movement." In *Climate Change and Society: Sociological Perspectives,* edited by Riley Dunlap and Robert Brulle, 300–332. New York: Oxford University Press.

Farrell, Justin. 2016. "Network Structure and Influence of the Climate Change Counter-movement." *Nature Climate Change* 6 (December): 370–374.

Fink, Richard. 1996. "From Ideas to Action: The Role of Universities, Think Tanks, and Activist Groups." *Philanthropy Magazine* 10(1): 10–11; 34–35. Available at https://archive.org/stream/TheStructureOfSocialChangeLibertyGuideRichardFinkKoch/The+Structure+of+Social+Change+_+Liberty+Guide+_+Richard+Fink+_+Koch_djvu.txt.

Freudenberg, William, Robert Gramling, and Debra Davidson. 2008. "Scientific Certainty Argumentation Methods (SCAMs): Science and the Politics of Doubt." *Sociological Inquiry* 78 (1): 2–38.

Friedman, Milton. 1962. *Capitalism and Freedom.* Chicago: University of Chicago Press.

Fuchs, Doris, and Berenike Feldhoff. 2016. "Passing the Scepter, Not the Buck: Long Arms in EU Climate Change Politics." *Journal of Sustainable Development* 9(6): 58–74.

Gale, Richard. 1986. "Social Movements and the State: The Environmental Movement, Counter-movement, and Government Agencies." *Sociological Perspectives* 29 (2): 202–240.

Goozner, Merrill and Eryn Gable. 2008. *Big Oil U.* Washington, DC: Center for Science in the Public Interest, 25 pages.

Greenberg, Josh, Graham Knight, and Elizabeth Westersund. 2011. "Spinning Climate Change: Corporate and NGO Public Relations Strategies in Canada and the United States." *International Communication Gazette* 73(1–2): 65–82.

Habermas, Jürgen. 1962. *The Structural Transformation of the Public Sphere: An Inquiry into a Category of Bourgeois Society.* Cambridge: MIT Press.

Habermas, Jürgen. 1989. "The Public Sphere: An Encyclopedia Article." In *Critical Theory and Society: A Reader*, edited by Stephen Bronner and Douglas Kellner, 136–144. New York: Routledge.

Hansen, James. 1988. "The Greenhouse Effect: Impacts on Current Global Temperature and Regional Heat Waves." In *Testimony to the U.S. Senate Committee on Energy and Natural Resources*, June 23.Washington, DC: U.S. Congress.

Harrison, E. Bruce. 1992. "A Quality Approach to Environmental Communication." *Total Quality Environmental Management* 1(Spring): 225–231.

Helvarg, David. 1994. *The War against the Greens: The "Wise-Use" Movement, the New Right, and Anti-Environmental Violence.* New York: Random House.

Jacques, Peter, Riley Dunlap, and Mark Freeman. 2008. "The Organization of Denial: Conservative Think Tanks and Environmental Skepticism." *Environmental Politics* 17(3): 349–385.

Jenkins, Vance. 1954. "The Petroleum Industry Sponsors Air Pollution Research." *Air Repair* 3(3): 144–149.

Kerr, Robert. 2005. *The Rights of Corporate Speech: Mobil Oil and the Legal Development of the Voice of Big Business.* New York: LFB Scholarly Publishing.

Ledbetter, James. 1997. *Made Possible By … The Death of Public Broadcasting in the United States.* London: Verso Books.

Lepore, Jill. 2012. "The Lie Factory: How Politics became a Business." *The New Yorker*, September 24.

Lewandowsky, Stephan, Naomi Oreskes, James Risbey, Ben Newell, and Michael Smithson. 2015. "Seepage: Climate Change Denial and Its Effect on the Scientific Community." *Global Environmental Change* 33 (July): 1–13.

Lo, Clarence Y. H. 1982. "Countermovements and Conservative Movements in the Contemporary United States." *Annual Review of Sociology* 8: 107–134.

Magan, A. 2006. "Refeudalizing the Public Sphere: 'Manipulated Publicity' in the Canadian Debate on Genetically Modified Foods." *Canadian Journal of Sociology* 31(1): 25–53.

Manheim, Jarol. 2011. *Strategy in Information and Influence Campaigns.* New York: Routledge.

Mayer, Jane. 2016. *Dark Money: The Hidden History of the Billionaires behind the Rise of the Radical Right.* New York: Anchor Books.

Meyer, David S., and Suzanne Staggenborg. 1996. "Movements, Countermovements, and the Structure of Political Opportunity." *American Journal of Sociology* 101(6): 1628–1660.

Miller, David and William Dinan. 2008. *A Century of Spin: How Public Relations became the Cutting Edge of Corporate Power.* Ann Arbor: Pluto Press.

Murphy, Priscilla Coit. 2005. *What a Book Can Do: The Publication and Reception of* Silent Spring. Amherst: University of Massachusetts Press.

National Research Council. 1983. *Changing Climate: Report of the Carbon Dioxide Assessment Committee.* Washington, DC: The National Academies Press. doi.org/10.17226/18714.

NEDA. 1979. "Invitation to the National Environmental Development Association." Conference on Regulatory Issues. http://ufdc.ufl.edu/WL00002412/00001/print?options=1JJ.

Oreskes, Naomi, and Erik Conway. 2011. *Merchants of Doubt: How a Handful of Scientists Obscured the Truth on Issues from Tobacco Smoke to Global Warming.* New York: Bloomsbury Publishing.

Potter, Stephen. 1990. *The American Petroleum Institute: An Informal History (1919–1987).* Washington, DC: American Petroleum Institute, 77 pages.

Powell, Jr., Lewis. 1971. "Attack on the Free Enterprise System." Memorandum. Lewis F. Powell, Jr., *Archives,* August 23, 34 pages. http://law2.wlu.edu/deptimages/Powell%20Archives/PowellMemorandumTypescript.pdf.

*Publicist.* 1982. "D. C. Agency Created First Client." *The Publicist*, March/April, pp. 1–4.

Savitt, Sara, Dan Barry, Allison Daly, and Mike Shelhamer. 1997. "A CLEAR View: Special Kyoto Issue." *Clearinghouse on Environmental Advocacy and Research (CLEAR)* 4(16), 10 pages.

Schmertz, Herb. 1986. *Goodbye to the Low Profile: The Art of Creative Confrontation*. New York: Little Brown and Co.

Seidel, Stephen, and Dale Keyes. 1993. *Can We Delay a Greenhouse Warming? The Effectiveness and Feasibility of Options to Slow a Build-Up of Carbon Dioxide in the Atmosphere*. Washington, DC: U.S. Environmental Protection Agency.

Shabecoff, Philip. 1993. *A Fierce Green Fire: The American Environmental Movement*. New York: Hill and Wang.

Sievers, Bruce. 2010. *Civil Society, Philanthropy, and the Fate of the Commons*. Lebanon, NH: Tufts University Press.

Stefancic, Jean, and Richard Delgado. 1996. *No Mercy: How Conservative Think Tanks and Foundations Changed America's Social Agenda*. Philadelphia, PA: Temple University Press.

Teles, Steven. 2008. *The Rise of the Conservative Legal Movement: The Battle for Control of the Law*. Princeton, NJ: Princeton University Press.

Walker, Edward. 2014. *Grassroots for Hire: Public Affairs Consultants in American Democracy*. Cambridge: Cambridge University Press.

Washburn, Jennifer. 2010. *Big Oil Goes to College: An Analysis of Ten Research Collaboration Contracts between Leading Energy Companies and Major U.S. Universities*. Washington, DC: Center for American Progress.

Wright, Christopher, and Daniel Nyberg. 2017. "An Inconvenient Truth: How Organizations Translate Climate Change into Business as Usual." *Academy of Management Journal* 60(5): 1633–1661.

# 18

# A critique of techno-optimism

## Efficiency without sufficiency is lost

*Samuel Alexander and Jonathan Rutherford*

## Introduction

Across human history technology has played a significant role in the development of civilisation – a role that has accelerated dramatically over the last few centuries due to the interrelated emergence of capitalism and the scientific and industrial revolutions. Through technological advancement humans have been able to produce electricity, cure diseases, split the atom, travel into space, invent computers and the Internet, and map the human genome, among an unending list of things that often seem like miracles. Notably, these technological advancements have also assisted in the unprecedented expansion of our productive capacities through harnessing the energy in fossil fuels and developing machines to augment human labour. This has allowed many people, primarily in the developed nations, to achieve lifestyles of material comfort that would have been unimaginable even a few generations ago. Increasingly it seems that all 7.7 billion people on the planet are set on achieving these high consumption lifestyles for themselves, and at first consideration the universalisation of affluence indeed seems a coherent and plausible path of progress.

But all that glitters is not gold. No matter how awesome the advancement of technology has been as a means of raising material living standards, there are also well known social and environmental dark sides that flow from this mode of development. Economic activity depends on nature for resources, and as economies and populations have expanded, especially since the industrial revolution, more pressure has been placed on those natural and finite resources, ecosystems, and waste sinks. Today we face a series of overlapping crises owing to the heavy burden our economies are placing on the planet (Ehrlich and Ehrlich, 2013; Ripple et al., 2017). The global economy now exceeds the sustainable carrying capacity of the planet (Global Footprint Network, 2017), with deforestation, ocean depletion, soil erosion, biodiversity loss, nitrogen and phosphorous pollution, water shortages, and climate change being just a sample of these acute, unfolding problems (Rockström et al., 2009). Recent publications from the IPCC (2014) reiterate the immense challenge of climate change in particular, with the necessity of rapid emissions reductions becoming ever more pressing as carbon budgets continue to shrink through lack of committed action. At the same time, great multitudes of people around the planet still live in material destitution, and the global population continues to grow (UN, 2017), suggesting the environmental burden is only going to be exacerbated as the global development agenda – the goal of promoting growth in global economic output – is pursued into the future (Turner, 2012).

Technological optimists believe, however, that just as the application of technology has been a primary cause of environmental problems, so too does it provide the primary solution (Lomborg, 2001; Lovins, 1998, 2011). Techno-optimism, in this sense, can be broadly defined as the belief that science and technology will be able to solve the major social and environmental problems of our times, without fundamentally rethinking the structure or goals of our growth-based economies or the nature of Western-style, affluent lifestyles. In other words, techno-optimism is the belief that the problems caused by economic growth can be solved by more growth (as measured by GDP), provided we learn how to produce and consume more efficiently through the application of science and technology.

There is debate, it should be stressed, among techno-optimists regarding environmental policy solutions. While some believe that problems can be solved by "free-markets" (see e.g. Beckerman, 2002; Simon and Kahn, 1984), most recognise a strong role for government in regulating markets, commonly through either tax or cap-and-trade market mechanisms (see e.g. Hatfield-Dodds et al., 2015) designed to incentivise the uptake of innovative technologies. Common to all techno-optimists, however, is an assumption that environmental problems can be solved via the applications of technology within the framework of the growth economy (Purdey, 2010).

This chapter presents an evidence-based critique of such techno-optimism, arguing that the vision of progress it promotes is unrealisable due to the limits of technology and the inherent structure of growth economics (see also, Hickel and Kallis, 2019). The focus of this critique, however, is not on specific technological solutions (see Hamilton, 2013; Huesemann and Huesemann, 2011) but rather the subtler faith techno-optimists place in "efficiency" as the environmental saviour. The critical analysis begins in section "Techno-optimism and the ideology of growth" by placing techno-optimism in theoretical context. It is important to understand the structure of techno-optimism and see why it forms a central part of the ideology of growth. In section "Are economies decoupling growth from impact?" and "Efficiency and the rebound effect" the poor historical record on "decoupling" GDP from environmental impacts is examined, and this analysis is used to explain why efficiency improvements have not produced sustainable economies despite extraordinary technological advance in recent decades. It turns out efficiency improvements have not often been able to keep up with continued economic and population growth, meaning that overall environmental impact continues to grow, despite efficiency improvements. Section "The 'growth model' has no techno-fix" unpacks the arithmetic of growth to expose how deeply implausible techno-optimism really is, in ways the optimists themselves seem scarcely aware. The central conclusion of this critique is that technology cannot and will not solve environmental problems so long as it is applied within a growth-based economic model. The degree of decoupling required is too great.

## Techno-optimism and the ideology of growth

In 1971, Paul Ehrlich and John Holdren published an article that greatly advanced the understanding and communication of environmental problems and their potential solutions (Ehrlich and Holdren, 1971). In this chapter they developed what has become known as the IPAT equation. This equation holds that environmental impact (I) equals, or is a function of, Population (P), Affluence (A), and Technology (T). While this equation is not without its limitations and drawbacks, it nevertheless made it easy for environmentalists to talk about the nature of the unfolding environmental crisis (Meadows et al., 2004). With the IPAT equation, it could be shown in clear terms that environmental impact could be mitigated by the various means of reducing population, reducing per capita income, and reducing the energy or resource intensity of production and consumption through technological development and better design. Put otherwise, the equation showed that continuous population and consumption growth would exacerbate environmental problems, unless technological advancement could outweigh those impacts through efficiency gains.

One of the attractions of the IPAT equation was the way in which it highlighted how individuals and policy-makers had various options available to them for tackling environmental problems. People who cared about the environment could try to lessen impact either by trying to reduce population, by trying to reduce consumption, or by trying to produce and consume as efficiently as possible.

Nevertheless, the fact that there were options turned out to be a mixed blessing. After all there was (and still is) extreme social, economic, and political resistance – by governments, business, and across much of civil society – to taking concerted action on two of the three IPAT variables: population growth and per capita income. With respect to population growth, while we know it is the multiplier of everything (Alcott, 2010), population control is obviously a thorny issue, in that procreation seems like a very private and intimate issue that governments should not try to regulate. For this reason, population has been, and to a large part remains, one of the great taboo subjects of the environmental debate (see Coole, this volume). A similar dynamic could explain the marginalisation of consumption. Since a higher income is almost universally considered better than a lower income, governments, and indeed the voting public, have looked for other ways to lessen environmental impact. To borrow a phrase from George Monbiot (2006), people do not "riot for austerity."

The IPAT equation, however, had within it the win–win solution that people seemed to be seeking: efficiency improvements. Even if a nation was unable to reduce population, and even if it was unwilling to reduce its income, the equation provided a theoretical framework that showed that it was nevertheless possible to reduce environmental impact through technological advancement (Simon and Kahn, 1984). At the core of this claim was the idea of "decoupling," according to which economic growth could be divorced from growing environmental impacts.

This "techno-fix" approach was a much more politically, economically, and socially palatable way to address environmental problems. It provided governments, businesses, and individuals with a means of responding to environmental problems (or being seen to respond to environmental problems), without confronting population growth or questioning affluent lifestyles. As the next sections show, however, the empirical support for decoupling is lacking, which is a most inconvenient truth for those consciously or unconsciously committed to the ideology of growth (Hamilton, 2003; Kallis, 2017).

## Are economies decoupling growth from impact?

As noted, "decoupling" is the idea that GDP growth can be, over time, progressively divorced from environmental impacts. In assessing the recent decoupling record, it is imperative to distinguish between "relative" and "absolute" decoupling (Jackson, 2016). Relative decoupling refers to a decline in the ecological impact *per unit* of economic output. Absolute decoupling refers to a decline in the *overall* ecological impact of total economic output. While relative decoupling may occur, making each commodity less materially intensive, if the total consumption of commodities increases then there may be no absolute decoupling; indeed, the absolute ecological impact of total economic activity may increase.

Given that the global economy already exceeds the planet's sustainable carrying capacity by 70% (Global Footprint Network, 2017), large-scale absolute decoupling is what is needed. The problem is, the record to date suggests very little *absolute* decoupling is occurring, let alone at the rates that would be needed for long-term sustainability – an issue we will return to below.

Consider the example of carbon emissions. There is no doubt that significant *relative* decoupling – i.e. emissions per unit of GDP – has taken place. Tim Jackson (2016: 88) reports that the amount of carbon released per unit of world's economic output has declined continuously over several decades, from 760 grams of carbon dioxide per dollar in 1965 to just under 500 grams of carbon dioxide per dollar in 2015. That is an average decline in carbon intensity of a little under 1% per year. Nevertheless, despite these efficiency gains, global carbon emissions have continued to rise in absolute terms, more than *trebling* over the same period. It is true carbon emissions from fossil fuels and industry (excluding land-use change) were flat from 2014 to 2016 at about 36 billion tonnes, suggesting that emissions might have peaked and could soon start to decline. Unfortunately, however, global emissions have since recommenced their upward trajectory, with indications that record levels were reached in 2017 (Global Carbon Budget, 2017). This shows that – even 30 years after the Intergovernmental Panel on Climate Change (IPCC) was established – the significant relative decoupling of carbon (and energy) intensities has so far failed to translate into actual absolute declines. To date, technological advance is not fulfilling its promise to reduce overall impact (Hickel and Kallis, 2019).

A similar story holds with respect to global resource consumption, a measure which includes aggregate consumption of biomass, fossil fuels, metal ores, and minerals. A review of the evidence found that resource efficiency improvement for the global economy between 1980 and 2009 averaged 0.9% per annum (Giljum et al., 2014). This, however, represented a per annum efficiency improvement that was less than one third of the rate that would have been needed for "absolute" decoupling (Giljum et al., 2014: 328), i.e. growth of GDP without any increase in materials use. As such, over the same period global materials use more than doubled. Furthermore, as a UNEP (2016) report found, this efficiency improvement rate masks a more recent efficiency *decline* since the turn of the century, from 1.2 kg per one US$ of GDP in 2000 to almost 1.4 kg per US$ by 2010 (UNEP, 2016: 40). Far from decoupling – even in relative terms – this report showed that, from the turn of the century, the global economy has undergone a process of material "recoupling." Given the fact that increasing material consumption use "is one of the key drivers for environmental problems and is directly or indirectly responsible for problems such as climate change, water scarcity or biodiversity loss" (Giljum et al., 2014: 332–333), it should be no surprise that these problems, far from improving at the global level, continue to get worse (Ripple et al., 2017).

It is true that some limited absolute decoupling is underway in certain sectors of some nations, specifically as some developed economies move towards "service," "information," or "post-industrial" modes of production and consumption (see e.g. Steinberger et al., 2013). This is especially the case for localised pollutants, such as wastewater discharge, sulphur dioxide emissions, and carbon monoxide emissions (Bo, 2011; Dinda, 2004). Some of these nations have reduced domestic carbon emissions (i.e. emissions released within the national territory) in absolute terms (Carbon Tracker, 2016).

However, while these reductions are positive steps in the right direction, the achievement is often less impressive on deeper interrogation (see Hickel and Kallis, 2019). Often a large fraction of the decoupling taking place in rich nations is a result of environmental "leakage" – that is, the process whereby wealthy nations have, throughout the globalisation era, increasingly externalised environmental damage via mechanisms such as pollution trading and the outsourcing of environmentally intensive production to developing countries, especially China. While it may be possible to "externalise" impacts from a given nation, the planet, of course, is a closed system in this regard. Accordingly, when "externalised" manufacturing or agricultural commodities – and their associated environmental harms – are "internalised" from an accounting perspective, much of the apparent environmental progress of high consuming countries disappears. For example, it is no good claiming a reduction in national deforestation, say, if a nation is simply importing more wood from abroad rather than cutting down its own trees (Asici, 2013). Similarly, it has hardly environmental progress if the rate of species loss is reduced within a nation if, at the same time, the net import of luxury agricultural crops is driving accelerated species extinction across the globe (Lenzen et al., 2012). The Organisation for Economic Co-operation and Development's (OECD) aggregated carbon efficiency improvements between 2000 and 2013 reduce by about half when a consumption-based methodology is used and the emissions embedded in imports from "pollution havens" in China and other industrialising nations are fully accounted for (Carbon Tracker, 2016). To the extent that some nations have achieved absolute decoupling in carbon emissions, the problem remains that this process has been too slow and too minor to provide much solace, and as noted earlier, from the global perspective that ultimately matters, carbon emissions remain on the rise. Absolute decoupling, in itself, would not be enough. What is needed is "sufficient absolute decoupling," which is certainly not occurring (Kallis, 2017; Raworth, 2017).

The effect of environmental leakage is most pronounced with respect to overall material usage. Wiedmann et al. (2015) have shown that absolute decoupling within OECD nations is only evident when using the "domestic material consumption" (DMC) accounting measure, which fails to factor in upstream raw materials embedded in imports that originate outside the focal country. If instead one uses the recently developed material footprint (MF) – a measure which includes all the raw materials associated with final demand of a given economy – the picture looks very different. The study found that between 1990 and 2008 the MF of the OECD (as well as the EU) "kept pace with increases in GDP and no improvements in resource productivity at all are observed when measured as the GDP/MF" (Wiedmann et al., 2015: 6273). When a proper accounting is applied, many OECD nations have not even achieved relative decoupling in MF over the last two decades, let alone absolute reductions. Accordingly, the idea that wealthy nations have reached "peak stuff" (Goodall, 2011) seems to be no more than a widely promoted accounting error that only serves to entrench the economics of growth.

In the final analysis, any rich world decoupling to date remains far from adequate to put humanity on a sustainable pathway – a point we will elaborate on below. As a rule, the richest nations remain, on a per capita basis, by far the most environmentally impactful (Global Footprint Network, 2017). Certainly, the environmental impacts of the rich nations are not showing an "inverted U" shape claimed by advocates of the so-called Environmental Kuznets Curve, which holds that getting rich is the best way to live lightly on the planet (Beckerman, 2002). At the global level, decades of extraordinary technological development have failed to decouple GDP from environmental impact. Despite isolated examples of success, decade after decade more carbon emissions are sent into the atmosphere and more renewable and non-renewable resources are extracted from our finite Earth. It seems, therefore, that the "optimism" in "techno-optimism" lacks evidential foundation. And if, in coming years, absolute reductions in global emissions and resource extraction are achieved, at that stage one must still ask questions about whether this is occurring at *sufficient rates* to avoid or seriously mitigate the ecological crises that are, in many cases, already unfolding.

## Efficiency and the rebound effect

The poor historical record of decoupling reviewed earlier is counter-intuitive, perhaps, because one might ordinarily think that efficiency gains would lead to substantial reductions in energy and resource demands. It is plausible to think that as the world gets better at producing commodities more efficiently, the absolute impacts of our economic activity would naturally decline. But we have seen that this assumption has not played out in reality. In part this is due to sheer growth in economic throughput and energy services, which has overwhelmed the efficiency gains that have been made. But another critical part of the explanation is due to what are known as "rebound effects" (Alcott, 2005; Polimeni et al., 2009; Saunders, 2013). A rebound effect is said to have occurred when the benefits of efficiency improvements are partially or wholly negated by consumption growth – either by consumers or by industry – that was made possible by the efficiency improvements (Alcott and Madlener, 2009; Herring and Sorrell, 2009). For example, a 5% increase in energy efficiency may only reduce energy consumption by 2% if the efficiency improvements incentivise consumers or industry to increase demand for energy (meaning 60% of anticipated savings are lost or "taken back"). In other words, efficiency improvements can provoke behavioural or economic responses ("rebounds") that end up reducing some of the anticipated benefits of the efficiency improvements. When those rebounds are significant enough they can even lead to *increased* resource or energy consumption, which is sometimes called "back-fire" or the "Jevons paradox," in reference to the classical economist William Stanley Jevons who first observed the phenomena. There are two main categories of rebound effects – direct rebound and indirect rebound.

A direct rebound occurs when an efficiency gain in production results in increased consumption of the same good or service (Frondel et al., 2012; Khazzoom, 1980). For example, a more fuel-efficient car can lead people to drive more often, or drive further, since the costs of fuel per kilometre have gone down; a more efficient heater can lead people to warm their houses for longer periods or to hotter temperatures, since the relative costs of heating have gone down; energy-efficient lighting can lead people to leave the lights on for longer, etc. (Sorrell, 2009). Similar dynamics operate in industry whereby the introduction of more energy efficiency methods either reduces the price of a commodity (since it makes production less resource-intensive or time-intensive), or else enables the development of new product lines, thus acting to incentivise higher consumption, meaning that some or all the efficiency gains are lost.

An indirect rebound occurs when efficiency gains in one good or service lead to increased consumption of other goods or services. For example, insulating one's home might reduce the annual consumption of energy for electricity, but the money saved from reduced energy costs is often spent on other commodities that require energy (e.g. a plane flight or a new computer). This can mean that some or all the energy saved from insulating one's house is consumed elsewhere, meaning overall energy dependence can stay the same or even increase.

While the basic mechanism of rebound effects is widely acknowledged, and, indeed, beyond dispute, there is an ongoing debate over the magnitude of the various rebound phenomena (Chakravarty et al., 2013). There is, however, a strong body of literature arguing that the rebound effect is larger than has been previously assumed (Saunders, 2013; Sorrell, 2015).

Saunders (2013), for example, finds that there was an average *direct* energy rebound (thus excluding potential indirect rebound) of 62% across 30 sectors of US industry from 1980 to 2000. Without entering further into the intricacies of the complex empirical and theoretical debates, it is fair to say that despite the uncertainties, there is broad agreement that rebound effects exist and that they are significant. The benefits of technology are almost always less than presumed, and, in fact, at times efficiency improvements can lead to more, not fewer, resources being consumed overall.

## The "growth model" has no techno-fix

The forgoing lines of argument should be enough to cast doubt on the faith of the techno-optimists. But faith dies hard, even in the face of compelling evidence. Thus, the techno-optimist is likely to double down and proclaim that past behaviour is not a reliable guide to future success. As Nordhaus and Shellenberger (2011) argue: "The solution to the unintended consequences of modernity is, and has always been, more modernity – just as the solution to the unintended consequences of our technologies has always been more technology." While this can be accepted as a theoretical possibility, there are dynamics at play – including the laws of physics – that suggest that decoupling through efficiency gains will not reduce the overall ecological impacts of economic activity if global growth remains the primary economic goal (Ward et al., 2016).

This uncomfortable reality was highlighted by a recent study (Ward et al., 2016), which extrapolated out the implication of Australia pursuing economic growth at the modest rate (by historical standards) of 2.41% per annum each year until the end of the 21st century. The study assumed rapid technological development and proactive policy settings (i.e. a steeply rising global carbon price), resulting in historically unprecedented energy and resource efficiency gains. Nevertheless, given that by 2100 the Australian economy would have undergone an eightfold expansion, the study found that by 2100 Australian material and energy use, instead of declining, would have risen by 29% and 256%, respectively, on current levels (Ward et al. 2016: 9). The authors conclude that this demonstrates "categorically that GDP growth cannot be sustained indefinitely" (Ward et al 2016: 10).

Even this scenario, however, understates the true magnitude of the challenge for those advocating techno-optimism. First, these scenarios assume that efficiency gains will be just as easy to achieve in coming decades as they have in the past. But this assumption is highly questionable, given the reality of diminishing returns from investment, which we are already witnessing today, and which are likely to accelerate as ecological conditions deteriorate. This is most obvious with respect to the resource sector, which includes, of course, fossil fuels accounting for over 80% of the global economy's primary energy (IEA, 2017). For basic economic reasons, humans tend to extract the cheapest and easiest resources first while leaving the more difficult, dangerous, and expensive resources for later. What this means is that, over coming decades, the energy and financial costs of resource extraction will tend to rise. Indeed, this trend is already apparent; several decades ago, for example, the energy in one barrel of oil could be used to extract 30 barrels of oil, but the ratio is now estimated to be around 18 and the ratio will only decrease in coming decades (Hall et al., 2014). The impact of depletion is also evident in the mining sector with declining mineral ore grades and increasing mining waste rock and tailings resulting in higher energy, water, and emission costs for mining and ore separation (Mudd, 2009). The upshot is that, even if there are rapid efficiency gains in, say, manufacturing processes, in coming decades they are likely to be increasingly counteracted by efficiency *declines* in resource extraction. This situation evokes the challenge faced by the Red Queen in Lewis Carroll's *Through the Looking Glass*, who must run faster and faster simply to stay in the same place.

Second, the Ward et al.'s (2016) scenario only looks at the implication for decoupling in one economy (Australia); but the challenge for techno-optimism is far more daunting if we look at the wider global context (Hickel and Kallis, 2019). This is especially true if we take seriously the long-term goal of the global development agenda (UNEP, 2016) which aims to bring the poorest parts of the world up to the living standards enjoyed by the developed world – a goal that many techno-optimist scenario's neglect to take on board (see e.g. Hatfield-Dodds et al., 2015). After all, from a moral perspective, it is difficult to argue that one section of the global population is entitled to a certain income per capita while denying a similar level to others. If the global development agenda of universalised affluence were to be achieved over the next 80 years, how big would the global economy be relative to the existing economy?

The figures are confronting, to say the least. Let's assume, as with the Ward et al.'s (2016) scenario, that continuous economic growth at a modest 2.41% growth rate leads today's developed nations (i.e. OECD) to expand their economies eightfold by 2100. Let us also assume that by this time the world population will have reached 11 billion, in line with median UN projections. Let us finally assume that this population has by the end of the century caught up to the per capita incomes of the OECD. If this scenario were ever to be achieved, the global economy would end up approximately 28 times larger than it is today!

Needless to say, ecosystems are already trembling under the pressure of one "developed world" at the existing size. Who, then, could seriously think our planet could withstand the equivalent of a 28-fold increase in the size of the global economy? The very suggestion is absurd, and yet this very absurdity defines the vision of the global development agenda. It is the elephant in the room. If we remember that humanity is already in ecological overshoot by 70%, then to achieve long-term sustainability humanity would need to achieve a factor 48 reduction in overall environmental impact (i.e. resource use, carbon emissions) per unit of GDP. Compare this 48-factor reduction with the 5-factor reductions that some techno-optimists think might be achievable via an efficiency revolution which has historically failed to fulfil its promise (Lovins, 1998; Von Weizsacker, 2009). Accordingly, even if these figures are overstated by an order of magnitude, the point would remain that efficiency gains could not possibly be expected to make the projected amount of GDP growth sustainable. The levels of decoupling required would simply be too much (Huesemann and Huesemann, 2011; Trainer, 2012). To think otherwise is not being optimistic but delusional.

The question, therefore, must not be: "How can we make the growth model sustainable?" The question should be: "What economic model is sustainable?" And the answer, it seems, must be: "Something other than the growth model" (see Lange, this volume).

## Efficiency without sufficiency is lost

The central message of this critical analysis is that efficiency gains that take place within a growth-oriented economy tend to be negated by further growth, resulting in an overall increase in resource and energy consumption, or at least no reduction. To take advantage of efficiency gains – that is, for efficiency gains to reduce resource and energy consumption to sustainable levels – what is needed is an economics of sufficiency, an economics that directs efficiency gains into reducing ecological impacts rather than increasing material growth (Alexander, 2012a; Herring, 2009). Once the limits of technology (and thus the limits to growth) are recognised, however, it becomes clear that embracing an economics of sufficiency is necessary if we are to create an economic model that is ecologically sustainable.

Exploring what a sufficiency paradigm might entail, however, is beyond the scope of this chapter. Let it simply be stressed that it would not mean rejecting the use of innovative technologies in relevant fields, nor more efficient ways of doing things. Rather, it would place such efforts in the context of a wider societal goal of achieving steady-state economies within the sustainable limits of a finite planet (Czech, 2013). For the most developed nations this will mean not just producing and consuming more *efficiently*, but also producing and consuming *less* and *differently*. This follows from the critique of techno-optimism detailed earlier. This can be achieved partly by cultural change, through which people practice "voluntary simplicity" by exchanging superfluous consumption of resource-energy intensive products and services, for a variety of non-material pursuits that promise to bring greater life satisfaction (Alexander, 2012b). But lifestyle changes at the individual and household level will also need to be accompanied by profound changes in macro-economic policy, political systems, and settlement design (see e.g. Trainer, 2010; van den Bergh, 2011). Other chapters in this volume explore some of these issues in more depth (see e.g. Di Giulio & Defila; Fuchs; Hayden; Lange; Larsson, Nässén & Lundberg, Philipsen, this volume).

## Conclusion

This chapter has reviewed the evidence in support of techno-optimism and found it to be wanting. This is significant because it debunks a widely held view, even amongst many environmentalists, that "green growth" is a coherent path to sustainability. Perhaps it would be nice if affluence could be globalised without damaging the planet. It would certainly be less confronting than rethinking cultural and economic fundamentals. But there are no credible grounds for thinking that technology is going to be able to protect the environment if economic growth is sustained and high consumption lifestyles continue to be globalised. The levels of decoupling required are simply too great. More efficient growth in GDP, therefore, is not so much "green" as slightly "less brown" (Czech, 2013: Ch. 8), which is a wholly inadequate response to the crises facing humanity (see Higgs, this volume).

Since there are no reasons to think that more efficient growth is going to reduce humanity's ecological footprint within sustainable bounds, it follows that we must consider alternative models of economy – alternative models of progress – even if these challenges conventional economic wisdom. To draw on the dictum often attributed to Einstein: we cannot solve our problems using the same kinds of thinking that caused them. This may not be a popular message, and it may already be too late for there to be a smooth transition beyond the growth model. But on a finite planet, there is no alternative. The sooner the world realises this, the better it will be for both people and planet.

We must embrace life beyond growth before it embraces us.

## References

Alcott, B. "Jevons' paradox." *Ecological Economics* 54 (2005): 9–21.
Alcott, B. "Population matters in ecological economics." *Ecological Economics* 80 (2010): 109–120.
Alcott, B. and R. Madlener. "Energy rebound and economic growth: A review of the main issues and research needs." *Energy* 34 (2009): 370–376.
Alexander, S. "The optimal material threshold: Toward an economics of sufficiency." *Real-World Economics Review* 61 (2012a): 2–21.
Alexander, S. "The sufficiency economy: Envisioning a prosperous way down." *Simplicity Institute Report* 12s (2012b): 1–31.
Asici, A.A. "Economic growth and its impact on environment: A panel data analysis." *Ecolgical Economics* 24 (2013): 324–333.

Beckerman, W. *A poverty of reason: Sustainable development and economic growth*. Oakland: Independent Institute, 2002.

Bo, S. "A literature review on environmental Kuznets curve." *Energy Procedia* 5 (2011): 1322–1325.

Carbon Tracker. "The 35 countries cutting the link between economic growth and emissions." 2016. Accessed 5 April 2016. www.carbonbrief.org/the-35-countries-cutting-the-link-between-economic-growth-and-emissions

Chakravarty, D., S. Dasgupta and J. Roy. "Rebound effect: How much to worry?" *Current Opinion in Environmental Sustainability* 5 (2013): 216–228.

Czech, B. *Supply shock: Economic growth at the crossroads and the steady state solution*. Gabriola Island: New Society Publishers, 2013.

Dinda, S. "Environmental Kuznets curve hypothesis: a survey." *Ecological Economics* 49, No. 4 (2004): 431–435.

Ehrlich, P. and A. Ehrlich. "Can a collapse of civilization be avoided?" *Simplicity Institute Report* 13a (2013): 1–19.

Ehrlich, P. and J. Holdren. "Impact of population growth." *Science* 171 (1971): 1212–1217.

Frondel, M., N. Ritter and C. Vance. "Heterogeneity in the rebound effect: Further evidence from Germany." *Energy Economics* 34 (2012): 461–467.

Giljum, S., M. Dittrich, M. Lieber and S. Lutter. "Global patterns of material flows and their socio-economic and environmental implications: A MFA study on all countries world-wide from 1980 to 2009." *Resources* 3 (2014): 319–339.

Global Carbon Project. Carbon budget and trends 2017, 2017. Accessed 13 November 2017. www.globalcarbonproject.org/carbonbudget.

Global Footprint Network. Ecological Footprint, 2017. Accessed 9 February 2017. www.footprintnetwork.org/our-work/ecological-footprint/.

Goodall, C. 'Peak stuff: Did the UK reach a maximum use of material resources in the early part of the last decade?' Carbon Commentary, Research Paper, 2011.

Hall, C.S., G.J. Lambert and B.S. Balough. "EROI of different fuels and the implications for society." *Energy Policy* 64 (2014): 141–152.

Hamilton, C. *Growth fetish*. Crows Nest, NSW: Allen & Unwin, 2003.

Hamilton, C. *Earthmasters: Playing God with the climate*. Crows Nest, NSW: Allen & Unwin, 2013.

Hatfield-Dodds S., H. Schandl, P.D. Adams, T.M. Baynes, T.S. Brinsmead, B.A. Bryan, F. Chiew, P.W. Graham, M. Grundy, T. Harwood, R. McCallum, R. McCrae, L.E. McKellar, D. Newth, M. Nolan, I. Prosser, A. Wonhas. "Australia is 'free to choose' economic growth and reduced environmental pressures." *Nature* 527 (05 November 2015): 49–53. doi:10.1038/nature16065.

Herring, H. "Sufficiency and the rebound effect." In *Energy efficiency and sustainable consumption: The rebound effect,* edited by H. Herring and S. Sorrell. London: Palgrave Macmillan, 2009.

Herring, H. and S. Sorrell. *Energy efficiency and sustainable consumption: The rebound effect*. London: Palgrave Macmillan, 2009.

Hickel, J. and G. Kallis. "Is green growth possible?" *New Political Economy* (2019, in press). doi:10.1080/13563467.2019.1598964

Huesemann, M. and J. Huesemann. *Techno-fix: Why technology won't save us or the environment*. Gabriola Island: New Society Publishers, 2011.

IEA. "Key World Energy Statistics 2017." Accessed 12 February 2017. www.iea.org/publications/freepublications/publication/key-world-energy-statistics.html.

IPCC. Climate Change 2014: Synthesis Report. *Contribution of working groups I, II and III to the fifth assessment report of the intergovernmental panel on climate change* [Core Writing Team, R.K. Pachauri and L.A. Meyer (eds.)].Geneva: IPCC, 2014.

Jackson, T. *Prosperity without growth: Foundations for the economy of tomorrow*. London: Earthscan, 2016.

Kallis, G. "Radical dematerialization and degrowth." *Philosophical Transactions of the Royal Society A*. 375, No. 2095 (2017). doi:10.1098/rsta.2016.0383.

Khazzoom, J.D. "Economic implications of mandated efficiency in standards for household appliances." *Energy Journal* 1, No. 4 (1980): 21–40.

Lenzen, M., D. Moran, K. Kanemoto, B. Foran, L. Lobefaro and A. Geschke. "International trade drives biodiversity threats in developing nations." *Nature* 486, No. 7401 (2012): 109–112.

Lomborg, B. *The sceptical environmentalist: Measuring the real state of the world*. Cambridge: Cambridge University Press, 2001.

Lovins, A. *Factor four: Doubling wealth – halving resource use*. London: Earthscan, 1998.

Lovins, A. *Reinventing fire: Bold business solutions for new energy era.* White River Junction, VT: Chelsea Green Publishing, 2011.

Meadows, D., J. Randers and D. Meadows. *Limits to growth: The 30-year update.* White River Junction, VT: Chelsea Green Publishing, 2004.

Monbiot, G. *Heat: How we can stop burning the planet.* Penguin: London, 2006.

Mudd, G.M. *The sustainability of mining in Australia: Key production trends and their environmental implications for the future.* Research Report No. RR5. Clayton: Department of Civil Engineering, Monash University and Mineral Policy Institute, 2009.

Nordhaus, T. and M. Shellenberger. 2011. "Evolve: A case for modernization as the road to salvation." *Orion Magazine* (September/October), 2011. www.orion magazine.org/index.php/articles/article/6402/.

Polimeni, J., K. Mayumi, M. Giampietro and B. Alcott. *The myth of resource efficiency: The Jevons paradox.* London: Earthscan, 2009.

Purdey, S. *Economic growth, the environment, and international relations: The growth paradigm.* New York: Routledge, 2010.

Raworth, K. *Doughnut economics: Seven ways to think like a 21st-century economist.* White River Junction, VT: Chelsea Green Publishing, 2017.

Ripple, William J., C. Wolf, T.M. Newsome, M. Galetti, M. Alamgir, E. Crist and M.I. Mahmoud, et al. "World scientists' warning to humanity: A second notice." *BioScience* 67, No. 12 (1 December 2017): 1026–1028. doi:10.1093/biosci/bix125.

Rockström, J., W. Steffen, K. Noone, Å. Persson, F.S. Chapin, III, E. Lambin, T.M. Lenton, et al. "Planetary boundaries: Exploring the safe operating space for humanity." *Ecology and Society* 14, No. 2 (2009): 32. [Online] www.ecologyandsociety.org/vol14/iss2/art32/.

Saunders, H.D. "Historical evidence for energy efficiency rebound in 30 US sectors and a toolkit for rebound analysis." *Technological Forecasting and Social Change* 80, No. 7 (2013): 1317–1330.

Simon, J. and H. Kahn. *The resourceful earth: A response to Global 2000.* London: Blackwell Publishing, 1984.

Sorrell, S. "The evidence for direct rebounds." In *Energy efficiency and sustainable consumption: The rebound effect,* edited by H. Herring and S. Sorrell, 23–46. London: Palgrave Macmillan, 2009.

Sorrell, S. "Reducing energy demand: A review of issues, challenges and approaches." *Renewable and Sustainable Energy Reviews* 47 (July 2015): 74–82.

Steinberger, J., F. Krausmann, M. Getzner, H. Schandl, and J. West. "Development and dematerialization: An international study" *PLoS One* 8(10) (2013): e70385.

Trainer, T. *The transition to a sustainable and just world.* Sydney: Envirobook, 2010.

Trainer, T. "But can't technological advance solve the problems?" *Simplicity Institute Report* 12g (2012): 1–7.

Turner, G. "Are we on the cusp of collapse? Updated comparison of *The Limits to Growth* with historical data." *Gaia* 21, No. 2 (2012): 116–124.

UNEP, Schandl, H., M. Fischer-Kowalski, J. West, S. Giljum, M. Dittrich, N. Eisenmenger, A. Geschke, et al. "Global Material Flows and Resource Productivity." An Assessment Study of the UNEP International Resource Panel. United Nations Environment Programme, Nairobi, 2016.

United Nations, Department of Economic and Social Affairs, Population Division. "World Population Prospects: The 2017 Revision, Key Findings and Advance Tables." 2017. ESA/P/WP/248.

Van den Bergh, J. "Energy conservation more effective with rebound policy." *Environmental Resource Economics* 48, No. 1 (2011): 43–58.

Von Weizsacker, E.U., C. Hargroves, M.H. Smith, C. Desha and P. Stasinopoulous. *Factor five: Transforming the global economy through 80% improvements in resource productivity.* London: Routledge, 2009.

Ward, J.D., P.C. Sutton, A.D. Werner, R. Costanza, S.H. Mohr and C.T. Simmons. "Is decoupling GDP growth from environmental impact possible?" *PLoS One* 11, No. 10 (2016): e0164733. doi:10.1371/journal.pone.0164733.

Wiedmann, T., H. Schandl, M. Lenzen, D. Moran, S. Suh, J. West and K. Kanemoto. "The material footprint of nations." *PNAS* 112, No. 20 (May, 2015): 6271–6276. doi:10.1073/pnas.1220362110.

# 19

# Consumer values and consumption

*Naomi Krogman*

## Introduction to the problem of consumption for the planet

At the core of all of our environmental problems is consumption. To be human is to consume. Human survival and moreover human dignity fundamentally require that we each have food, shelter, and clothing. Certainly of profound importance for governance of consumption is the disproportionate contribution of more-developed countries' material impacts to the planet compared to the lesser-developed world. Medium-developed countries such as China, India, and Korea have also dramatically increased their material demands on the earth. This disproportionality is a matter of international debate in regards to how energy and material goods are needed to maintain or advance development, and the fairness of access to energy, material goods, and the safe disposal of waste. Moreover, given the inextricable relationship between energy use and consumption, consumption is an issue for a number of justice issues in regards to costs and responsibilities for climate change adaptation, mitigation, and disaster recovery.

Consumer values are often the focus of public attention on drivers of overconsumption. Overconsumption results in a myriad of wastes along the supply chain that causes extensive harm to the living systems of the planet. Overconsumption, or consuming well beyond one's needs, is characteristic of the elite in all countries and more commonly across the population in medium- and more-developed countries. Societies arrange material production from the environment for the end result of consumption, be it consumption of energy, water, food, clothing, shelter, and a host of other material items for convenience and comfort (e.g. cell phones, cars). The final disposal of materials moving through these stages poses a particular challenge. The world's garbage crisis is predicted to grow exponentially in the coming decades as people become richer and increasingly move to urban areas (Hoornweg et al., 2013). A poignant example is the plastic that clogs many rivers (Ridgwell, 2017) – leading to floating islands of plastic garbage in the ocean (Zuckerman, 2017). Often the greatest cost to municipal budgets is solid waste disposal to address burgeoning waste streams (Hoornweg et al., 2013).

The human population has already crossed three of the nine thresholds associated with the planetary boundaries that define a "safe operating space for humanity" (Rockström et al., 2009). These planetary boundaries are those of climate change, rate of biodiversity loss, and "human interference with the nitrogen cycle" (Rockström et al., 2009, p. 472), all of which can cause significant environmental change and human suffering. The Union of Concerned Scientists recently published an article in *BioScience* where they asserted that in order to avoid widespread suffering, humanity must drastically reduce its per capita consumption of fossil fuels, meat, and other resources. Consequently, to provide a good life for the world's population within ecological limits, we will need to transform consumption patterns, especially among more- and medium-developed countries (Ripple et al., 2017).

A challenge in writing this chapter, which focuses on consumer values and consumption, is to situate human values in a larger system that perpetuates overconsumption by structurally supporting extrinsic values (e.g. material wealth and status) more than intrinsic values (e.g. community relationships). The impersonal nature of many market mechanisms that promote overconsumption (e.g. advertisement, easy money lending, a ubiquitous commercial presence across the globe coupled with a permeation of the cash economy) is topics that are rarely addressed in the scholarship on values and consumption. Most discussions of values, often held in disciplines such as economics, psychology, and business, conjure up images of individuals with values, and individuals who have choices (for a more critical perspective see Fuchs et al., 2016). In that individualistic mindset, the focus to change overconsumption becomes changing values that drive buying too much stuff or harmful stuff. Alternatively, the focus on individual values may lead to scholarship that examines the human values that prompt reduced or green consumerism, or greater participation in a sharing economy. If the public discussion on consumption focuses on individual actions, discussions about power, structural change, and overall reductions in global material consumption may be missed.

Thus, a caveat to the reader is that while understanding human values is critical for understanding consumption trends, it is more important to understand the role of various forces in society that reinforce certain values in the discursive realm of production and consumption, and moreover the structural barriers to addressing consumerism. The focus on the individual ends up being on the tail end of the commodity chain that examines the "buyer beware" or "vote with your dollar" approach rather than addressing many front-end consumption impacts of extraction, production, distribution, and waste disposal. The focus on values may also obscure the ways in which consumer choice is highly constrained by the rules of the market (e.g. regulations) and the power influences (e.g. lobbyists) on key decision-makers in society. Thus, as scholars of governance and sustainability, our task is to understand human and consumer values as proximal rather than direct drivers of overconsumption, given the economic, political, and cultural systems that perpetuate overconsumption.

## Human values and consumer values

### Human values

Values can be defined as what one holds to be desirable, and relates to the world of thought. Rokeach (1973), a seminal theorist of human values, separated out values into those that are terminal, or tied to desirable end-states of existence, such as family security, happiness, or freedom, in contrast to instrumental values, or values that are preferable modes of behaviour, such as being capable, helpful, or obedient. Human values literature suggests there is a core set of values that are shared by most human beings across cultures (Maslow, 1943; Schwartz, 2012). One commonly cited list of human values includes self-direction, stimulation, hedonism, achievement, power, security, conformity, tradition, benevolence, and universalism (Schwartz and Bilsky, 1987; Crompton et al., 2010).

Maslow, another leading figure in scholarly discussions about how material realities are tied to human values, tied material conditions of one's life to an individual's hierarchy of needs, where the fewer the material needs (satisfying hunger and the ability to feel safe) one had the more a person was free to pursue ideal needs such as belongingness and love, greater self-esteem, and self-actualisation (Maslow, 1943, 1954). Maslow's hierarchy of needs has been challenged by many theorists who argue that material needs in one culture may be perceived differently from another culture (or individual), and thus the hierarchy does not always apply (e.g. you may have less physical security but feel at peace with the world and yourself, especially if you're a Buddhist monk).

Many studies have established correlations between peoples' values and their corresponding behaviours. Some values are compatible with each other and other values are in psychological opposition to each other. Behaviours generally favour one set of values over another, where compatible values make it easier for people to pursue certain behaviours, and incompatible values make it more difficult. For example, the intrinsic value of having a "community feeling" often manifests itself in helping behaviours, nothing expected in return, and the way one behaves to "make the world a better place" (Crompton et al., 2010, p. 27). This value may be in opposition to the value of financial success, an extrinsic value. Crompton et al. (2010) explain that people simultaneously hold extrinsic values such as popularity and financial status, and intrinsic values such as affiliation and self-acceptance. These two sets of values exist in everyone, but intrinsic and extrinsic values are emphasised and manifest differently, in terms of what people consume in particular contexts and cultures. In the more-developed world, ubiquitous advertising, higher incomes, speed of transport and communication, and positive affiliation with extra-version, physical attractiveness, efficiency, and convenience induce higher consumption levels.

The priority people attach to these different values will vary by culture, life stage, and specific context (Schwartz, 2012), but there appears to be a universal relationship between value families, so to speak, that are in opposition to or compatible with each other. To address the relationship between consumption and the environment, an important finding of Crompton et al.'s (2010) work is that self-transcendent values are generally at odds with self-enhancement values. Self-transcendent values refer to the value placed on "understanding, appreciation, tolerance and protection for the welfare of all people and nature" and benevolence is the value placed on "preserving and enhancing the welfare of those with whom one is in frequent personal contact (the in-group)." These values are found to be in opposition to self-enhancement values: power (the value placed on "social status and prestige, control or dominance over people and resources") and achievement (the value placed on "personal success through demonstrating competence according to social standards"). Those who score strongly on self-enhancement values are found to be less concerned about environmental damage, and less likely to behave in environmentally friendly ways than those who score more strongly on self-transcendent values. Similarly, other research suggests that those who place more emphasis on values of financial success show lower engagement with bigger-than-self problems such as biodiversity loss and environmental pollution (Crompton et al., 2010 pp. 32–33). The implication of this research summary is that to build concern about the impacts of overconsumption, individuals and organisations need to activate, reinforce, and demonstrate self-transcendent values.

## Consumer values

Consumer values need to be understood in the broader context of human values because of the integration of consumer behaviour into our daily lives. Values, as personal guidelines that help people make personal choices, are of greater focus in the work on consumer values than are social values. Marketing and consumer studies literature often categorises the kinds of materialist values people hold, and further breaks down consumer values into "yearning to acquire and consume" (Cushman, 1990) and the "ceaseless pursuit of the 'good life' (Fromm, 1967). Richins and Dawson (1992), for example, identify three key kinds of consumer values. First, "acquisition centrality" is held when material items are so important to someone it takes the place of religion in structuring and orienting their behaviours, or where "consumption for the sake of consumption becomes a fever that consumes all the potential energy it gets access to" (p. 305). The second consumer value they identify is "acquisition as the pursuit of happiness" where a person's pursuit of happiness is through possessions and money. Third, the consumer value of "possession-defined success" attributes materialist choices to the social comparison with others, where consumption provides a proxy for "evidence of success and right-mindedness" (p. 304).

Consumers will be looking for products or services that express values that are important to them, often looking for signals that are implied by a product. These signals or "flags" help consumers to make choices, and thus brands often profile certain values such as in ethical consumption where "care, solidarity and collective concern" are part of the brand (Barnett et al., 2005, p. 45). Ethical consumption can be demonstrated, for example, by buying products like Tom's shoes (see http://www.toms.ca/what-we-give). In fact, many companies spend a great deal of money associating their brands with certain values. Heineken advertisements, for example, profile the core values of "respect, enjoyment and a passion for quality."

In the 2018 documentary "Advertising at the Edge of the Apocalypse," Jhally, the narrator, asserts that more money has been spent on advertising than any other collective human effort, and that advertising is the major storyteller of our time, reinforcing the collective belief that happiness and satisfaction are tied to consumerism. Advertisers especially effectively use price competition, tailoring to one's individuality, and gaining status, as reasons to buy their goods (Bell and Ashwood, 2016). Schor (2010) maintains "post modern accounts of consumer culture argue that what we care about as we consume is not the products themselves, but the signs and symbols they connect to" (p. 40). In sum, signals of products induce consumers to behave in a certain way (Kostelijk, 2017).

## Why people consume so much despite increased environmentalism

There are a number of trends in North America and across the world that suggest there is increasing awareness of our environmental conditions. The US environmental movement in 1970s, a global environmental movement in the 1990s (Carolan, 2013), and the recent attention on the drivers and impacts of climate change have heightened general awareness. Schor (2010), for example, found that the US public is aware that the American way of life is not sustainable. The majority of those surveyed from 2004 to 2009 agreed with the statement "protecting the environment would require most people to make major changes in the way they live" (p. 10). Pro-environmental attitudes have become more widely spread across the world, although people's habits have not necessarily followed (Stone, 2014). In addition, political polarisation in the United States has made the environment, especially climate change mitigation, a divisive issue (McCright and Dunlap, 2011) where conservative leaders and many of their constituents discount the connections between consumerism and environmental decline.

Schor (2010) notes that in the United States, the average person in 1960 consumed just a third of what they did in 2008. Overall material use has increased, before accounting for imports (Schor, 2010). For example, in the United States the average home size has almost doubled in the past 20 years to nearly 3,000 square feet. While the number of people in the home has shrunk to less than three people on average, nearly half of all new homes built have four or more bedrooms, and 38% have three bathrooms or more (US Census Bureau, 2016). Fast fashion (cheap clothing and high turnover of styles) has also resulted in people purchasing far more clothing than they did in the 1960s – Schor notes that as of 2007, Americans were purchasing a new piece of clothing every 5.4 days. Nearly one in four American households owns four or more TVs (Stone, 2014) and most people own a cell phone. Americans buying habits are highest in the category of electronics (Consumer Reports, 2014) and Americans over 30 years of age continue to buy cars and homes like the baby boomers (those born between 1946 and 1964). Thus, despite the widespread recognition that protecting the environment would require most people to make major changes in the way they live, North Americans have not reduced their absolute material throughput since the emergence of many phases of the environmental movement.

## Public understanding of pathways to action to address a macro-level problem

While environmentalism has challenged consumerism, there is public uncertainty about how to change current practices related to consumption to improve environmental conditions. The link people make between environmental values in particular and pro-environmental behaviours, such as consuming less, consuming green products, or consuming more durable, well-made items, is murky (Kennedy et al., 2009, 2015). Part of the reason why the relationship between environmental values and consumerism is murky is because of the poor understanding of the actual behaviours that can remediate environmental conditions (McBeth and Volk, 2010). Norgaard (2010) also makes the compelling argument that fear, guilt, and helplessness are psychological barriers to putting knowledge into transformative action, particularly in relation to making connections between consumption and climate change. Thus, while there appears to be an increased understanding of the role humanity has on the environment and the reliance humanity has on a healthy environment for survival and human dignity, there is not a corresponding sense of empowerment and efficacy to change the status quo. In this same volume, Maniates makes the point that the value–action gap is also driven by a sense of resignation that small actions of "responsible," "conscientious," "green," or "environmental" living are the only ways to drive change, given the lack of vision and social support to pursue collective, structural change in how citizens work together to reorganise the material world. This value–action gap is exacerbated by a context in which high consumption is normalised, so that persons in more developed countries especially must deliberately eschew the patterned behaviours supported by these structural and contextual factors described below.

## Easy spending

Obtaining credit cards became easier in the 1980s, and taking out loans became easier in the 2000s, resulting in far more purchases and a higher debt load in the United States (*Euronews*, 2018). Laird (2009) notes that during the early 2000s, countries such as Italy, Canada, and France actually outpaced the United States in the growth of consumer debt. Credit cards have become consumers' safety net in an era of wage stagnation (Thorne cited in *Euronews*, 2018), and are a pathway to instant gratification in a societal context of increased time pressures and stress (American Psychological Association, 2011).

## Enchantment

George Ritzer in *Enchanting a Disenchanted World: Revolutionizing the Means of Consumption* (2010) argues that "cathedrals of consumption" or spectacle shopping spaces have replaced the awe and wonder citizens once had with religious cathedrals and other spaces of beauty. He points to examples such as Dubai, Disneyworld, or Times Square, where shopping spaces might even simulate landscapes and other man-made wonders to entice the shopper to buy things or experiences. For example, Disneyworld might simulate what it is like to ride down class four rapids of a Western US river, and then sell you a $25 offered picture of you as you experienced your last splash of the ride. Other shopping areas in malls are often designed to mimic quaint areas of Monte Carlo or Paris, and thus enchant the consumer with a sense they are on a luxury holiday. Consequently, people shop for the experience of enchantment rather than for obtaining goods and services. Costs to re-enchant these spaces with new designs and spectacles fall to the consumer, Ritzer argues, because enchantment soon requires re-enchantment, once the novelty of any spectacle is gone. Ritzer holds that given the value consumers have for novelty, excitement, and entertainment, even athletic stadiums, educational settings, and some small towns embrace spectacle consumption. He predicted the rise of seeking enchantment through virtual means, which encourages consumers to live out fantasies via artificial intelligence, such as through Second Life, and fantasy vacations such as that depicted in the HBO TV series West World.

## Discount culture

Another recent trend that has aided and abetted high consumption is the discount culture that has blossomed in North America since the 1970s. Shell (2009) in *Cheap: The High Cost of Discount Culture* explains the history of going from dime stores in the 1960s and 1970s to cathedrals of consumption like shopping malls in the 1970s and 1980s into a growth phase of vast low price-promising Walmart and IKEA department stores in the 1980s, to even cheaper dollar and outlet stores thereafter. While the gross national product tripled in the United States from 1970 to 2008, wages flattened and job benefits declined. Between 2011 and 2016 this trend continued as wages increased by 1% or less in many Western economies such as the United States, Germany, the United Kingdom, and Japan between 2011 and 2016 (Angus, 2018). Eighty per cent of the net income gains went to the top 1% of earners, and so the ability to buy things cheap for the vast majority of the population became even more attractive. The subprime mortgage crisis of 2008 also reduced citizens' purchasing power, although employment trends in the United States have improved since then. Currently Amazon online sales are attracting buyers looking for cheap prices and quick delivery. Further, those living in poverty have increasingly purchased their needs from resale shops, grocery discounters, and value-based retailers (e.g. Aldi, Lidl and Daiso). In the United States, the traditional thrift store market is reportedly growing by 8% (Angus, 2018).

## Time crunch

In addition, numerous scholars have documented the time crunch many people feel themselves under due to long work hours, commuting in traffic, and distractions from being online, via texting, tweeting, Facebook, Netflix, or email much of the day (Shulte, 2014). This time crunch makes the importance of convenience even stronger. Amazon delivery has appeal due to the reduced time it takes to shop for items. Thus, consumerism that saves time, i.e. is convenient, has been the key reason for the huge waste stream from packaged food, bottled water, to-go coffees, and electronic items that demand easy servicing at particular school and work settings. People also buy more and buy new items because of the lack of time to find the same item one has already purchased, hidden away in their homes, offices, or garages. Consumers may also replace material goods because the item was poorly made in the first place, it takes too much time to get the item fixed, or due to the time required to figure out if there is a repair shop available.

Whether it is readily acknowledged or not, many people value their time more than their money. Mullainathan and Shafir (2013) eloquently identify this in their book *Scarcity: Why Having Too Little Means So Much* in which they illustrate how scarcity of time limits or distorts human skills. We may have intentions to consume deliberately, but our time scarcity may lead to mindless consumption. These authors suggest, like Thaler and Sustein (2008) in their popular book *Nudge: Improving Decisions about Health, Wealth and Happiness*, that the more resource light consumption choices are built into our daily lives, the more likely we can take them up, because they don't require energy-depleting vigilance. Time scarcity contributes to the predominance of the small-things approach that, as Maniates argues in this volume, is environmentally inadequate. The time crunch likely contributes to this short term immediate gratification one can have over "personal responsibility" actions versus engaging in collective political action that has no certain outcome.

## Inconspicuous consumption

Elizabeth Shove (2003) points out that much of the consumption in more-developed countries is embedded in inconspicuous routines and habits. In other words, a great deal of consumption in the more-developed world is invisible as it is built into tools, appliances, and household infrastructure that people take for granted. Shove points out the proliferation of products, arrangements, and devices that promise consumer comfort, cleanliness, and convenience. For example, the normalisation of owning a cell phone, and having multiple bathrooms in a home, has carried with it a large environmental footprint. These material features become normalised in everyday environments, and carry habits with them that have resource-intensive consequences, and ultimately cause environmental harm (Shove, 2003). Similarly, even if the ecological rucksack of an item is lower over time (e.g. cell phones become smaller, TVs have less hardware) the ubiquitous purchase of them makes the overall impact of these items higher. As resource-intensive features of societies increase, inconspicuous consumption increases, making it harder to tie environmental values to consumption given the unnoticed normalisation of these consumption patterns.

For example, Fuchs et al. (2016) point out the inconspicuous structural features of the supply chain of meat consumption in much of the Global North. While there are many efforts to make meat consumption more sustainable, they argue business as usual prevails in meat production. Hidden from the meat consumer may be the cheap land used to raise cows, the complicit government that prioritised privatisation of that range land over subsistence farming, the externalisation of environment and public health risks associated with large-scale, capital-intensive animal farming, the highly concentrated ownership of meat processing plants, and the few supermarket chains worldwide that push prices downward given their disproportionate power in what reaches the urban consumer. Opaque to the consumer are all of these exercises of power and attendant social and environmental impacts because all the consumer sees is the normalisation of the convenient availability of meat at cheap prices.

The structure of one's built environment also significantly influences consumer behaviours. You may care a lot about your personal carbon footprint, but you may drive to work, even though you could take public transit, ride share, bike to work, or have rented an apartment closer to work. The mode you travel to work may be in relation to other values—such as values to spend time with your family instead of time on the bus, or the value you have to arrive at work looking clean and well groomed (which biking to work may not make easy), as expected on the job. Thus, the way we organise our work and family lives has a great deal to do with how we consume, particularly in relation to car travel (Barr, 2015). This also begs the question, "how can we integrate multiple human values into city planning and design for where people live, work and play to reduce harmful elements of consumption?"

## Shifts in prioritising values for mindful living and building community

There are hopeful trends that suggest shifts are occurring, but as Maniates points out, these would be far more impactful if accompanied by collective political action. For example, a recent Euromonitor International report summarises consumer trends for 2018, suggesting that people in the 20–29 age group are assuming greater social responsibility for how they impact the world as consumers. Angus (2018) asserts, "Consumers are adopting clean-living, more minimalist lifestyles, where moderation and integrity are key" (p. 3). Clean Lifers feel they can make a difference through their spending habits, and feel less need to impress through ownership. Rather, those in the 20–29 year old range are more likely to spend money on experiences they can share with friends such as weekend trips, festivals, and restaurants. They are more likely to spend less on alcohol and other drugs, and more on "genervacations" or choosing to holiday with family. Millennials are less likely to desire belongings like vehicles, houses, and clothing than previous generations. There is greater comfort among this generation in being self-employed, and they are delaying owning a home or having children (Angus, 2018).

The report also asserts that the new generation of consumers is more likely to be community-minded sharers, renters, and subscribers to sharing economy arrangements such as Uber, Rent the Runway, and Airbnb. The younger generation is renting equipment, cars, consumer electronics, and clothing. Similarly, millennials are more interested in "co-living and sharing spaces and mutual facilities to save money and inspire collective ideas or provide comfortable, more acceptable living conditions" (Angus, 2018, p. 30).

They are also much more likely to use their collective voice to call brands to account, trace the full product journey, and use social media to highlight injustice associated with particular products. "They look for evidence of Fair Trade procurement, environmentally-friendly production, fair wages, FSC certified paper packaging and energy-efficient distribution" (Angus, 2018, p. 25). These discriminating shoppers are more interested in shopping locally (e.g. supporting local brew pubs) and are more concerned about using their purchases to be distinct from others over wanting to be considered by others that they are doing well (Angus, 2018).

Two other trends in the last 15 years that are intersecting with consumer values are the growing values of "mindfulness" and "simplicity." There has been a recent proliferation of popular literature on how to be more deliberate in the use of one's time, especially in regards to building healthy practices into one's life like yoga and meditation, with the ultimate goal of living more in the present. Books like *The Power of Now* (Tolle, 1999) and *The Not So Big Life* (Sushanka, 2007) are exemplars. Links are being made between mindfulness and less need for consumption, where time spent shopping or consuming un-needed items is rather spent outside, doing exercise, or in communal acts of provision such as in the slow food movement and neighbourhood revitalisation efforts (Krogman and Kennedy, 2016). While the voluntary simplicity movement has been around since the 1970s (Elgin, 1981) it is now manifested in multiple strands of efforts to simply life, often through buying less, and holding on to less stuff. Popular reading that takes this on wholeheartedly is *Essentialism: The Disciplined Pursuit of Less* (McKeown, 2014) and *The Japanese Art of Decluttering* (Kondo, 2014). Popular TV shows such as "The Secret Lives of Hoarders," "Storage Wars," and "How Clean is your House?" emphasise how clutter has costs to human well-being, and inadvertently make the case to own fewer but higher quality items.

Schor (2010) documented a shift in thinking among "downshifters" or those people who choose to work less hours and spend less money in exchange for more time to do what they want such as spend time with friends and family, pursue hobbies, and ultimately self-fulfilment. A sharp increase in people to subscribe to this "plentitude" way of living occurred after the 2008 economic downturn in the United States. Schor contends many of these people discovered how much happier they were when they got off the work and spend treadmill. Given income is the strongest predictor of consumption, Schor makes a powerful argument that working less and earning less are key to more moderate consumption in the more-developed world. Downshifters are more likely to be frugal rather than cheap in their purchases, and learn to have moderate consumption of items that have traditionally been luxuries. They are more likely to buy high quality items that they can repair, if need be, and buy products and create spaces in their homes that have multi-functionality. Like the millennials, down shifters are interested in the sharing and collaborative economy and bartering with those with whom they have affinity.

Jaeger-Erben and Ruckert-John (2016) have categorised a number of innovations in sustainable consumption led by mostly local groups who share these new consumer values. Local groups may focus on systems of sharing in a particular setting, through clothing swapping parties, home swapping, and private car sharing. Other groups are focusing on the competence of residents to fix their own items and build sewing cafes and repair shops. More serious sustainable consumption community-led efforts have focused on altering material arrangements by supporting professional car sharing organisations, bicycle renting systems, purchasing solidarity groups (Forno et al., 2016), and working with financial institutions to set up systems for local investment (Gismondi et al., 2016) in cooperatively held wind farms, forests, or other kinds of energy delivery. While intentional villages such as Eco villages have been around for 50 years, more recently the number of tiny home (homes that are from 65 to 400 square feet) communities is growing (Shaw, 2011; Kaufmann, 2016).

These new forms of practice mentioned earlier have created some new rules of the game to provide for oneself and community in a less resource-intensive way. Regulations have changed to allow co-housing developments in many cities, and provide support for craft-businesses, community gardens, and community-supported agriculture. While often controversial, new policies are being developed to govern the sharing economy, driven by such companies like Airbnb, Zipcar, Etsy, and Uber (Botsman 2015). Botsman (2010) asserts that the sharing economy is uniquely placed to reflect consumers' desires to connect directly and see their consumer selves as part of a collective act that can save money and reduce environmental harm.

Consumer social movements are seeking to "transform various elements of the social order surrounding consumption and marketing" (Kozinets and Handelman, 2014, p. 691). For example, some consumer activist groups are lobbying government to control the flow of junk food in schools, remove genetically modified foods from groceries and restaurants, and eliminate or reduce advertising (e.g. as suggested by Klein (1999) in her book *No Logo*). However, anti-consumer ideology has not been a strong force at mobilising social movements against global capitalism (Sklair, 1995).

## Governance to support new consumer values

Given levels of concern and awareness are high, the real question may be how the public demands that values are manifested into collective action. To address strongly held intrinsic and self-transcendent values, citizens need to see how inconspicuous consumption requires such collective efforts. As Lacasse (2016) and Maniates (this volume) have pointed out, a danger in focusing on individual pro-environmental behaviour is that the environmentally concerned population may satisfy their desire to influence change by making lifestyle changes alone. By choosing to focus on consumption choices alone, key segments of the population delude themselves into thinking that eventually "everyone will get on board" voluntarily, missing the complexities of the significant material shifts needed that can only come about through transformative environmental governance.

Consumers rely on governments to protect them with consumer safety laws, safety inspections, and develop new certification systems to verify features of products people want. For example, Fair Trade certification, Eco-labelling, and Organic certification continue to be a key way in which governments interact with consumer values. Less available are government supports such as tax breaks, business incubators, research, development, and demonstration funding to reduce negative environmental and social impacts along the entire supply chain. Infrastructure investments are needed to reduce waste of water, energy, material throughput, and minimise the negative environmental impacts of waste streams. Consumption could also be reduced by governments that invested in better public services that obviate the need for so many privatised services (Lertzman, 2012). Our governance systems have far more potential to impact the environment by developing and enforcing regulations in building codes, energy systems, soil and water management, transportation requirements, and the absorptive capacity of ecosystems, than in appealing to the individual shopper to buy less or buy green.

Given the majority of people in the world live in or are moving to cities, cities will continue to be the most important generators of innovative consumption practices and changes in governance systems that reduce environmental harm and inequity. The way consumer values are embedded into citizen muscle, political decision-making about collective goods and institutions that engage citizen participation are important areas for future scholarship.

## Acknowledgements

I would like to acknowledge Julian Faid, MSc in Environmental Sociology from University of Alberta, who did background literature searches for this chapter.

## References

American Psychological Association. 2011. "Stressed in America." *Monitor on Psychology*. 42(1): 60. www.apa.org/monitor/2011/01/stressed-america.aspx [Accessed March 25, 2018].

Angus, A. 2018. *The Top 10 Global Consumer Trends for 2018*. London: Euromonitor International.

Barnett, C., N. Clarke, P. Cloke and A. Malpass. 2005. "The Political Ethics of Consumerism." *Consumer Policy Review*. 15(2): 45–51.

Barr, S. 2015. "Beyond Behaviour Change: Social Practice Theory and the Search for Sustainable Mobility." In *Putting Sustainability into Practice: Applications and Advances in Research on Sustainable*

*Consumption*, edited by Emily Huddart Kennedy, Maurie J. Cohen and Naomi T. Krogman, 91–108. Cheltenham: Edward Elgar.

Bell, M. and L. L. Ashwood. 2016. *An Invitation to Environmental Sociology*. Los Angeles: Sage.

Botsman, R. 2010. *What's Mine Is Yours: The Rise of Collaborative Consumption*. New York: Harper Collins.

Botsman, R. 2015. "Defining the Sharing Economy: What is Collaborative Consumption – And What Isn't?" *Fast Company Newsletter*. www.fastcompany.com/3046119/defining-the-sharing-economy-what-is-collaborative-consumption-and-what-isnt [Accessed March 25, 2018].

Carolan, M. 2013. *Society and the Environment: Pragmatic Solutions to Ecological Issues*. Boulder, CO: Westview Press.

Consumer Reports. 2014. "How Americans Consumers Shop Now: After Seven Years of Cutting Back, Consumers Are Finally Opening Their Wallets Again. But the Recession Has Changed This Country's Buying Habits – Big Time." *Consumer Reports*. www.consumerreports.org/cro/magazine/2014/11/how-america-shops-now/index.htm [Accessed March 23, 2018].

Crompton, T., J. Brewer, P. Brewer and T. Kasser. 2010. *Common Cause: The Case for Working with Our Cultural Values*. Surrey: WWF-UK. http://assets.wwf.org.uk/downloads/common_cause_report.pdf [Accessed March 25, 2018].

Cushman, P. 1990. "Why the Self Is Empty." *American Psychologist*, 45(May): 599–611.

Elgin, D. 1981. *Voluntary Simplicity: Toward a Way of Life That Is Outwardly Simple*. New York: Inwardly Rich.

Euronews. 2018. "America's Debt Load Is Hitting Record – and Risky – Territory." January 12. www.euronews.com/2018/01/12/america-s-debt-load-hitting-record-risky-territory-n837251 [Accessed March 25, 2018].

Forno, F., C. Grasseni and S. Signori. 2016. Italy's Solidarity Purchase Groups and 'citizenship labs. In *Putting Sustainability into Practice: Applications and Advances in Research on Sustainable Consumption*, edited by Emily Huddart Kennedy, Maurie J. Cohen and Naomi T. Krogman, 67–90. Cheltenham: Edward Elgar.

Fromm, E. 1967. "The Psychological Aspects of the Guaranteed Income." In *The Guaranteed Income*, edited by Robert Theobald, 183–193. Garden City, NY: Anchor.

Fuchs, D., A. Di Guilo, K. Glaab, S. Lorek, M. Maniates, T. Princes and I. Ropke. 2016. Power: the missing element in sustainable consumption and absolute reductions and action. *Journal of Cleaner Production*. 132: 298–307.

Gismondi, M., J. Marois and D. Straith. 2016. "'Unleashing Local Capital': Scaling Cooperative Local Investing Practices." In *Putting Sustainability into Practice: Applications and Advances in Research on Sustainable Consumption*, edited by Emily Huddart Kennedy, Maurie J. Cohen and Naomi T. Krogman, 204–230. Cheltenham: Edward Elgar.

Hoornweg, D., P. Bhada-Tata and C. Kennedy. 2013. "Environment: Waste Production Must Peak in This Century." *Nature*. 502(October 31): 615–617.

Jaeger-Erben, M. and J. Ruckert-John. 2016. "Researching Transitions to Sustainable Consumption: A Practice Theory Approach to Studying Innovations in Consumption." In *Putting Sustainability into Practice: Applications and Advances in Research on Sustainable Consumption*, edited by Emily Huddart Kennedy, Maurie J. Cohen and Naomi T. Krogman, 204–230. Cheltenham: Edward Elgar.

Kaufmann, C. 2016. "Tiny Houses Are Becoming a Big Deal." AARP Liveable Communities. www.aarp.org/livable-communities/housing/info-2015/tiny-houses-are-becoming-a-big-deal.html [Accessed March 28, 2018].

Kennedy, E. H., T. Beckley, B. McFarlane. 2009. "Why We Don't 'Walk The Talk': Understanding the Environmental Values/Behaviour Gap in Canada." *Human Ecology Review*. 16(2): 151–160.

Kennedy, E. H., H. Krahn and N. T. Krogman. 2015. "Are We Counting What Counts? A Closer Look at Environmental Concern, Pro-Environmental Behaviour, and Carbon Footprint." *Local Environment*. 20(2): 220–236.

Klein, N. 1999. *No Logo*. New York: Picador.

Kondo, Marie. 2014. *The Life-Changing Magic of Tidying Up: The Japanese Art of Decluttering and Organizing*. New York: Random House LLC.

Kostelijk, Erik. 2017. *The Influence of Values on Consumer Behaviour: The Value Compass*. London: Routledge.

Kozinets, R. V. and J. M. Handelman. 2014. "Adversaries of Consumption: Consumer Movements, Activism, and Ideology." *Journal of Consumer Research*. 31(4): 691–704.

Krogman, N. and E. H. Kennedy. 2016. "The Potential of Emotional Energy and Mindfulness to Expand Sustainable Consumption Practices." *Tvergastein: Interdisciplinary Journal of the Environment*. 7(8): 74–84.

Lacasse, K. 2016. "Don't Be Satisfied, Identify! Strengthening Positive Spillover by Connecting Pro-Environmental Behaviours to an 'Environmentalist' Label." *Journal of Environmental Psychology.* 48: 149–158.

Laird, G. 2009. *The Price of a Bargain: The Quest for Cheap and the Death of Globalization.* Toronto, ON: Emblem.

Lertzman, Renee. 2012. Book Review of *Living in Denial: Climate Change, Emotions and Everyday Life.* Cambridge: MIT Press. *Organization and Environment.* 24(4): 491–492.

Maslow, A. H. 1943. "A Theory of Human Motivation." *Psychological Review.* 50(4): 370–396.

Maslow, A. H. 1954. *Motivation and Personality.* New York: Harper and Row.

McBeth, W. and T. L. Volk. 2010. "The National Environmental Literacy Project: A Baseline of Middle Grade Students in the United States." *Journal of Environmental Education.* 41(1): 55–67.

McCright, A. and R. Dunlap. 2011. "Cool Dudes: The Denial of Climate Change among Conservative White Males in the United States." *Global Environmental Change.* 21(4): 1163–1172.

McKeown, Greg. 2014. *Essentialism: The Disciplined Pursuit of Less.* New York: Random House LLC.

Mullainathan, S. and E. Shafir. 2013. *Scarcity: The New Science of Having Less and How It Defines Our Lives.* New York: Times Books.

Norgaard, K. M. 2010. *Living in Denial: Climate Change, Emotions and Everyday Life.* Cambridge: MIT Press.

Richins, M. L. and S. Dawson. 1992. "A Consumer Values Orientation for Materialism and Its Measurement: Scale Development and Validation." *Journal of Consumer Research.* 19: 303–316.

Ridgwell, H. 2017. "Most Ocean Plastic Pollution Carried by 10 Rivers." Voice of America News (VOA), November 24. www.voanews.com/a/ninety-percent-of-ocean-plastic-pollution-carried-by-10-rivers-/4134909.html [Accessed March 22, 2018].

Ripple, W. J., C. Wold, T. M. Newsome, M. Galetti, M. Alamgir, E. Crist, M. L. Mahmoud, W. F. Laurance, 2017. 15,364 Scientist Organizations from 184 Countries. "World Scientists' Warning to Humanity: A Second Notice." *BioScience.* 12(1): 1026–1028.

Ritzer, G. 2010. *Enchanting a Disenchanted World: Revolutionizing the Means of Consumption.* London: Pine Forge Press.

Rockström, J., et al. 2009. "A Safe Operating Space for Humanity." *Nature.* 461(24): 472–475.

Rokeach, M. 1973. *The Nature of Human Values.* New York: Free Press.

Schor, J. 2010. *Plenitude: The New Economics of True Wealth.* New York: Penguin Press.

Shaw, N. 2011. "The Tiny House Movement." *Synergy Magazine,* April 14. http://synergymag.ca/the-tiny-house-movement/ [Accessed March 28, 2018].

Shell, E. R. 2009. *Cheap: The High Cost of Discount Culture.* New York: The Penguin Press.

Shove, E. 2003. *Comfort, Cleanliness and Convenience: The Social Organization of Normality.* Oxford: Berg.

Shulte, B. 2014. *Overwhelmed: Work, Love and Play When No One Has the Time.* New York: Harper Collins.

Sklair, L. 1995. "Social Movements and Global Capitalism." *Sociology* 29(August): 495–512.

Stone, A. 2014. "8 Surprising, Depressing, and Hopeful Findings from Global Survey of Environmental Attitudes." *National Geographic,* September 27. https://news.nationalgeographic.com/news/2014/09/140926-greendex-national-geographic-survey-environmental-attitudes/ [Accessed March 22, 2018].

Sushanka, S. 2007. *The Not So Big Life: Making Room for What Really Matters.* New York: Random House.

Schwartz, S. H. 2012. "An Overview of the Schwartz Theory of Basic Values." Online Readings in *Psychology and Culture.* 2(1). https://scholarworks.gvsu.edu/orpc/vol2/iss1/11/ [Accessed March 22, 2018].

Schwartz, S. H. and W. Bilsky. 1987. "Toward a Universal Psychological Structure of Human Values." *Journal of Personality and Social Psychology.* 53: 550–562.

Thaler, R. H. and C. R. Sustein. 2008. *Nudge: Improving Decisions about Health, Wealth, and Happiness.* London: Penguin Books.

Tolle, E. 1999. *The Power of Now: A Guide to Spiritual Enlightenment.* Novato, CA: New World Library.

United States Census Bureau. 2016. "Characteristics of New Housing." https://www.census.gov/construction/chars/completed.html [Accessed March 24, 2018].

Zuckerman, C. 2017. "Garbage Swell." *National Geographic.* 231(4): 14.

# 20

# The population challenge

*Diana Coole*

Population change is a dynamic, complex process. It has environmental significance because of its implications for natural and social resources, including for land, space, and biodiversity. While alterations in the distribution or composition of peoples (such as urbanisation, mass migrations or an age bulge at either younger or older ages) can have a substantial negative impact, policy interventions in these areas tend to be practically and politically very challenging. Arguably the trend that has the biggest impact on sustainability, and which presents the greatest challenge for sustainability governance in the 21st century, however, is the continuing expansion of human numbers. This is the main issue discussed in this chapter. The discussion focuses on two questions: should population growth be brought under sustainability governance and if it is, what sort of measures is appropriate (that is, effective *and* ethical)?

The human population is an aggregate of embodied individuals manifesting measurable patterns and trajectories. Mortality and fertility are the most important variables since their relationship determines natural population growth (or decline). During the modern age they have been progressively wrenched from the natural realm to become increasingly amenable to biomedical interventions, personal lifestyle choices, and political decision-making. While modern individuals have been empowered to manage their own bodies and life chances, their health and fertility especially, the state has also acquired biopolitical capacities for and interests in managing aggregate outcomes. As such, population matters present a febrile mixture of intimate individual experiences and anonymous general trends. In a field where ethical considerations and biophysical consequences seem ineluctably bound, the population question presents policymakers with a particularly knotty form of collective action problem.

Reducing growth rates essentially means curtailing fertility. Collectively, individuals' reproductive behaviour may result in dangerous outcomes for society and the planet. Yet within this intimate area, personal autonomy is staunchly defended as a freedom belonging to the private realm. Since 1967, couples' right to decide when, whether, and how many children to procreate has been recognised as a human right by the United Nations (UN). Since 1994, women's reproductive rights have been at the centre of the dominant transnational discourse (the "Cairo consensus"), which is overtly anti-Malthusian in its denunciation of anti-natalist population programmes. The latter's opponents reject policies whose goals mean treating women's bodies instrumentally, while asking *whose* fertility and growth rates are targeted (Corrêa 1994). Any suggestion of managing fertility in order to achieve broader environmental outcomes, including its incorporation within sustainability governance, is controversial because it is freighted with gender, race, and class significance.

Not everyone agrees, moreover, that ongoing population growth is a problem and even among those who do, not all protagonists accept that political intervention is necessary to resolve it. Even when numbers are acknowledged to be problematic and susceptible to policy interventions, perceptions of political risks and ethical hazards may rule it out. Yet over the past decade, there has been a noticeable resurgence of interest in the contribution world population growth is making to an unfolding environmental crisis (e.g. Alexander and Rutherford, this volume; Higgs, this volume), with a growing number of empirical studies and modelling exercises demonstrating the environmental benefits of swiftly stabilising or reducing human numbers. Perhaps the most immediate challenge for a new generation of limits-to-growth or planetary-boundaries thinkers is simply, then, to get the issue back onto the political agenda by addressing objections, reframing the issue as one of sustainability that concerns everyone, and promoting an informed public debate that takes into account the material conditions and ethical sensibilities of the 21st century.

## The demographic challenge

When considering whether population growth should be brought within the orbit of sustainability governance, some background information about demographic trends is instructive. Until the industrial revolution, the Earth's human population increased extremely slowly and precariously since births and deaths largely cancelled one another out. With better diets, higher living standards, and improved public health, Europe's population began to rise after the mid-18th century: a process subsequently replicated worldwide as mortality rates everywhere declined. Rapid population growth ensued. According to demographic transition theory, which is the principal model for understanding this process, this is the first stage of a transformative process contemporaneous with modernisation (Szreter 1993; Dyson 2010). It took until 1804 for the first billion people to appear but the two billion threshold had been crossed by 1927 and the three billion one by 1960. In a context of postwar economic growth and new environmental concerns about planetary limits, fears of a population explosion engendered calls to halt population growth (Ehrlich 1968; Meadows and Meadows 1972). Since then, the total has more than doubled: to over 7.6 billion currently and with approximately 83 million additional people each year. A world total of 9.7 billion by 2050 and 11.2 billion by 2100 is projected (UN 2017a, medium-variant).

Opinion is divided about the significance of this ongoing increase. Some environmentalists are alarmed anew by the prospect of adding more human beings to a finite planet already manifesting chronic signs of systemic stress and exceeding safe operating boundaries. Their concerns are explained further below. On the other hand, economists and technological optimists may suggest that free markets and technological innovation can unlock the elasticity and substitutability of resources, thereby allowing the Earth to accommodate additional, wealthier inhabitants. This looks more feasible inasmuch as their case is supplemented by demographic optimists who point out that the annual growth rate is slowing (Dorling 2013). Between 2005 and 2017 it fell from 1.24% to 1.10% (UN 2017a). These optimists generally welcome this downward trend but mostly agree with economists that political intervention is unnecessary, since fertility decline is occurring as a consequence of development. Allowing this process freely to run its course has obvious libertarian and ethical appeal. If numbers at the end of the journey are far higher than during the pre-industrial period, a supplementary assumption is that modernisation will also have procured the technological advances needed to maintain a balance between more, wealthier people and the planet's carrying capacity.

From these latter perspectives, recent concerns represent an unwarranted return of Malthusian pessimism with its attendant implication of (coercive) population control (Coole 2013). Certainly, the kind of massive totals once anticipated no longer look credible because world fertility is indeed declining almost everywhere. This defines the second stage of demographic transition. Yet some outstanding questions remain. Will world population stabilise quickly enough, and at a sufficiently low level, for the planet to support future generations at a level compatible with Sustainable Development Goals (SDGs)? Demographic transition theoretically ends when a new balance between low mortality and fertility rates restores equilibrium, with fertility hovering around replacement level and numbers stabilising. But in a context of globally rising living standards and the Earth's degraded life-support capacities, can completing transition be guaranteed to be sufficiently universal and fast to avoid unacceptable existential risk of systems overload and collapse? Crucially, might some form of policy direction not be required if a sustainable equilibrium is to be achieved?

For sceptics unpersuaded by the laissez-faire route, completing transition is not a destiny but an urgent project for governance. While all nations have successfully achieved the first stage of transition (for example through Green Revolution biotechnology and global immunisation programmes) and virtually all have started the second, far fewer have completed the process and in some places, fertility rates remain well above replacement level. The duration of the lag between the two stages of transition is critical in shaping a nation's demographic expansion and its contribution to the global total. During this phase especially, political will and good governance seem to be crucial for completing transition and stabilising numbers (Coole 2018). Centralised welfare states and international support for family planning programmes after all played important roles when transition was being completed in Europe and Asia.

When making the case for active population governance, the latest projections are informative. Projections are not predictions, but estimates based on existing trends and assumptions about their likely evolution. Uncertainty is represented by calculating three main variants. The medium version is generally cited but the high and low versions are statistically only relatively less probable. The key variable determining actual outcomes, which also defines the difference between each of the three variants, is the total fertility rate (TFR). Basically, this is the number of children that women will, on average, bear during their lifetime. "Future population growth is highly dependent on the path that future fertility will take, as relatively small changes in fertility behaviour, when projected over several decades, can generate large differences in total population" (UN 2015, p.8). The difference between projected totals is just an average half a child more or less per woman. A TFR of 2.1 (replacement-level fertility) is equated with population stabilisation. The assumption built into the UN's relatively optimistic medium-variant projection is that worldwide fertility rates will converge at this level (rising a little in more developed countries; plummeting in the least developed) by the end of this century.

In 2010–2015 the global TFR stood at 2.5. But this occludes significant regional differences. In the 48 least developed (mainly African) countries, the average TFR is 4.3 and the annual population growth rate 2.4%. As life expectancy rises, they are experiencing the kind of increase associated with the first stage of transition: populations here are expected to double by mid-century, with three- to fivefold increases projected for some already populous countries and Africa's total population projected to rise from 1.26 to 4 billion by 2100. Should expected fertility decline falter, the number would be correspondingly higher (unless death rates rise again). The high-variant global total for 2100 is 16.5 billion. The projections carry a clear message: achieving the medium-variant requires "substantial reductions in fertility" for which "it will be essential to support continued improvements in access to reproductive health care services, including family planning" (UN 2017a, p.6). In sum, population stabilisation peaking at around 11 billion during the 21st century is an achievement not a prediction.

An ideal scenario is that the average TFR will actually decline faster than expected, effectively realising the low-variant projection. Based on 2017 estimates, this yields a peak population of 8.8 billion in 2050, declining to 7.3 billion by 2100. An advantage of peaking earlier is that the peak is lower. This outcome – of swift population stabilisation – is in fact quietly endorsed by most transnational organisations, including the UN and the World Bank, despite their reluctance explicitly to endorse Malthusian principles. One reason for this reluctance is that uneven demographics complicate any global commitment to stabilisation because of its diverse regional implications. "In all plausible scenarios of future trends, Africa will play a central role in shaping the size and distribution of the world's population over the next few decades." (UN 2017a, p.4). Since most developed, and some emergent, economies already have a TFR of 2.1 or less, the principal target of "shaping" will be the least developed countries. This fuels suspicions that environmental rationales for population policies are de facto alibis for racism and eugenics, in which poor and marginalised peoples are the target (Hardt and Negri 2004).

Three rejoinders might nonetheless be made. First, it is by no means obvious that the sort of rapid population growth projected for Africa is in the continent's own interests, especially inasmuch as it is undermining development prospects. Although individuals here mostly have small ecological footprints, moreover, the environmental impact of multiplying them significantly (on deforestation, for example) is not negligible and Africa is at the sharp end of climate change. Second, many African countries do not appear to be following the same transition path as other continents: stagnation at stage one, with stalling or even reversed fertility decline, is causing concern (Guengant and May 2013; World Bank 2015). Third, it could (also) be argued that the "over-developed" nations of the Global North are already over-populated since they no longer live sustainably within their own territories. Per capita, their peoples' large ecological footprints contribute disproportionately to exhausting and polluting the planet. From this perspective there may be an argument for a double-pronged strategy: increasing (anti-natalist) population governance in least-developed countries and ending the (pronatalist) population governance that is striving to rejuvenate fertility trends in more developed nations.

## The environmental argument

When making an environmental case for urgent stabilisation, certain characteristics of population growth are salient: its exponential rate; its multiplication of individuals' unsustainable practices; adverse effects from rising densities and absolute size on well-being and biodiversity. Exponential growth means that even apparently small growth rates can quickly yield substantial increases as numbers keep doubling. A population growing by 2% per annum will double about every 35 years. There exist many illustrations of this statistical phenomenon. They show how initially modest increases may be regarded with equanimity until opportunities for intervening in measured ways, as problems associated with over-population emerge, may already be foreclosed.

It is nonetheless rare to come across reductionist arguments that attribute environmental degradation to population expansion alone. Its impact relative to other changes is expressed in the IPAT equation.

$$I(\text{impact}) = P(\text{population}) \times A(\text{affluence}) \times T(\text{technology})$$

This accommodates geographical and historical variability while recognising changing relationships between key variables (Holdren and Ehrlich 1974; see also Alexander and Rutherford, this volume). In some accounts, increasing affluence and technological innovation are credited with capacities to compensate for population growth (Chertow 2001). Recent sustainability studies reject this move: the Royal Society's *People and the Planet* is indicative of renewed emphasis on the P factor. It concludes that "a gradual and equitable decline in numbers will serve humanity best," particularly if accompanied by reduced consumption by richer people and by "changes to the current economic model" of persistent economic growth (Royal Society 2012, p.83). The importance of population is affirmed by modelling exercises that show the environmental benefits of reducing numbers relative to other sustainability activities. Regarding climate change, for example, increasing the number of greenhouse gas emitters is found substantially to outweigh benefits from reducing per capita emissions: an effect amplified once descendants are also taken into account (e.g. Murtaugh and Schlax 2009; O'Neill et al. 2010). Producing one child fewer emerges as by far the most effective way to reduce one's carbon footprint, its benefits massively exceeding other green lifestyle changes such as cycling, recycling, or becoming vegetarian (Wynes and Nicholas 2017).

A premise of such studies is that population serves as a multiplier of individually unsustainable practices. If potentially catastrophic changes to the geo-biosphere are primarily anthropogenic (human-caused), then it is unsurprising that more humans will, other things being equal, amplify the risk of irreversible damage. Johan Rockström's Planetary Boundaries approach is influential here. It estimates safe operating spaces for human development and warns that the "exponential growth of human activities is raising concern that further pressure on the Earth System could destabilize critical biophysical systems and trigger abrupt or irreversible environmental changes that would be deleterious or even catastrophic for human well-being" (Rockström 2009, p.2). When advising on setting the 2015 SDGs, Rockström's team accordingly prioritised "decelerating population growth more rapidly" (SDSN 2013, p.11), although this advice went unheeded.

Adverse impacts of population growth are distributed across the full range of an impending environmental crisis, from climate change and soil degradation to food, land, and freshwater shortages. To this one might add externalities from increasingly high-density living that reduce well-being, such as loss of non-humanised nature and green amenities, shrinking per capita dwelling sizes, traffic congestion, and air pollution. While these are linked to poor mental and physical health, "lifestyle" impacts from over-crowding are generally less well researched because they are perceived as either subjective and unmeasurable or as politically treacherous. More incontrovertible are the harmful effects of absolute size on biodiversity loss (Ceballos et al. 2017). As spaces shared with other species are colonised by urban development and the hinterlands that service it, corresponding mass extinctions and declines among non-human populations are well documented. Environmentalists suggest that tackling any social and environmental challenge becomes more difficult, if not impossible, with ever more people since infrastructure, social justice initiatives, and capacity-building must struggle just to keep up with rising demand (Attenborough 2011).

## Population and development

If stabilising numbers by reducing fertility rates can make a substantial contribution to sustainability, what kind of strategies are appropriate and where are they most effectively targeted? To address these questions it is helpful to visit a longstanding disagreement among demographers. This centres on the principal causes of fertility change and has important policymaking implications. Simplifying greatly, if rapid population growth hinders development, it makes sense to focus directly on contraceptive technologies and on generating small-family norms. If, on the other hand, fertility-decline occurs as a consequence of development, then it can be achieved indirectly through broader development strategies that do not require overt population governance.

Views on this cause/effect relationship have varied over time and are politically charged. Initially, development was prioritised. When it remained sluggish in a context of significant (mainly Asian) population growth, however, a theoretical reversal occurred and high fertility was blamed for persistent under-development. From the 1950s to 1970s, family planning was promoted as a panacea (Notestein 1983/1964). This direct approach was however denounced by poor countries as a technological fix that avoided restructuring global inequalities and justified coercion. A backlash followed that still influences current debates. Development was subsequently re-prioritised, with particular emphasis since the mid-1990s on women's empowerment and reproductive health. The premise is again that development will indirectly resolve the population problem, but pursuing overt demographic objectives is disavowed.

This complicates the situation for environmentalists wishing to advocate a demographic component to sustainability governance. Should they, controversially, call for a further reversal of causality in order to justify more direct interventions in the reproductive sphere? Or should interventions remain indirect and highly circumspect regarding their demographic ambitions? In which case does including the population factor among sustainability challenges actually add anything specific to the existing repertoire of sustainable development strategies? The problem here is that evidence challenges an assumption which underpins hostility to population governance, namely that development automatically causes fertility decline and that once started, transition only moves in one direction (O'Sullivan 2018). Historically, no country has developed sufficiently to eliminate poverty without first reducing its fertility rate (All Party Parliamentary Group on Population, Development and Reproductive Health [APPG] 2007). The UN still warns that rapid population growth makes it harder for governments "to eradicate poverty, reduce inequality, combat hunger and malnutrition, expand and update education and health systems, improve the provision of basic services and ensure that no-one is left behind." (UN 2017a, p.5).

A theoretically attractive solution is to recognise a bi-directional process, in which population and development are co-dependent while their precise relationship is context-dependent. There are, for example, instances of less developed countries (like Bangladesh) achieving significant fertility decline prior to development through well organised, voluntary family planning programmes. Conversely, there is no guarantee that having embarked on development, a country's fertility rate will keep falling until it reaches replacement level. Recent experience in Egypt, Algeria, Ghana, Kenya, and Indonesia suggests otherwise. Sluggish or stagnant fertility decline in some populous countries where women typically bear more than five children implies a significant disparity between embarking upon and completing transition. Optimistic global projections predicated on substantial fertility falls from the 1970s to 1990s (in Asia especially) may therefore be premature (Bongaarts 2008).

The relevant question for this chapter is whether population governance can be the contingency that makes the difference.

There are many risk factors that can suppress fertility decline. Among them are cultural forces such as the ascendancy of conservative values and religious orthodoxies that reject gender equality and embrace pro-family norms. Economic factors, such as faltering development or privatisation and austerity policies, may reduce universal access to reproductive health services. Political contingencies associated with state capacity and ideological orientation can hinder or reverse transition, too. Wars and state failure may cause family planning services to collapse. Nationalist regimes may regard a larger population as a tactical advantage (Turkey and Iran have both recently called on women to bear more children) and thus reverse former demographic objectives. Cuts to foreign aid for family planning that followed the demise of a demographic rationale for population policies (an estimated 80% drop since the 1980s) are widely blamed for stultifying transitions, too. Since 1984, every Republican Administration in the United States, from Reagan to Trump, has banned American funding for any organisation that even in principle supports abortion.

But if failings and pronatalist commitments of governments can hinder transition, other political interventions can surely hasten it. An international will to support and fund comprehensive family planning programmes, mobilised by evidence-based transnational recognition of shared risks to the global commons from ongoing population growth, is arguably a prerequisite of fulfilling low- or medium-variant projections and thus a condition of sustainable development. From the perspective of the current international consensus, governments' role in supporting family planning services within a wider context of reproductive health is uncontroversial, provided their use is voluntary, since it increases gender equality and individual choice. Merely by eliminating the estimated 40% of unplanned conceptions, the goals of women's reproductive freedom and fertility decline could be advanced in tandem. Similarly, meeting the "unmet need" of over 140 million women who say they would use contraception if it were available combines demographic efficacy with an ethical way to satisfy voluntary choice (UN 2017c). Yet, in the absence of any broader population stabilisation goal, evidence suggests that incentives to prioritise family planning services or actively encourage replacement-level fertility are greatly diminished. While, then, acknowledging the population factor marks an important addition to current modelling exercises, an important next step is to legitimise population policies by reframing the issue in sustainability terms and advertising well-funded family planning services as a cheap, egalitarian contribution to sustainability governance. If a new 21st-century population narrative is to gain legitimacy, however, devising effective and ethical instruments of population governance that are compatible with human rights and reproductive choice is crucial.

## Population governance

The final challenge is to identify non-coercive instruments of population governance. Supplying comprehensive family planning services has emerged as relatively uncontroversial in principle (although in practice comprehensive services often remain aspirational). But this is not necessarily the case for demand-side approaches that must actively persuade couples to plan and choose smaller families, where birth control may be discouraged for cultural reasons or large families are valued. How is a low-fertility norm to be engendered while respecting voluntary consent? If policies are to be both effective and ethical, the public good and private choices have to be aligned. Educating people about the environmental benefits of procreative parsimony within a revised population narrative is undoubtedly important, but it is probably insufficient for inculcating universally responsible conduct (Conly 2016). Fortunately, contemporary states have an additional toolkit of policy measures at their disposal, since behaviour modification has become a principal instrument of contemporary governance.

According to rational choice theories and neoliberal economics, fertility decline is achieved by free markets rather than governments since markets both spur development and disincentivise reproduction as children become an economic burden (Becker 1960). Arguably this opposition between state intervention and markets (and between public and private, too) becomes anachronistic, or at least blurred, however, to the extent that economic (dis-)incentives have become the stuff of the public sphere. It is illuminating from this perspective to approach the issue of population governance in terms of competing models of government and changing conceptions of the modern (welfare) state, including their implications for policymaking.

Political scientists identify the emergence of a new model of governing during the 1980s, as government became governance. This has implications for the way fertility is managed. The former model of (vertical) bureaucratic, top-down government is replaced, here, by dispersed practices of (horizontal) governance in which a variety of stakeholders includes non-political actors and sectors (public, private, voluntary). Whereas the previous command-and-control model was associated with centralised state planning that focused on achieving collective goals identified with the public good, postmodern governance makes greater use of mechanisms that are less visible as modes of power precisely because they are adept at structuring individual choices prior to action and thereby obtain voluntary compliance.

This helps explain an apparent contradiction regarding population policy. Although governments everywhere speak the language of the Cairo consensus and its disavowal of demographic goals, it is also the case that almost all contemporary states do endorse policy interventions for purposes of achieving preferred demographic outcomes. The new model of governance allows them to proceed in quiet, fine-grained ways that are politically uncontentious to the extent that individuals either feel they are freely consenting or they blame anonymous economic, rather than accountable political, forces for constraining their (fertility) choices.

In 2017, 42% of governments had policies for lowering fertility. This includes all 22 countries defined as high fertility (with a TFR over 5) and almost two thirds of the 96 manifesting intermediate fertility (a TFR between 2.1 and 5). Among the policy measures identified are increasing the minimum marriage age; providing access to reproductive health care that includes affordable, safe, and effective contraception; integrating the latter into primary health care structures; improving education and employment opportunities for girls and women (UN 2017b). Notable about this list is the way it mixes more and less direct fertility-reducing strategies while manifesting a fairly conventional model of government (the law, public health, and education services) whose aims are framed in terms of SDGs. The list focuses on supply-side approaches (facilitating fertility management) commended for widening individual choice and satisfying unmet need. The discourse through which such interventions are encouraged tends to rely, however, on an economic rubric of behaviour modification in which development objectives (such as cutting infant mortality rates or providing social security) are advertised as strategies for disincentivising high fertility and large families.

This approach draws on a model of governance in which reproductive compliance is achieved not by coercion, but by reconfiguring fertility choices through recalibrating the costs and benefits of procreation. With economic development and urbanisation, couples are both incentivised to make such rational choices and to calculate that children have become a financial burden. Large families equate with household poverty. Educated women with jobs are more likely to postpone marriage and hence first births, take control of their own bodies and their fecundity, and appreciate the opportunity costs of numerous offspring. Better infant survival rates as a result of reproductive health services discourage producing extra children as an insurance against child deaths. Social security reduces dependency on children, especially in old age. The ability to control one's fertility is a necessary condition for responding to the rewards and sanctions the new governance perspectives entail, but the purposes for which individuals control it can be directed through subtle behavioural nudges.

This mode of population governance is actually more expressly practised in developed countries, where it is most obviously congruent with a neoliberal model of policymaking and mobilised as a solution to post-transitional population ageing. Here the New Public Management relies on market (price) mechanisms to (dis)incentivise behaviour even in the public sector. This works in conjunction with a sophisticated application of behaviouralist psychology that relies on unconscious cues to "nudge" people into willing compliance (McMahon 2015). These countries, too, endorse population policies where fertility comprises a core component. In this case, however, the rationale is sustained economic growth and the orientation is pronatalist. In 2013, 20% of nations worldwide were trying to raise their population growth rates and 27% to increase their TFR (UN 2014). In 2017, the 28% striving for fertility increase included 66% of 44 European nations and 38% of 48 Asian ones, representing nearly two thirds of nations with a TFR below 2.1. A summary of their pronatalist measures mentions baby bonuses, family allowances, various permutations of parental leave, tax incentives, and flexible work schedules (UN 2017b). Their success relies on engineering reproductive decisions by appealing to individuals' economic rationality when making important lifestyle decisions. Tweaking tax and benefit systems, for example by providing a range of subsidies and tax breaks that encourage larger families, has been successful in countries like Sweden and France. EU policy documents reveal the underlying demographic purpose as a response to labour market challenges from population ageing, although policies are typically justified as contributions to gender equality (Repo 2015). Contrary to a popular misconception, many of the world's wealthiest countries (such as the United States, Australia, most western European countries including the United Kingdom and France) are projected to expand their numbers quite considerably by mid-century, despite their inhabitants already having the world's largest and most unsustainable footprints. Might a more sustainable option not be to allow these low-fertility, post-transitional regions naturally to lose population, despite temporary fiscal problems as a generation of baby boomers retires (Coleman and Rowthorn 2011)?

As a policy regime, reproductive behaviour modification strikes me as an ambiguous mode of sustainability governance. On the one hand, it exemplifies the current repertoire of liberal governance and its measures are no more or less coercive when applied to reproductive conduct than when used elsewhere. Because they are framed by, and practised at the level of, individuals' self-interest, personal choice, and human rights, these measures provide legitimate means for pursuing demographic goals that conform with liberal criteria of efficacy and ethical acceptability. They therefore refute objections that population stabilisation goals require coercive control, offer a justifiable approach to any government wishing to encourage low-fertility behaviour, and avoid other more egregious ways of tackling persistent population growth, such as compulsory sterilisation or abortion, by engendering voluntary consent. Inasmuch as anti-natalist goals remain unpopular, this method defuses opposition by deploying apparently apolitical micro-measures to achieve a desirable collective outcome (a sustainable planet for current and future generations of humans and non-humans).

On the other hand, whether this last characteristic of population governance is regarded as helpful or harmful will to some extent depend on one's evaluation of its underlying goals.

When behaviour modification operates in the service of pronatalism and regenerated birth rates, environmental critics might be especially concerned about a lack of democratic accountability and public debate that stems from policies whose operation is largely invisible and whose broader aims remain opaque. Since, moreover, fertility interventions are not specifically directed at environmental sustainability in either less or more developed countries, but are rather pursued for purposes of sustained economic development and growth, critics might be wary of an approach that relies on the economic rationality of the unsustainable business-as-usual model. In affluent countries, especially, more transparency and candour about their population policies could facilitate critical discussion about their merits in light of sustainability goals and environmental crises. Finally, critics might harbour ethical concerns about instruments of behaviour modification that blur the line between voluntary and coerced consent. From this perspective, policies that use positive incentives ("soft incentive-changing policies") seem preferable to punitive disincentives (Cripps 2015).

## Conclusion

In this chapter, ongoing world population growth has been identified as a major contributory cause of ecological peril. This suggests that stabilising human numbers as quickly as possible, provided the measures used comply with liberal standards of voluntary consent, is an important aspect of environmental mitigation and adaptation strategies and may enhance communities' resilience to existential risk. Bringing reproduction under the aegis of sustainability governance is commended inasmuch as the new political regime of governance furnishes non-coercive ways to manufacture compliance with a low-fertility norm. While this approach greatly improves on coercive population control measures and conforms with the wider regime of liberal governance, it is not however ideal. Precisely because its subtle interventions depoliticise the population issue, they foreclose public debate about the role of demographics within the larger sustainability picture. I have suggested in this chapter that it is incumbent on environmentalists who find population growth endangering sustainability to initiate such a discussion and publicise their findings. If publics thereby become convinced that population stabilisation in all regions is beneficial for all peoples and species, they are more likely intentionally to abide by low-fertility norms and to support funding and research for better reproductive technologies and health services.

Putting the population question back on the political agenda and reopening public debate on the topic might seem a modest suggestion in the context of sustainability governance; yet it faces prodigious challenges given that the topic has been taboo for the past four decades and that many powerful interests are invested in sustained population growth. Since the prevailing consensus was forged in the mid-1990s, however, considerably more has become known about the dangers inhering in the Anthropocene, environmental indicators have mainly deteriorated and the world's human population (and per capita footprints) has continued to expand. On the horizon there still lurks the Malthusian spectre that rising mortality may check our numbers if we do not. Should life expectancy rise dramatically, however, replacement-level fertility would fall below 2.1. Alternatively, other possibilities exist outside the circuit of current projections and sustainable governance regimes. Further denaturalisation of reproduction, with gestation no longer reliant on wombs or age and disconnected from sexual activity or identity, could usher in an entirely different procreative context from the one in which demographic transition occurs. Likewise, the replacement of humans by artificially intelligent robots in labour markets could substantially alter personal motivations and public inducements to procreate. It is unpredictable what impact either might have on fertility choices. Meanwhile, it is surely timely to reconsider the benefits of actively pursuing a stable, and preferably a smaller, population globally and of persuading the world's peoples about the collective and planetary benefits of doing no more than replace themselves.

## References

All Party Parliamentary Group on Population, Development and Reproductive Health [APPG]. (January 2007) *Return of the Population Growth Factor. Its Impact upon the Millennium Development Goals. Report* of Hearings. www.appg.popdevrh.org.uk

Attenborough, Sir. D. (2011) 'RSA President's Lecture 2011: People and the Planet.' Royal Society, London. https://populationmatters.org/sites/default/files/Sir%20David%20Attenborough%20RSA%20speech%202011.pdf

Becker, G.S. (1960) 'An Economic Analysis of Fertility', chapter in *Universities-National Bureau, Demographic and Economic Change in Developed Countries* (New York: Columbia University Press): 209–240.

Bongaarts, J. (2008) 'Fertility Transitions in Developing Countries: Progress or Stagnation?' *Studies in Family Planning* 39.2: 105–110.

Cebellos, G., Ehrlich, P. and Dirzo, R. (2017) 'Biological Annihilation via the Ongoing Sixth Mass Extinction Signaled by Vertebrate Population Losses and Declines', *PNAS* online, 8pp.

Chertow, M.R. (2001) 'The IPAT Equation and Its Variants. Changing Views of Technology and Environmental Impact', *Journal of Industrial Ecology* 4.4: 13–29.

Coleman, D. and Rowthorn, R. (2011) 'Who's Afraid of Population Decline? A Critical Examination of Its Consequences', *Population and Development Review* 37 (Supplement): 217–248.

Conly, S. (2016) *One Child: Do We Have a Right to More?* (Oxford: Oxford University Press).

Coole, D. (2013) 'Too Many Bodies? The Return and Disavowal of the Population Question', *Environmental Politics* 22.2: 195–215.

Coole, D. (2018) *Should We Control World Population?* (Cambridge and Medford, MA: Polity).

Corrêa, S. (1994) *Population and Reproductive Rights. Feminist Perspectives from the South* (London: Zed Books).

Cripps, E. (2015/2016) 'Population and the Environment: The Impossible, the Impermissible, and the Imperative', in S. Gardiner and A. Thompson eds, *The Oxford Handbook of Environmental Ethics* (Oxford, OUP. Downloaded from Oxford handbooks online, 2015: 1–14).

Dorling, D. (2013) *Population Ten Billion. The Coming Demographic Crisis and How to Survive It* (London: Constable and Robinson Ltd).

Dyson, T. (2010) *Population and Development: The Demographic Transition* (London: Zed Books).

Ehrlich, P. (1968) *The Population Bomb* (London: Pan/Ballantine).

Guengant, J.-P. and May, J. (2013) 'African Demography' *Journal of Emerging Market Economies* 5.3: 215–267.

Hardt, M. and Negri, A. (2004) *Multitude. War and Democracy in the Age of Empire* (New York: Penguin Press).

Holdren, J. and Ehrlich, P. (1974) 'Human Population and the Global Environment', *American Scientist* 62: 282–292.

McMahon, J. (2015) 'Behavioral Economics as Neoliberalism: Producing and Governing Homo Economicus', *Contemporary Political Theory* 14.2: 137–158.

Meadows, D. and Meadows D. (1972) *The Limits to Growth. A Report for the Club of Rome's Project on the Predicament of Mankind* (New York: Universe Books).

Murtaugh, P.A. and Schlax, M.G. (2009) 'Reproduction and the Carbon Legacies of Individuals', *Global Environmental Change* 19: 4–20.

Notestein, F. (1983/1964) 'Frank Notestein on Population Growth and Economic Development', *Population and Development Review* 9.2: 345–360.

O'Neill, B. et al. (2010) 'Global Demographic Trends and Future Carbon Emissions', *PNAS* 107.41: 17521–17526.

O'Sullivan, J. (2018) 'Synergy between Population Policy, Climate Adaptation and Mitigation', in M. Houssain et al. eds, *Pathways to a Sustainable Economy* (New York: Springer International Publishing): 103–125.

Repo, J. (2015) *The Biopolitics of Gender* (Oxford: Oxford University Press).

Rockström, J. et al. (2009) 'Planetary Boundaries: Exploring Safe Operating Space for Humanity', *Ecology and Society* 14.2. www.ecologyandsociety.org/vol14/iss2/art32/.

The Royal Society. (2012) *People and the Planet* (London: Royal Society).

Sustainability Development Solutions Network [SDSN] (Report to the General-Secretary, United Nations). (2013) *An Action Agenda for Sustainable Development*. https://unstats.un.org/unsd/broader progress/pdf/130613-SDSN-An-Action-Agenda-for-Sustainable-Development-FINAL.pdf.

Szreter, S. (1993) 'The Idea of Demographic Transition and the Study of Fertility Change: A Critical Intellectual History', *Population and Development Review* 19.4: 659–701.

United Nations (2014) *World Population Policies 2013* (New York: UN).

United Nations (DESA). (2015) *World Population Prospects: The 2015 Revision, Key Findings and Advance Tables.*

United Nations (DESA). (2017a) *World Population Prospects: The 2017 Revision, Key Findings and Advance Tables.* https://esa.un.org/unpd/wpp/Publications/Files/WPP2017_KeyFindings.pdf.

United Nations (DESA). (2017b) *Population Facts No 2017/10, Government Policies to Raise or Lower the Fertility Level.* www.UN.org/en/development/desa/population/publications/pdf/popfacts/PopFacts_2017-10.pdf.

United Nations. (2017c) *World Family Planning. Highlights.* www.un.org/en/development/desa/population/publications/pdf/family/WFP2017_Highlights.pdf.

World Bank (2015) *Africa's Demographic Transition: Dividend or Disaster?* (Washington, DC: World Bank).

Wynes, S. and Nicholas, K. (2017) 'The Climate Mitigation Gap: Education and Government Recommendations Miss the Most Effective Actions', *Environmental Research Letters* 12: 074024.

# Part 4

# Transformative approaches

# 21

# Beyond magical thinking

*Michael Maniates*

Transformational global governance faces two fundamental challenges. One is the unravelling of the natural world, which grows warmer, less biologically diverse, and increasingly hostile to human flourishing. The other is the funnelling of environmentally concerned publics into cul-de-sacs of political irrelevance. Addressing the first challenge means confronting the second.

This chapter explores the dimensions of such confrontation. It argues that "magical thinking"[1] about the inevitable aggregation of small living-green actions into powerful social change is a fundamental impediment to transformative global governance. Such thinking short-circuits the political potency of growing public concern about climate change and other environmental ills. It also fuels "the trinity of despair," a mutually reinforcing dynamic of cynicism, misanthropy, and preoccupation with crisis (Maniates 2016). When it comes to individual agency and transformative social change, magical thinking is deeply corrosive.

Given this volume's emphasis on *collective* change in the global allocation of power and authority, this chapter's focus on environmentally concerned *individuals* may seem odd. It shouldn't. If one lesson emerges from decades of struggle for transformative global governance, it is this: institutional change necessary to the restoration of critical environmental systems will not be driven by governments or corporate elites, not initially at least, and not while economic growth remains the yardstick for state performance and corporate success. Strategic, synergistic mobilisation of environmentally committed populations must come first.

Hopeful assessments notwithstanding (e.g. Hawken 1993/2010), dominant economic organisations will not lead the way to transformative governance. Corporations are built to maximise short-term returns and externalise costs, often via stubbornly opaque global commodity chains that separate production from consumption and exploit the environment and poor alike.[2] Too often, corporate sustainability efforts are exercises in greenwashing or, as Robert Reich (2007) observes, an umbrella for efficiencies in production and distribution rebranded as "green."

Governments are similarly hobbled. On their own, nation-states systematically discount the future – future generations and non-human species cannot vote or threaten civic unrest (Lawrence, this volume; Levin et al. 2012) – and GDP growth remains the litmus test for policy success and good leadership (Philipsen, this volume). Acting collectively, national governments are constrained by a global order that impedes sustained international cooperation. "Blocking" or "laggard" states opposed to environmental safeguards enjoy disproportionate power and they use it, often to distort or disregard compelling scientific knowledge, while ambitious international agreements signed in far-flung cities inevitably face the wrath of producer groups and polluting industries at home, to the detriment of policy vitality (Hempel 1996).

Absent catastrophic crisis, the only solution to these equations of lethargy is the strategic mobilisation of environmentally anxious citizens, who exist in sufficient number around the world to force necessary change in global governance.[3] In some instances, citizens could join together to compel corporate and state elites to embrace elements of lasting environmental sustainability, or what Sylvia Lorek and Doris Fuchs call "strong sustainable consumption" (Lorek and Fuchs 2013). In other settings, citizen pressure can provide political cover for elites aspiring to lead. Either way, the majorities around the world for whom "the environment" is a salient issue (Dunlap and York 2012) must rediscover, in the words of Annie Leonard (2010), their "citizen muscle" – their capacity, working together as individuals, to drive transformative change.

Such transformation won't be easy, and it won't be incremental. But it can happen. Cass Sunstein gets it right when he observes that "large-scale changes" occur in "an astoundingly short time…Stunning surprises are nearly inevitable" (2017, par. 17). Instances of norm cascades, where issue-specific environmental action in one corner of the world spreads across borders, speak to the power of sustained local action in a global society (Clapp and Swanston 2009). Humans are social animals, predisposed to collaboration around goals kept aloft by strong normative claims. Civic action, moreover, is often its own reward: struggling with others of like mind towards grand social reform offers deep intrinsic benefits. And evidence of mounting ecological crises that challenge everyday understandings of prudence and decency is on full display. As Tom Princen (2005) observes in his treatise on sufficiency, the materiality of biogeophysical decline is laying the groundwork for a rapid, unexpected shift in the dominant social organising principles of efficiency, extraction, and externalisation of costs.

Amidst these forces of possibility and transformational change, it is only fair to ask: what are the vast numbers of the environmentally concerned doing? Why aren't they flexing their citizen muscles?

The following sections offer one explanation. As these nascent eco-activists wrestle with the gap between their morals and their practices, they do what most overextended, frequently distracted, occasionally stressed-out people do – they gravitate to established behaviours that are straightforward, affirmed by credible sources, and promise meaningful agency. In practice, this means, for many, an embrace of small acts of "responsible," "conscientious," "green," or "environmental" living, with the expectation that these small gestures will sum with millions of others around the planet to produce deep institutional change. In other words, magical thinking.

It is silly to believe that a cabal of evildoers is working overtime to disempower environmentally concerned publics. But if such a cabal existed, it would quietly promote notions of magical thinking, which took root in the 1980s without much notice, as the next section describes.

## The rise of magical thinking

I have an "Earth Day" rubbish bin from the early 1970s. It features colourful drawings of environmental protesters with bell-bottoms, long hair, and women in knee-high boots shaking signs with phrases that resonate with my own recollection of the time: "This is where it's at: fight pollution," "Pollution is a bummer," "Don't Cop Out – Get Active," and "Fight the System – Save the Earth." One young man is waving a large Earth Day flag. Everyone looks to be marching off to protest. The bin sits near my desk in reminder of a time when citizen engagement and political mobilisation were natural expressions of anxiety about environmental abuse.

Whenever I bring the bin into my environmental-studies seminars, my students are alternately amused and flummoxed by the scene it depicts. They typically view their 1970s-counterparts as idealistic, naïve, even counterproductive. My pupils wonder aloud if "all that protesting" really made a difference, forgetting the major environmental laws that were adopted in the United States in the early 1970s under Richard Nixon, a president not known for his progressive leanings, and that fostered similar reforms around the world.[4]

But it is not just the seeming futility of public protest that troubles my students, at least not lately. Charges of hypocrisy also fly with an intensity that suggests deeper anxieties at work. Those protesters, say some students, undoubtedly drove cars, ate meat, and flew. Why should they command respect given the chasm between their words and their deeds?[5] Such preoccupation with environmental hypocrisy becomes disabling – how can any of us plead innocent to all acts of environmental impact, living as we do in a globalised, consumeristic society? There are echoes here of Paul Loeb's (1985) chronicle of "the perfect standard," which infected politically disempowered college students of the 1980s who frequently asserted their inability to engage controversial issues until they had *complete* information – an utterly immobilising threshold for action.

While this "perfect standard" is familiar, its insinuation into mainstream environmentalism is recent, and dangerous. It seems connected to the burgeoning norm that "living environmentally" is among the best mechanisms, both politically and ethically, for addressing environmental ills. Buy green and live lean to influence business decisions and persuade, by lived example, others to do the same. Avoid confrontational protest and work within the system. After all, in this view, only a tsunami of public outrage will alter dominant institutions and realign personal values, both necessary to save the planet. And you don't get everyone on board with protests by individuals who themselves are despoiling the planet. That will only alienate the super-majorities upon which fundamental change is thought to depend. Far better to walk one's talk, lead by example, and reward "green" companies with consumer loyalty, all while hoping for the best.

It is understandable that my students, and so many like them, see the world this way. Indeed, it would be surprising if they did not. Everyday life is awash with messaging that we "save the world" and our own souls one small eco-act at a time. So awash, in fact, that it is hard to notice until one actually looks.

On a recent business trip I tried tallying the many ways I was asked to address environmental problems via small acts meant to aggregate with the good deeds of others. I gave up after 17 instances in the first five hours of my trip. The recycling bins in my condominium implored me to recycle to save the world. My taxi featured a placard explaining that by using a little less water I'd be joining thousands of others to create real impact. The seat-back screen during my flight flashed "A simple act can save the planet – Please lower your window shade before leaving the aircraft." And there were not one but three reminders in my hotel room that I could stop climate change if I reused my towels and acceded to the intermittent change of bed linens. The list goes on in surely familiar ways. This messaging isn't just ubiquitous; it stands largely unopposed across the everyday landscape. No rival assertions emerge with any consistent force about how one best translates concern for the planet into meaningful action.

Of course, faith in the spontaneous aggregation of good deeds isn't new to environmental thinking. The voluntary simplicity movement, which valorises low-consumption living, was a potent agent of this message in the 1970s and early '80s (e.g. Maniates 2002; Ballantine and Creery 2010). The appropriate technology movement, flourishing at the same time, advanced a similar sensibility. Its "Cuisinart theory of social change" (if everyone owned a Cuisinart we'd all become great home chefs) asserted that transformative institutional change could be achieved through individual embrace of small-scale, environmentally friendly technologies. To drive fossil-fuel companies out of business, or to at least bring them to the point of political malleability, bolt a solar collector to your roof, persuade your neighbour to do the same, and wait for the social power of aggregation to emerge.[6]

Both movements were swept aside in the early 1980s by engines of neo-liberalism. Writing about appropriate technologists, Langdon Winner was exactly right when he noted that "they were lovely visionaries, naïve about the forces that confronted them" (1986, 80). As important as living a simple life or thinking about technological choice can be, focusing only on these elements ultimately constitutes a flight from power rather than engagement with it.

One might have expected an alternate ideology to emerge to fill the void, but this was not to be. Instead, three self-reinforcing elements produced a deepening of magical thinking even as its shortcomings were becoming evident. One was escalating public concern over global environmental ills. The sudden and starkly visual discovery of the ozone hole in 1983 crystallised such apprehension; events culminating in the 1992 Earth Summit gave it full form. But what was one to do with all this worry?

The primary answer: join environmental lobby groups. The 1980s saw a surge in support for major environmental groups like the Sierra Club, Greenpeace, and the World Wildlife Fund (Bosso 2005). But this windfall created its own headaches. As these organisations expanded their staff and programs, they fretted about how to engage and retain their new members. Launching another political-action campaign wouldn't be enough. The politics of the day were hostile to new environmental initiatives – this was the time of Reagan and Thatcher, and the best that most lobby groups could do was resist the rollback of cherished environmental policies. A new angle of engagement was necessary, one distant from the toxic politics of environmental policymaking.

Through trial and error, environmental groups hit upon a strategy, launched in the mid-1980s, of celebrating personal responsibility over collective political action. The new story of change coming from key environmental organisations went something like this: "If you care about the environment but are frustrated with governmental short-sightedness and corporate malfeasance, don't despair – you still can make a difference through small acts of ecological living, and we'll show you how." Soon thereafter, lists of simple ways to save the planet proliferated, and books like *50 Simple Things You Can Do to Save the Earth* (1989) became bestsellers. Almost overnight, a contentious politics of environmental protection became a tidy, feel-good process of smart shopping and doing with less.

This environmental-group strategy shift was the second of three progenitors of magical thinking. The global recession in the early 1980s, and the concomitant collapse of corporate profits, was the beginning of a third. As the economy began to recover in 1983, major corporations, scrambling to restore their bottom line, began experimenting with "green marketing" earlier shunned as too niche. They were startled by the success of these initial attempts to grow consumption. For the rest of the decade, with increasing sophistication punctuated by occasional overreach, business rolled out an array of so-called planet-friendly products, each wrapped in a story of consumer power to effect political change (Crane 2000; Peattie 2001; Peattie and Crane 2005). "Buying green" and "conscientious consumption" took off like few marketing initiatives before or since, simultaneously meeting the needs of business, environmental groups, and a public in desperate search of agency (e.g. Mendleson and Polonsky 1995).

These three elements, converging amidst a neo-liberal celebration of markets and individual choice, have done great damage to the environmental imagination (Lukacs 2017). Indeed, if a rubbish bin portraying environmental activism were produced today, it would show environmentally minded shoppers in a checkout line, or perhaps concerned environmentalists installing energy- and water-saving devices in their residences. It would be purchased in droves by anxious people of conscience looking to display their eco-credentials. The message of this contemporary bin – that we are at our best as agents of change when we modify our lifestyles and consume differently, battling an enervating sense of hypocrisy all the while – undermines any hope for transformative governance, for reasons laid bare in the next section.

## Not just distracting or delaying, but demobilising: the trinity of despair (TOD)

Even if unseemly forces are fostering a deepening faith in small acts of eco-living, that doesn't necessarily make these acts worthless. Living simply and striving to purchase environmentally friendly products are pillars of mindful living. As Karen Litfin (2014) powerfully observes, these choices can become daily personal reminders of the urgency of environmental decline, helping us act with grace in the midst of the biological unravelling of the planet. The problem is that, alone, such behaviours are mismatched to the imperative of transformative change in social structures. It is important to walk an elderly neighbour across the street when you're both standing on the corner. Doing so cultivates inner decency and community connection. Just don't assume that your good deed will solve the pension crisis, no matter how many people follow your example.

For some, assertions about the political impotence of eco-living are troubling. Couldn't shifts in everyday behaviour become on-ramps to the collective citizen work of redistributing power and reforming institutions? Wouldn't persuading a neighbour to buy organic food today prime her to become a food activist tomorrow? Since we all know how to be consumers, couldn't people be drawn into environmental activism through accessible, "first-step" acts of enlightened consumption (e.g. Lorenzen 2014)?

Alas, there is scant empirical evidence that individual acts of environmental stewardship lead people to meaningful political action.[7] The two frequently coexist: citizen activists troubled by biodiversity loss may purchase rainforest-friendly coffee, and owners of energy-efficient appliances might participate in climate-action rallies. But any straight-line causality looks to run in one direction. Environmental citizen activists often embrace elements of green living (though perhaps not enough to satisfy their critics), while conventional forms of green consumption fail to activate Annie Leonard's citizen muscle (e.g. Johnston 2008; Webb 2012). Intuitively, this makes a certain kind of sense: the skills and rewards associated with conscientious consumption are worlds apart from those associated with engaged citizenship.

More may be at work than mere disconnect, however; green living may actually undermine citizen action. When individuals enact their environmental concerns through small lifestyle changes and concerted green consumption, researchers observe a weakening of pro-environmental behaviours, including citizen mobilisation (e.g. Lacasse 2016). Imagine a well-intentioned environmentalist recycling their rubbish, grilling their organic vegetables on their eco-friendly charcoal grill, then putting their feet up on their sustainably sourced wooden ottoman and congratulating themselves for a job well done. For many scholars (e.g. Fridell 2007;

Princen 2010) this scenario isn't far-fetched: it is a daily fact that impedes more muscular citizen responses to environmental decline. Since green consumption and simple living cannot, on their own, meaningfully address our most pressing environmental ills – public policy must also change, and it's not for sale at the check-out counter (Sanne 2002) – the prospect of millions of deeply committed but politically complacent eco-consumers is unnerving, especially if citizen restiveness is a critical engine of transformational governance

Faced with these research findings, advocates of aggregation urge patience (e.g. Schudson 2007; Middlemiss 2014; Zamwel, Sasson-Levy, and Ben-Porat 2014; Atkinson 2015). The small-and-easy path to social change *is* a politics of transformative governance, they insist, if only we'd give it more time. Three lines of argument emerge. One centres on social norms, where consumption choices that visibly challenge the dominant culture are read as fostering new norms of restraint and sufficiency, making subsequent political change easier and more enduring. Another suggests that personal struggle with lifestyle choices slowly cultivates citizen capacities necessary to later political struggle. A third defence insists that market pressure from conscientious consumers heightens big business's sensitivity to political pressure for transformational change, leading to big payoffs when citizens eventually mobilise.

Each argument has merit, but all overlook a devastating effect of magical thinking. Rather than distracting or delaying individuals from their obligations and capacities as citizens, faith in spontaneous aggregation can transform the environmentally concerned into potent agents of cynicism and citizen demobilisation. The forces at work cohere as "the trinity of despair" (see Figure 21.1)

The TOD begins with the belief that humans are short-sighted creatures who focus narrowly on their own prosperity and security. We are those *homo economicus* creatures described in economics courses, or in environmental science textbooks that view environmental degradation through a "tragedy of the commons" lens (Hardin 1968) that reifies the same *homo economicus* caricature. Pundits asserting that "people will never sacrifice" for environmental sustainability reflect this monochromatic view of human nature. In fact, while humans can be narrowly selfish, they can also be magnanimous and altruistic; we live within community where loyalties, passions, group association, and stories of belonging and allegiance frequently privilege the better angels of our nature (Stone 2012). These promising complexities are ignored, however, in conversation about environmental decline where "human nature" receives the blame (Maniates and Meyer 2010).

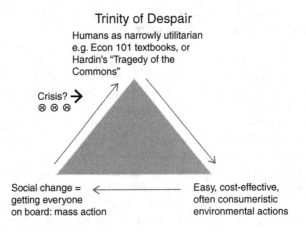

*Figure 21.1* Trinity of despair

If we are short-sighted and sacrifice averse, then attempts to engage us in struggles for environmental sustainability and social justice must cater to self-interest, per the strategies promoted in top "easy ways to save the world" lists and "guides to sustainable living" books.[8] Familiar measures include adoption of eco-efficient technologies that produce economic savings (e.g. new light bulbs), simple behaviour changes that confer personal benefits (e.g. eating less meat for health reasons), and cost-effective lifestyle changes or consumer products that signal a commitment to the environment, thereby earning the admiration of others. Each is tailor-made for rational, sacrifice-allergic actors, and relentlessly promoted by corporate marketers and environmental groups.

The social-change narrative that connects these practices is by now obvious, but is worth emphasising here and goes something like this: if small groups of individuals adopt some of the above measures, others will notice and join in. This social mimicking scales as eco-behaviours slowly become normal, and as alarming information about environmental problems spreads. As more people climb aboard the bandwagon, the cumulative environmental benefits of these small acts grow apparent, prompting laggards and late-adopters to hop aboard for fear of missing out or being ostracised. Inspired by these changes, some will become politically active, and policymakers will feel the pressure. Major corporations will feel pressure too, since consumers are now clamouring *en masse* for clean and green products. The outcome is a more sustainable and just planet achieved not by a counterproductive politics of confrontation (recall the concerns of my students), but rather by seemingly innocuous individual decisions that became an unstoppable force for good.

The linchpin of this story is mass participation. Personal acts of green living morph into political potency only if nearly everyone participates, since each act alone is incommensurate to looming planetary threats. Ubiquitous advertisements for environmental action acknowledge as much: "If *everyone* recycled their newspapers, we'd save 10,000 trees a year" or "If *we all* bought LED lights, five coal-fired power plants would close." This story says that corporations won't change their ways until they see massive change in buying patterns. Governments won't alter policy unless most people shift from "uncaring consumer" to "eco-shopper." My neighbour won't start composting until all his neighbours do, at which point he'll awkwardly realise that he's the odd man out. All the gears of this process – more environmental education, savvy information campaigns, labels and ratings that communicate the environmental consequences of individual choice, eco-consumption by environmentalists – centre on the goal of mass voluntary participation. Social change won't happen without it.

It's seductive, this narrative of institutional transformation. But it is wrong, and it is debilitating. It is wrong when it imagines that past environmental successes – dolphin-safe tuna; bans on leaded petrol, dichlorodiphenyltrichloroethane (DDT), and chlorofluorocarbons (CFCs); a solar-electric revolution; and more – arose from the mass mobilisation of individual consumers, and that future successes must too. And it is debilitating by obscuring potent possibilities for citizen mobilisation and social change, reinforcing the fiction that citizen action is too messy or loud to pursue. Instead of celebrating, for example, the roughly 20% of Americans who say they act on their environmental concerns "all of the time" (Pew Research Center 2016), and strategising about how to turbocharge the political salience of this minority, those seduced by this story obsess over the absent 80% as evidence of environmentalism's failure, and work harder to bring the wayward onboard. In doing so, advocates forget their history lessons. During critical moments of political transformation, majorities are disengaged, or alarmed by the prospect of change. Abolition, women's suffrage, the New Deal, the rise of economic liberalism and right-wing macroeconomics, gay marriage, the US civil rights movement, the banning of ozone-destroying CFCs, the impact of the so-called alt-right – in each of these instances and others like them, determined and strategic minorities made change happen. Mass acceptance comes later or, at times, not at all.

In the end, this "all aboard" catechism breeds cynicism and misanthropy. When advocates of small and easy grasp that large majorities aren't being drawn into a whirlwind of behaviour change, as their story predicts, they logically double-down with flashier messaging, further appeals to self-interest, and more information about the virtues of "being green." When these strategies falter – and they always falter, since super-majority participation is a fantasy, and appeals to immediate self-interest are paradoxically counterproductive (e.g. Hurst et al. 2013) – guilt, blame, and fear (a TOD all its own) become the favoured prods to action. But these fail too, since enduring political movements thrive by appealing to the best in us, rather than resorting to guilt and shame.

Deceived by cartoonish notions of social change and human nature, but not yet realising their error, those ensnared by the TOD are left with one option: return to blaming humans and their selfish, self-destructive nature, and wait for crisis – deep, broad, Old Testament sort of stuff – to force social change. Once optimistic and engaged, these individuals morph into disappointed and misanthropic observers of global environmental collapse, only dimly aware that their starting assumptions about environmental action and social change, methodically reinforced by marketers and others, were their undoing. Their fear and anger become infectious, pushing transformational governance further out of reach.

Magical thinking isn't so magical after all.

## Escape

In theory, magical thinking and the TOD it spawns undermine new forms of global governance. If the TOD is more than a thought experiment – if it truly captures conditions on the ground – then transforming the distribution of power and authority across the planet demands puncturing the attitudes and assumptions that give this trinity life.

Two questions arise. Is the TOD, fuelled by magical thinking, more than an instructive heuristic? If so, how might the TOD best be opposed in service of transformational governance? Tentative answers to the first question reveal useful insights into the second.

On this first question, abundant anecdotal evidence points to the realities of the TOD. In the classroom, students speak with confidence about the power of small choices and big crises. Worldwide, faith in naïve aggregation appears to be rising (with a backlash brewing against plastic straws at the time of this writing). And far too many environmental events still end with audience commentary about public ignorance of environmental issues, the necessity of education, and the urgency of making environmental action convenient and economic in order to drive rapid transformation of individual and household behaviour. In this respect, Elizabeth Shove's (2010) concerns about the discredited "ABC" path to environmental sustainability, where altering environmental **a**ttitudes is mistakenly thought to change individual **b**ehaviours, which then supposedly yield powerful **c**hange in consumption patterns and public policy, are on full display. Again and again, human nature, or at the very least ignorance and "bad values," is the problem. The solution lies in fashioning digestible bits of enticing environmental action that, in ways never fully interrogated, will spread to save the world.

Research on the TOD reinforces these anecdotal impressions. In 2010, Allegheny College researcher Samuel Rigotti's analysis of a non-randomised sample of 400+ undergraduate environmental-studies students uncovered several TOD-elements in play, including preoccupation with individual consumption, a cynical view of human nature, and a profound faith in crisis among respondents (Rigotti 2010). A three-year research project completed in 2018 offers even deeper evidence of the TOD in everyday life (Chee, Kaur, and Maniates 2018; Maniates 2018). Like Rigotti's earlier work, this project focused on students of the environment as especially representative of individuals primed for citizen action. More than 1,200 undergraduates at 73 randomly selected US colleges and universities completed a 31-item questionnaire, usually in association with an environmentally focused course in which they were enrolled.

Although a complete reporting of research results is forthcoming (Maniates, forthcoming), some conclusions can be confidently reported here. The general picture from the study is distinctly neo-liberal and individualistic; debilitating assumptions about human nature and social change, typically expressed as natural truths, abound. For instance, "consumers" are the most frequently identified actor capable of generating meaningful social change, rising above other choices with statistical significance. Super-majorities of the sample, exceeding 80% of all respondents, characterise small and easy environmental measures as the central mechanism for marshalling public support ("green consumption as an on-ramp"), blame inherent deficiencies in human nature for our environmental ills ("people only respond to what is immediately best for them"), and identify crisis as the singular driver of important change ("nothing changes without a crisis" and "humans are short-sighted when it comes to environmental problems").

These views, moreover, become entangled with an "everyone on board" view of social change that confuses the cultural benefits of environmental awareness with the fundamentals of political transformation. When asked to speculate about levels of public commitment necessary to spur change in environmental policy, more than half the sample insisted that citizen majorities must deeply engage with environmental issues before policy change could occur. Strikingly, more than a quarter of the sample indicated that social change is impossible absent the enduring commitment of 70%, 80%, or even 90% of their fellow citizens. Less than one respondent in ten recognised that persistently strategic minorities typically drive social change.

This multi-year study, the most comprehensive of its kind, assesses the beliefs and attitudes of young people who, by virtue of their education, are exposed to diverse accounts of the cause of and cure for environmental ills. That so many in the sample subscribe to magical thinking and the disabling understandings of social change that follow was unexpected. The effects of the ubiquitous "save the world one small consumption choice at a time" narrative appear overwhelming.

But perhaps not, for at least three reasons. One is that a majority of respondents express discontent with the limitations of green consumption as a primary vehicle for change, even as they extoll its virtues. They embrace the notion that voting with their purchases makes a difference while acknowledging the power of economic actors to edit consumer choice and shape consumption preferences. There is a dissonance here that awaits cultivation, not just among undergraduates but also among the environmentally committed at large.

Another reason is what Yale-NUS College researchers Stephanie Chee and Sonia Kaur call "the all-of-the-above problem" (Chee, Kaur, and Maniates 2018). In their analysis of the 1,200+-response data set, Chee and Kaur note that students reassess their views about individual agency and social change as they move through the survey. Drawing on qualitative and quantitative responses, they emphasise that respondents seemed unaccustomed to examining their notions of social change, and often welcomed the opportunity for reflection. There is, say Chee and Kaur, a palpable uncertainty among undergraduates about how to be powerful in the world, an openness to new ways of thinking about social change, and a yearning for alternatives beyond the familiar guideposts to action. Perhaps the magical-thinking narrative is less powerful than it appears.

A final reason: outliers. In both studies – Rigotti's and the more recent inquiry – arguably the most interesting population is the 10–15% who are sceptical of magical thinking, who understand that mobilised minorities can make a difference, and who see great possibility for future social change. If theorists of social norming are correct, this minority is likely hiding in the shadows, waiting until their understandings of action and change become less oppositional to the prevailing views of their peers. And, as Cass Sunstein (2017) notes, once it becomes more acceptable to think and act in ways consistent with this minority, an uptick of interest and action – a so-called norm cascade – around strategic activism and community mobilisation is not just possible, but likely.

In a world where the problem is less a lack of environmental concern among the public than how this concern is enacted, the implications for proponents of transformative governance become clear. The top priority must be to reverse the cooptation of environmental movements by narratives that privilege magical thinking. Advocates of transformational governance often embrace this narrative, thinking that it aids their cause. This practice must end, and the marketing of magical-thinking solutions must be systematically opposed, for they do more than distract. They are viruses of immobilising notions of social change and human nature. By trivialising the challenges before us they spread diminished expectations about what our fellow citizens can be called upon to do to avert environmental catastrophe. Too often, resulting appeals to immediate self-interest prime the very behaviours that make lasting progress to sustainability impossible (Crompton and Kasser 2009; Hurst et al. 2013; Kasser 2016).

Another imperative is to move from arguments about the need for transformative governance to stories about the potency of everyday acts of ecological citizenship. To this end, a narrative of painless aggregation around consumption choices must be supplanted by stories of threshold and feedback, where diligent struggle for social change suddenly, and often unexpectedly, succeeds. These stories exist, stories of norm cascades and punctuated-equilibria and policy windows, around topics as diverse as smoking in public, plastic bags, seatbelt use, gay marriage, and #MeToo. They offer theoretical lessons and empirical proof of the joys of working with others to slowly drive governance systems towards thresholds of abrupt policy change. They illustrate too the importance of optimistic persistence in the face of uncertainty around how close those thresholds might be.

From here, it is a short hop to outlining everyday acts of ecological citizenship that are effective and empowering, and that put green consumption in its proper place. For example, rather than buying fair-trade/organic coffee at the market, hoping that someone in the corporate office notices, why not gather a few friends to speak directly to the manager about stocking more socially responsible products on the best shelf space, and relegating less enlightened products to the nooks and crannies of the store? Joining with others to reconfigure the architecture of consumer choice, if only in small ways, is a more promising on-ramp to active citizenry than cajoling a neighbour to buy green, and more effective too. By making these paths of individual action more natural and normal, advocates of transformative governance embolden the minorities already thinking like environmental citizens while creating space for others struggling with dissonance around green consumption.

Most important is the need to revive a sense of adventure and joy around the daunting task of birthing new forms of transformative governance. The late sociologist Robert Bellah and his colleagues had it right when they wrote that "it isn't enough to exhort people to participate. . . We must build institutions that make participation possible, rewarding, and challenging" (Bellah et al. 1992, 15). The sad fact is that contemporary environmentalism has become a movement of guilt and shame, where individual consumption choices become the measure of one's commitment to social justice and environmental sustainability. That nothing challenging or rewarding lies down that path is further argument for cultivating new models of individual action that tie citizenship to consumption, and focus on small-wins of institutional change (e.g. challenging the architecture of consumer choice at the local market) that make rewarding activism the norm rather than the exception.

For Bellah, institutions are patterned ways of doing things – ways of living, individually and collectively, that feel natural and normal and that, at their inception at least, met a need or offered a solution. Magical thinking and the forces around it are one such institution. They for a time met a need, served a purpose, and solved a problem, and for some interests they still do. But as the false promise of naïve aggregation increasingly hobbles those who care deeply about environmental degradation, and upon whom the promise of transformative governance rests, it is time for a change. Challenging the forces swirling about the TOD must now rise to the top of the governance agenda.

## Notes

1  My thanks to Simon Nicholson for "magical thinking."
2  As Jennifer Clapp (2002) notes, the distancing of production from the end use is among the most pernicious drivers of environmental decline.
3  Reports from the Yale Program on Climate Communication (e.g. Leiserowitz et al. 2017) are especially illustrative. This latest reporting (as of this writing) of climate-change views in the United States reports that 22% of Americans (with an error of +/− 3% with 95% confidence) are "deeply worried" about climate change, with another 42% reporting as "somewhat worried." Leaders of past social movements – abolition, women's' suffrage, India's struggle for independence, the US civil rights movement, and efforts to legalise same-sex marriage, to name a few – would celebrate such levels of public support.
4  My students today are not unlike their parents who may have attended college in the late 1980s, as filmmaker James Klein (1990) effectively demonstrates.
5  Protesters from the 1970s aren't the only ones subject to this gaze; climate scientists are on the hook too, it seems (Attari, Krantz, and Weber 2016).
6  The "Cuisinart theory of social change" is Langdon Winner's (1986) phrase. Meyer (2015) offers a contemporary treatment of the same phenomenon.
7  "Meaningful" means sustained engagement with others to alter social rules, policies, norms, and/or patterned ways of doing things, as per Bellah et al. (1992) and, more recently, Steinberg (2015).
8  For a biting take on the proliferation of these lists, see Tom Friedman's (2008) "205 Easy Ways to Save the Earth."

## Works Cited

Atkinson, Lucy. 2015. "Locating the 'Politics' in Political Consumption: A Conceptual Map of Four Types of Political Consumer Identities." *International Journal of Communication* 9: 2047–2066.

Attari, Shahzeen, David Krantz, and Elke Weber. 2016. "Statements about Climate Researchers' Carbon Footprints Affect Their Credibility and the Impact of Their Advice." *Climatic Change* 138, no. 1–2: 325–338.

Ballantine, Paul, and Sam Creery. 2010. "The Consumption and Disposition Behaviour of Voluntary Simplifiers." *Journal of Consumer Behaviour* 9, no. 1: 45–56.

Bellah, Robert, et al. 1992. *The Good Society*. New York: Vintage.

Bosso, Christopher. 2005. *Environment, Inc.: From Grassroots to Beltway*. Lawrence: University Press of Kansas.

Chee, Stephanie, Sonia Kaur, and Michael Maniates. 2018. "'All of the Above?'" Confusion and Disempowerment among Change-Seeking ESS Undergraduate Students." Conference Poster Presented at the Annual Meetings of the Association of Environmental Studies and Science, Washington DC, June.

Clapp, Jennifer. 2002. "The Distancing of Waste: Overconsumption in a Global Economy." In *Confronting Consumption*, eds. Thomas Princen, Michael Maniates, and Ken Conca, 155–176. Cambridge: MIT Press.

Clapp, Jennifer, and Linda Swanston. 2009. "Doing Away with Plastic Shopping Bags: International Patterns of Norm Emergence and Policy Implementation." *Environmental Politics* 18, no. 3: 315–332.

Crane, Andrew. 2000. "Facing the Backlash: Green Marketing and Strategic Reorientation in the 1990s." *Journal of Strategic Marketing* 8, no. 3: 277–296.

Crompton, Tom, and Tim Kasser. 2009. *Meeting Environmental Challenges: The Role of Human Identity.* Godalming: WWF-UK.

Dunlap, Riley, and Richard York. 2012. "The Globalization of Environmental Concern." In *Comparative Environmental Politics: Theory, Practice, and Prospects*, eds. Paul Steinberg and Stacy VanDeveer, 89–112. Cambridge: MIT Press.

The Earthworks Group. 1989. *50 Simple Things You Can Do to Save the Earth.* Berkeley, CA: Bathroom Readers Press.

Fridell, Gavin. 2007. "Fair-Trade Coffee and Commodity Fetishism: The Limits of Market-Driven Social Justice." *Historical Materialism* 15, no. 4: 79–104.

Friedman, Thomas. 2008. "205 Easy Ways to Save the Earth." In *Hot, Flat, and Crowded.* New York: Farrar, Straus and Giroux.

Hardin, Garrett. 1968. "The Tragedy of the Commons." *Science* 162, no. 3859: 1243.

Hawken, Paul. 1993/2010. *Ecology of Commerce: How Business Can Save the Planet.* New York: Harper Business.

Hempel, Lamont. 1996. *Environmental Governance: The Global Challenge.* Washington, DC: Island Press.

Hurst, Megan, Helga Dittmar, Rod Bond, and Tim Kasser. 2013. "The Relationship between Materialistic Values and Environmental Attitudes and Behaviors: A Meta-Analysis." *Journal of Environmental Psychology* 36: 257–269.

Johnston, Josée. 2008. "The Citizen-Consumer Hybrid: Ideological Tensions and the Case of Whole Foods Market." *Theory and Society* 37, no. 3: 229–270.

Kasser, Tim. 2016. "Materialistic Values and Goals." *Annual Review of Psychology* 67: 489–514.

Klein, James. 1990. *Letter to the Next Generation* (Documentary). Blooming Grove, NY: New Day Films.

Lacasse, Katherine. 2016. "Don't Be Satisfied, Identify! Strengthening Positive Spillover by Connecting Pro-Environmental Behaviors to an 'Environmentalist' Label." *Journal of Environmental Psychology* 48: 149–158.

Leiserowitz, Anthony, Edward Maibach, Connie Roser-Renouf, Seth Rosenthal, Matthew Cutler, and John Kotcher. 2017. *Climate Change in the American Mind: October 2017.* New Haven: Yale Program on Climate Change Communication.

Leonard, Annie. 2010. *The Story of Stuff: How Our Obsession with Stuff Is Trashing the Planet, Our Communities, and Our Health.* New York: Simon and Schuster.

Levin, Kelly, Benjamin Cashore, Steven Bernstein, and Graeme Auld. 2012. "Overcoming the Tragedy of Super Wicked Problems: Constraining Our Future Selves to Ameliorate Global Climate Change." *Policy Sciences* 45, no. 2: 123–152.

Litfin, Karen. 2014. *Ecovillages: Lessons for Sustainable Community.* Cambridge: Polity Press.

Loeb, Paul. 1995. *Generation at the Crossroads: Apathy and Action on the American Campus.* Chicago: Rutgers University Press.

Lorek, Sylvia, and Doris Fuchs. 2013. "Strong Sustainable Consumption Governance – Precondition for a Degrowth Path?" *Journal of Cleaner Production* 38: 36–43.

Lorenzen, Janet. 2014. "Convincing People to Go Green: Managing Strategic Action by Minimising Political Talk." *Environmental Politics* 23, no. 3: 454–472.

Lukacs, Martin. 2017. "Neoliberalism Has Conned Us into Fighting Climate Change as Individuals." *The Guardian*, July 17.

Maniates, Michael. 2002. "In Search of Consumptive Resistance: The Voluntary Simplicity Movement." In *Confronting Consumption*, eds. Thomas Princen, Michael Maniates, and Ken Conca, 199–235. Cambridge: MIT Press.

Maniates, Michael. 2016. "Make Way for Hope: A Contrarian View." In *A New Earth Politics*, eds. Simon Nicholson and Sikhina Jinnah, 135–154. Cambridge: MIT Press.

Maniates, Michael. 2018. "Just Give Us a Good Crisis: Deconstructing the Incapacitating Faith in Crisis among ESS Undergraduates." Paper Presented at the Annual Meetings of the Association of Environmental Studies and Science, Washington DC, June.

Maniates, Michael. Forthcoming. "A Mirror of the Movement? Diversity, Despair, and Indecision among Environmentally Focused Undergraduates in the United States."

Maniates, Michael, and John Meyer, eds. 2010. *The Environmental Politics of Sacrifice.* Cambridge: MIT Press.

Mendleson, Nicola, and Michael Polonsky. 1995. "Using Strategic Alliances to Develop Credible Green Marketing." *Journal of Consumer Marketing* 12, no. 2: 4–18.

Meyer, John. 2015. *Engaging the Everyday: Environmental Social Criticism and the Resonance Dilemma.* Cambridge: MIT Press.

Middlemiss, Lucie. 2014. "Individualised or Participatory? Exploring Late-Modern Identity and Sustainable Development." *Environmental Politics* 23, no. 6: 929–946.

Peattie, Ken. 2001. "Towards Sustainability: The Third Age of Green Marketing." *The Marketing Review* 2, no. 2: 129–146.

Peattie, Ken, and Andrew Crane. 2005. "Green Marketing: Legend, Myth, Farce or Prophesy?" *Qualitative Market Research: An International Journal* 8, no. 4: 357–370.

Pew Research Center. 2016. "The Politics of Climate: 3. Everyday Environmentalism." www.pew research.org/science/2016/10/04/everyday-environmentalism/ Last accessed 2 May 2019.

Princen, Thomas. 2005. *The Logic of Sufficiency.* Cambridge: MIT Press.

Princen, Thomas. 2010. *Treading Softly: Paths to Ecological Order.* Cambridge: MIT Press, 2010.

Reich, Robert. 2007. *Supercapitalism: The Transformation of Business, Democracy, and Everyday Life.* New York: Vintage.

Rigotti, Samuel. 2010. *Environmental Problem Solving: How Do We Make Change?* Meadville, PA: Department of Environmental Science, Allegheny College. https://michaelmaniatesblog.files. wordpress.com/2018/03/rigotti_environmentalproblemsolving.pdf.

Sanne, Christer. 2002. "Willing Consumers—or Locked-In? Policies for a Sustainable Consumption." *Ecological Economics* 42, no. 1: 273–287.

Schudson, Michael. 2007. "Citizens, Consumers, and the Good Society." *The Annals of the American Academy of Political and Social Science* 611, no. 1: 236–249.

Shove, Elizabeth. 2010. "Beyond the ABC: Climate Change Policy and Theories of Social Change." *Environment and Planning A* 42, no. 6: 1273–1285.

Steinberg, Paul. 2015. *Who Rules the Earth? How Social Rules Shape Our Planet and Our Lives.* Oxford: Oxford University Press.

Stone, Deborah. 2012. *Policy Paradox: The Art of Political Decision Making.* New York: W. W. Norton.

Sunstein, Cass. 2017. "Unleashed: The Role of Norms and Law." *Harvard Law Review Blog*, October 17.

Webb, Janette. "Climate Change and Society: The Chimera of Behaviour Change Technologies." *Sociology* 46, no. 1 (2012): 109–125.

Winner, Langdon. 1986. "Building the Better Mousetrap." In *The Whale and the Reactor: A Search for Limits in an Age of High Technology*, 61–84. Chicago: University of Chicago Press.

Zamwel, Einat, Orna Sasson-Levy, and Guy Ben-Porat. 2014. "Voluntary Simplifiers as Political Consumers: Individuals Practicing Politics through Reduced Consumption." *Journal of Consumer Culture* 14, no. 2: 199–217.

# 22

# Democracy in the Anthropocene

*Ayşem Mert*

The Anthropocene emerged as a novel and powerful paradigm in the last two decades. The concept signifies changing perceptions of humanity and its relationship with the non-human environment, by asking if we live in a new geological era in which humans are the main force altering planetary ecosystems. The speed and intensity with which the Anthropocene captured popular imagination show its power as a contemporary transformative approach to global sustainability governance. Simultaneously, the transformations it entails in human societies and in our conceptions of self, other, and interconnectedness reveal a new set of challenges and problematiques. This chapter introduces the Anthropocene concept and the debates surrounding what it means in terms of these transformations. It specifically focuses on how democracy can be imagined in this new era. Rather than theorising about the potential workings of a planet-wide representative democracy (for a recent take on this debate see Leinen and Bummel 2018), or possibilities of democratic deliberation in existing governance schemes (for this debate see Niemeyer 2014; Eckersley 2017; Dryzek and Pickering 2019), it focuses on the conditions for democratic governance under a new set of restraints. This is necessary, because there is little evidence to suggest that the institutions of the earlier (Holocene) era can tackle the problems herein, or that this would be desirable. In what follows, I identify the conditions and challenges for democracy in the Anthropocene, problematise the nation-level bias in global governance and certain strains of democratic theory, and indicate some of the conceptual tools in our disposal to understand and shape the type of democracy that can emerge and address the issues in the Anthropocene era.

## The emergence of the Anthropocene and contestations around the concept

Although Eugene Stoermer coined the term Anthropocene earlier, it gained critical attention when Nobel laureate Paul Crutzen popularised it (Steffen et al. 2011: 843). Crutzen (2002) suggested that the stable biogeochemical conditions of the Earth, which were present for the last 11,000 years (the Holocene) were no longer present. We now lived in a new geological era characterised by the unprecedented impact of a single (human) species on the planet, which should hence be named the Anthropocene. In this new era, the Earth's biological fabric, surface energy balance, and physio-chemical cycles were being altered by the industrial and economic activities of humans to the extent that its effects were visible in stratigraphy and geomorphology (Steffen et al. 2007). The resulting toxic pollution, biodiversity loss, and global warming are irreversible and life threatening, and human behaviour had to drastically change at all levels (Crutzen 2002).

Following Crutzen's initial publication, contestations emerged about the legitimacy and definitions of the term Anthropocene. In the making of previous geological epochs, scientists sought recognisable breaking points in the planet's long history. Accordingly, the time frames were as long as millions of years. The Anthropocene marks a new way of looking at geological time. Marcel Wissenburg (2016, 16–19) explains that the Anthropocene is "named after the (alleged) origin of the break [whereby it] denotes an artificial break in geological and climate history." Scientists asked, therefore, if it was a new geological era at all. Does the Anthropocene signify an epoch or a mere event in the long history of our planet? Is human influence on the planetary ecosystems critical enough to be called a new era? Should the term be Anthropocene or one of the various other terms connoting the impact of political and economic systems on the planet (see Gumbert, this volume, Haraway 2015; Moore 2015)?

These debates are far from settled: in August 2016, the Anthropocene Working Group has voted in favour of designating the Anthropocene as a subdivision of geological time, and sent a formal recommendation to the International Geological Congress (SQS 2016). Yet, in July 2018, the International Union of Geological Sciences (IUGS) ratified a different proposal, which states that the Holocene continues, and the stage we are currently in was to be called the Meghalayan Stage. This surprising turn of events was regarded by Anthropocene proponents as "a strange coup to downplay humans' impact on the environment [by] a small group of scientists – 40 at most" (Maslin and Lewis 2018). Even if the term receives official approval at a later stage, contestations will likely continue regarding when the Anthropocene has started, what its markers are, and how the relevant natural science disciplines should respond (cf. Wissenburg 2016; Castree 2019).

It is not of great significance that the term was turned down by IUGS. Many geologists still use the term; the popular imagination has been captured by the Anthropocene narratives; and the concept has gained widespread recognition in daily life. In fact for many, "the Anthropocene has become an important framework for thinking about the processes and consequences of worldwide environmental change" (Ogden et al. 2013: 341). In short, we have already entered the Anthropocene and started organising our thoughts, policies, and institutions around this paradigm (also see Inoue et al., this volume).

## A radical break in scientific thought and its repercussions

The scholarly debate on the Anthropocene is important because it points to a radical break in the way technoscientific pursuit is organised and perceived. Regardless of their stance on the questions about the Anthropocene, natural scientists are now operating in a new field. Since the Enlightenment, modernist approaches to and definitions of "nature" have prevailed in scientific discourse "from the molecular to the cosmic scale" (Castree 2019: 38). Technoscience has been the dominant paradigm in predicting, controlling, and transforming natural cycles to ensure continuous production and processing of resources for human use (see also Alexander and Rutherford, this volume). This forecasting power was to a large extent due to the systematic measuring and recording of "natural" phenomena and making relatively accurate predictions on the basis of this data. Although it was always already a political process, the legitimacy of technoscience depended largely on the agreement that scientific facts were universal and objective. On the other hand, the same predictive power has provided technoscience with unprecedented legitimacy, funding, and authority over the way life is conducted in modern societies (Lane, this volume). In the Anthropocene, the stable, predictable conditions of the Holocene can no longer be expected and therefore the forecasting power of technoscientific endeavour is also at risk. The theories of knowledge and the value attached to scientific truth will necessarily change. For instance, as the number and intensity of extreme weather events grow, ecosystemic and geophysical changes become less predictable. For human systems that rely heavily on technoscientific calculation (e.g. on average climatic conditions as basic assumption for agricultural production, on algorithms to make decisions, on statistical models to plan and build infrastructure), this means increasingly risky public investments and private undertakings. Furthermore, without a stable steady state, interventions in socioecological systems are also riskier than usual. In sum, modern paradigms of technoscience are no longer as accurate in foreseeing and dominating nature, or securing continuous production and distribution of resources.

This is *the first condition of the Anthropocene* democratic institutions and imaginaries have to accommodate. The reliance of modern human societies and institutions on the existing technoscientific model will decrease significantly in the absence of stable ecologies and economies, with all the political transformations and contestations it entails (for a detailed discussion see Mert 2019a). The way science is conducted, technological achievement is perceived, findings are introduced into policy making need to be reassessed (for a detailed discussion see Latour 2015, also see Bulkeley 2019; Paterson 2019). More pluralist, flexible, and adaptable forms of knowledge and technologies befit the conditions of swift and fundamental change and such research is necessarily reflexive, normative, and empirical at once (Beck et al. 2014; Biermann 2014a, Lövbrand et al. 2015; Dryzek and Pickering 2017).

Inversely, imagining a democratic Anthropocene requires a commitment to pluralism in the sources, methods, and paradigms of knowledge, as well as the recognition that knowledge production has always already been political. At present, the hegemonic knowledge structures produced by the past and present relations of domination are largely regarded universal. In contrast, the less industrialised, the less modernised, and the less westernised knowledge-holders in the global South, and the bottom-up, holistic, non-patriarchal knowledge-bases in the global North are marginalised. The hegemony of a particular (Western, white, male, modernist, capitalist) knowledge over all others has socioecological repercussions. This dominant paradigm often proposes technofixes, large-scale technological interventions, which promise solutions to the symptoms of environmental problems rather than addressing their root causes. Such proposals often involve employing unsafe technologies to intervene in ecosystems in irreversible ways, with unpredictable potentially catastrophic results. For instance, carbon sequestration aims to address the problem of carbon emissions *after* they have been emitted to the atmosphere, and aims to inject carbon captured from the atmosphere back into rocks and soil, without the knowledge of how this might affect land ecosystems. The underlying assumption of technofixes is that activities resulting in climate change do not have to change, and industrial and economic causes of the problem, such as commitment to fossil fuels and continuous economic growth, need not be addressed (for a critique of growth and development see Higgs, this volume). Furthermore, whether and how to engage with climate engineering technologies are shaped by a small set of experts in the global North, marginalising the countries in the global South that are most vulnerable to climate change as well the potential consequences of these technologies (Biermann and Möller 2019).

The first step towards democratic governance in the Anthropocene is, then, to step back from quick fixes, which promise unrealistically easy and efficient solutions to difficult problems without deep alterations in contemporary socio-economic structures. Not only do these arguments pave the way to technocratic authoritarianism and risky and irreversible interventions in large-scale ecosystems, but such projects often fail in delivering their promises, too (Huesemann and Huesemann 2011). By claiming that the current social, political, and economic structures (such as patriarchy, capitalism, inequality, industrial, and economic growth paradigms) are unchangeable, they legitimate and normalise highly risky policies with little chance of success.

If the concern is the effectiveness of environmental policies, they are found to increase with more reflexive and inclusive governance models (Voss et al. 2006). Democratic and diverse processes also produce better outcomes than exclusive and homogeneous ones at the planetary level (Stevenson 2016). Political institutions should therefore maximise the ways and capacities for participation of various knowledge-holders. However, such a policy recommendation is not without its problems and there are many complexities associated with public participation. Participation can be insufficient in generating meaningful sustainability transitions, and obscure possibilities of direct action. This can generate disappointment; and participation is at times regarded as a way to maintain the environmental crisis (see Blühdorn and Deflorian, this volume). Moreover, participation in transnational and global environmental governance has become increasingly instrumentalised since the 1990s and used to conceal other (often neoliberal) types of institutional change (Mert 2019b).

## The context of the Anthropocene: neoliberal globalisation and international relations

The Anthropocene concept has emerged in a specific conjuncture that requires elaboration. Since the 1990s, the world is regarded as increasingly interdependent and densely connected. With globalisation, individuals frequently acquire the same sets of information, respond to the same set of stimuli, identify with similar political projects, and therefore often act in similar ways.[1] Taken together, their identical reactions can have overwhelming and at times disastrous effects on economies, ecologies, and societies. The Anthropocene is the realisation of this aggregate effect of individual actions on planetary ecosystems. In other words, the Anthropocene paradigm recognises that individual action under the conditions of globalisation affects nature/cultures at a planetary scale. It points to a contradiction in political agency: on the one hand, individuals often act in certain ways because their various identities (as citizens, shareholders, consumers, etc.) present them with only a few viable options in a given situation with the available information. On the other hand, a single individual can rarely influence an economic crisis or ecological disaster by acting intentionally towards a certain goal (i.e. to avoid a certain disaster) (see also Maniates, this volume). The causal chain between individual behaviour and planet-level catastrophe is too long and the impact of singular deeds on the bigger scheme of things is too small to establish accountability, let alone liability.

In this sense, the Anthropocene highlights the disconnect between agency, cumulative effects of individual actions, and planetary consequences. Thus, being a morally responsible individual, an ethical consumer, a good citizen (however constructed) no longer implies that a healthy political community can be formed and sustained. A democratic imaginary for the Anthropocene also has to account for *this second condition*: the Anthropocene represents an end to the modernistic perceptions of the world and the place humans occupy in these cosmologies both as morally responsible individuals and as political agents. The emergence of the concept alone dislocates the previously stable constructions of structure, hierarchy, morality, identity, interdependence, and interconnectedness. Furthermore, as the human subject is decentred, what is public and what is community are reformulated to include non-human actants, inter-actions across species, ecosystems, institutions, and networks (Gumbert, this volume, also see Haraway 2003, 2007; Schlosberg 2012; Burke et al. 2016, Latour 2016). With this conceptual change, the political agency in a democratic Anthropocene has to build on the complex inter-connectedness between human/non-human, self/other, and nature/society.

Finally, the particular form globalisation has taken in the context of neoliberal capitalist ideology also delimits the democratic imaginary in the Anthropocene (cf. Swyngedouw 2015; Moore 2017). Asymmetries in economic resources and other structural sources of power have shaped current societies, institutions, and imaginaries. Constructing conditions of equal participation among nation-states let alone individual citizens would be to deny the historical conditions in which the Anthropocene emerged. While I discuss the repercussions of this context for democracy below, it is critical to remember that the economic structures and power relations in the international systems have been hindering far-reaching democ-ratisation reforms of international institutions, such as the United Nations or the World Bank, for decades. With globalisation and the expansion of the corporate realm, the power and responsibilities of the corporations are augmented, whereas their liabilities in global environmental governance have become increasingly elusive (Mert 2012). Simultaneously, the discursive power of business is intensified, allowing business interests to redefine the political order and the corporation to become the main provider of public goods (Fuchs and Kalfagianni 2010), and thus challenging especially (but not only) democracies (Fuchs 2013). The very logics of global environmental and climate governance have become increasingly financialised (Clapp and Stephens, this volume), economized, and depolicitised (e.g. Büscher 2010; Methmann et al. 2013; Swyngedouw 2013).

Global power asymmetries will continue to constrain the change that is required to democratise governance and address exploitation and inequality. However, there is in-creasing interest among international relations (IR) scholars to respond to this constraint in various ways. The dislocation of the main concepts of IR in the context of the Anthropocene debate instigated new debates in the discipline. On the one side, there are calls for a political programme to reform "Earth system governance" institutions with an emphasis on the polit-ical nature of the Anthropocene (Biermann 2014b). There is at least one manifesto that calls for increasing control of human enterprise and reliance on international law to limit human influence on the planet, ranging from turning coal into a controlled substance to sustainably critical due diligence (Burke et al. 2016). On the other side, there are calls for bottom-up anti-capitalist emancipatory projects for the Anthropocene, which regard interventionism, and such reliance on international law problematic, de-politicising, and fundamentally un-democratic (Chandler et al. 2018). There are also calls to rethink sovereignty in the face of ecological crisis without falling back on modernistic dichotomies (Latour 2016), and for a "postmodern global imaginary" that prioritises neither the whole (the world) nor the parts (the states) (Falk 2016: 134). Each party claims its project to be political, empowering, crit-ical, and democratic. This already reveals the centrality of the Anthropocene debate for democracy at all levels of politics, as well as in ontological and epistemological frameworks.

## The Anthropocene question: challenges to democracy

The Anthropocene is at once a crisis and an opportunity to rethink and (re-)construct dem-ocratic imaginaries, and correct some failures of the existing regimes. This goal is critical for any discussion of democracy today, considering the recent disillusionment conveyed by the citizens of liberal democracies. Jedediah Purdy (2015: 257) explains that restructuring our lives as citizens has become impossible with the decline in our political imagination. "Calling for more democracy when democracy seems a formula for failure" can cause further alienation and frustration; therefore the democracy that can address the Anthropocene question cannot be more of the same (Purdy 2015: 267). This is also the case for the contemporary global governance architecture and IR. Global environmental problems have emerged at the so-called "end of history," in an era where liberal democracies have supposedly triumphed over alternatives. As Simon Niemeyer (2014: 16) notes "the failure to anticipate and address the anthropocenic challenge [has been] viewed as a failure of democracy."[2] In other words, on many levels of governance, democracy is being associated with dissatisfaction and failure while the stress on socioecological structures is ever increasing.

The greatest threat to democracy, therefore, will be the conditions under which most societies will live in the Anthropocene, and how they will respond to these circumstances. Extreme climatic change and the impossibility of predicting the conditions of the near future will bring with them feelings of deep uncertainty and insecurity. These feelings often translate into othering, polarisation, and a desire for faster and more decisive action, often associated with authoritarianism. This is because a situation where order itself becomes impossible represents much more than the "contingency found in all empirical reality: it is the very definition of the state of nature" (Laclau 1990: 70). In such cases, *establishing* order becomes urgent and important, whereas the content of that order is less significant. Ensuring survival takes precedence over how democratically decisions are reached. By the time most of humanity experiences the ecological crisis as radical disorder, it will be difficult if not impossible to argue for democratisation. It is critical to note, however, that we are not living in that reality just yet, which makes it even more urgent to demand deeply democratic procedures to be put in place and radically democratic regimes to be established at all levels.

As we enter the Anthropocene, we already notice the increasing dislocations and stress on socioecological systems. Under extreme duress, however, democratic institutions can only survive if they provide citizens with the following: (1) basic security and order, characterised by flexible and adaptable plans and processes; (2) the political narratives that justify the difficult transformations required for such flexibility; and (3) inclusive public platforms on which the required transformations are debated, contested, and agreed upon. However, we enter the Anthropocene with the institutions that were created for a fundamentally different set of problems, institutions that have emerged in steady conditions, where decision-making could take a very long time. Contemporary institutions and global environmental governance architecture emerged in response to human societies' experiences in the Holocene, to tackle Holocenic environmental problems. To be able to protect the citizens and the public good, to prepare for potential calamities and rebuild resilient communities in their aftermath requires these institutions to change and adapt drastically and swiftly.

While societies opting out of democracy are a potential problem, another challenge emerges when experts (often scientists) actually call for undemocratic solutions and processes.[3] These post-political arguments highlight the urgency and severity of the ecological crisis (climate exceptionalism) while framing democracy as too slow and ineffectual (Lövbrand et al. 2015). Calls for eco-authoritarianism have been present since concern for environmental issues began in modern societies. They follow the same line of argument for instance with the neo-Malthusian eco-authoritarianism debates of the 1970s based on severity and complexity, and similarly propose technofixes as solutions and technocracy as an ideal regime. They confirm Purdy's concern about the decline in political imagination and indirectly show the importance of imagining democracy in the Anthropocene. Second, they have been

287

proven wrong empirically at state level and at large (Gilley 2012; Purdy 2015; Shahar 2015) and repudiated conceptually (Niemeyer 2014; Stevenson and Dryzek 2014). Even the imperfect democratic models in contemporary societies prove more environmentally friendly than authoritarian regimes.[4]

Another set of challenges result not from the conditions of the Anthropocene, but from the way it has been framed. *The positivist framing* of the Anthropocene, particularly regarding the post-social and post-political ontologies it rests upon (Lövbrand et al. 2015), is not conducive to a new democratic imaginary to flourish for several reasons (Mert 2019a). Critics highlight that a multiplicity of cultures and social relations is subsumed under the abstract conception of a single humankind (see Inoue et al., this volume). As a result, social categories such as nationality, age, class, race, gender, power, and capital (and the historical inequalities therein) are overlooked. Furthermore, differences in responsibilities for environmental change (causes) and differences in experiences of it (results) are absent, and arguably muted in this framing.

However, an imaginary for a democracy in the Anthropocene does not necessarily rest on its positivist framing which originated in the initial scientific debates among natural scientists. A more pluralist narrative, which I call the *deconstructivist framing*, has emerged shortly afterwards where a democratic imaginary of the Anthropocene is encouraged and enabled (Mert 2019a). This framing allows for alternative interpretations and narratives of the Anthropocene and focuses often on its underlying dilemmas and contradictions (e.g. the impossibility of maintaining coherent social structures such as a capitalist world economy, global industrial production, and institutions such as traditional family structures, large-scale political organisations).

## Unlearning the state-level bias: an all-inclusive democracy

What kind of a democracy, then, could anticipate, mitigate, and adapt to the transformations invoked by the Anthropocene? While there might be new material restrictions on decision-making process (e.g. the need for faster decisions made on less accurate data), it is reasonable to assume some continuity and some change: some of the current practices of democracy will be a part of the Anthropocene governance architecture, and others will fade or be fundamentally reinterpreted. The fundamental differences between democracy in the Anthropocene and the Holocene will be in the scale of the problems, the decisions made, and their repercussions on the affected human and non-human stakeholders. If we understand the Anthropocene as a new scale, what kind of democracy could address its tensions? Which current practices in democratic theory and practice impede a democratic imaginary for the Anthropocene?

One of the practices obstructing the emergence of a democratic imaginary for the Anthropocene is the state-level bias in IR and global governance scholarship (Mert 2019a). Traditionally, in IR literature, citizens are assumed to be represented by their respective governments and the relations between nation-states are considered anarchic with little democratic potential. Archibugi and Held (2011: 433) observe that most IR textbooks prior to 1989 do not even contain the word "democracy." Academic literature on democracy beyond the nation-state reflects time and again that state-centred democratic concepts are not applicable at the international and transnational levels, and particularly for non-electoral, non-territorial, or hybrid modes of governance that characterise global environmental governance (Bäckstrand 2012: 170). While there are attempts to go beyond this paradigmatic restraint and reinterpret certain democratic concepts in the face of transnationalisation, hybridisation, or polycentricity (Nanz and Steffek 2004; Bäckstrand 2006; Dingwerth 2007; Scholte 2011) IR research generally assumes that states have a high threshold of democracy, whereas democracy beyond the state cannot be expected to have such high standards (Keohane 2011). As a result, scholars tend to apply nation-level democratic principles less stringently to the international level.

One of the reasons for this bias is that nation-states have been and arguably still are the main actors in the present international society. This bias structures and limits the democratic imaginary. For instance, in a recent debate, Robert Keohane (2016: 938) wrote that global governance involves the difficult task of production of public goods; however, there is no global government which could "harness the emotional support of nationalism." The infrastructure that is needed for democracy, he argues, is absent at the global level. If we do not want to support a hypocritical nominal democratisation that is void of substance, we should aim to maintain some key features of democracy (such as accountability of elites to publics, some stakeholder participation, protection of minority rights, and deliberation within civil society) rather than insist on more radical ideas such as deliberation or representation.

It is the hegemony of state-level democratic imaginaries that focuses the IR debates on the impossibility of a global democracy. When all of humanity constitutes the *demos*, it is argued, the representative model cannot be operationalised:

> While [political theorists] want to democratize global institutions such as the UN or the WTO, there is not yet a people or *demos* on the global level to undertake this task. [They try] to speak *for* the people by constructing a theory *without* the people.
>
> *(Näsström 2010)*

This is, however, a misleading argument for three reasons: first, as Robert Dahl (1989: 3) observes, founding a democracy always presupposes that "a people" already exists while the accuracy of this argument is almost always dubious. Second, there have been numerous examples of extraordinarily diverse and large populations creating democratic governance such as the United States, Brazil, and India.

Third, and most critically, defining the *demos* always involves bias and exclusion. Democracy cannot be established without defining a *demos*, which cannot be democratically determined. This constitutes the founding paradox of democracy (Mouffe 2000, also see Benhabib, 2008; Honig, 2009). The Anthropocene can perhaps solve this paradox to some degree: if there were the slightest possibility of delimiting "a people" without bias and exclusion, this would entail the inclusion of all humanity. (The problem remains unchanged, however, about excluding future generations, non-human life, and so forth.) In a most original contribution to this debate, Hans Agné (2010, fc.) argues that not only is it possible to imagine a global demos that includes all humanity and find a new political order, but this premise also resolves the foundational dilemma of democracy without producing a boundary problem. As such, he concludes, there is no paradox of democratic founding in a global democracy.

Other scholars argue that a loosely connected albeit largely unorganised global *demos* already exists and is taking part in global environmental governance (e.g. Dingwerth 2007). To imagine democratisation of *planetary* governance, this all-inclusive conception of *demos* is central, particularly when it comes to understanding its political activity: Masses of people make demands regarding multilateral agreements at environmental summits; at G8 meetings, they gather to protest the structures that allow the financial elites to govern the world's largest economies; and they take action on the global networks of social media every day. They change and re-structure global politics and discourses and deem international agreements illegitimate with their actions (e.g. the 2009 Copenhagen Climate Accord). Often the local, global, and transnational levels are more critical for their actions, by-passing more structured and rigid nation-level institutions. If we follow their lead, it would be possible to note that there is no inherent reason for democracy in the Anthropocene to be based on the same assumptions as democracy at the scale of the nation-state. In its historical development as a regime, democracy has been fundamentally transformed when it was being applied to a new political scale. During the American and French revolutions, the principles of an ancient, small-scale regime were reinterpreted and significantly transformed to suit the needs of larger nation-states, with an extensive *demos*. Most importantly, representation was introduced as a democratic procedure, which was previously a monarchic tradition. For Aristotle, elections belonged to the logic of oligarchy, for Montesquieu and Rousseau that of aristocracy. Nevertheless, elections began to serve the purpose of deciding "who gets the office," generating representative councils, and ensuring accountability. Today, every democratic nation-state employs representation in its practice of democratic rule. In other words, the tension between direct participation and representation acts as "the restrictive myth for democratic governance" (Mert 2015: 189): direct participation of all affected citizens in decision-making is "the participatory *ideal* in democratic discourses," as a result many inclusive and participatory practices emerge across the scale. On the other hand, this ideal serves as "the ultimate and unreachable fantasy" (ibidem). There is almost complete consensus that participation is not possible and representation has long become its substitute. This was the first scalar revolution in the history of democracy, and it can be summarised as follows: since participation of all citizens has been regarded as impossible, representation has become equated to democracy.

With the end of the Cold War and intensifying globalisation, democratic criteria have become even more abstract than representation. This time, representation was regarded as impossible and therefore second-order democratising principles and criteria such as transparency, accountability, and consultation (e.g. with civil society, stakeholders, or industry) have become proxies for democracy. There are three problems with this practice: first and foremost, this type of democratisation understands representation as a goal in itself. However, as discussed earlier, representation was devised as an imperfect remedy for the difficulties of democracy for the newly forming nation-states of the long 19th century. Its relationship with the democratic ideal is problematic at best. Second, these criteria are often encouraged as desirable but not obligatory principles. In other words, the actors are invited to be accountable and transparent, or to voluntarily report on democratic quality of their activities at regular intervals. Nevertheless, employing nation-level criteria as non-binding ground rules fails to address the democratic deficit at the transnational and global levels where there is no *shadow of hierarchy* (Mert 2009). Most transparency measures are geared towards disclosing information without any behavioural change on the side of the governance institutions and the corporations, whereas accountability at the global level only means accountability to a limited number of actors, often experts and the international development elite (Hardt and Negri 2005). Currently, even these second-order principles are regarded as too difficult to achieve and are regarded desirable rather than obligatory for democratisation of governance institutions.

Finally, as Karin Bäckstrand (2010) notes, the intense academic and corporate interest in accountability and transparency indicates that these measures have, to some degree, been institutionalised into the dominant system of rules. They have ended up legitimising and maintaining existing power imbalances, rather than fundamentally democratising these institutions. This is also true for participation measures in global and transnational environmental governance. The power imbalances between North and South are often nominally concealed by including a like-minded Southern actor without much say over the decisions. It is more difficult to ensure the inclusion and participation of the more vulnerable groups, less represented demands, and the rights of non-national minorities (Pattberg and Wiederberg 2016). A nominal democracy, where the appearance of citizens' participation is ensured, can easily obscure the deeper challenge of the democratic production of actual solutions to fundamentally political questions (Mert 2019b).

In sum, the state-level bias haunting imageries of democracy at the global level derives from the (conceptually and normatively misleading) assumption that a global *demos* does not and cannot exist. Taking the counter-position allows us to ask the critical question: how can the issues that concern all (i.e. all of humanity and non-human environment) be governed in a democratic fashion?

## Imagining democracy in the Anthropocene: some convivial tools

With the Anthropocene, democracy must be reinterpreted once more. A second scalar revolution is needed to fundamentally transform the practice and conception of democracy for the planet. It is possible to juxtapose some of the conditions emerging from the earlier discussion with some of the already existing tools at the disposal of social sciences, humanities, and legal studies. This would help co-construct a new democratic imaginary for the coming era.

The first instrument of critical value is *deconstruction*. Modern, nation-level traditions and experiences of democracy do not produce democratic governance mechanisms that address the challenges of the Anthropocene, or democratise older governance institutions. Thus, existing practices must be deconstructed with the explicit aim of seeking and delineating democratic principles and institutions that respond to the uncertain conditions of the Anthropocene. This requires the acceptance that certain institutions no longer serve their purpose nor can they be reformed to democratise governance. For instance, does sustainable development, the liberal compromise that seemingly reconciles social, economic, and ecological demands, serve our democratic political purposes in the Anthropocene with its emphasis on a less ambitious and versatile ecological project? Policy questions like this one are difficult to address: institutions are agents that aim to perpetuate themselves, produce their own gatekeepers, and develop strategies to maintain control (Illich 1971). Even after obsolete institutions are phased out, new institutions should only be endorsed if they have a clear purpose and a time limit to fulfil their policy goals. This would reduce the number of counterproductive institutions, provide more space and resources for convivial practices, and promulgate innovative governance approaches (for a more radical approach see Illich 1971).

However, phasing out counterproductive institutions would require *institutional and ecological reflexivity* – the second core concept of democratic governance in the Anthropocene: by critically analysing the results of past decisions and policies from the viewpoint of socioecological change, institutions of the Anthropocene must continuously reform themselves. This requires deliberation, an ongoing public debate on how to address the socioecological problems (Dryzek and Pickering 2017). Furthermore, it requires a new understanding of democratic accountability and political success. In today's representative democracies, it is assumed that public officials are held accountable for the results of their decisions in the next election cycle. The unpredictability of Anthropocenic change demands that politicians notice and swiftly change direction when a policy fails to address a crisis. Accepting the imperfection of policy, backtracking and adapting it to new conditions in open public deliberation can take place with a new framing and novel practices of policy making and evaluation. Some of these practices are present in adaptive governance experiments, whereas others, such as new conceptions of a healthy policy cycle or a successful politician, will have to be reinvented.

A third relevant tool is *democratic experimentation*: ways of including non-human actants that cannot represent themselves and traditionally underrepresented groups are continuously invented. For instance, the non-human environment increasingly has been given rights and legal persona in the last decade (e.g. Whanganui and Ganges rivers, Pachamama). While the agency of non-human nature is discussed by Gumbert in this volume, it is also necessary to note that non-human entities have been treated as persons for limited legal purposes for several centuries. Most recently, multi-national corporations have enjoyed this privilege, resulting in deepening social and economic inequality as well as ecological injustice. Providing parts of non-human nature with similar legal rights potentially counterbalances some of these issues.

Participation of underrepresented groups in global environmental governance is replete with problems such as instrumentalisation and green-washing. However, there are also new practices that involve transfer of decision-making power to the affected citizens. These practices can and should be transferred to other levels of governance. The goal of *democratic transfer*, however, is neither problem-solving by designing the most effective policy-options, nor is it rebranding the traditional representative institutional model with a façade of bottom-up, civic innovation. As Henk Wagenaar (2016: 114) aptly puts, democratic transfer entails "design[ing] institutions that incorporate the deliberative, non-monetized practices of collective problem solving that have emerged in civil society (Avritzer 2002)."

New, radical, emergent practices in the margins of contemporary democratic imaginary should be reconsidered and evaluated carefully and reflexively before being taken up at the Anthropocenic scale. It can also be fruitful to return to the earlier principles and practices of direct democracy and innovatively rethink ways to bring back meaningful participation, deliberation, and even sortition. If some of these can be revived and tried at different platforms, a repertoire of democratic experiments would develop, which the less powerful, the less represented, and the more vulnerable agents could employ. Of critical importance in this context are those already affected by the Anthropocenic change, particularly climate change, to the extent that they cannot be regarded as traditional stakeholders. Climate refugees who lost their livelihoods, populations of low-lying small islands who increasingly are/have been deprived of the places they live, know, and co-produce have a different kind of agency, one that emerges from their loss and sentiments of "planetary justice" (Dryzek and Pickering 2019).

## Conclusions

The idea of the Anthropocene affected scientific discourse and captured the popular imagination in an unforeseen fashion. The role of the human species in transforming the planet; the perils, risks, and limits of such transformative power; and its implications for morality, politics, and identity are some of the most intensely debated issues of global environmental governance. A continuum of co-existing rationales is replacing the modern dichotomies such as nature/culture, (hu)man/non-human, self/other, rational/irrational, agency/structure, and being/becoming. In other words, the contestation around the concept of the Anthropocene is a hegemonic struggle about how to understand, signify, and respond to a radical threat, and at what scale.

At once, the Anthropocene threatens and potentially invigorates the practice of democratic governance, and forces us to think innovatively about democracy, to deconstruct traditions and learn from the peripheral and the marginalised knowledge-bases and to relate in new ways with the non-human environment. A reflexive, all-inclusive, and innovative approach to science *and* decision-making, which allows for flexible and adaptable experimentation with democracy, is the most likely starting point for a democratic imaginary for the Anthropocene.

## Notes

1 This applies in particular to large segments of the population in the global North as well as the middle and upper socio-economic classes in the global South. See Inoue et al.'s chapter in this volume on differences in experiences and conditions.

2 Niemeyer (2014: 16–17) aptly questions this conclusion, since the liberal paternalism that characterises most modern democracies precludes a deeper transformation towards deliberative democracy. It is also necessary to question the misleading equivalence often established between democracy and liberal capitalism (Mouffe 2000). Francis Fukuyama's (1989: 3) argument was not about democratisation but about the triumph of Western liberalism as the only viable political system that ensures prosperity. It is important to note the difference between liberalism and democracy, and that its critique is made in democratic theory even though it is largely uncited in global environmental governance literature.

3 This is a different challenge: in the earlier example, where I discuss the likely responses to radical dislocation, there is no reason to assume that the emergent authoritarian regime will be ecologically concerned.

4 This is the case even in the face of China's recent commitment to curb air pollution, which is at times offered as the proof that authoritarianism is the solution to environmental crises. Time and again, even most nominal democracies are found more environmentally friendly than authoritarian regimes (e.g. Bernauer and Bättig 2009), and that environmental policies in authoritarian states tend to produce policy outputs than outcomes (see Gilley 2012 on China).

## References

Agné, H. (2010). "Why democracy must be global: self-founding and foreign intervention", *International Theory*, 2, no. 3: 381–409.

Archibugi, D., & Held, D. (2011). Cosmopolitan democracy: paths and agents. *Ethics & International Affairs*, 25, no. 4: 433–461.

Avritzer, L. (2002). *Democracy and the public space in Latin America*. Princeton, NJ: Princeton University Press.

Bäckstrand, K. (2006). "Democratizing global environmental governance? Stakeholder democracy after the World Summit on Sustainable Development." *European Journal of International Relations* 12, no. 4: 467–498.

Bäckstrand, K. (2010). "From rhetoric to practice: The legitimacy of global public-private partnerships for sustainable development". In *Democracy and public-private partnerships in global governance* (pp. 145–166). London: Palgrave Macmillan.

Bäckstrand, K. (2012). *Are partnerships for sustainable development democratic and legitimate*. Public-private partnerships for sustainable development: emergence, influence and legitimacy. Cheltenham.

Beck, S., Borie, M., Chilvers, J., Esguerra, A., Heubach, K., Hulme, M., Lidskog, R., et al. (2014). "Towards a reflexive turn in the governance of global environmental expertise. The cases of the IPCC and the IPBES." *GAIA-Ecological Perspectives for Science and Society*, 23, no. 2: 80–87.

Benhabib, S. (2008). *Another cosmopolitanism*. Oxford: Oxford University Press.

Biermann, F. (2014a). *Earth system governance: World politics in the Anthropocene*. Cambridge: MIT Press.

Biermann, F. (2014b). "The Anthropocene: A governance perspective." *The Anthropocene Review* 1, no. 1: 57–61. doi:10.1177/2053019613516289.

Biermann, F., & Möller, I. (2019). "Rich man's solution? Climate engineering discourses and the marginalization of the Global South." *International Environmental Agreements: Politics, Law and Economics*, 19: 151–167.

Bernauer, T., & Bättig, M. (2009). "National institutions and global public goods: Are democracies more cooperative in climate change policy?" *International Organization*, 63, no. 2: 281–308.

Bulkeley, H. (2019). "Navigating climate's human geographies: Exploring the whereabouts of climate politics." *Dialogues in Human Geography*, 9, no. 1: 3–17.

Burke, A., Fishel, S., Mitchell, A., Dalby, S., & Levine, D. J. (2016). "Planet Politics: A Manifesto from the End of IR". *Millennium*, 44(3), 499–523. https://doi.org/10.1177/0305829816636674

Büscher, B. (2010). "Anti-politics as political strategy: Neoliberalism and transfrontier conservation in southern Africa." *Development and Change*, 41, no. 1: 29–51.

Castree, N. (2019). "The "Anthropocene" in global change science: Expertise, the earth, and the future of humanity." In *Anthropocene encounters: New directions in green political thinking*, eds. F. Biermann & E. Lovbrand, 25–49. Cambridge: University of Cambridge.

Chandler, D., Cudworth, E., & Hobden, S. (2018). "Anthropocene, capitalocene and liberal Cosmopolitan IR: A response to Burke et al.'s 'planet politics'." *Millennium*, 46, no. 2: 190–208.

Crutzen, P. J. (2002). "Geology of mankind: The Anthropocene." *Nature*, 415: 23.

Dahl, R. A. (1989). *Democracy and its critics*. New Haven, CT: Yale University Press.

Dingwerth, K. (2007). *The new transnationalism: Transnational governance and democratic legitimacy*. Berlin: Springer.

Dryzek, J. S., & Pickering, J. (2017). "Deliberation as a catalyst for reflexive environmental governance." *Ecological Economics*, 131: 353–360.

Dryzek, J. S., & Pickering, J. (2019). *The politics of the Anthropocene*. Oxford: Oxford University Press.

Eckersley, R. (2017). Geopolitan democracy in the Anthropocene. *Political Studies*, 65, no. 4: 983–999.

Falk, R (2016). *Power Shift: On the New Global Order*. London: Zed.

Fuchs, D. (2013). "Theorizing the power of global companies." In *The handbook of global companies*, ed. J. Mikler, 77–95. Hoboken, NJ: Wiley-Blackwell.

Fuchs, D., & Kalfagianni, A. (2010). The causes and consequences of private food governance. *Business and Politics*, 12, no. 3: 1–34.

Fukuyama, F. (1989). "The end of history?" *The National Interest* 16: 3–18.

Gilley, B. (2012). "Authoritarian environmentalism and China's response to climate change." *Environmental Politics* 21, no. 2: 287–307.

Haraway, D. J. (2003). *The companion species manifesto: Dogs, people, and significant otherness* (Vol. 1, 3–17). Chicago: Prickly Paradigm Press.

Haraway, D. J. (2015). Anthropocene, capitalocene, plantationocene, chthulucene: Making kin. *Environmental Humanities*, 6, no. 1: 159–165.

Hardt, M., & Negri, A. (2005). *Multitude: War and democracy in the age of empire*. New York: Penguin.

Honig, B. (2009). *Democracy and the foreigner*. Princeton, NJ: Princeton University Press.

Huesemann, M., & Huesemann, J. (2011). *Techno-fix: Why technology won't save us or the environment*. Gabriola Island, BC: New Society Publishers.

Illich, I. (1971). *Deschooling society*. New York: Harper & Row.

Keohane, Robert O. (2011). "Global governance and legitimacy." *Review of International Political Economy*, 18, no. 1: 99–109.

Keohane, R. O. (2016). "Nominal democracy?: A rejoinder to Gráinne de Búrca and Jonathan Kuyper and John Dryzek". *International Journal of Constitutional Law* 14, no. 4: 938–940.

Laclau, E. "New reflections on the revolution of our time: Ernesto Laclau". *Verso*, 1990.

Latour, B. (2015). "Telling friends from foes in the time of the Anthropocene." In *The Anthropocene and the global environmental crisis: Rethinking modernity in a new epoch*, eds. C. Hamilton, C. Bonneuil, & F. Gemenne, 145–155. Abingdon: Routledge.

Latour, B. (2016). "Onus Orbis Terrarum: About a Possible Shift in the Definition of Sovereignty". *Millennium* 44, no. 3: 305–320. https://doi.org/10.1177/0305829816640608

Leinen, J., & Bummel, A. (2018). *A world parliament: governance and democracy in the 21st century*. Berlin: Democracy without Borders.

Lövbrand, E., Beck, S., Chilvers, J., Forsyth, T., Hedrén, J., Hulme, M., Lidskog, R., & Vasileiadou, E. (2015). "Who speaks for the future of Earth? How critical social science can extend the conversation on the Anthropocene." *Global Environmental Change*, 32: 211–218.

Maslin, M., & Lewis, S. (2018). https://theconversation.com/anthropocene-vs-meghalayan-why-geologists-are-fighting-over-whether-humans-are-a-force-of-nature-101057.

Mert, A. (2009). "Partnerships for sustainable development as discursive practice: Shifts in discourses of environment and democracy." *Forest Policy and Economics*, 11, no. 5: 326–339.

Mert, A. (2012). "The privatisation of environmental governance: On myths, forces of nature and other inevitabilities." *Environmental Values*, 21, no. 4: 475–498.

Mert, A. (2015). *Environmental governance through partnerships: A discourse theoretical study.* Cheltenham: Edward Elgar Publishing.

Mert, A. (2019a). "Democracy in the Anthropocene: A new scale." In *Anthropocene encounters: New directions in green political thinking,* eds. F. Biermann & E. Lövbrand, 128–149. Cambridge: Cambridge University Press.

Mert, A. (2019b). "Participation(s) in transnational environmental governance: Green values versus instrumental use." *Environmental Values,* 28, no. 1: 101–121.

Methmann, C., Rothe, D., & Stephan, B., eds. (2013). *Interpretive approaches to global climate governance: (de) Constructing the greenhouse.* London: Routledge.

Moore, J. W. (2015). *Capitalism in the web of life: Ecology and the accumulation of capital.* New York: Verso Books.

Moore, J. W. (2017). "The Capitalocene, Part I: On the nature and origins of our ecological crisis." *The Journal of Peasant Studies* 44, no. 3: 594–630.

Mouffe, C. (2000). *The democratic paradox.* New York: Verso.

Nanz, P., & Steffek, J. (2004). "Global governance, participation and the public sphere." *Government and Opposition,* 39, no. 2: 314–335.

Näsström, S. (2010). "Democracy counts: Problems of equality in transnational democracy." In *Transnational actors in global governance,* eds. C. Jonsson & J. Talberg, 197–217. Basingstoke: Palgrave Macmillan.

Niemeyer, S. (2014). "A defence of (deliberative) democracy in the anthropocene." *Ethical Perspectives,* 21, no. 1: 15–45.

Ogden, L., Heynen, N., Oslender, U., West, P., Kassam, K. A., & Robbins, P. (2013). Global assemblages, resilience, and Earth Stewardship in the Anthropocene. *Frontiers in Ecology and the Environment,* 11, no. 7: 341–347.

Paterson, M. (2019). "Climate-as-condition, the origins of climate change and the centrality of the social sciences." *Dialogues in Human Geography,* 9, no. 1: 29–32.

Purdy, J. (2015). *After nature: A politics for the Anthropocene.* Cambridge, MA: Harvard University Press.

Schlosberg, D. (2012). "Justice, ecological integrity, and climate change". Ethical adaptation to climate change: human virtues of the future, pp. 165–183.

Scholte, J. A., ed. (2011). *Building global democracy? Civil society and accountable global governance.* Cambridge: Cambridge University Press.

Shahar, D. C. (2015). "Rejecting eco-authoritarianism, again." *Environmental Values,* 24, no. 3: 345–366.

SQS [The Subcommission on Quaternary Stratigraphy]. (2016). Working group on the 'Anthropocene'. http://quaternary.stratigraphy.org/workinggroups/anthropocene/.

Steffen, W., Crutzen, P. J., & McNeill, J. R. (2007). The Anthropocene: Are humans now overwhelming the great forces of nature." *AMBIO: A Journal of the Human Environment,* 36, no. 8: 614–621.

Steffen, W., Grinevald, J., Crutzen, P., & McNeill, J. (2011). The Anthropocene: Conceptual and historical perspectives. *Philosophical Transactions of the Royal Society A: Mathematical, Physical and Engineering Sciences,* 369, no. 1938: 842–867.

Stevenson, H. (2016). "The wisdom of the many in global governance: An epistemic-democratic defense of diversity and inclusion." *International Studies Quarterly,* 60, no. 3: 400–412.

Stevenson, H., and Dryzek, J. S. (2014). *Democratizing global climate governance.* Cambridge: Cambridge University Press.

Swyngedouw, E. (2013). "The non-political politics of climate change". *ACME: An International Journal for Critical Geographies* 12, no. 1: 1–8.

Swyngedouw, E. (2015). "Depoliticized environments and the promises of the Anthropocene". In International Handbook of Political Ecology, eds. E. Swyngedouw and R. Bryant. London: Edward Elgar (2015).

Voss, J.-P., Bauknecht, D., & Kemp, R. eds. (2006). *Reflexive governance for sustainable development.* Basingstoke: Edward Elgar Publishing.

Wagenaar, H. (2016). "Democratic transfer: Everyday neoliberalism, hegemony and the prospects for democratic renewal". In *Critical Reflections on Interactive Governance,* eds. J. Edelenbos & I. van Meerkerk, 93–119. Cheltenham: Edward Elgar.

Wissenburg, M. (2016). "The Anthropocene and the body ecologic." In *Environmental politics and governance in the Anthropocene: Institutions and legitimacy in a complex world,* eds. P. Pattberg & F. Zelli, 15–30. London: Routledge.

# 23

# Living well within limits

## The vision of consumption corridors

*Doris Fuchs*

Living well within limits has emerged as the core global challenge of our time. Planetary *Limits to Growth* seemed to be a future threat taken seriously only by a small group of experts in the 1970s, when the Club of Rome (Meadows et al. 1972) first published its monumental study. Today, we know that human resource consumption already reaches beyond planetary boundaries in a number of dimensions (Rockström et al. 2009). Indeed, Earth Overshoot Day[1] arrives worryingly early every year (Global Footprint 2018).

The central driver behind this development is consumption. Overconsumption by the global consumer class[2] fosters a reckless exploitation of global ecological resources (see also Krogman, this volume). It is also paired with the social exploitation of and continuing under-consumption by the global poor. Scholars have described our lifestyle as "the imperial mode of living" and our societies as "exploitation societies," accordingly (Brand 2017; Lessenich 2016). These concepts capture the situation well: we clearly would not be able to continue to live the way we live if we were to stop degrading and destroying people and planet.

It does not have to be this way, one would think. We can provide consumers with better information about the environmental and social consequences of their consumption, we can subsidise environmentally superior products and tax inferior ones, or we can use a combined tool set to nudge consumers in the desired direction, right?

Unfortunately, decades of sustainable consumption governance have shown such approaches at weak sustainable consumption governance[3] to fail in fostering a sustainability transformation in consumption (Fuchs 2017). Efforts to improve the energy efficiency of products and services and get consumers to buy them have not altered the unsustainability of consumption in any substantial way. Technological household appliances may have become more efficient, but also grown in size and numbers used (Brohmann et al. 2012). Individual automobile models may use less gas per mile or kilometre, but sales in SUVs have skyrocketed and air travel has exploded. In general, dynamics such as the attitude-implementation gap and the rebound effect have meant that both technological efficiency gains and improvements in consumer awareness, fostered in part by sustainable consumption governance, have not yet achieved absolute reductions in resource use (Wiedmann et al. 2015). In addition, research has shown that even consumers intending to drastically reduce consumption face challenges in realising a substantial reduction in their ecological footprint due to lock-in effects in infrastructures and other systemic conditions (Csutora and Zsóka 2016). More fundamentally, structural forces, including business and economic models based on mass consumption, the tight connections between money and politics, the social embeddedness of consumption, and challenges associated with collective action mean that any attempt to individualise responsibility for the sustainability of consumption can only fail (Fuchs et al. 2019; Isenhour 2017; Maniates 2001 and in this volume). After decades of political and societal attention to the idea of sustainable consumption, GDP remains the number one predictor of a society's ecological footprint (Boucher 2017).

In contrast to efforts to improve consumption at the margins, some scholars have engaged with the limits Western societies have to face in terms of their economic activities and lifestyles in a more direct and fundamental way. This chapter depicts a number of such approaches to then focus in on consumption corridors as a particularly pertinent one. It delineates the concept, discusses typical trepidations associated with the notion of limits, and reflects on pathways towards consumption corridors and supportive structural changes. The chapter concludes by drawing linkages to ongoing research and normative debates.

## Limits to growth, limits to consumption?

The idea that the Earth provides humans with limits to what they can consume most prominently reached the scene with the *Limits to Growth* report, published by the Club of Rome in the 1970s (Meadows et al. 1972). At that time, the authors still hoped that their forecast would foster the recognition and acceptance of the need for limits to economic activities and resource consumption and thereby induce changes towards a more sustainable world. Critics, however, suggested that the model was faulty and made the future look too bleak. Yet, recent research including a "30 year update" largely confirms the predictions of the 1972 model (Meadows et al. 2009; Randers 2012; Turner 2008, 2014).

Likewise drawing on the basic notion that ecological resources provided by the Earth are limited, scholars developed the concept of the "environmental space" in the 1980s and 1990s (Opschoor 1987). "The concept reflects that at any given point in time, there are limits to the amount of environmental pressure that the Earth's ecosystems can handle without irreversible damage to these systems or to the life support processes that they enable" (Wetering and Opschoor 1994, as cited by Hille 1997). These limits result both from the stock of natural resources and the sink capacity of ecosystems. On that basis, the concept indicates an ecological "ceiling" and has been used to delineate the idea of an environmentally sustainable space of each person within ecological and social[4] boundaries (Hille 1997; Spangenberg 2002). According to the scholars working with this concept, this environmental space can be quantified by calculating the amount of resources and ecosystem services per head of the population.

More recently, a group of scholars has used the terminology of "planetary boundaries" to draw attention to the questions to what extent the planet can support current human activity (Rockström et al. 2009). The group of 28 scientists identified nine processes regulating the earth system's resilience to human activity. Below those planetary boundaries lies a "safe space" where humankind can develop without altering the earth system, according to the model. In their research, scholars have found, however, that anthropogenic perturbation already exceeds four planetary boundaries: climate change, biosphere integrity, biogeochemical flows, and land system change. Thus, their research reveals the necessity of limits to human (economic) activity, as surpassing the proposed thresholds increases the risk of generating large-scale abrupt or irreversible environmental changes (op. cit.).

A similarly illustrative model based on and communicating the idea of necessary limits is Raworth's "doughnut economy." Inherent to the concept is a critique of economics as a science of growth, which Raworth replaces with a model balancing human needs and planetary boundaries. The "doughnut" defines an ecologically safe and social just area for human activity (Raworth 2017, 2018). Social foundations, derived from internationally agreed social standards, delimit this space on the inner side, while the ecological ceiling defines the outside of the doughnut. The "doughnut economy" has become quite popular as a tool to reconceptualise sustainable development (e.g. Dearing et al. 2014; Stopper et al. 2016).

Starting their inquiry not from the results of unsustainable human economic activity in terms of resource overuse but from its drivers in terms of income and wealth, other scholars have begun to discuss limits to those drivers. Highlighting the correlation between wealth, political influence, and social exploitation, Neuhäuser (2018), for instance, enquires into the ethical dimension of income and wealth. Arguing for the immorality of wealth accrued in the context of exploitative relationships and practices, he proposes measures that limit the accumulation of wealth such as a maximum net income of 4,200€ per month. According to Neuhäuser, such measures would permit everybody to satisfy their basic needs without hurting the dignity of others. In a similar manner, Gough (2017) explores the idea of a maximum income (as well as reduction in working time, see also Larsson, Naessén, and Lundberg, this volume), due to concerns about people's willingness to voluntarily reduce their consumption.

The earlier concepts are similar in that they are all interested in the pursuit of well-being in a world of limits. They differ, for instance, in their attention to ecological and/or social resources. Most fundamentally, however, they could be stronger in the focus on consumption as one of the core drivers of unsustainability. The next section, therefore, details the concept of consumption corridors, which similarly enquires into the potential for well-being in a world of limits, integrating notions of a good life, justice, and planetary boundaries. In doing that, however, it focuses squarely on consumption, linking it to the foundations of a good life, on the one side, and the use of ecological and social resources, on the other side. Accordingly, this concept promises to be particularly powerful as a transformative approach in (global) sustainability governance.

## Consumption corridors

### The concept

Similar to the approaches identified previously, the concept of consumption corridors pursues the aim of living well within limits. Importantly, well-being, specifically the idea of a 'good life' for all humans living now and in the future, is the concept's very starting point. Its premise is further that the purpose of consumption is to allow individuals to live a good life (Di Giulio and Defila, this volume; Fuchs 2017). It is not to acquire materials, foster economic growth, or absorb surplus production. The purpose of consumption is to allow individuals to secure what they need to live a good life, in other words to meet their fundamental needs.

Importantly, the concept of consumption assumes that certain needs are innate to all human beings, thus adopting a perspective based on natural law and more specifically anthropological approaches. We may satisfy our needs in different ways, depending on our historical and cultural contexts and socio-economic and political conditions. But the assumption is that all humans have these needs, and some scholars, accordingly, call them "objective needs" to distinguish them from subjective wants (Doyal and Gough 1991. In a similar manner, we need to distinguish between needs and satisfiers.

Numerous thinkers and scholars have pondered the exact circumscription of these fundamental human needs for thousands of years. Since the 1990s, such inquiries have received new attention and scholars from different disciplinary backgrounds like the philosopher Martha Nussbaum (1992) and the ecological economist Robert Costanza (Costanza et al. 2007) have suggested concrete lists of needs that individuals need to be able to fulfil, if they are to be able to live a good life (see also Max-Neef 1991; Sen 1996). Recently, Di Giulio and Defila (2018 and in this volume) have empirically analysed the societal acceptance of a list of selected needs in Switzerland. Importantly, they have not just asked what needs the interviewees perceived to be fundamental, but specifically to whom the possibility to satisfy these needs should be provided by the state and society. Di Giulio and Defila have thus explicitly included the question of societal responsibility for an individual's ability to satisfy fundamental needs. This perspective is shared by Nussbaum, who also claims that the state has a responsibility to allow its citizens to fulfil their needs. Di Giulio and Defila suggest using the term "protected needs" rather than "objective needs" to highlight this question of responsibility. Sustainable development extends this question of responsibility beyond the national level and today's temporal context, of course.

Given the premise that individuals need to be able to satisfy certain needs if they are to be able to live a good life, the question has to be: what is the quality and the quantity of ecological and social resources that they need to be able to consume? If sustainable development is about allowing all humans living now or in the future to live a good life, then, access to these resources needs to be guaranteed to them. In other words, one can identify a minimum consumption standard.

In a world of limited resources, the identification of such a minimum consumption standard for all individuals, living now and in the future, implies the need to define maximum consumption standards as well. Such standards would need to be set at the level beyond which consumption by one individual or group would hurt the chances of other individuals or groups to consume at least at the minimum consumption standards. In other words, maximum consumption standards would safeguard individual consumption from destroying the ability of others to live a good life. The space between minimum and maximum consumption standards represents a sustainable consumption corridor (Blättel-Mink et al. 2013; Di Giulio and Fuchs 2014). Such corridors could be defined for example by resource or sector, and thus there would be a number of such corridors.

Within the corridors, that is between the floor of minimum consumption standards and the ceiling of maximum consumption standards, individuals are free to consume as they wish. Thus, consumption corridors organise consumption in the interest of opportunities for a good life for all, thereby addressing the challenges of justice and planetary boundaries. They do not prescribe how individuals design their lives. Similarly, the concept of consumption corridors neither implies a plea for asceticism nor a general rejection of consumption. Its core aim is to reconceptualise and reorganise consumption so as to allow all humans living now and in the future to not just survive, but to live a good life.

## But what about my freedom (of choice)?

Should I not be allowed to consume whatever and however much I want and can afford? It is my money, after all, so why should I not be allowed to freely decide how to spend it? In other words, is the intervention in our lives involved in the setting of maximum consumption standards justifiable and acceptable?

Presently, freedom is a value that appears to be prioritised in many Western societies, and a sense of entitlement often prevails. But it is a very specific idea of freedom that opponents of state intervention on behalf of environmental and social objectives tend to hold up. It is a focus on freedom as "freedom from" rather than "freedom to." Specifically, we are supposed to be free from the state interfering in our lives. Such arguments ignore, however, how much interference there already is, most importantly from market actors. Today, we can witness the powerful setting of consumption norms by enterprises with immense discursive power. Our world of "limitless" consumption opportunities is based on the normalisation of very specific lifestyles, which are made possible by the systematic hiding of the implications of our consumption and stabilised by power asymmetries in national and global governance. Indeed, the systematic commercialisation of satisfiers for all of our needs becomes comprehensible only against the background of the opportunities this presents for economic actors and the associated continuous efforts to activate materialistic (in contrast to non-materialistic) impulses in individuals (Kasser 2016).

Rather than focus on our freedom from state intervention, we should retake our freedom in building the societies we want to live in, specifically our freedom to pursue a good life for all and our freedom to set limits to individual action for the sake of the joint pursuit of a collective goal. After all, freedom is a value that societies always have balanced and will continue to have to balance against other values such as security and justice (see also Kalfagianni et al., this volume). It is the very task of governments to constrain individual freedom in those cases in which it endangers and hurts others. In this sense, setting limits to resource consumption by individuals and groups with the aim of protecting the opportunities of others to be able to satisfy their protected needs is indeed justifiable and necessary.

The setting of limits, thus, is in no way a new idea for democratic societies. Some of these limits are legally imposed, others reflect cultural norms, and this can vary across cultures and times. In one way or another, societies have come to agree on these limits. They result, therefore, from the institutions and priorities created by the society individuals envision (Sahakian and Lorek 2018). Limits, moreover, tend to be accepted when they are deemed just, i.e. when their relevance is understood and they apply to everybody (see also Di Giulio and Defila, this volume). The latter aspect thereby addresses one of the pitfalls of consumption governance solely relying on value-based appeals to the individual consumer: the possibility of freeriding on others' efforts.

Even in the context of consumption, we already are used to dealing with limits. We know lower consumption limits in the form of consumption baskets or budgets, used in welfare policy to determine social security payments at the national level. In a similar manner, societies have agreed on minimum levels of education to be "consumed," frequently considering minimum amounts of school years, levels of public funding, and student-teacher ratios in this context. At the international level, the International Labour Organization's (ILO) and United Nations Educational, Scientific and Cultural Organization's (UNESCO) Social Protection Floor similarly captures the idea to identify what is necessary to guarantee basic income security and access to basic social services such as health care.

Maximum consumption limits frequently exist with respect to drugs. But more interesting ones can be found as well. They result from an observed scarcity of resources as well as the societal consequences of their use. Driving was restricted in many places during the oil crises, and is restricted in some places due to concerns about safety, congestion, or emissions, for example. Speed limits exist primarily for safety reasons, but are also discussed as a means to reduce energy consumption. Due to a lack of progress with emission reductions on the part of the automobile industry, courts in some German cities have started to prohibit certain cars from entering the city centre or using certain roads. In Singapore, the overall number of automobile licenses is limited. Governments also frequently set limits on what one can do with one's property via building codes, planning regulations, or environmental laws. In social contexts, countries identify upper limits in terms of the number of years parents receive support for children enrolled in educational institutions or health treatments covered, for example.

Neither minimum nor maximum consumption limits are new then. The idea of consumption corridors is to think of these upper and lower limits together and to thereby use corridors as a means to rethink and reorganise our system of consumption (and thereby production and distribution) in pursuit of a good life for all.

## Pathways towards consumption corridors

If the very aim of consumption corridors is to allow all individuals living now or in the future to live a good life, and if protection of others and the pursuit of other societal norms such as justice are reasons for imposing limits on an individual's freedom, then society clearly needs to play a pivotal role in defining those limits (Fuchs and Di Giulio 2016; Sahakian and Lorek 2018). In consequence, deliberative participatory processes are highly desirable as a basis for the design of consumption corridors. Such processes will need to focus first and foremost on generating agreement on protected needs and corresponding satisfiers. Participants from different socio-economic, demographic, and cultural backgrounds need to ponder and discuss the question of what matters most to them as a condition for living a good life. Importantly, these participatory processes need to be inclusive in engaging with as representative a sample of society as possible, making sure that each voice is heard and fostering capacities for political judgement formation (Bohn and Fuchs 2018).[5] Linking these needs to relevant satisfiers provides the basis for identifying qualitative and/or quantitative minimum consumption standards. In a third step, experts can then identify maximum consumption standards by relating minimum consumption standards to population size and available resources. As the concrete nature of needs and their satisfaction, the relationship between satisfiers and resource use, and (knowledge about) the availability of resources can change over time, lower and upper consumption limits will need to be reconsidered periodically and readjusted when necessary.

Such participatory processes can take place at different levels of governance and their organisation as well as their ability to prepare binding decisions would vary accordingly. Implementing corridors may well have to take place step by step at different scales and in different arenas. Given the complexity of the task involved in reorganising consumption (and production) as guided by minimum and maximum consumption standards and the embedded nature of consumption in societal practice and interconnected politico-economic systems, hoping for immediate all-encompassing change does not seem advisable. Thus, local initiatives may well provide promising stepping stones as can national or regional ones.

One may wonder how consumption corridors could be institutionalised at a local level, however. Would it not be easy to side-step them by moving to the neighbouring city/county/ region? Indeed, if the population of a town or city were to decide to experiment with consumption corridors with respect to residential floor space per person (see Box 23.1) or water consumption, for instance, moving outside the borders of that town or city would be an option for those rejecting the idea. In a similar manner, implementing corridors at a national level could be challenged, of course, as economic elites in particular tend to be sufficiently mobile these days to be able to choose to live abroad. We are familiar with similar arguments being made in the context of taxation. A number of reasons can be named for the meaningfulness of corridors at local or national levels, however. First, introducing corridors as an idea will have to involve societal dialogue in any case, and such dialogue still is the most feasible at the local level. In addition, even if some inhabitants of a city or town were to move away due to conflicting ideas about quality of life, that does not mean that the remaining inhabitants could not benefit from the chosen corridors. Limits on maximum residential floor space per person are likely to translate into protection of the natural environment in the city and on its outskirts. Moreover, cities pursuing progressive policies in terms of environmental protection and social equity tend to be very attractive, especially to younger generations, and thus may well draw new residents to them. Finally, we also have to be aware that we simply cannot wait for international negotiations and agreement on this. Just think about how long it has taken from the first negotiations about the United Nations Framework Convention on Climate Change to the (still extremely weak) Paris Agreement. Accordingly, momentum will most likely have to come from the local and national levels.

---

### Box 23.1 Consumption corridors and residential space

In many Western societies, residential space per person has greatly increased over the last half century (Sahakian and Lorek 2018). This is partly a result of demographic changes, in particular a rise in the number of single person households. But the perceived need for more space also results from increasing possessions of household goods and appliances, cultural expectations regarding correlations between home and status, and politico-economic regulations and practices such as the role of real estate as an investment vehicle (Cohen 2018). While governments have addressed household energy efficiency with a broad range of policies, per capita living space has received little attention (Wilhite 2016). We have become more efficient in heating or cooling our homes, but we have used the energy saved to heat or cool larger spaces (not to mention seal more ground and use more building materials) – the rebound effect is in full swing!

Defining a consumption corridor in this context would mean relating the human need for shelter to a minimum standard for residential space per person. Ideally, communities would deliberate on the precise nature of this need and appropriate requirements for its satisfaction. In the absence of such deliberations, however, we may take existing minimum standards developed by the International Code Council (ICC), the organisation tasked with the development of engineering standards for the design and construction of buildings, as guideposts. The ICC has identified 13.9 $m^2$ (149.6 $ft^2$) for the first and 9.3 $m^2$ (96.88 $ft^2$) for each additional occupant as a minimum requirement for residential space (Fuchs et al. 2018). A two-person household thus would have a right to 23.2 $m^2$ (249.72 $ft^2$) floor space, a four-person household to 41.8 $m^2$ (449.93 $ft^2$).

Taking such minima as a starting point and planetary boundaries and population size into account, some research projects and initiatives have developed suggestions for appropriate maximum standards for per capita floor size. They suggest that such maxima would need to lie around 20 $m^2$ (215.28 $ft^2$) per capita (Lorek and Spangenberg 2019). These calculations have not considered the rather convincing differentiation between the first occupant and additional ones, as in the ICC calculations, and also need further substantiation. However, they provide a first insight into what a consumption corridor with respect to home size could look like.

> The implications of such a corridor for per capita residential space would include an increasing focus on co-living spaces, vertical villages, or communal or multi-generational housing projects, i.e. projects combining reduced individual flat size with generous areas for common use (cafés, gyms, recreation areas) and short-term rent options (for guest rooms, for instance) for city planners. Such developments are already taking place in niches in some cities around the world, but would need to be scaled-up and become mainstream.

As pointed out earlier, developing corridors via participatory processes is highly desirable. Once fundamental agreement on the value of consumption limits is reached in a society, there may be instances in which further corridor development by governments is possible without complex and costly participatory processes. Even with such participatory processes, governments, of course, are not relieved from their responsibility to work towards effective solutions to the environmental and social crises faced by (their) populations in any case.

## Facilitating the transition to consumption corridors

Beyond the question of the immediate processes in the design of corridors, there is the question of what broader structural changes may support a transition to a world of sustainable consumption corridors. Clearly, our current system is set up both in terms of its ideational and material structures to foster overconsumption by those with sufficient economic resources. To allow for an effective transformation, then, structural changes in both of these dimensions are likely to be necessary (Fuchs et al. 2019; Isenhour 2017).

On the ideational side, the vision of a good life for all can provide a powerful replacement for narratives linking the ownership of material goods to happiness (see Di Giulio and Defila, this volume). In a similar manner, paradigms measuring well-being in terms of GDP per capita and correspondingly prioritising growth (as well as extractivism and technological optimism) and locating its basis in consumption need to be transformed (see also the chapters by Alexander and Rutherford, Higgs, Philipsen, and Princen, in this volume). Likewise, the hegemony of efficiency concerns needs to be challenged and the notion and relevance of sufficiency as a paradigm need to be advanced (Princen 2005, Hayden, this volume).

In terms of material structures, the entanglement of politics and business interests, or – to put it differently – the role of money in politics needs to be dramatically reduced (Gumbert and Fuchs 2018, Fuchs et al., this volume). In a similar manner, societies need to critically reflect on the implications of today's media dependence on business funds and the role of advertising in general. To open up space for a reconsideration of the role of consumption in pursuit of a good life and with respect to justice and sustainability, the constant bombardment of each and every individual with messages encouraging material consumption needs to be stopped. This applies, in particular, with respect to children and teenagers, of course. Given the diversity of channels through which such messages are communicated today, including advertorials or influencers on social media, ending this bombardment will be a challenging task, of course.

Combining material and ideational dimensions, the insufficient and often symbolic nature of much of what is pursued in terms of sustainable consumption governance needs to be highlighted and contested. Policies as well as research promoting consumption of "greener" or "fairer" products need to be, at a minimum, balanced, with policies and research focusing on absolute reductions in resource use. Indeed, critical research has pointed out the extent to which such weak sustainable consumption governance legitimises and maintains unsustainable systems and practices (Blühdorn 2013 and this volume). Accordingly, the need for fundamental change as a basis for absolute reductions in resources use must be communicated (see also Fuchs et al., this volume). At the same time, measures supporting collective consumption opportunities and the provision of synergic[6] satisfiers should be promoted to aid individuals in Western societies in efforts to avoid resource consumption at unsustainable levels.

## Conclusion

If the objective of sustainable development is to allow human beings to live a good life, we must strive for a situation in which every individual, living now and in the future, has access to the amount of ecological and social resources necessary to satisfy her or his protected needs (see Di Giulio and Defila, in this volume). Thus, we need to define minimum consumption standards. The existence of planetary boundaries, in turn, means that we also need to address the question of maximum consumption levels. Specifically, we need to define maximum consumption standards at the level of consumption beyond which consumption by one individual or group would hurt other individuals' chances to meet their needs, i.e. reach minimum consumption levels. Sustainable consumption corridors encompass the space between minimum consumption levels needed to be able to live a good life and maximum consumption levels not to be overstepped in order not to hurt other people's chances to live a good life (Blättel-Mink et al. 2013; Di Giulio and Fuchs 2014). Only within this corridor can sustainable consumption take place.

This chapter depicted a number of concepts focusing on the pursuit of human well-being and justice in a world of limited resources, before detailing the concept of consumption corridors and pondering the justifiability of limits to consumption, pathways towards corridors and facilitating accompanying ideational and material structural changes. Consumption corridors, according to the chapter's core argument, provide a transformative approach allowing the pursuit of a good life for all within planetary boundaries while making consumption and its role with respect to both a core concern.

To make this argument, the concept of consumption corridors takes a number of normative stances. It starts from the assumption that every human being living now and in the future should have access to a minimum level of ecological and social resources necessary to be able to live a good life just because he or she is a human being. In making this assumption, the concept also identifies justice and responsibility as important normative foundations for human society, which can, if necessary, present a foundation for constraining individual freedom.

The corridors concept is at the centre of a number of current research debates and strands of inquiry. On the empirical side, it is tied to analyses of societal consensus on protected needs (Di Giulio and Defila 2018), limits to income (Gough 2017), and the relationship between well-being/needs satisfaction and resource consumption (Syse and Mueller 2015, see also Brand-Correa and Steinberger 2017 on energy provisioning systems). On the normative side, the concept is linked to research on responsibility and justice, as well as their relationship to freedom, in global governance in general and global sustainability governance and sustainable consumption governance in particular (Bohn and Gumbert 2018, see also both, Kalfagianni et al. and Pellizzoni, in this volume). Moreover, pondering pathways towards consumption corridors necessarily means considering foundations and processes of political participation and democratic legitimacy (Bohn and Fuchs 2018).

Clearly, designing and implementing sustainable consumption corridors will be a complex task. Defining objective needs, defining satisfiers and translating them into minimum consumption standards, is challenging in itself and likely to involve considerable debate. Likewise, debating and identifying maximum consumption standards are challenging tasks. In the end, however, a world of consumption corridors, a world in which consumption is intended to allow each and every individual now and in the future to live a good life, would be a wonderful world to live in, and more fundamentally, a sustainable world, something worth striving for!

## Notes

1 Earth Overshoot Day is calculated by the Ecological Footprint Network and captures the date by which we have used up more than the resources provided to us by our planet for the year. In other words, it identifies the day by which start living from the stock rather than the interest.
2 The "global consumer class" can be identified via the adoption of the diets, transportation systems, and lifestyles that were once mostly limited to the rich nations of Europe, North America, and Japan (Worldwatch Institute 2018).
3 Such measures can be summarised as weak sustainable consumption (WSC) governance, in contrast to strong consumption governance (SSC) which addresses not just the resource efficiency of products but also fundamental patterns and levels of consumption (Fuchs and Lorek 2005).
4 Captured in terms of minimum sustainable use rates.
5 Political judgement formation involves addressing participants as citizens rather than consumers and based on intrinsic rather than extrinsic motivations, valuing practical knowledge next to expert knowledge, considering emotions and empathy next to rationality, pursuing societal well-being as the central goal, sharing responsibility between society and the state, and finally, the three steps of deliberation, decision, and action (Bohn and Fuchs 2018).
6 Following Max-Neef (1991), synergic satisfiers are those that can satisfy several needs at the same time.

## References

Blättel-Mink, Birgit, Bettina Brohmann, Rico Defila, Antonietta Di Giulio, Daniel Fischer, Doris Fuchs, Sebastian Gölz, Konrad Götz, Andreas Homburg, Ruth Kaufmann-Hayoz, Ellen Matthies, Gerd Michelsen, Martina Schäfer, Kerstin Tews, Sandra Wassermann, and Stefan Zundel. 2013. *Konsum-Botschaften. Was Forschende für die gesellschaftliche Gestaltung nachhaltigen Konsums empfehlen.* Stuttgart: S. Hirzel Verlag.
Blühdorn, Ingolfur. 2013. "The Governance of Unsustainability: Ecology and Democracy beyond the Post-Democratic Turn". *Environmental Politics* 22 (1): 16–36. doi:10.1080/09644016.2013.755005.
Bohn, Carolin, and Doris Fuchs. 2018. "Transformation through Participation: The Pivotal Role of Political Judgement Formation." Paper presented at the Third International Conference of the Sustainable Consumption Research and Action Initiative, Copenhagen, June 27–30.
Bohn, Carolin, and Tobias Gumbert. 2018. "Die Regierung der Freiheit innerhalb biophysischer Grenzen – Zur Notwendigkeit der Rekonzeptualisierung eines Begriffs „grüner liberaler Freiheit" im Kontext neoliberaler Nachhaltigkeitspolitiken." Paper presented at the 27th Congress of the Deutsche Vereinigung für Politikwissenschaft (DVPW) "Grenzen der Demokratie/Frontiers of Democracy", Frankfurt a.M., September 27.
Boucher, Jean. 2017. "Culture, Carbon, and Climate Change: A Class Analysis of Climate Change Belief, Lifestyle Lock-In, and Personal Carbon Footprint." *Socijalna ekologija: Journal for Environmental Thought and Sociological Research* 27 (1–2): 53–80. doi:10.17234/SocEkol.25.1.3.
Brand, Ulrich. 2017. *Imperiale Lebensweisen: Zur Ausbeutung von Mensch und Natur in Zeiten des globalen Kapitalismus.* München: oekom.
Brand-Correa, Lina and Julia Steinberger. 2017. "A Framework for Decoupling Human Need Satisfaction from Energy Use". *Ecological Economics* 141: 43–52. doi:10.1016/j.ecolecon.2017.05.019.
Brohmann, Bettina, Christian Dehmel, Doris Fuchs, Wilma Mert, Anna Schreuer und Kerstin Tews. 2012. "Bonus Schemes and Progressive Electricity Tariffs as Instruments to Promote Sustainable Electricity Consumption in Private Households". In *The Nature of Sustainable Consumption and How to Achieve It*, edited by Rico Defila, Antonietta Di Giulio and Ruth Kaufmann-Hayoz, 411–420. München: oekom.
Cohen, Maurie. 2018. "Sustainable Consumption and New Conceptions for Sufficient Home Size." Paper presented at the World Social Science Forum, Fukuoka, September 25–28.

Costanza, Robert, Brendan Fisher, Saleem Ali, Caroline Beer, Lynne Bond, Roelof Boumans, Nicholas L. Danigelis, Jennifer Dickinson, Carolyn Elliott, Joshua Farley, Diane Elliott Gayer, Linda MacDonald Glenn, Thomas Hudspeth, Dennis Mahoney, Laurence McCahill, Barbara McIntosh, Brian Reed, S. Abu Turab Rizvi, Donna M. Rizzo, Thomas Simpatico, and Robert Snapp. 2007. Quality of Life: An Approach Integrating Opportunities, Human Needs, and Subjective Well-Being. *Ecological Economics* 61: 267–276. doi:10.1016/j.ecolecon.2006.02.023.

Csutora, Mária and Ágnes Zsóka. 2016. "Breaking through the Behaviour Impact Gap and the Rebound Effect in Sustainable Consumption". Paper presented at Sustainable Consumption and Social Justice in a Constrained World, Budapest, August 29–30.

Dearing, John, Rong Wang, Ke Zhang, James G. Dyke, Helmut Haberl, Md. Sarwar Hossain, Peter G. Langdon, Timothy M. Lenton, Kate Raworth, Sally Brown, Jacob Carstensen, Megan J. Cole, Sarah E. Cornell, Terence P. Dawson, C. Patrick Doncaster, Felix Eigenbrod, Martina Flörke, Elizabeth Jeffers, Anson W. Mackay, Björn Nykvistk, and Guy M. Poppy. 2014. "Safe and Just Operating Spaces for Regional Social-Ecological Systems". *Global Environmental Change* 28: 227–238. doi: 10.1016/j.gloenvcha.2014.06.012.

Di Giulio, Antonietta, and Doris Fuchs. 2014. Sustainable Consumption Corridors: Concept, Objections, and Responses. *GAIA* 23: 184–192. doi:10.14512/gaia.23.S1.6.

Di Giulio, Antonietta and Rico Defila. 2018. "How the Concept of Consumption Corridors Is Received in Switzerland". Paper presented at the Third International Conference of the Sustainable Consumption Research and Action Initiative, Copenhagen, June 27–30.

Doyal, Len, and Ian Gough. 1991. *A Theory of Human Need*. London: Macmillan.

Fuchs, Doris. 2017. "Consumption Corridors as a Means for Overcoming Trends in (Un-) Sustainable Consumption". In *International Conference on Consumer Research 2016: The 21st Century Consumer: Vulnerable, Responsible, Transparent*, edited by Christian Bala und Wolfgang Schuldzinski, 147–159. Düsseldorf: Verbraucherzentrale NRW.

Fuchs, Doris and Antonietta Di Giulio. 2016. "Consumption Corridors and Social Justice: Exploring the Limits." In *Sustainable Consumption and Social Justice in a Constrained World*, edited by Sylvia Lorek and Edina Vadovics, SCORAI Europe Workshop Proceedings, August 29–30. Sustainable Consumption Transition Series, Issue 6.

Fuchs, Doris and Sylvia Lorek. 2005. "Sustainable Consumption Governance. A History of Promises and Failures". *Journal of Consumer Policy* 28 (3): 261–288. doi:10.1007/s10603-005-8490-z.

Fuchs, Doris, Sylvia Lorek, Antonietta Di Giulio, and Rico Defila. 2019. "Sources of Power for Sustainable Consumption: Where to Look." In *Politics, Power, and Ideology in Sustainable Consumption*, edited by Mari Martiskainen, Lucie Middlemiss and Cindy Isenhour. London: Routledge.

Fuchs, Doris, Sylvia Lorek, Tobias Gumbert, Marlyne Sahakian, and Antonietta Di Giulio. 2018. Consumption Corridors: Sufficiency and Wellbeing in a World of Limits. Paper presented at the Degrowth Conference, Malmö, Sweden, August 21–25.

Global Footprint Network. 2018. National Footprint Accounts, Edition 2018.

Gough, Ian. 2017. *Heat, Greed and Human Need. Climate Change, Capitalism and Sustainable Wellbeing*. Cheltenham: Edward Elgar.

Gumbert, Tobias and Doris Fuchs. 2018. "The Power of Corporations in Global Food Sector Governance". In *Handbook of the International Political Economy of the Corporation*, edited by Andreas Nölke and May Christian, 435–447. Cheltenham: Edward Elgar.

Hille, John. 1997. *Concept of Environmental Space*. Copenhagen: European Environmental Agency.

Isenhour, Cindy. 2017. "When "Gestures of Change" Demand Policy Support: Social Change and the Structural Underpinnings of Consumption in the United States." In *Social Change and the Coming of Post-Consumer Society: Theoretical Advances and Policy Implications*, edited by Maurie Cohen, Halina Szejnwald Brown, and Philip Vergragt. New York: Routledge.

Kasser, Tim. 2016. Materialistic Values and Goals. *Annual Review of Psychology* 67: 489–514.

Lessenich, Stephan. 2016. *Neben uns die Sintflut: Die Externalisierungsgesellschaft und ihr Preis*. München: Hanser Berlin.

Lorek, Sylvia and Joachim Spangenberg. 2019. "Identification of Promising Instruments and Instrument Mixes to Promote Energy Sufficiency. EUFORIE - European Futures for Energy Efficiency". sites.utu.fi/euforie/wp-content/uploads/sites/182/2019/05/649342_EUFORIE_D5.5.pdf.

Maniates, Michael. 2001. "Individualization: Plant a Tree, Buy a Bike, Save the World?" *Global Environmental Politics* 1 (3): 31–52.

Max-Neef, Manfred. 1991. *Human-Scale Development — Conception, Application and Further Reflection*. London: Apex.

Meadows, Donella, Dennis Meadows, Jørgen Randers, and William Behrens. 1972. *The Limits to Growth*. New York: Universe Books.

Meadows, Donella, Jørgen Randers, Dennis Meadows. 2009. *The Limits to Growth. The 30-year Update*. London: Earthscan.

Neuhäuser, Christian. 2018. *Reichtum als moralisches Problem*. Berlin: Suhrkamp.

Nussbaum, Martha C. 1992. "Human Functioning and Social Justice: In Defense of Aristotelian Essentialism." *Political Theory* 20 (2): 202–246.

Opschoor, J. 1987. Duurzaamheid en verandering: over Ecologische inpasbaarheid van Economische ontwikkelingen. Oratie, Amsterdam: VU Boekhandel/Uitgeverij.

Princen, Thomas. 2005. *The Logic of Sufficiency*. Cambridge: MIT Press.

Randers, Jørgen. 2012. *2052. A Global Forecast for the Next Forty Years: A Report to the Club of Rome Commemorating the 40th Anniversary of the Limits to Growth*. White River Junction, VT: Chelsea Green Publishing.

Raworth, Kate. 2017. "A Doughnut for the Anthropocene: Humanity's Compass in the 21st Century". *The Lancet Planetary Health* 1 (2): 48–49. doi:10.1016/S2542-5196(17)30028-1.

Raworth, Kate. 2018. "What on Earth Is the Doughnut?" www.kateraworth.com/doughnut/, accessed December 18, 2018.

Rockström, Johan, Will Steffen, Kevin Noone, Åsa Persson, F. Stuart Chapin, III, Eric F. Lambin, Timothy M. Lenton, Marten Scheffer, Carl Folke, Hans Joachim Schellnhuber, Björn Nykvist, Cynthia A. de Wit, Terry Hughes, Sander van der Leeuw, Henning Rodhe, Sverker Sörlin, Peter K. Snyder, Robert Costanza, Uno Svedin, Malin Falkenmark, Louise Karlberg, Robert W. Corell, Victoria J. Fabry, James Hansen, Brian Walker, Diana Liverman, Katherine Richardson, Paul Crutzen, and Jonathan A. Foley. 2009. "Planetary Boundaries: Exploring the Safe Operating Space for Humanity". *Ecology and Society* 14 (2): Art. 32.

Sahakian, Marlyne and Sylvia Lorek. 2018. "Laying the Foundations for Consumption Corridors: The Case of Heating Bigger Homes". Paper presented at the Third International Conference of the Sustainable Consumption Research and Action Initiative, Copenhagen, June 27–30.

Sen, Amartya. 1996. "Capability and Well-Being". In *The Quality of Life*, edited by Amartya Sen and Martha Nussbaum, 30–54. Oxford: Clarendon.

Spangenberg, Joachim. 2002. "Environmental Space and the Prism of Sustainability: Framework for Indicators Measuring Sustainable Development." *Ecological Indicators* (2): 295–309. doi:10.1016/S1470-160X(02)00065-1.

Stopper, Markus, Anja Kossik, and Bernd Gastermann. 2016. "Development of a Sustainability Model for Manufacturing SMEs Based on the Innovative Doughnut Economics Framework." In *Proceedings of the International MultiConference of Engineers and Computer Scientists 2016 Volume II, IMECS 2016*, edited by S.I. AO, Oscar Castillo, Craig Douglas, David Dagan Feng and A. M. Korsunsky, Hong Kong, March 16–18.

Syse, Karen Lykke and Martin Lee Mueller, ed. 2015. *Sustainable Consumption and the Good Life: Interdisciplinary Perspectives*. Oxon and New York: Routledge.

Turner, Graham. 2008. "A Comparison of the Limits to Growth with 30 Years of Reality". *Global Environmental Change* 18 (3): 397–411. doi:10.1016/j.gloenvcha.2008.05.001.

Turner, Graham. 2014. "Is Global Collapse Imminent? An Updated Comparison of the Limits to Growth with Historical Data". Melbourne Sustainable Society Institute. https://sustainable.unimelb.edu.au/__data/assets/pdf_file/0005/2763500/MSSI-ResearchPaper-4_Turner_2014.pdf, accessed December 11, 2018.

Weterings, R., and Opschoor, J.B. 1994. *Towards Environmental Performance Indicators Based on the Notion of Environmental Space*. The Netherlands: Advisory Council for Research on Nature and Environment (RMNO).

Wiedmann, Thomas, Heinz Schandl, Manfred Lenzen, Daniel Moran, Sangwon Suh, James West, and Keiichiro Kanemoto. 2015. "The Material Footprint of Nations." *PNAS* 112 (20): 6271–6276. doi:10.1073/pnas.1220362110.

Wilhite, Harold. 2016. *The Political Economy of Low Carbon Transformation: Breaking the Habits of Capitalism*. London: Routledge.

Worldwatch Institute. 2018. "Chapter 1: The State of Consumption Today". www.worldwatch.org/node/3816, accessed September 30, 2018.

# 24

# Beyond GDP

## The economics of well-being

*Dirk Philipsen*

As other chapters in this book demonstrate, the world of economic activity is on a collision course with the planet's ecosystem. The dominant global business model is to consume and degrade at rates long known to be unsustainable. It is no longer a model that can deliver on the promise of greater prosperity, opportunity, and freedom. The challenges of a necessary shift from growth to well-being are significant.

This chapter briefly explores the nature and logic of the reigning economic paradigm with the intent of identifying both reasons and consequences behind today's preeminent measure of economic success. It concludes by arguing that to go beyond the measure requires going beyond the paradigm.

## An economics based on growth

At the core of modern societies lies something we can broadly describe as the economy – the way people make a living, create goods and services, and, in the process, create rules that regulate things like property rights, market exchanges, and access to entitlements such as nature, capital, or income.

Today, the global economy is governed by a system that provides a specific, and narrow, definition of economic activity: monetised transactions in the marketplace. As such, it also provides a clear definition of economic activity that informs most standard economic metrics: to increase financial wealth by growing output (Philipsen, 2015/17; Hickel, 2018).

What concerns us here is this particular economy's defining assumption. At its core lies the claim that growth of output[1] is not only good, but that it constitutes a vital precondition for progress and well-being. It is a wide-ranging claim that deserves close attention.

Who are the participants in this economy? They are people who exchange goods, services, and labour in the marketplace – mostly people who work and produce, people who own and manage, and people who consume. Activities outside market transactions are not part of the economy. Parenting, volunteer work, political activism, friendships, and communal services all happen, for the most part, outside of the economy.

Second, our natural environment, the source of all wealth, becomes narrowly conceptualised as "natural capital," a subsystem of the economy. Nature counts only if it can be monetised through the extraction of resources and the production of marketable goods. Depletion counts, by the standards of prevailing economic logic, as a plus. The permanent loss of resources is counted no more than the loss of resilience, diversity, and health of ecosystem services. In this topsy-turvy world, nature as necessary foundation is degraded to convenient appendage.

Both people and nature, in short, are counted improperly. They matter only as contributors to monetised market transactions. Within this economic logic, people turn into self-interested consumers pursuing something called utility maximisation; nature becomes a supply store of resources and a container for waste. As such, the richness and complexity of life disappear behind the logic of the economic model (Polanyi, 1944/2011). Essential choices about broader social values and priorities shrivel to the purely instrumental logic of cost benefit analyses. Hidden behind graphs, charts, and econometric models is a profound ethical choice: the central focus is on "what sells" (revealed consumer preferences), not what is good, or right, or sustainable. The much vaunted hope that market transactions would be value-neutral representations of free and informed choices reveals itself as a delusion (Daly, 1996; Costanza et al., 2014; Felber and Hagelberg, 2017).

Modern economics applies a wide range of short-term metrics to gauge success or failure – stock prices, quarterly returns, net profit margins. As such, it values only the here and now, not the future. An outlook beyond the next couple of years is rare, usually discounted into insignificance. A perspective on the next generation, much less the seventh generation, largely disappears from view.

As a system based on private property, not the commons, and on financialised commodities, not use value, modern economies also undermine democratic governance. First, concentrated wealth and power increasingly define policy. Second, as ever more components of life are commodified, financialised, and marketed, politics itself inevitably begins to function within the logic of the economic paradigm – what matters is what sells. Third, though nation-states initially developed, regulated, and enforced markets, they now find themselves increasingly hamstrung by the imperatives of their own creation – governments in the service of market logic. During the Great Recession, hundreds of billions were spent bailing out the corporations who caused the crash, not the real people drowning in the consequences (Yarrow, 2010; Higgs, 2014).

Within this system, the one overriding purpose of all governments today is economic growth – how to stimulate it, build infrastructure for it, incentivise it, safeguard it, and instruct young people how best to contribute to it. Consistently, social needs and ecological boundaries are sacrificed at the altar of economic growth. The logic is actually quite simple: within global capitalism, government can't function without economic growth (de Vogli, 2013; Schmelzer, 2016, Higgs, this volume). Not surprisingly, dominant political parties from right to left, across the world, may occasionally disagree on how best to facilitate growth, but they all agree on its overwhelming need.

In the aggregate, we have a system of necessary exponential growth of a largely indiscriminate array of goods and services, transacted in the market for money, for the purpose of expanding profit, wealth, and power – a market logic that reveals itself as fundamentally at odds with human and ecological well-being.

## The measure of success – reflection and goal

How is this particular growth measured? Gross Domestic Product (GDP) has been by far the most powerful economic performance measure since the middle of the last century. Economists have called it "one of the greatest inventions of the 20th century" (Landefeld, 2000). GDP emerged as a response to the challenges of the Great Depression. It became the key indicator of a System of National Accounts (SNAs) that, by the end of the 20th century, every single national economy had adopted in order to track its performance and growth (Mitra-Kahn, 2011; Philipsen, 2015/17).

The initial focus on output made sense. During the Great Depression, every new tractor built, every new field ploughed, every new machine fired up created jobs and incomes. During the Second World War, rapid growth not only made sense, it propelled the United States and its allies to victory. Providing vital information, GDP numbers helped organise and streamline production of jeeps, tanks, and guns to defeat fascism. Indeed, the greatest victory of growth-centred market capitalism happened under a centralised command economy.

Today, modern economies still follow the logic of winning a war. Yes, jobs and incomes are as important as ever – perhaps more so, as families find it increasingly difficult to get by on one income. But every new job, every new product also adds to material throughput, carbon emissions, waste, overwork, stress. Homes are filling up with junk, waste disposal sites are overflowing and leaking toxins into soil. Vital resources, meanwhile, are depleting at accelerating rates. At the same time, stress and loneliness are pressing mental health issues (*The Economist*, 2018). "Inequalities and lack of social cohesion," in the end, "are becoming unbearable, ... generating conflicts, migrations, social unrest, and political instability" (Fleurbaey, 2018). We are managing 21st-century problems with a 1930s toolkit. GDP has become what one might call "Grandpa's Definition of Progress."

The greatest strength of GDP continues to lie in its narrow simplicity. By counting only market transactions, it allows economists and policymakers to track volume of economic activity over time and across regions and sectors, and provides information for investment, finance, and development decisions (Marcuss and Kane, 2007). The system is straightforward. It has the advantage of being easily quantifiable – income, output, or expenses, all come with a dollar figure.

Yet the sole focus on aggregate market activity provides no information on quality or purpose of output, much less on consequences or sustainability of economic activity. As a flow measure of economic activity, GDP entirely fails to consider wealth – both human and natural – creating a "clash between economic growth and growth in wellbeing" (Dasgupta, 2014).

Ignoring this design flaw, policymakers nevertheless elevated GDP growth as the key indicator for success. Today, the logic of GDP informs every other key economic metric – income, wealth, employment, investment, standard of living. Despite this unrefined reduction of life to a monetary figure – defining global poverty, for instance, as living with less than $1.90 a day, or evaluating investment by its contribution to GDP – economic growth is now widely accepted, if not as a reliable proxy, then as necessary precondition for an increase in prosperity and welfare (see also Higgs, this volume).

How did we get here? At the end of the Second World War, the promise of prosperity inherent in the logic of GDP growth served multiple purposes. It offered the prospect of more profits to businesses, better incomes to workers and employees, more tax revenues to governments, more funds to academic research – in short, more tools to improve life for nearly everyone. It is precisely this grand promise that led GDP growth, in fairly short order, to become the single most important political goal of governments around the globe.

The creators of GDP understood the dangers of this logic. Simon Kuznets famously warned in 1936 that "the welfare of a nation can scarcely be inferred from a measurement of national income." On the question of what to count as legitimate economic output, he eventually lost the argument (he wanted to exclude expenditures such as the military, commuting, or job training, as "the evils necessary in order to be able to make a living") (Kuznets, 1937).

The shortcomings of a metric that merely tracks market exchanges point to a larger problem. Modern economics as both discipline and lived reality prioritises the quantity of market exchanges over their quality. "Leisure, inequality, mortality, morbidity, crime, and the natural environment" are essentially absent from GDP (Jones & Klenow, 2016). Life-saving immunisations or a new generation of smartphones count the same way as oil spills or cigarettes. As such, improved health or better communication becomes equivalent in its presumed contributions to collective well-being as dead oceans or millions of unhealthy addicts.

Of course, even if one were to make a distinction between goods and bads exchanged in the market, the transaction still provides insufficient to no information about results – the computer is counted, but not the quality of education; the pill, but not health; the security lock, but not safety. Oil spill clean-ups, while adding to GDP, do not add to welfare. My current Apple MacBook costs less than my 1986 Atari ST, yet is of vastly higher quality. GDP accounts for neither welfare nor quality.

The main problem is not what we count, but what we do not count. As many welfare theorists have pointed out, a theory "that can address neither aggregate welfare nor inequality seems of little practical or conceptual use" in representing human well-being (Stanton, 2007). The key economic performance indicator simply ignores all things that do not get exchanged on the market – fresh air, democracy, vital resources, vibrant communities, or healthy ecosystems. It is a world stood on its head: modern societies value what they measure, rather than measure what they value.

Based on the assumption that growth of economic activity provides a social good, the measure turns into goal. As GDP is blind to what makes life possible and good, the costs of GDP growth may long have begun to outpace its benefits (Daly, 1996; Stiglitz et al., 2010; Philipsen, 2015/17).

GDP is not just a useless measure of welfare (as most economists are ready to admit), but also an inaccurate indicator of economic performance.

The Great Recession of 2007/2008 and its aftermath starkly revealed the inadequacy of GDP as a representation of lived reality. The number is up, but America is still down. Ten years after the crash, both stock markets and GDP growth are hitting record territory; yet people experience a real loss in net worth, suffer through a decline in homeownership, face a precarious job market, and watch climate change ravage the ecosystem. All the while, inequality rises as a tiny minority at the top hordes the financial gains of the last decade (Gabler, 2016; Gilens and Page, 2017; Jordan and Sullivan, 2017). According to GDP, all is good.

The logic of GDP effectively defines the parameters of policymaking: standard of living, membership in the exclusive G-7 club of countries, even the very determination of an economy's overall performance (a recession/depression is defined by nothing other than GDP). Core concepts of popular understanding – concepts that are inherently political such as wealth, income, or inequality – are reduced to mere components of GDP.

The consequences are often quite direct. The EU Fiscal Compact of 2012, for instance, requires member states to respect a fixed ratio between GDP and public deficit and sovereign debt, which "have tied the hand of democratically elected governments, overriding elections and referenda, and resulting in the enforcement of austerity policies" (Fioramonti, 2013). Likewise, in the United States, lax financial and environmental regulations, or free trade agreements that do little to protect labour standards and local communities, are sold to the public as effective policies to boost GDP growth. Meanwhile, policymakers routinely portray living wage campaigns or improved environmental regulations as detrimental to economic health. All are politics in the service of GDP growth.

Based on what economists consider a desirable growth rate, the economy would have to double in GDP output roughly every 25 years. If such exponential growth is difficult to imagine, that's because it is absurd (imagine economies with 16 times the output in 100 years,

256 times in just 200 years, or 5,000 times in as little as 300 years). It is true: there is one diagram in economic theory that "is so dangerous that it is never actually drawn: the long-term path of GDP growth" (Raworth, 2017).

Still, business and political leaders alike preach GDP's indefinite exponential growth as the presumed precondition for addressing our mounting environmental and social problems. The solution is to speed up. The systemic need for endless growth is what authors like Richard Smith (2016) consider the central dilemma of modern capitalism: "we have to destroy our children's tomorrow to hang on to our jobs today."

In summary, hidden behind the elaborate apparatus of data, estimates, and formulas lies a simple truth: while GDP may be an adequate representation of market activities, it is a wholly inadequate representation of success, for it cannot capture economic performance, social welfare, or sustainability (Wilkinson and Pickett, 2011; Jackson, 2016).

## The well-being of GDP

When describing GDP's narrow focus, it's important to recall "that the current indicators are not a naturally occurring phenomenon. They are political creations, with the flaws, limitations and choices that politics usually involves" (Leonhardt, 2018). It's not "the economy, stupid," but a stupid economy. We mark escalating concentrations of wealth at the top, but fail to generate gains in well-being for the many, or a sustainable path forward for all.

In the United States, the first major politician to alert the broader public was Robert F. Kennedy. During his presidential campaign he concluded that GDP "measures everything, in short, except that which makes life worthwhile." Though his critique was trenchant, it largely died with him in 1968.[2] During a period of robust (and much coveted) GDP growth under Donald Trump, the issue made a brief resurgence. Senate majority leader Chuck Schumer said: "For too long we've relied on GDP alone as a bellwether for how Americans are faring in the economy when we should be looking as well at income data across the economic spectrum." The reason was simple, as the economist Paul Krugman stressed: "If Jeff Bezos walks into a bar, the average wealth of the bar's patrons suddenly shoots up to several billion dollars—but none of the non-Bezos drinkers have gotten any richer" (Krugman, 2018).

Grave concerns about the validity of GDP, meanwhile, have penetrated the mainstream. China, the European Union, the Organisation for Economic Co-operation and Development (OECD), the UN, the World Bank all have initiated efforts to move "beyond GDP." Perhaps most significantly, the UN succeeded in getting most nations to sign on to a new set of broader performance standards. Called the Sustainable Development Goals (SDGs), the UN initiative covers 17 "global goals" with 169 target measures between them. Hundreds of possible Sustainable Development Indicators (SDIs) are currently debated as possible metrics for the SDGs.

If awareness has shifted, why does GDP continue to be the predominant metric of success? The short answer is that GDP has become the operating system of modern economies. As a recent UN report remarked, countries that have developed alternative metrics have effectively given up on the prospect of "replacing" GDP (UN Global Sustainability Development Report, 2015). Not only is it nearly impossible "to get rid of an indicator as emblematic" as GDP, but "despite its many limitations, GDP retains several powerful features," such as standardisation and global reach, that give it as yet unchallenged power (Chancel, 2015).

As a consequence we have entire economies informed and guided by a metric so flawed that the chief economist of the *World Economic Forum* came to the devastating conclusion that GDP manages to overlook the three key areas of economic growth: "is growth fair, is it green, and is it improving our lives?" (Blanke, 2016). As a European Environmental Bureau petition put it in 2018: the "aggressive pursuit of growth at all costs divides society, creates economic instability, and undermines democracy." The urgency to go beyond GDP would seem present and clear.

## Beyond GDP: well-being

"Beyond GDP" is a broad term. Logically, it encompasses every characteristic not captured by GDP's narrow lens of economic output – from inequality to technological advances to carbon footprint to human well-being.

There are, today, hundreds of initiatives attempting to measure a world beyond GDP. They variously attempt to

- highlight issues completely ignored by GDP – human development and sustainability in general, or, at times, things like ecosystem services or inequality in particular;
- devise broader metrics of human development and sustainable progress.

All have in common that they shift focus beyond marketed commodities and towards a broader understanding of human well-being. Each has to wrestle with the fundamental dilemma between conceptual rigour and immediate public policy relevance.

The market provides information about means, not ends. One way to summarise the history of capitalism, modernity, and Westernisation is to say that it attempted to separate "the economy" from the rest of social life. The magic of the market (the invisible hand) presumably entailed that individual preferences expressed in the market inevitably lead to a social good. The means (market choices, utility maximisation, money, profit), presumably translates into a positive end (consumer satisfaction, social welfare, wealth, progress).

The various fallacies of this assumption are well documented. For one, as Martha Nussbaum and Amartya Sen, among many others, have shown, consumer choices do not equal quality, much less happiness or freedom. While a sufficient amount of food is necessary for basic health, for instance, volume or price say nothing about quality, fairness of distribution, individual differences in needs, choices about its use, or about consequences for people or environment. It says nothing, to use Sen's and Nussbaum's terminology, about how goods and services function in society; how they affect human capabilities to find meaning or stability, to express values, or how to secure future generations.

A concept like utility, in Sen's words, "at best captures part of the good life, but at worst justifies severe deprivation and inequality." Goods and services have value only, so the claim, to the extent to which they improve people's well-being, and provide them the freedom to pursue the good life (Sen, 1999; Nussbaum, 2011).

Increasingly, our lives are governed by numbers. Yet we currently possess no numbers globally recognised by business or policymaking as essential to the one thing that matters most to people: their personal and collective well-being. The same is, by and large, true for the biophysical foundation of life – our natural environment. We seem able to agree on how to count market transactions, but not meaning or happiness or sustainability.

A wide range of scholars now believe that the interrelated challenges of ecosystem destruction, political and social instability, and escalating inequality need to be addressed on at least three planes at the same time: equity (making sure that real opportunities/capabilities exist for everyone); freedom (both political and economic, including the guarantee of extensive human and democratic rights); and a broad definition of sustainability (preserving and protecting healthy and vibrant ecosystems as much as social and political communities) (Fleurbaey, 2018).

Consequently, the focus of most beyond-GDP initiatives is on outcomes, rather than merely on means/flows. This shift de-emphasises economic growth in favour of human well-being and sustainability. It necessitates, in turn, inclusion of qualities that lie beyond the realm of monetised and marketed commodities.

A direct consequence of all "Beyond GDP" initiatives is to make visible the explicitly political choices of all metrics. They move the focus from transactions to outcomes, from numbers to qualities, from dollars to consequences. Such initiatives dive into real challenges of human life – questions about justice and equality as much as purpose, sustainability, and a deeper meaning of progress and development.

Beyond GDP thus faces many challenges. For future generations, it is much like a person whose health has long been evaluated by her consumption of calories. The more calories, the better, everyone tells her. Yet she also realises that all those calories are beginning to make her sick, and certainly don't address her desire for health or meaning. Sure, she knows that some caloric intake is essential (a basic level of GDP output), but she also begins to suspect that adding ever more calories to the diet will eventually kill her (exponential economic growth).

The biggest challenge is likely the question of purpose: what particular goals should a measure encompass? If, as Emanuele Felice argues, GDP actually does a very good job reflecting "the dominant values of our capitalist market economies and their prevailing interests," the question of metrics becomes, at base, a question of power. Who or what does the economic system serve? Who are the primary beneficiaries (Felice, 2016)?

Since the 1970s, well-being in most high-growth economies, as measured by the Genuine Progress Indicator, has actually declined through uneconomic growth – things like fixing social decay, stealing resources from future generations, or simply moving functions previously provided by families and communities into the realm of the monetised economy (Kubiszewski et al., 2013).

Above all, new metrics have to address the world's collective ability to survive. Without food to eat, air to breathe, water to drink, and land to live on, conversations about well-being and progress become moot. In the dry language of natural scientists, "the various components of the environmental footprint of humanity must be reduced to remain within planetary boundaries" (Hoekstra and Wiedmann, 2014).

Our collective train can't avoid the cliff if all of our gauges are pushing us towards it. "Green growth" or "sustainable growth" is tantamount to equipping the train with solar panels, making the ride more comfortable and more equitable, and introducing vegetarian options in business class. Such efforts can be useful, but they do not change the direction of the train. Additional economic growth, if accounted properly, generates more costs than benefits. Speeding up the train will simply bring us closer to the inevitable crash (Schandl et al., 2016; UNEP, 2017).

Given the available information, a disturbing characteristic of this historical moment is the question: why do so many leaders in politics and business do the opposite of what they know to be necessary and right?

"What we measure informs what we do. And if we're measuring the wrong thing, we're going to do the wrong thing" (Stiglitz et al., 2010). Measuring output will continue to be important, but only if it is embedded in a framework of performance indicators that actually measure what people want, and does so within the inevitable limits of strong sustainability.

What are the alternatives to GDP? The chart below illustrates the most prominent "beyond GDP" metrics, and some sample nation rankings. Some strive to replace GDP, others see themselves as supplements. All provide valuable data essential to an informed public debate. None have, thus far, come close to dislodging GDP as the world's central gauge for economic decision-making (Table 24.1).

| Indicator | What it covers | Main challenges | Possible relevance | Ranking |
|---|---|---|---|---|
| Gross Domestic Product (GDP)<br><br>Unit: Monetary | Value of all final goods and services | Too many to list: see Fioramonti, Philipsen, Higgs, Mitra-Kahn | Globally dominant metric of economic performance | Per capita GDP<br>• United States: 7<br>• Costa Rica: 60<br>• Denmark: 9 |
| Inclusive Wealth Index (IWI)<br><br>Unit: Monetary | Attempt to measure all assets upon which human well-being is based (produced, human, and natural) to evaluate well-being. Created in connections with the UN Sustainable Development Goals | Guesswork on "shadow prices," especially on ecosystem services, leading to likely serious under-valuations; also a "weak" sustainability indicator; includes produced, natural, and human capital; claims that 81 out of 140 countries analysed are on a sustainable path | UN effort to replace GDP and HDI; more aligned with SDGs than HDI | By IWI growth rates<br>• United States: 60<br>• Costa Rica: 19<br>• Denmark: 84<br><br>All three have seen declines in natural capital |
| Human Development Indicator (HDI)<br><br>Unit: Index | Combination of three metrics: GDP per person; years of education; and life expectancy | Narrow focus ignores inequality and sustainability – efforts underway to include both in an updated HDI version (a I-S-HDI) | Global coverage, widely consulted, though little impact on economic policymaking | • United States: 13<br>• Costa Rica: 63<br>• Denmark: 11<br><br>Adjusted for inequality:<br>• US: 25<br>• Costa Rica: 64<br>• Denmark: 9 |
| Genuine Progress Indicator (GPI)<br><br>Unit: Monetary | Similar to GDP in covering personal consumption expenditures, but accounts for income inequality, adds volunteer and household work; subtracts environmental and social costs | Inadequately covers ecosystem services and future challenges; like GDP reduces everything to monetised units; possible incommensurability of components; only calculated for very few countries/outdated coverage | Conceptualised to replace GDP; shows how economic growth does not lead to equal progress in well-being | No country rankings available |

(Continued)

| Indicator | What it covers | Main challenges | Possible relevance | Ranking<br>• United States<br>• Costa Rica<br>• Denmark |
|---|---|---|---|---|
| Genuine Savings (or Adjusted Net Savings)<br><br>Unit: Monetary | Combines the net investment in physical and human capital with the depletion of natural capital and pollution | Covers "weak" sustainability only; many issues with discounting of future effects; very limited ecosystem degradation metrics; only updated sporadically | Likely the most established indicator to measure sustainability in monetary terms – possible complement to other common economic indicators. | United States: 79<br>Costa Rica: 76<br>Denmark: 28*<br><br>* Data for Denmark and United States are from 2015; for Costa Rica from 1990 |
| Legatum Prosperity Index<br><br>Unit: Variety of metrics | Based on 104 different variables for 149 nations around the world, including growth, education, health, well-being, and quality of life | A sophisticated prosperity metric that undervalues long-term sustainability and international inequality | Reputational scorecard for nations, unclear direct policymaking relevance | United States: 17<br>Costa Rica: 31<br>Denmark: 5 |
| Ecological Footprint<br><br>Unit: Index | Amount of biologically productive area used by people for their consumption and waste, and compare that to the biologically productive area available within a region or the world | Methodology not entirely clear; exclusively focused on human need; does not address the need for biodiversity | Easy to use, very accessible. Of increasing reputational value; helpful guidepost for policymakers | 188 countries; highest ranking equals worst performance<br><br>United States: 5<br>Costa Rica: 90<br>Denmark: 34<br><br>(based on US consumption, we would need five planets to survive) |
| Happy Planet Index<br><br>Unit: Index | Well-being × life expectancy × inequality of outcomes, all divided by ecological footprint | Relies on imprecise ecological footprint | Clear, easy to use, relatively comprehensive, focus on sustainable well-being | United States: 108<br>Costa Rica: 1<br>Denmark: 32 |
| World Happiness Report<br><br>Unit: Surveys | Personal life evaluations, based on Cantril ladder (ranking from 1 to 10 between best and worst imaginable); recent attempts to attribute to factors such as GDP per capita, social support, life expectancy, freedom of choice, etc. | Predominant emphasis on individual perceptions; not sufficiently accounting for structural inequities | Relatively easy to survey and rank most nations/localities; unclear direct policymaking impact; meets increasing global attention on happiness (similar to Gross National Happiness) | United States: 18<br>Costa Rica: 13<br>Denmark: 3 |

| | | | | |
|---|---|---|---|---|
| Sustainable Society Index<br>Unit: Variety of metrics | Covering most countries in the world on three separate scored categories (human well-being; environmental well-being; economic well-being) comprised of 21 indicators | High human and economic well-being mask vital shortcoming in environmental well-being; very little use by policymakers; graphs not intuitive | More accurate sustainability measure than HDI or GPI – can inform scholarship and outreach | **Human well-being:**<br>• United States: 47<br>• Costa Rica: 75<br>• Denmark: 9<br>**Environmental well-being:**<br>• United States: 140<br>• Costa Rica: 26<br>• Denmark: 97<br>**Economic Well-being:**<br>• United States: 87<br>• Costa Rica: 45<br>• Denmark: 7 |
| OECD Better Life Index<br>Unit: Index | Each country is graded on a scale of 1 to 10 in each of BLI's 11 well-being dimensions: (1) *income and wealth*, (2) *jobs and earnings*, (3) *work-life balance*, (4) *housing*, (5) *environmental quality*, (6) *health status*, (7) *education and skills*, (8) *social connections*, (9) *civic engagement and governance*, (10) *personal security*, and (11) *life satisfaction* | Most comprehensive and by far the most flexible of alternative indicators – each user decides on weight of dimension; outside of OECD recommendations, very little policymaking application; very weak sustainability criteria (present-day pollution) | Further refinements could lead to significant educational relevance; only metric with built-in regional and personal variability | Rankings can easily be obtained based on weights and preferences: www.oecdbetterlifeindex.org/#/55555555555 |
| Environmental Performance Index (EPI) | Global coverage; two overarching issues provide the foundation for EPI: Ecosystem Vitality and Environmental Health, which, in turn, are divided into nine issue categories, each of which are comprised of one or more of 20 different indicators. | Currently the most credible environmental indicator – includes no categories for social or economic well-being (though a good supplement to metrics like HDI or GPI) | Able to inform environmental policies; significant attention in academic circles | • United States: 27<br>• Costa Rica: 30<br>• Denmark: 3 |

It should be noted that those countries considered most successful by traditional standards (the G7, for instance) routinely do not make the top ten list of sustainable progress indicators. On the contrary, only those countries that have begun to make serious efforts linking economic progress to human and ecological well-being, like Costa Rica or Denmark, make the cut.

With all alternative performance indicators, the challenges of emphasis, clarity, and empirical quality are vast. Is health more important than education? Is environmental sustainability a precondition, or just a factor among others? How to compare or merge quantitative with qualitative indicators? Do we assume substitutability – i.e. can technological innovations, say, stand in for natural resources, or are some things essential, and therefore non-substitutable?

The key purpose of alternative indicators – and what makes them preferable to GDP despite their flaws – is the shared focus on human well-being. What outcomes of economic activity are desirable? Should emphasis be on ecological or human prosperity? By what process can the world community decide what to include, how to prioritise, and how to measure? Per capita carbon footprint calculations, for instance, are of vital importance for our collective survival. How to measure them accurately and apportion them fairly, on the other hand, runs up against serious obstacles.

Within the world of growth-centred economics, the Achilles heel of all alternative metrics is how to be relevant without falling prey to the logic of markets, profits, and accumulated wealth. There are, for instance, a total of 17 UN SDGs (such as "end hunger"). Every goal comes with several different targets (like end malnourishment, double agricultural productivity, ensure agricultural sustainability, etc.), each of which, in turn, can be measured in multiple ways. Assuming one could, through democratic decision-making, agree not just on goals and categories, but also on ways to measure outcomes, goals such as "effective action on climate change" (SDG 13) or "significant reduction of inequality" (SDG 10) inevitably challenge deeply entrenched power. Within the current GDP system, oil companies profit from ruining our delicate web of life; the top 1% benefit from tax breaks that deplete the public coffers. Historical experience indicates that neither will give up their extraordinary privileges voluntarily.

Beyond GDP metrics don't just represent technical fixes or marginal improvements of an otherwise well-functioning system. In order to be effective, they question the very logic of a system narrowly and irrationally focused on output and growth, rather than prosperity and well-being of people and planet. They variously do so by

1    providing a coherent vision of performance that lies beyond merely addressing the various blind-spots of GDP;
2    breaking down artificial boundaries between qualitative and quantitative, as well as between so-called objective and subjective measures;
3    offering specific tools and incentives to inform policymaking and business decisions;
4    attempting to provide built-in flexibility to reflect changes over time, and cultural differences across and within societies.

Whether such metrics provide a single indicator or a dashboard, a monetary figure or a performance indicator, likely matters less than that they offer clarity and coherence in their focus on people and planet. On topics ranging from inequality to health to happiness to sustainability, alternative indicators consistently provide insights that go far beyond the simplistic, and increasingly anachronistic, story of GDP growth. "Too much and for too long," said Robert Kennedy over half a century ago, we "have surrendered personal excellence and community values in the mere accumulation of material things" (Kennedy, 1968). The struggle to correct this fatal flaw has just begun. Better metrics provide essential guidance.

The real challenge, in the end, lies in giving them political power in a policy world shackled to GDP growth – and then to break the shackles.

## Notes

1 It is important to note that this growth is, by definition, always exponential, not linear.
2 Within mainstream American politics, the elemental shortcomings of GDP did not appear again until two congressional sub-committee meetings – one in 2001, the next in 2008 (Senate Hearing, 2008). Here, again, GDP was deemed flawed and insufficient. As entrenched and powerful as the underlying rationale of blind growth was, however, the conversation stayed safely contained behind the doors of the committee chambers. GDP continued to reign supreme.

## Bibliography

Ackerman, Frank, and Lisa Heinzerling. 2002. *Pricing the Priceless – Cost-Benefit Analysis of Environmental Protection*. Washington, DC: Georgetown University.

Anielski, Mark, and Jonathan Rowe. 1999. *The Genuine Progress Indicator*. San Francisco: Redefining Progress.

Arrow, Kenneth J., Partha Dasgupta, et al. 2012. "Sustainability and the Measurement of Wealth." *Environment and Development Economics* 17 (3): 317–353.

Blanchet, Didier, and Marc Flaubert. 2013. *Beyond GDP – Measuring Welfare and Assessing Sustainability*. Oxford/New York: Oxford University Press.

Blanke, Jennifer. 2016. "What Is GDP, and How Are We Misusing It?" *World Economic Forum*, April 13.

Boroway, Iris, and Matthias Schmelzer. 2017. *History of the Future of Economic Growth-Historical Roots of Current Debates on Sustainable Degrowth*. London/New York: Routledge.

Büchs, Milena, and Max Koch. 2017. *Postgrowth and Wellbeing: Challenges to Sustainable Welfare*. New York: Palgrave.

Chancel, Lucas. 2015. Brief for GSDR 2015 beyond GDP Indicators: To What End? Global Sustainable Development Report.

Costanza, Robert, Ida Kubiszewski, et al. 2014. *Creating a Sustainable and Desirable Future*. Singapore: World Scientific Publishing.

Coyle, Diane, and Benjamin Mitra-Kahn. 2017. "Making the Future Count." *Indigo Prize Entry*. Accessed May 2018. http://global-perspectives.org.uk/wp-content/uploads/2017/10/making-the-future-count.pdf.

Dale, Garth. 2012a. "The Growth Paradigm: A Critique." *International Socialism* 134: 55–58.

Daly, Herman. 1996. *Beyond Growth – The Economics of Sustainable Development*. Boston: Beacon Press.

Dasgupta, Partha. 2014. "Measuring the Wealth of Nations." Barcelona Graduate School of Economics, Opening Lecture. Accessed August 2018. www.youtube.com/watch?v=D7igTrNPOW4.

De Vogli, Roberto. 2013. *Progress or Collapse – The Crises of Market Greed*. New York: Routledge.

Dietz, Robert, and Daniel O'Neill. 2013. *Enough Is Enough: Building a Sustainable Economy in a World of Finite Resources*. San Francisco: Berrett-Koehler Publishers.

Easterlin, Richard A. 2013. "Happiness, growth, and public policy." Discussion Paper Series, *Forschungsinstitut zur Zukunft der Arbeit, No. 7234*. Bonn: Institute for the Study of Labor (IZA).

Ekins Paul, and Nick Hughes, et al. 2016. *Resource Efficiency: Potential and Economic Implications*. Report of the International Resource Panel, United Nations Environment Program (UNEP), Paris.

Felber, Christian, and Gus Hagelberg. 2017. "Economy for the Common Good: A Workable, Transformative Ethics-Based Alternative." *Next Systems Project*. https://thenextsystem.org/the-economy-for-the-common-good.

Felice, Emanuele. 2016. "The Misty Grail: The Search for a Comprehensive Measure of Development and the Reasons for GDP Primacy." *Development and Change* 47 (5): 967–994.

Fioramonti, Lorenzo. 2013. *Gross Domestic Problem. The Politics Behind the World's Most Powerful Number*. London/New York City: Zed Books.

Fioramonti, Lorenzo. 2017. *The World after GDP*. Cambridge: Polity Press.

Fletcher, Robert, and Crelis Rammelt. 2017. "Decoupling: A Key Fantasy of the Post-2015 Sustainable Development Agenda." *Globalizations* 14 (3): 450–467.

Fleurbaey, Marc. 2015. "Beyond Income and Wealth." Review of Income and Wealth 61 (2): 199–219.

Fleurbaey, Marc. 2018. *A Manifesto for Social Progress – Ideas for a Better Society*. Cambridge: Cambridge University Press.

Gabler, Neil. 2016. "The Secret Shame of Middle-Class Americans." *The Atlantic*, May.

Gilens, Martin, and Benjamin Page. 2017. *Democracy in America? What Has Gone Wrong and What We Can Do About It*. Chicago: University of Chicago Press.

Hickel, Jason. 2018. *The Divide – Global Inequality from Conquest to Free Markets*. New York/London: W.W. Norton &Co.

Higgs, Kerryn. 2014. *Collision Course: Endless Growth on a Finite Planet*. Cambridge: The MIT Press.

Hoekstra, Arjen, and Thomas Wiedmann. 2014. "Humanity's Unsustainable Environmental Footprint." *Science* (June) 344/6188: 1114–1117.

Jackson, Tim. 2016. *Prosperity without Growth: Foundations for the Economy of Tomorrow*. New York: Earthscan.

Jones, Charles I., and Peter J. Klenow. 2016. "Beyond GDP? Welfare across Countries and Time." *The American Economic Review* 106 (9): 2426–2457.

Jordan, Mary, and Kevin Sullivan. 2017. "The New Reality of Old Age in America." *Washington Post*, September 30.

Kahneman, Daniel, and Alan B. Krueger. 2006. "Developments in the Measurement of Subjective Well-Being." *Journal of Economic Perspectives* 20 (1): 3–24.

Kennedy, Robert F. 1968. "Remarks at the University of Kansas." Accessed March 2018. www.jfk library.org/Research/Research-Aids/Ready-Reference/RFK-Speeches/Remarks-of-Robert-F-Kennedy-at-the-University-of-Kansas-March-18-1968.aspx.

Klein, Naomi. 2014. *This Changes Everything – Capitalism vs. the Climate*. London: Penguin.

Krugman, Paul. 2018. "For Whom the Economy Grows." *NYT*, August 30.

Kubiszewski, Ida, Robert Costanza, Carol Franco, Philip Lawn, John Talberth, Tim Jackson, and Camille Aylmer. 2013. "Beyond GDP: Measuring and Achieving Global Genuine Progress," *Ecological Economics*, 93 (September): 57–68.

Kuznets, Simon. 1937. Discussion Section. In M.A. Copeland, "Concepts of National Income," *Studies in Income and Wealth*, I: 37.

Landefeld, J. Steven. 2000. "GDP: One of the Great Inventions of the 20th Century." *Survey of Current Business* 80 (1): 6–9.

Laurent, Eloi. 2017. *Measuring Tomorrow: Accounting for Well-Being, Resilience, and Sustainability in the Twenty-First Century*. Princeton: Princeton University Press.

Layard, Richard. 2005. *Happiness: Lessons from a New Science*. New York: Penguin Books.

Leonhardt, David. 2018. "We're Measuring the Economy All Wrong." *NYT*, September 14.Maddison, Angus. 2007. *Contours of the World Economy, 1–2030 AD. Essays in Macroeconomic History*. Oxford: Oxford University Press.

Marcuss, Rosemary D., and Richard E. Kane. 2007. "U.S. National Income and Product Statistics: Born of the Great Depression and World War II." *Survey of Current Business* 87 (2): 32–46.

Meadows, Donella H., et al. 1972. *The Limits to Growth: A Report for the Club of Rome's Project on the Predicament of Mankind*. Washington: Potomac Associates.

Mitra-Kahn, Benjamin H. 2011. *Redefining the Economy: How the 'Economy' Was Invented 1620*. (Unpublished Doctoral thesis, City University London). Accessed 2015. http://openaccess.city.ac.uk/1276/.

Nordhaus, William and James Tobin. 1972. "Is Growth Obsolete?" In *Economic Growth*, New York: NBER.

Nussbaum, Martha C. 2011. *Creating Capabilities: The Human Development Approach*. Boston: Belknap Press.

OECD Observer. 2004/05. No 246/247.

O'Neill, Daniel W. 2015. "The Proximity of Nations to a Socially Sustainable Steady-State Economy." *Journal of Cleaner Production*, 108, Part A: 1213–1231.

Paech, Niko. 2016. *Liberation from Excess – The Road to a Post-Growth Economy*. Munich: oekom.

Philipsen, Dirk. 2015/17. *The Little Big Number – How GDP Came to Rule the World and What to Do About It*. Princeton: Princeton University Press.

Polanyi, Karl. 1944/2011. *The Great Transformation – The Political and Economic Origins of Our Time*. Boston: Beacon Press.

Raworth, Kate. 2017. *Doughnut Economics: Seven Ways to Think Like a 21st Century Economist*. White River Junction, VT: Chelsea Green Publishing.

Schandl, Heinz, S. Hatfield-Dodds, T. Wiedmann, et al. 2016. "Decoupling Global Environmental Pressure and Economic Growth: Scenarios for Energy Use, Materials Use and Carbon Emissions." *Journal of Cleaner Production* 132: 45–56.

Schmelzer, Matthias. 2016. *The Hegemony of Growth: The OECD and the Making of the Economic Growth Paradigm*. Cambridge: Cambridge University Press.

Sen, Amarty. 1999. *Commodities and Capabilities* (3rd ed). New York: Oxford University Press.

Senate Hearing. 2008. "Rethinking the Gross Domestic Product as a Measurement of National Strength," Senate Hearing 110–1141, Subcommittee on Interstate Commerce, Trace, and Tourism; Senator Daniel K. Inouye, Chairman (12 March).

Smith, Richard. 2016. "Six Theses on Saving the Planet." *The Next Systems Project*. Accessed September 2018. https://thenextsystem.org/six-theses-on-saving-the-planet.

Stanton, Elizabeth A. 2007. "The Human Development Index – A History. *Political Economy Research Institute, University of Massachusetts*, Working Paper 127.

Stiglitz, Joseph E., Amartya Sen, and Jean Paul Fitoussi. 2010. *Mismeasuring our Lives: Why GDP Doesn't Add Up*. New York: The New Press.

Talberth, John, C. Cobb and N. Slattery (2006), "The Genuine Progress Indicator 2006 – A Tool for Sustainable Development," *Redefining Progress*, http://rprogress.org/publications/2007/GPI%20 2006.pdf.

*The Economist*, "Loneliness is a Serious Public Health Problem." 1 September 2018. www.economist.com/international/2018/09/01/loneliness-is-a-serious-public-health-problem

UN Global Sustainability Development Report (GSDR). 2015. "Beyond GDP Indicators: To What End?" Accessed March 2019. https://sustainabledevelopment.un.org/content/documents/5769Beyond%20 GDP%20Indicators%20to%20what%20end_rev.pdf.

Wallis, Stewart. 2016. "Five Measures of Growth That Are Better than GDP." *World Economic Forum*. Accessed May 2018. www.weforum.org/agenda/2016/04/five-measures-of-growth-that-are-better-than-gdp.

Washington, Haydn and Paul Twomey, eds. 2016. *A Future Beyond Growth: Towards a Steady-State Economy*. New York: Routledge.

Wilkinson, Richard and Kate Pickett. 2011. *The Spirit Level: Why More Equal Societies Almost Always Do Better*. London: Penguin Group.

World Commission on Environment and Development. 1987. "Our Common Future." Accessed Mai 2018. www.un-documents.net/our-common-future.pdf.

Yarrow, Andrew L. 2010. *Measuring America: How Economic Growth Came to Define American Greatness in the Late Twentieth Century*. Amherst: University of Massachusetts Press.

# 25

# Beyond a-growth

## Sustainable zero growth

*Steffen Lange*

The debate on the relationship between economic growth and the environment has a long and controversial history (see also Higgs this volume). Three positions can be identified: (1) the view that (the right kind of) economic growth is beneficial for sustainability (green growth) (Ekins, 2000; Fay, 2012; Jacobs, 2016; OECD, n.d.); (2) it should be focused on reducing environmental throughput and whether this goes along with positive, zero, or negative growth is to be seen (a-growth) (Petschow et al., 2018; van den Bergh, 2011; van den Bergh & Kallis, 2012); (3) that the necessary reduction of environmental throughput has to go along with zero growth or a reduction in production and consumption (degrowth) (Alexander, 2012; D'Alisa, Demaria, & Kallis, 2014; Demaria, Schneider, Sekulova, & Martinez-Alier, 2013; Kallis, 2011, 2015;van den Bergh & Kallis, 2012). This chapter relates to the latter two positions and goes beyond the question "can economic growth and sustainability be reconciled?" The chapter instead investigates the question, "how would sustainable economies without growth look like at the macroeconomic level?" It starts by laying out why going beyond the position of a-growth helps to understand how sustainable economies of the future could be constituted (section "From a-growth to sustainable zero growth"). In the following, neoclassical, Keynesian and Marxian theories are used to investigate conditions for sustainable zero growth economies. These three have been chosen due to their prominence in the history of economic thought. However, the analysis could certainly be extended by additional schools of thought. Neoclassical theories focus on aggregate supply and in particular the role of technological change (section "Redirected technical change – based on neoclassical growth theories"). Marxian theories highlight the role of capital accumulation, firm expansion, and incentives connected to ownership (section "Diseconomies of scale and types of firms – based on Marxian economics"). Keynesian approaches connect supply with demand-side aspects of the economy (section "Constant consumption and zero net investments – based on Keynesian approaches"). The insights from these three schools of thought are used to describe central elements of sustainable economies without growth (section "Sustainable zero growth economies").

## From a-growth to sustainable zero growth

Posing this question seems implausible at first sight. Why should we presuppose that sustainable economies cannot grow? Why not rather ask, along the lines of an a-growth position, how the environmental throughput can be reduced to a sustainable level – without assuming that such economies do not grow? However, if it is true that reducing the environmental throughput to a sustainable level goes along with zero or negative economic growth, this has consequences for several economic and societal systems. Therefore, solutions need to be found for how such systems can function without growth (Petschow et al., 2018; Seidl & Zahrnt, 2010a, 2010b, 2012; Strunz & Schindler, 2018).

The discussion between a-growth and degrowth positions typically goes something like this (cf. Petschow et al., 2018; van den Bergh & Kallis, 2012): sustainable economies by definition imply lower environmental throughput. The essential tool to reduce environmental throughput within a market economy is to internalise environmental costs – mostly operationalised by introducing taxes or caps (so far, most agree). However, if such measures were to be implemented to the degree necessary to achieve sustainability (for example, to keep global warming below 1.5 degrees), this would have massive impacts on economic activities (most still agree). The economic effects would be three-fold. First, more clean[1] products would be produced and consumed that have a lower environmental throughput, because they become relatively cheaper compared to dirty[2] products. At the same time, less dirty products are consumed (Brock & Taylor, 2005). Second, clean technologies (which typically have a higher labour intensity[3]) would be used for production across the diversity of products (Lange, 2018). Third – and this is the crucial point here – production would be lower than today (this is where disagreement takes place). The reason is that the increases in labour productivity the world has experienced since the industrial revolution have been based on massive increases in environmental throughput. If less resources would be available and less emissions possible (due to caps), it would in all likelihood not be possible to produce as much as today, let alone to further increase labour productivity (Ayres, 2003; Ayres & Warr, 2010; Common & Stagl, 2005; Kümmel, 2011; Kümmel & Lindenberger, 2014).

Hence, strict environmental regulation implies an end to economic growth. But – and this is crucial – economic growth is said to be necessary for the functioning of important aspects of the economy (Seidl & Zahrnt, 2010a). Economic growth is supposedly indispensable to keep unemployment low (Antal & van den Bergh, 2013), to facilitate profits (Foster, 2011), and even for the stability of the economy as a whole (Binswanger, 2013). Below, it is argued that some of these fears are justified only under certain circumstances. However, the fears themselves already prevent policy-makers and decision-makers at large to implement strict environmental regulation. Some of the fears are very likely to come true. Therefore, even if strict environmental regulation would be implemented, it seems likely that policy-makers reverse course when experiencing economic instabilities.

This is why we need to understand how economies could be stable without growth. If an economy can be stable and generate high social welfare without growth, strict environmental regulation is more likely to be implemented. Therefore, in the following it is summarised how "sustainable economies without growth" could be constituted. Referring to the argument so far, "sustainable" does not only mean environmental sustainability but also economic stability and high social welfare.

# Redirected technical change – based on neoclassical growth theories

One manner to describe an economy on a macroeconomic scale is by looking at the supply side. In fact, the majority of theories on economic growth and environmental throughput take a supply-side perspective. In such theories, economic growth is determined by the development of the supply of production factors and of their productivities. As the latter depend on the speed and direction of technological change, technology plays a crucial part regarding economic growth. Technology is also decisive for environmental throughput, as it determines the emission and resource intensity of production (see, however, Alexander and Rutherford, this volume, for a critical perspective on placing too much hope in technology).

## Determinants of economic growth

By far the most important determinant of economic growth is technological change. It increases labour productivity and therefore allows not only an increase in GDP but also GDP per person. It is to a major degree responsible for increasing environmental productivity, and thereby facilitates an improving (i.e. decreasing) relation between emissions and economic output. Many models of economic growth do not take into account environmental factors but rather focus on capital and labour. Early models, such as the famous Solow-model (Solow, 1956), include technological change as an exogenous factor (a factor that is not explained by the model but taken as a given). In such theories, technological change is the only determinant of labour productivity and – combined with population growth – of economic growth. However, these theories were incapable of explaining what factors are responsible for the speed of economic growth. Also, they could not explain different types of technological change (Barro & Sala-i-Martin, 2004). The first of such shortcomings was addressed by so-called endogenous growth theories. Such theories argue that knowledge and ideas are the drivers behind technological change and ask what determines the growth of such knowledge. The central answer is that firms need the right incentives to invest into developing new techniques (e.g. Aghion, Howitt, & Peñalosa, 1998). The second issue has been the topic of various theories from environmental economics. The question usually asked is how environmental productivity can be increased. In many approaches, this environmental productivity is independent from the question of labour productivity (Brock & Taylor, 2010). Others argue that by increasing the price for resources and emissions, firms get incentives to switch to technologies using more labour and less resources and also invest in the development of such technologies (Acemoglu, 2001; Acemoglu, Aghion, Bursztyn, & Hemous, 2012). Hence, the type of technological change depends on the accessibility and prices of production factors. The abundant access to natural resources and in particular fossil fuels since the industrial revolution has led to a focus on labour-saving technological change. The relation between the costs of using natural resources and labour is decisive. If the usage of natural resources was, relative to labour, more expensive than it is today, technological change would shift towards increasing environmental productivity instead of labour productivity.

## Sustainable economies without growth

There are in principle two ways an economy can be organised without growth from such supply-side perspectives. If technological change continues to increase labour productivity, a reduction in labour supply (at the same rate as the increase in labour productivity) would lead to zero growth. Alternatively, technological change ceases to increase labour productivity. In this case, no change in labour supply is needed for a zero growth scenario (Lange, 2018).

It is insightful to connect this result to the starting point of the discussion. As outlined earlier, the a-growth position argues that strict environmental regulation should be implemented and whether economic growth still takes place is to be seen (van den Bergh, 2011). This point of view makes sense within the neoclassical framework. Strict environmental regulation here leads to the implementation of a different set of technologies and to a redirection of innovations, so that future technologies alter the relations between capital, labour, and natural resources differently than past innovations. Such changes would certainly lead to cleaner production. Whether such developments would still go along with increases in labour productivity is difficult to predict. The a-growth position makes sense within the neoclassical framework because independent of whether the economy grows, stagnates, or shrinks under strict environmental regulation, there are no economic instabilities (this is due to common neoclassical assumptions such as perfect substitutability and market clearing).

Depending on whether economic growth still takes place within such strict environmental regulation, one of two manners to organise economies with zero growth applies. If the environmental regulation leads to technological change that does not increase labour productivity (contrary to technological change in the past that has increased labour productivity), no further changes are needed. If labour productivity still increases, a reduction in labour supply would be necessary for zero growth to take place. (It should be noted, however, that the reduction in labour supply would not be necessary to achieve environmental sustainability in the neoclassical framework.)

Neither of the two manners to organise zero growth leads to economic instabilities, nor do they go along with unemployment or (at least in prominent models) with increasing poverty. Economic instabilities, in the sense of economic crises, do not take place because they cannot occur in neoclassical theories of economic growth or environmental economics, due to the design of the underlying models. Unemployment cannot be a problem because substitutability between production factors is assumed, so that the amount of labour supplied is always employed (Irmen, 2011). Issues of poverty or income inequality do not occur because the models assume a representative household – implying equal distribution of income (Lange, 2018).

In sum, within neoclassical supply-side theories, the key element is to implement strict environmental regulations, in order to alter technological change. This does not lead to economic or social problems. If such regulations still go along with increases in labour productivity, a reduction in labour supply would be needed to achieve zero growth. However, the environmental regulation is sufficient to achieve environmental sustainability – the reduction in labour supply is not needed for that goal. This also explains why many theorists using neoclassical frameworks support an a-growth position.

## Diseconomies of scale and types of firms – based on Marxian economics

Another very different perspective from the supply side is to analyse the dynamics within and between firms and how they are related to economic growth and the environment. Such analyses are usually conducted by authors who refer to Marxian economics.

### Determinants of economic growth

The analysis starts from the question of how firms act and how financial capital is used: why do firms try to maximise profits and why is capital continuously invested into expanding production? There are two major reasons why profits are reinvested and not consumed by its owners – in Marxian terminology, the capitalists. First, capitalists are interested in reinvestment in order to accumulate wealth. Harvey (2010) argues that "[c]apitalists […] are necessarily interested in and therefore motivated by the accumulation of social power in money-form" (p. 257). Or as Marx (1990) puts it: The capitalists' "motivating force is not the acquisition and enjoyment of use-values, but the acquisition and augmentation of exchange-values" (p. 739). It is important to note though that this is not due to the attitude of capitalists, but the function the capitalists take within the capitalist system. And even if capitalists wanted to use their income differently, they are coerced to reinvest due to the following reason.

Second, capitalists are coerced to reinvest due to price competition. They stand in competition with each other and can only sell products when they are able to offer them at the market price. This is done by introducing new, cost-saving technologies. If they do not follow this logic, their firms go bankrupt and other firms take over their market share. Capitalists have both an incentive and an imperative to apply newly available technologies that allow production at lower costs per unit of production. The incentive is that those capitalists who introduce the cost-reducing technologies can earn extra profits, that is, profits above the normal profit rate: "The innovative capitalist gains an extra profit, extra surplus-value, by selling at or close to the social average while producing at a rate of productivity far higher than the social average" (Harvey, 2010, p. 167). These capitalists can sell the products at the prior price, while having lower costs until the other capitalists also introduce the new technologies and the average price falls. The imperative to apply new technologies rests upon the fact that when an increasing share of capitalists introduces the new technologies, the market price falls. The capitalists who do not introduce cost-reducing technologies are not able to offer products at the reduced price and are therefore pushed out of the market (Harvey, 2010).

When many or all capitalists reinvest profits, overall investments are high. In addition, new, more cost-efficient technologies are invented and implemented. These are – from a supply-side perspective – the crucial reasons for economic growth. Investments plus increases in labour productivity lead to economic growth.

The described analysis of the dynamics between private ownership, competition, and capital accumulation also explains the increasing environmental degradation of the capitalist system within Marxian theories. Two mechanisms are central. First, the described dynamics lead to continuous expansion – economic growth – on a macroeconomic level. Increasing levels of output go – ceteris paribus – along with more environmental throughput (Schnaiberg, 1980). Second, the dynamics lead to an incentive for firms to externalise costs to the environment. As firms have to compete, they have to reduce costs. This can be achieved by introducing more efficient technologies; however, it can also be obtained by exploiting the environment where possible (Foster, Clark, & York, 2010).

## Sustainable economies without growth

Various Marxian authors have argued that zero growth is incompatible with capitalism (Blauwhof, 2012; Magdoff & Foster, 2010; Smith, 2010). According to Magdoff and Foster (2010) "No-growth capitalism is an oxymoron […]. Capitalism's basic driving force and its whole reason for existence is the amassing of profits and wealth through the accumulation (savings and investment) process" (p. 8). As I have argued elsewhere (Lange, 2018), this conclusion depends on how capitalism is defined and whether one looks only at economic aspects in a narrow sense or also at the political economy of capitalism.

Regarding economic aspects, two conditions are necessary for a zero growth economy within a Marxian framework: (1) the coercion to invest in order to stay competitive and

(2) the profit-motive of capitalists need to be addressed. Concerning the first issue, an additional analytical point needs to be made: the Marxian argument rests upon the assumption of economies of scale. Only when large-scale production is more efficient (in the sense that it allows firms to supply products at a lower price) than production at smaller scales, firms have to reinvest in order to stay competitive. Therefore, if diseconomies of scale would prevail over economies of scale, the need to expand (on the firm level) would not exist. Such diseconomies can, in particular, be introduced by introducing environmental regulation that would increase the price of trading products and intermediate products globally – which is closely related to the emergence of global companies. In addition, resource-intensive production tends to go along with large-scale production, while production methods with lower resource-intensity are often conducted on smaller scales. Therefore, an increase in the price of environmental throughput would be an essential step towards implementing diseconomies of scale. However, diseconomies of scale can also be introduced beyond the price of environmental throughput – for example by focusing state expenditure on regional production, changing the relative tax-burdens on small and large companies or governmental investments in local rather than global transport systems (cf. Gebauer, Lange, & Posse, 2017). Such changes of the economic framework would dampen or even reverse the coercion to expand on a firm level.

However, capital is not tied to one firm but can be invested anywhere. Even if owners of specific firms do not have the incentive to expand production of such firms, they have an incentive to use profits to invest elsewhere – for the "amassing of profits and wealth" (Magdoff & Foster, 2010, p. 8). The interest to reinvest goes along with the profit-motive as explained earlier. Within high-income countries, such investments beyond the coercion of price competition often go along with the so-called sales effort – firms try to sell products despite satisfied markets by using advertising, inventing new products, or planned obsolescence (Baran & Sweezy, 1966). In order to prevent such motives and strategies to expand consumption and (thereby) also production, firm ownership would need to be collectivised. Firms that are owned by stakeholder-groups, for example employees, follow different logics and in particular follow different goals. Rather than focusing on the interests of individual owners such as firm growth and profit maximisation, they can concentrate on the interests of the stakeholders, for example good working conditions, democratic participation and high wages (cf. Blauwhof, 2012; Gebauer et al., 2017; Lange, 2018).

In sum, the analysis of Marxian theories shows that diseconomies of scale and collective firm ownership would be crucial aspects of non-growing market economies. In addition, Marxian theories also take into account aspects of political economy – including the power-relations within a society. It is argued that political decisions are influenced by powerful societal actors, in particular the representatives of the interest of capital (Sweezy, 1942). This implies that introducing changes such as diseconomies of scale or collective firm ownership are unlikely to be implemented without significant political struggle, as they contradict such interests.

## Constant consumption and zero net investments – based on Keynesian approaches

The supply is only one side of the economic story. Demand is the other. That is why other theories – in particular Keynesian types - emphasise the role of demand. While focusing on it, macroeconomic dynamics are always understood as an interplay between aggregate demand and aggregate supply in those theories.

## *Determinants of economic growth and environmental throughput*

Keynesian analyses often refer to the equation of aggregate demand $(Y_d)$, which is determined by the sum of private consumption $(C)$, government spending $(G)$, and investments $(I)$: $Y_d = C + G + I$. This is an identity and always holds for closed economies. However, it helps to illustrate the interconnections between the components of aggregate demand and how such dynamics lead to economic growth. Keynesian authors also see technological change and capital accumulation as central pieces in explaining economic growth (Kalecki, 1971). The key driver behind firms' investments is the demand for final goods by households and public bodies (Keynes, 2006). When firms experience high demand for their goods, they expand production, i.e. have high investments. Technological change is typically regarded as going along with investments (Hein & Tarassow, 2009). The faster production expands, the more technologies are implemented and therefore new technologies disseminate more rapidly. The demand for final goods depends on two major aspects. First, income inequality influences the amount of private consumption. As people with low income consume a larger share of income than richer households, low-income inequality goes along with a higher consumption share (of income). Second, the government can influence demand by the level of spending it undertakes – financed either by taxation (of richer households) or by borrowing (Hein, 2014).

The combination of consumption spending, investments, and technological change leads to a circular view on macroeconomic dynamics between aggregate demand and aggregate supply (Keynes, 2006). The level of private and government consumption influences the capacity utilisation of firms and is therefore the primary reason for the level of investments. Investments lead to an expansion of production, while at the same time introducing new technologies into the production process. This has opposing effects on the level of employment. More production implies larger employment, while new technologies typically go along with increasing labour productivity and hence less employment per unit of output. Which factor prevails is of major importance for the development of (un)employment. This is the primary determinant of the total wage level. And the total wage level, in turn, is the primary determinant of private consumption demand – as people with low or middle income primarily depend on wage income, rather than capital income.

There are two major arguments regarding the relation between economic growth and environmental throughput. The first is similar as in the supply-side theories. Robinson (1956) covers different choices of technique. Techniques with different sets of production factor proportions are chosen based on the price of capital and labour. Therefore, Robinson allows for different directions of technological change, based on different production factor prices. When, for example, the price of physical capital increases relative to the price of labour, firms are likely to switch to a different type of technique that uses less capital and more labour. As argued in Lange (2018), this approach can be extended to natural resources. The more expensive they relatively are, the more incentives exist for firms to use techniques saving on them. The second argument is a rather new one. Harris (2010, 2013) argues that economic activities can be divided into two categories: (1) those with high negative environmental effects and (2) those with minor negative, or even positive effects on the environment. While the former need to be limited, the latter can grow. The development of the two types of economic activities depends on the demand for them – by the government and by households.

## Sustainable economies without growth

The description of a sustainable economy without growth from such a Keynesian – demand side – point of view is similar to the one from supply-side perspectives (as is argued below in particular regarding working hours and technological change). However, while the description is similar, the causal links and therefore the way such an economy could be initiated are quite different.

In a zero growth economy, aggregate demand $(Y_d)$ has to stay constant over time. This implies that either all components of it $(C, G, I)$ need to stay constant or that while one increases, another one has to decrease. A stable zero growth economy is possible when each stays constant over time. We have seen earlier that investments depend on the demand from private and governmental consumption $(C + G)$. If total consumption stays constant over time, firms have the incentive to neither expand nor contract – investments would solely be used to replace depreciated capital or to modernise the capital stock. That implies that net investments are zero, while gross investments $(I)$ are positive and constant. In sum, aggregate demand would be constant. The other possibility – that one component shrinks while another one grows – is only possible in a certain case. Private consumption and government consumption could be substituted. For example, the government could increase taxation, reducing private consumption but using the additional revenues for government expenditures. However, the taxation would need to be of such kind that the sum of the shrinkage in private consumption and growth in government consumption is zero. On the other hand, it is not possible that private consumption and/or government consumption grows while investments decline (or the other way around). The reason is that investments are highly influenced by total consumption. If for example, private consumption would rise, investments would rise too – and if private consumption declines, so do investments. Therefore, the economy would go into a self-reinforcing feedback loop in case that one component rises continuously.

When technological change takes place and is labour-saving, such a zero growth economy would lead to continuously increasing unemployment: When less labour is needed per unit of production and total production stays constant, a smaller amount of workers is needed. However, this analysis assumes that hours per worker stay constant. If average working hours decrease, the amount of people employed stays constant, while overall hours worked decline (see Larsson and Nässén, this volume). Such working hours reductions would have to be at the same speed as technological change increases labour productivity. In addition, the reductions in working hours would have to go along with a constant real wage, implying rising hourly real wages. A constant real wage is necessary in order to keep consumption demand constant over time.

In sum, a zero growth economy would be initiated by working hours reductions at the speed of increases in labour productivity. These would lead to constant private consumption. In addition, government spending has to stay constant over time. Combined, these two aspects lead to constant demand of final goods so that firms have an incentive to have zero net investments. With consumption, government spending and investments staying constant over time, aggregate demand stays constant as well. Aggregate supply stays constant as well because the capital stock neither increases nor decreases and the labour applied decreases at the speed of labour productivity.

Note that the causal relation is very different to the one in section "Redirected technical change – based on neoclassical growth theories" on neoclassical approaches. There, a reduction in labour supply is the cause and zero growth is the effect. From a demand-side perspective, constant demand is the cause and zero growth the effect. Decreasing working hours are a necessary condition for this relation to stay stable over time. In addition, reductions in average working hours are not initiated due to changes in preferences, as was the case for the supply side. Instead, it is the outcome of a societal bargain, in particular between trade unions and firm unions (with a role of the government that sets the rules of the bargaining game).

Most Keynesian frameworks assume technological change that increases labour productivity. In this case, reductions in average working hours are necessary for stable sustainable economies with zero growth, as has just been depicted. However, as we have seen earlier, two developments could change this situation. First, a change in relative prices of production factors, in particular labour and natural resources, would incentivise firms to invest in resource-saving, rather than labour-saving technologies. This could lead to a slower increase in labour productivity, or even to a reversal of the trend, making reductions in average working hours less or even unnecessary. Second, a shift from dirty towards cleaner products would not only decrease the environmental intensity but could also increase labour intensity (in other words, decrease labour productivity). This would require, however, that the cleaner sectors have not only a lower environmental intensity but also a higher labour intensity (cf. Lange, 2018).

## Sustainable zero growth economies

Applying several theories to the question of sustainable economies without growth makes clear that asking the question for sustainable zero growth economies leads to quite different answers than taking an a-growth position. Environmental regulation is still necessary, but it does not suffice.

In sustainable zero growth economies, "getting the prices right" is still essential – in particular regarding the supply side. Environmental policies such as environmental taxation or limiting the exploitation of natural resources and emissions would make environmental throughput more expensive and the usage of labour (relatively) cheaper. This creates important incentives for households and firms. Households would consume less environmentally harmful products and firms would produce such a different set of products. In addition, firms would have incentives to develop and introduce technologies that focus on increases in environmental productivity. As long as there is a technological trade-off between environmental and labour productivity, the latter will grow more slowly, stagnate, or even decline. So far, the results are very near a-growth positions. In order to obtain a zero growth economy, average working hours need to be reduced, kept constant or increased, depending on how labour productivity develops (assuming a constant population size).

Preventing capital accumulation is the essential element for sustainable zero growth economies from the perspective of Marxian economics. Two aspects are central here. First, by introducing diseconomies of scale, the coercion to expand on a firm level can be avoided. Second, capital accumulation on the macroeconomic level can be countered by collective firm ownership, so that firms' revenues go into the hands of the stakeholders of firms, in particular employees. As such stakeholders have a wider set of interests than solely increasing sales and profits, collective ownership would also dampen the sales effort and its effect of fostering consumption.

From Keynesian perspectives, increases in labour productivity do not only need to be balanced out by decreases in average working hours to keep production constant but also need to keep income at the same level over time. Such a constant wage income enables constant private consumption. Combined with constant government spending, the demand for final goods stays constant as well. This incentivises firms to engage in zero net investments – so that production capacities do neither increase nor decrease.

The insights from the Marxian and Keynesian approaches show that a-growth positions do not suffice, if strict environmental regulations actually lead to very low, zero, or negative economic growth. If solely environmental regulations were implemented and they do lead to zero or negative growth, Marxian and Keynesian theories indicate economic instabilities and conflicts. From a Marxian perspective, the drive for capital accumulation would oppose a tendency for zero growth. Firms would have both an incentive and a pressure to find ways to expand production nonetheless – including strategies to circumvent or even prevent environmental regulation by influencing political decisions. Keynesian approaches show that private consumption, as well as government spending and firms' investments would have to stay constant over time. This has far reaching consequences regarding each of the three underlying economic and social systems. Consumers need to be satisfied with constant consumption and the question of distribution of income needs to take constant overall income into account. Governments need to keep spending constant, rather than increasing it – a major change compared to established conventions within modern governments and their bureaucracies. Firms have to become acquainted to the end of expansion, also implying a major change compared to current habits. The analysis of sustainable zero growth economies hence shows that an a-growth position of simply "getting the prices right" is not enough. Various economic institutions need to be adjusted if strict environmental regulation is to be implemented and future economies have to deal without growth.

## Notes

1 The term "clean" refers to products or technologies with relatively low emission and resource intensities.
2 The term "dirty" refers to products or technologies with relatively high emission and resource intensities.
3 And maybe more physical capital, this is an ongoing debate, see below.

## References

Acemoglu, D. (2001). Factor prices and technical change: From induced innovations to recent debates. *MIT Department of Economics Working Paper*, 01–39. doi: 10.2139/ssrn.290826

Acemoglu, D., Aghion, P., Bursztyn, L., & Hemous, D. (2012). The environment and directed technical change. *American Economic Review*, 102(1), 131–166. doi: 10.1257/aer.102.1.131

Aghion, P., Howitt, P., & Peñalosa, C. (1998). *Endogenous growth theory*. Cambridge: The MIT Press. Retrieved from http://books.google.com/books?hl=de&lr=&id=tLuqjIVJUcoC&oi=fnd&pg=PA1&dq=aghion+howitt+1998&ots=mvLZ_pQl2U&sig=9G2FR8cKzGkpdhg2gQpd-KlX5AVc

Alexander, S. (2012). Planned economic contraction: The emerging case for degrowth. *Environmental Politics*, 21(3), 349–368. doi:10.1080/09644016.2012.671569

Antal, M., & van den Bergh, J. C. (2013). Macroeconomics, financial crisis and the environment: Strategies for a sustainability transition. *Environmental Innovation and Societal Transitions*. Retrieved from http://www.sciencedirect.com/science/article/pii/S2210422413000038

Ayres, R. U. (2003). Exergy, power and work in the US economy, 1900–1998. *Energy*, 28(3), 219–273. doi:10.1016/S0360-5442(02)00089-0

Ayres, R. U., & Warr, B. (2010). *The economic growth engine: How energy and work drive material prosperity*. Cheltenham: Edward Elgar. Retrieved from https://books.google.com/books?hl=de&id=nLfJKVK9uJsC&oi=fnd&pg=PR1&dq=the+economic+growth+engine&ots=85zDbZbU-6P&sig=qnllEKRj_N1j0wNHJCyY55aUv6M

Baran, P. A., & Sweezy, P. M. (1966). *Monopoly capital: An essay on the American economic and social order.* New York: Monthly Review Press. Retrieved from http://philpapers.org/rec/BARMCA-10

Barro, R., & Sala-i-Martin, X. (2004). *Economic Growth* (2nd ed.). Cambridge: The MIT Press.

Binswanger, H. (2013). *The growth spiral: Money, energy, and imagination in the dynamics of the market process.* Heidelberg: Springer.

Blauwhof, F. B. (2012). Overcoming accumulation: Is a capitalist steady-state economy possible? *Ecological Economics, 84,* 254–261. doi:10.1016/j.ecolecon.2012.03.012

Brock, W. A., & Taylor, M. S. (2005). Economic growth and the environment: A review of theory and empirics. In *Handbook of Economic Growth* (Vol. 1, pp. 1749–1821). Amsterdam: Elsevier. Retrieved from http://www.ucalgary.ca/uofc/Others/iaprfiles/technicalpapers/iapr-tp-041007.pdf; http://www.sciencedirect.com/science/article/pii/S1574068405010282

Brock, W. A., & Taylor, M. S. (2010). The Green Solow model. *Journal of Economic Growth, 15*(2), 127–153. doi:10.1007/s10887-010-9051-0

Common, M., & Stagl, S. (2005). *Ecological economics: An introduction.* Cambridge University Press. Retrieved from https://books.google.com/books?hl=de&lr=&id=RYktw_SLIrQC&oi=fnd&pg=PT23&dq=stagl+ecological+economics&ots=SBz8wsUKF_&sig=FAuWryOyGFVrACPDcWqXYQB2bQ8

D'Alisa, G., Demaria, F., & Kallis, G. (Eds.). (2014). *Degrowth: A vocabulary for a new era.* Abington: Routledge. Retrieved from http://books.google.com/books?hl=de&lr=&id=ARxWBQAAQBAJ&oi=fnd&pg=PT35&dq=vocabulary+degrowth&ots=o2P2O_XcIy&sig=4A2EybhpKNMOiSq48jeh8VBOHAU

Demaria, F., Schneider, F., Sekulova, F., & Martinez-Alier, J. (2013). What is degrowth? From an activist slogan to a social movement. *Environmental Values, 22*(2), 191–215. doi:10.3197/0963271 13X13581561725194

Ekins, P. (2000). *Economic growth and environmental sustainability: The prospects for green growth.* London: Routledge.

Fay, M. (2012). *Inclusive green growth: The pathway to sustainable development.* Washington, DC: World Bank Publications.

Foster, J. B. (2011). Capitalism and degrowth – an impossibility theorem. *Monthly Review, 62*(8), 26–33.

Foster, J. B., Clark, B., & York, R. (2010). *The ecological rift: Capitalism's war on the earth.* New York: Monthly Review Press. Retrieved from http://books.google.com/books?hl=de&lr=&id=g5VECr8TjLYC&oi=fnd&pg=PP2&dq=The+Ecological+Rift:+Capitalism's+War+on+the+Earth&ots=gQZ1iUpnu7&sig=U61VtqYDfrx_slJiIVEBhw-ZT9Y

Gebauer, J., Lange, S., & Posse, D. (2017). Wirtschaftspolitik für Postwachstum auf Unternehmensebene. Drei Ansätze zur Gestaltung. In F. Adler & U. Schachtschneider (Eds.), *Postwachstumspolitiken. Wege zur wachstumsunabhängigen Gesellschaft.* München: oekom.

Harris, J. M. (2010). The macroeconomics of development without throughput growth. *Global Development and Environment Institute, Working Paper, 10–05.* Retrieved from http://books.google.com/books?hl=de&lr=&id=ClDyT8JroeMC&oi=fnd&pg=PA31&dq=the+macroeconomics+of+development+without+throughput&ots=QmxDcNxj4M&sig=DQohGpSHm0XpBPwtbTNH8eyJDxE

Harris, J. M. (2013). The macroeconomics of development without throughput growth. In M. Cohen, H. Brown, & P. Vergragt (Eds.), *Innovations in Sustainable Consumption. New Economics, Socio-technical Transitions and Social Practices* (pp. 31–47). Cheltenham: Edward Elgar. Retrieved from https://books.google.com/books?hl=de&lr=&id=ClDyT8JroeMC&oi=fnd&pg=PA31&ots=QmyFcLzk7J&sig=MXILoBtd1uO-9_Zrt9EGcYCjUPM

Harvey, D. (2010). *A companion to Marx's capital.* London: Verso. Retrieved from https://books.google.com/books?hl=de&lr=&id=u5N-Rrlz8FcC&oi=fnd&pg=PR7&dq=reading+marx+capital+harvey&ots=mSGHWeGoMS&sig=eLaU8IGVKoIv1eq9Y_i9NvsDxZs

Hein, E. (2014). *Distribution and Growth after Keynes: A Post-Keynesian Guide.* Cheltenham: Edward Elgar. Retrieved from http://books.google.de/books?hl=de&lr=&id=8oO1BAAAQBAJ&oi=fnd&pg=PR1&dq=Distribution+and+Growth+after+Keynes&ots=8KOKg4_859&sig=atXvAyiYvcqDYazlGXGf9zhRp94

Hein, E., & Tarassow, A. (2009). Distribution, aggregate demand and productivity growth: theory and empirical results for six OECD countries based on a post-Kaleckian model. *Cambridge Journal of Economics.* Retrieved from http://cje.oxfordjournals.org/content/early/2009/11/10/cje.bep066.short

Irmen, A. (2011). Ist Wirtschaftswachstum systemimmanent? *Department of Economics, University of Heidelberg, Discussion Paper Series, 509*. Retrieved from http://archiv.ub.uni-heidelberg.de/volltextserver/id/eprint/11538

Jacobs, M. (2016). Green growth. In R. Falkner (Ed.), *The Handbook of Global Climate and Environment Policy* (pp. 197–214). Chichester: John Wiley & Sons.

Kalecki, M. (1971). Selected essays on the dynamics of the capitalist economy 1933–1970. Retrieved from http://www.getcited.org/pub/101905393

Kallis, G. (2011). In defence of degrowth. *Ecological Economics, 70*(5), 873–880. doi:10.1016/j.ecolecon.2010.12.007

Kallis, G. (2015). The degrowth alternative. *A Great Transition Initiative Viewpoint*. Retrieved from http://www.greattransition.org/publication/the-degrowth-alternative.

Keynes, J. M. (2006). *General theory of employment, interest and money*. New York (original work published 1936). Retrieved from https://www.marxists.org/reference/subject/economics/keynes/general-theory/: Harcourt, Brace and Company. Retrieved from http://books.google.com/books?hl=de&lr=&id=xpw-96rynOcC&oi=fnd&pg=PR5&dq=general+theory+employment&ots=WWohBrlHFK&sig=26NW2HgijhEWA-oSM2GPKJhiNI

Kümmel, R. (2011). *The second law of economics: Energy, entropy, and the origins of wealth*. New York: Springer.

Kümmel, R., & Lindenberger, D. (2014). How energy conversion drives economic growth far from the equilibrium of neoclassical economics. *New Journal of Physics, 16*(12), 125008.

Lange, S. (2018). *Macroeconomics without Growth: Sustainable Economies in Neoclassical, Keynesian and Marxian Theories*. Marburg: Metropolis.

Magdoff, F., & Foster, J. B. (2010). What every environmentalist needs to know about capitalism. *Monthly Review, 61*(10), 1–30.

Marx, K. (1990). *Capital: Volume I*. London: Penguin Books (original work published 1867). Retrieved from http://158.69.150.236:1080/jspui/handle/961944/108713

OECD. (n.d.). *Towards Green growth*. Paris: Organisation for Economic Co-operation and Development.

Petschow, U., Lange, S., Hofmann, D., Pissarskoi, E., aus dem Moore, N., Korfhage, T., … Ott, H. (2018). *Gesellschaftliches Wohlergehen innerhalb planetarer Grenzen: Der Ansatz einer vorsorgeorientierten Postwachstumsposition* (No. UBA Texte 89/2018) (p. 194). Dessau-Roßlau: Umweltbundesamt. Retrieved from https://www.umweltbundesamt.de/publikationen/vorsorgeorientierte-postwachstumsposition

Robinson, J. (1956). *The accumulation of capital*. Basingstoke: Palgrave Macmillan.

Schnaiberg, A. (1980). *The environment: From surplus to scarcity*. New York: Oxford University Press. Retrieved from http://bases.bireme.br/cgi-bin/wxislind.exe/iah/online/?IsisScript=iah/iah.xis&src=google&base=REPIDISCA&lang=p&nextAction=lnk&exprSearch=145719&indexSearch=ID

Seidl, I., & Zahrnt, A. (2010a). Argumente für einen Abschied vom Paradigma des Wirtschaftswachstums. In I. Seidl & A. Zahrnt (Eds.), *Postwachstumsgesellschaft: Konzepte für die Zukunft* (pp. 22–36). Marburg: Metropolis. Retrieved from https://scholar.google.com/scholar?q=Konsum%3A+Der+Kern+des+Wachstumsmotors&btnG=&hl=de&as_sdt=0%2C5#0

Seidl, I., & Zahrnt, A. (Eds.). (2010b). *Postwachstumsgesellschaft: Konzepte für die Zukunft*. Marburg: Metropolis.

Seidl, I., & Zahrnt, A. (2012). Abhängigkeit vom Wirtschaftswachstum als Hindernis für eine Politik innerhalb der limits to growth. *GAIA, 21*(2), 108–115.

Smith, R. (2010). Beyond growth or beyond capitalism. *Real World Economics Review, 53*, 28–36.

Solow, R. M. (1956). A contribution to the theory of economic growth. *The Quarterly Journal of Economics, 70*(1), 65–94.

Strunz, S., & Schindler, H. (2018). Identifying barriers toward a post-growth economy–A political economy view. *Ecological Economics, 153*, 68–77.

Sweezy, P. M. (1942). *The theory of capitalist development: Principles of Marxian political economy*. London: Dennis Dobson Ltd. Retrieved from http://scholar.google.com/scholar?q=The+Theory+of+Capitalist+Development#0 http://philpapers.org/rec/SWETTO-2

van den Bergh, J. C. (2011). Environment versus growth – a criticism of 'degrowth' and a plea for 'a-growth.' *Ecological Economics, 70*(5), 881–890. doi:10.1016/j.ecolecon.2010.09.035

van den Bergh, J. C. J. M., & Kallis, G. (2012). Growth, a-growth or degrowth to stay within planetary boundaries? *Journal of Economic Issues, 46*(4), 909–920. doi:10.2753/JEI0021-3624460404

# 26

# Work-time reduction for sustainable lifestyles

*Jörgen Larsson, Jonas Nässén, and Erik Lundberg*

## Introduction

The relationship between lifestyle and ecological impact is strongly tied to consumption volumes. Conscious consumption choices within all lifestyle domains could make a difference of course, but studies involving larger samples of households have consistently found that disposable income and/or total consumption volume are the strongest predictors of both energy use and emissions (Lenzen et al. 2006; Roca and Serrano 2007; Kerkhof, Nonhebel, and Moll 2009; Nässén 2014; Nässén and Larsson 2015; Nässén et al. 2015). With an unchanged disposable income, reduced expenditure in one consumption category can always cause rebound effects through increased expenditure in other categories (Alfredsson 2002; Nässén and Holmberg 2009). Work-time reduction represents an important complementary strategy because it also incorporates the dimension of consumption volume in the opportunities to create more sustainable lifestyles.

Work-time reduction for reasons of sustainability has often been associated with more radical ideas such as degrowth (Kallis 2011; Victor 2012), but it is important to consider that last century, most countries already achieved substantial reductions in working hours without any major shift in the economic paradigm (Huberman and Minns 2007). Work-time reduction could also play an important role for sustainable development as part of a slower transformation where growth in private consumption is decelerated in favour of more leisure time (Lange, this volume). The aim of this chapter is to critically analyse the potentials and limits of work-time reduction as a form of sustainability governance.

Figure 26.1 presents historical data on working hours in European countries. We can see that the annual hours worked per employee continued to fall throughout the period albeit at a slower and slower rate (Figure 26.1). In recent decades, however, this reduction has been achieved primarily through increasing employment rates and job sharing, since the annual hours worked per capita in the total populations in these countries have remained almost unchanged. Hence, the substantial improvements in labour productivity (GDP per hour worked) have almost exclusively been channelled into increasing private and public consumption. A relative change in priorities where a larger share of these productivity improvements is used to reduce working hours would slow down economic growth, which

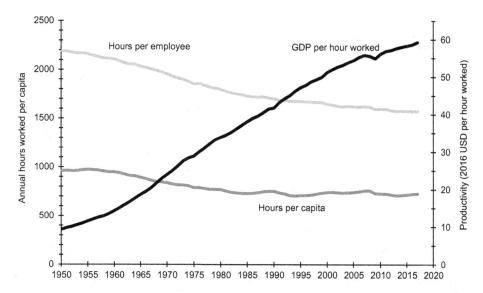

*Figure 26.1* Hours worked and labour productivity in 16 European countries 1950–2017. Countries were selected on the basis of the availability of complete data for the period: Austria, Belgium, Denmark, Finland, France, Germany, Greece, Ireland, Italy, Netherlands, Norway, Portugal, Spain, Sweden, Switzerland, and United Kingdom. Source: Total Economy Database (TCB 2017)

could potentially have benefits for the environment as well as for quality of life through more leisure time. Sections "Work-time reduction and the environment" and "Work-time reduction and quality of life" provide a review of the scientific evidence for such benefits. Section "Policy options for work-time reduction" describes different policy options for work-time reduction.

## Work-time reduction and the environment

A crucial question is what effect the number of working hours has on the environment. Our literature review has identified six quantitative studies on the relationship between working hours and environmental impact. The results are described in what are termed elasticities, which are estimates of how strong the link is between environmental indicators and annual hours worked per employee. For example, an elasticity of 0.8 can be interpreted as a 1 per cent reduction in working hours being associated with a 0.8 per cent reduction in environmental impact.

These studies have used a variety of indicators for environmental impact including ecological footprint, $CO_2$ emissions, and energy use. The studies are listed in Table 26.1 and are categorised as to whether they apply a territorial or a consumption-based perspective on emissions. Studies in the category "territorial perspective on emissions" have used data on emissions within country borders such as the official United Nations Framework Convention on Climate Change (UNFCCC) accounting systems for greenhouse gas (GHG) emissions. Studies in the category "consumption-based perspective on emissions" include data on emissions from imports and exclude data on emissions related to exports.

In our study (Nässén and Larsson 2015), we used micro-data on how time and money are used by individual households. The other five studies performed macro-level analyses based

*Table 26.1* Estimates of the elasticities of environmental indicators with respect to annual hours worked per employee

| | Indicator | Elasticity | Data set and method used |
|---|---|---|---|
| *Territorial perspective on emissions* | | | |
| (Rosnick and Weisbrot 2007) | Energy use | 1.33 | Cross-country comparison of 48 countries; 2003 |
| (Knight, Rosa, and Schor 2013) | $CO_2$ emissions | 0.50 | Panel of 29 OECD countries; 1970–2007 |
| (Fitzgerald, Jorgenson, and Clark 2015) | Energy use | 0.40 | Panel of 29 developed countries; 1990–2008 |
| (Shao and Rodríguez-Labajos 2016) | $CO_2$ emissions | 0.12 | Panel of 37 developed countries; 1980–2010 |
| *Consumption-based perspective on emissions* | | | |
| (Hayden and Shandra 2009) | Ecological footprint | 1.19 | Cross-country comparison of 45 countries; 2000 |
| (Knight, Rosa, and Schor 2013) | Ecological footprint | 1.37 | Panel of 29 OECD countries; 1970–2007 |
| | Carbon footprint | 1.30 | |
| (Nässén and Larsson 2015) | GHG emissions | 0.80 | Micro-data from Swedish Time Use Survey (N = 636) and Household Budget |
| | Energy use | 0.74 | Survey (N = 1,492); 2006 |

on averages of working hours per employee and environmental indicators for different countries. All the studies controlled for the number of employed in relation to the population as well as for economic output per hour of work (labour productivity). All of the macro-level studies utilised data sets from different versions of the Total Economy Database (TCB 2017), but the data sets differ as to whether they are panels (where individual countries are followed over time) or simply cross-country comparisons. They also differ in terms of the countries and time periods analysed, and there are some differences in the model specifications and econometric methods used.

These studies reveal a positive and statistically significant relationship between annual hours worked per employee and the environmental indicators studied. The average for the nine estimates mentioned earlier is an elasticity of 0.86, which can be interpreted as a 1 per cent reduction in working hours being associated with a 0.86 per cent reduction in environmental impact. However, the estimated elasticities differ substantially. In general, studies that used a consumption-based method to measure environmental indicators identified stronger relationships than studies that departed from the more traditionally applied territorial methods. This is no surprise, since we also know from other research that while GDP and $CO_2$ trends are partly decoupled when measured from a territorial perspective, the correlation is much stronger for consumption-based methods (Hertwich and Peters 2009; Davis and Caldeira 2010). The exception is the high estimate for territorial energy use found in Rosnick and Weisbrot (2007). A reason could be that this is a cross–country comparison with far fewer observations than in the three panel studies by Knight, Rosa, and Schor (2013), Fitzgerald, Jorgenson, and Clark (2015), and Shao and Rodriguez-Labajos (2016).

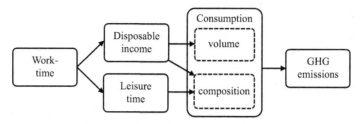

*Figure 26.2*  Links between work-time, consumption, and greenhouse gas emissions

The cross-country comparisons in Table 26.1 were all made with data sets from primarily developed countries, but two of the studies also attempted to make estimates for developing countries (Fitzgerald, Jorgenson, and Clark 2015; Shao and Rodríguez-Labajos 2016). None of these found any statistically significant relationships between work-time and environmental indicators. Possibly, this could be due to lower quality data or the fact that a smaller proportion of the population is employed in the formal labour market.

Our study (Nässén and Larsson 2015) offers a different empirical approach by studying time use and consumption patterns at the micro-level. A reduction in work-time is assumed to impact the GHG emissions of a household via two different pathways: (1) a reduction in disposable income that results in lower consumption volumes and a change in the relative composition of consumption (the income effect); and (2) an increase in leisure time that may also affect the composition of household consumption (the time effect). These pathways are illustrated in Figure 26.2.

The elasticity of GHG emissions with respect to work-time is estimated to be 0.8 in our study, meaning that a reduction in working hours of 1per cent could be expected to correlate with a reduction in GHG emissions of 0.8 per cent. It can be noted that this result puts our study somewhere in the middle of the five macro-level studies presented in Table 26.1. Our result is clearly dominated by the income effect of 0.82. The time effect lowers this estimate by −0.02 to a total elasticity of 0.8. Hence, leisure time becomes only slightly more GHG-intensive (kg $CO_2$-eq/cap/hour) with reduced work-time.

These results indicate that the reduction in consumption volume is the key mechanism by which reduced working hours could affect emissions and that the change in time use is of secondary importance. This is also corroborated by Melo et al. (2018) who did not find any statistically significant correlations between work-life balance and a number of pro-environmental behaviours related to home energy use, personal transport, recycling, and shopping. However, changes in time use could become important if they were to be shown to affect car travel in a substantial way, since this time use category stands out as having very high energy use per unit of time (Figure 26.3). This could also mean that the time effect is sensitive to how the reduction in working hours is organised, for example if people reduce the number of their working hours per day or the number of their work days per week (King and van den Bergh 2017).

A drawback of our micro-level approach is that it cannot capture the effects on the labour market. A reduction in the standard working week from 40 to 30 hours could also reduce unemployment through job sharing. This is one of the aspects reviewed in relation to quality of life in the next section. Reduced unemployment is clearly a positive outcome in itself, but it would simultaneously erode some of the environmental benefits of a working-hours reduction since consumption volume typically increases when a person gets a job. Another

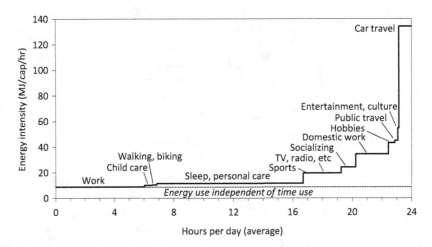

*Figure 26.3* The energy use of different activities on an average day. Car travel has the highest energy intensity of the activities analysed. Air travel has an even higher energy intensity but was not included because it is not generally an everyday activity. Energy use below the broken line is considered to be independent of how time is used (e.g. space heating). This is the level of private energy use while at work (Nässén and Larsson 2015)

uncertainty in relation to the environmental benefits of working hours reduction is that it may drive improved productivity at workplaces (increased output per hour). If this were to be the case, then reduced work-time may not result in a proportional reduction in production and consumption.

Despite these uncertainties, our summation of the previous research is that work-time reduction has relatively great potential for reducing environmental impact. However, major reductions in working hours are not likely to come about solely as a result of environmental concerns. Historically, reducing working hours has been primarily a matter of quality of life, which is the topic of the next section.

## Work-time reduction and quality of life

Reduction of work-time has wide-ranging and complex effects on quality of life due to its mirror-image effect on leisure time and its effects on economic factors. Table 26.2 is an attempt to summarise such potential effects at both the individual and the collective level.

For most types of jobs, a reduction of work-time would also be associated with a reduction in income or at least a slower growth in income over the long term. To what extent would this imply a negative outcome for well-being? Clearly, other types of determinants such as peoples' intentional activities and their social relationships are more important for well-being than material life circumstances (Lyubomirsky, Sheldon, and Schkade 2005). In developed economies, income has been found to explain as little as 2–3 per cent of the variance in subjective well-being between individuals (Ahuvia 2002). Kahneman and Deaton (2010) performed the most extensive cross-sectional analysis of this matter to date and found that the relationship depends on how well-being is measured. Life evaluation, where respondents rate their current life on a scale from "the worst possible life for you" to "the best possible life for you," was found to increase with the logarithm of income. Emotional well-being, assessed

*Table 26.2* Comparison of positive and negative effects from work-time reduction. The individual level refers to effects solely for those who reduce work-time, whereas the collective level refers to effects for the community at large

| | Effects on quality of life | |
| | Positive | Negative |
| --- | --- | --- |
| Individual | More time for leisure activities<br>More time for social relationships<br>Reduced or slower growth of income and consumption | Reduced time pressure and stress |
| Collective | Reduced unemployment through job sharing<br>Deceleration of economic growth makes it easier to reach common environmental targets | Reduced tax revenues for financing welfare services |

by the presence of various emotions (enjoyment, happiness, anger, sadness, stress, worry), also rose with the logarithm of income, but no further progress was found above a threshold level. Studies looking at the effects on well-being of actual changes in income level point towards no long-term effects of rising income since people tend to habituate to such changes, whereas reduced income can indeed have effects (Argyle 1999, 2013). Our interpretation of this literature is that a deceleration of people's growth in income is likely to have only minor effects on well-being whereas a sudden reduction could be negative, and particularly so if it is due to an involuntary work-time reduction.

The upside of reduced working hours of course is that it frees up time for activities that have been shown to be important for well-being including leisure activities and social relationships. It has been shown that people who experience the highest satisfaction with their time use as well as those with the lowest rates of experienced time pressure also work less than other people (Larsson and Björk 2017). Kasser and Sheldon (2009) also found that people who are experiencing so-called time affluence, i.e. the opposite of a life lived under time pressure, are less anxious and worry less about the past and the future than other people, have closer social relationships, and spend more time on physical activities. A meta-analysis by Albertsen et al. (2008) revealed that 26 out of the 30 studies reviewed showed that a higher number of working hours were associated with less balance between work and private life. Health research has primarily targeted individuals who work overtime, since this is known to be a risk factor. Bannai and Tamakoshi (2014) performed a meta-analysis of 19 previous studies of the association between long working hours defined as working more than 40 hours per week and several health outcomes. They concluded that long working hours are associated with depressive states, anxiety, sleep disorders, and coronary heart disease.

The most positive aspect of reducing working hours at the collective or societal level is probably the opportunities for job sharing, since unemployment has been found to be an important negative determinant for well-being (Helliwell 2003; Lucas et al. 2004; Pittau, Zelli, and Gelman 2010). The degree to which working hours reduction could lead to job sharing however is very uncertain and probably varies substantially between different sectors and types of jobs. A research overview by Bosch (2000) of the consequences of reduced working hours for employment showed that the estimated job sharing effects ranged from no effect or even negative effects up to 70 per cent of the theoretically possible effect. A limiting factor is that the unemployed often might not have the required skills and companies might then choose not to employ more people but rather reorganise their production in relation to the new circumstances (The Swedish Ministry of Finance 2015).

The biggest challenge in implementing working hours reduction probably lies in the funding of welfare services. With fewer hours worked, there would also be reductions in tax revenues. For example, the Long-Term Survey of the Swedish Economy prepared by officials at the Ministry of Finance highlights the challenge of a demographically conditioned increase in the demand for publicly funded welfare services and makes the assessment that the greatest potential for maintaining and developing the welfare state lies in increasing the number of hours worked in the economy (The Swedish Ministry of Finance 2015). Hence, the demands from the welfare system point in the opposite direction of work-time reduction. A large-scale implementation of work-time reduction would also need to take into account the impacts for the welfare system.

In the next section, we move on to how work-time reduction may be implemented in practice, including some results from a large-scale case study.

## Policy options for work-time reduction

There are many different ways to set up, or organise, a work-time reduction: the temporal level (day, week, etc.), implementation level (collective or individual) and "cost carrier" (employer, employee, etc.) (Spiegelaere and Piasna 2017). The way the reduction is organised influences what the effects will be on employment, well-being, gender equality, and public finances as well as environmental consequences.

### Temporal level

One aspect is whether the work-time reduction is per day, per week, per year, or for the whole of working life. Eight hours per day and 40 hours per week is the norm in most developed countries, but some countries have shortened this, for example Norway where the normal work day since the 1980s is 7.5 hours and France which reduced weekly working hours to 35 in 2002 (Hayden 2006). Reductions per year are often done through increasing the length of paid annual leave, ranging from zero days of statutory paid annual leave in the United States to 25 days in the Scandinavian countries, for example. Finally, there can be reductions in working hours over the whole of working life, including examples such as a lengthy period of paid parental leave (e.g. 16 months in Sweden), a low retirement age (e.g. in Russia, where retirement age is 60 for men and 55 for women), or possibilities to take a "career break," which in Belgium means that employees can take a full year off and receive a small but not insignificant benefit from the state (Pullinger 2014).

### Implementation level

Another aspect is at which level the work-time reduction is implemented, e.g. at the national level as in France where the statutory working week has been 35 hours for all companies since 2002 (Hayden 2006), but where in practice this has been partly eroded and working hours have increased again (Askenazy 2013). Work-time reductions can also be implemented at the sectoral level as a result of unions prioritising work-hour reductions, e.g. the German Metal Workers' Union IG Metall, which, after a seven-week long strike, reached an agreement on a 35-hour working week (Bispinck 2006). Another possible implementation level is within a specific company, with potential objectives such as avoiding lay-offs or to facilitate extended operating hours, e.g. to 12 hours per day with two 6-hour shifts. One example is an Austrian collective agreement that allows employees to choose between a wage increase of about three per cent or additional leisure time of around five hours per month (Gerold and Nocker 2018). Another example is the right to work part-time in the Netherlands, where all employees have the legal right to reduce their working hours if the individual employer is unable to show that it would have unreasonable consequences (Visser 2002).

## Cost carrier

A third fundamental aspect of how work-time reductions can be organised has to do with who will carry the downside, the "cost," of the work-time reduction. If the work-time reductions do not affect wage levels, then the cost is carried by the employer through higher hourly wage costs. For a private company, this could be expected to reduce profits. The main alternative is that the employee pays for the work-time reduction, such as when opting to work part-time with a cut in pay. The third potential cost carrier is the government, the tax-payers, who can be affected by lower tax revenues leading to a lower social security budget. Costs will be lower of course if the reform leads to higher productivity, reduced sick leave, or reduced unemployment benefits. In some work-time reduction schemes, the costs are carried by a combination of the above. For example, the French 35-hour week reform was financed through wage moderation (lower wage increases and, for some, wage freezes and decreased real wages) and increased productivity, as well as carried by the government through giving employers rebates on social insurance payments (Spiegelaere and Piasna 2017).

The options for how these three aspects of how working hours reductions are organised can be combined in many ways, all of which will have different consequences. In the next section, we describe and analyse one case of a reduction in working hours in detail. The case study is based on a right to work part-time that was formally implemented for all 50,000 employees of the City of Gothenburg in Sweden in 2015. This case has been chosen because it has not previously been described in the academic literature and because we have unique access to the data on this reform.

## Case study: right to work part-time

Recently, the City of Gothenburg changed its working hours policy to give its employees the right to request to reduce their working hours down to 50 per cent of full-time. The background to this reform was that the Swedish municipal sector has a long tradition of part-time work. However, its heavy use of part-time contracts had been criticised because it resulted in the City's employees, who are mainly women, having lower income and retirement income levels. The City of Gothenburg therefore decided to offer all its employees full-time contracts, but with good prospects of continuing to work part-time. In 2015, the right to request to work part-time was mandated for all employees in the City of Gothenburg. The policy stipulates that line managers are not allowed to decline a request to work part-time without first trying to find a solution with the aid of the central human resources (HR) department (Göteborg Stad 2015). In addition, all Swedish workers have the legal right to work part-time when studying and when they have children under eight years of age.

The right to work part-time is being studied in an ongoing research project led by Jörgen Larsson. The data presented here are based on a survey which was conducted in 2016 among part-time employees of the City of Gothenburg. Since the City has 50,000 employees, the sample frame was identified with the aid of the City's central HR department and consisted of those who had full-time contracts but had opted to work part-time. Employees who had worked part-time for more than 10 years were excluded from the sample frame, as well as those who had exercised their legal right to work part-time in order to study and those who were simultaneously receiving parental leave benefits. This resulted in 3,331 individuals and our analysis was based on employment data records from the HR department for these employees and survey responses. 994 respondents completed the web-based survey, which gives a response rate of 30 per cent.

The consequences of this reform regarding the effects on quality of life and gender equality, employment and public finances, as well as environmental impacts are briefly described here. The aim is to give a broad picture of the effects, but space precludes any deeper analysis.

## Effects on quality of life and gender equality

The survey included a question on the motives for choosing to work part-time. The two most common motives were that "full-time is too demanding, physically or mentally." This illustrates that for the majority in this sample, working part-time is a coping strategy for dealing with an unsatisfactory life situation. For a minority in the sample, working part-time is used as a means of managing their lives in a way that allows them more time for what really matters to them. The survey included questions on both positive and negative changes compared to when they worked full-time. Table 26.3 shows the proportions of respondents who agree with specific statements.

Table 26.3 shows that most of the respondents found that their health improved and that they had more time and energy. The negative side is mainly that some experienced financial worries or that their work had become more stressful. The survey also included questions on how their time use patterns had changed after reducing their working hours (Table 26.4).

Table 26.3 Positive and negative changes after starting to work part-time. Proportions who agree with the statements, including those who respond 4 or 5 on a 5-point Likert scale ranging from "do not agree at all" (1) to "agree completely" (5) (N = 966–982)

|  | Percentage that agree |
| --- | --- |
| *Positive changes* | |
| My health is better now compared to when I worked full-time. | 68 |
| It is now easier to find time to do everything I want, or need, to do in my private life. | 68 |
| I am now more well-rested when I come to work. | 57 |
| My reduced work-time makes me organise my work and prioritise tasks in a better way. | 40 |
| I now have more energy left for working effectively even at the end of the working day. | 53 |
| *Negative changes* | |
| I now feel obliged to work harder and take fewer breaks or to work involuntary overtime to get my tasks done. | 19 |
| I now feel more stress during working hours. | 18 |
| I now have a harder time making ends meet. | 33 |
| I am now more worried that my future retirement income will be very low. | 45 |

*Table 26.4* Changes in time use after reducing working hours, per cent (*N* = 825–931)

| Time use categories | Much more time | Slightly more time | Unchanged | Less time |
|---|---|---|---|---|
| Family | 42 | 41 | 17 | 0 |
| Friends | 10 | 34 | 53 | 3 |
| Exercise | 14 | 40 | 42 | 4 |
| Household chores | 8 | 40 | 49 | 3 |
| TV/computer | 1 | 11 | 80 | 8 |
| Shopping | 2 | 8 | 79 | 11 |
| Sleep/recovery | 15 | 45 | 36 | 4 |
| Social engagement (e.g. youth leader, volunteering, politics) | 3 | 8 | 82 | 7 |
| Other hobbies | 11 | 32 | 53 | 4 |
| Going for short holiday trips | 6 | 26 | 64 | 4 |
| Culture (e.g. reading, theatre, movies, museums) | 7 | 32 | 56 | 5 |

Previous research has shown that some of these time use categories correlate clearly with well-being such as time spent with family and friends and for exercise (Killingsworth and Gilbert 2010). The increased time use in these categories indicates that these altered time use patterns may have improved overall well-being.

From a gender perspective, there were both pros and cons of this reform. To date, part-time work is more widespread among women and this widens the gender gap in terms of wages and salaries, career development, and retirement income. If one looks at this kind of reform in isolation, then it could widen the gender gap in the future since, with existing gendered norms relating to work and care, it may be expected that more women than men would reduce their working hours. If the right to work part-time is combined with a right to work full-time, then this reform has a good chance of being positive with respect to gender equality.

## Effects on employment and public finances

The group of employees with full-time contracts who opted to work part-time constituted about 7 per cent of the employees and they work on average 79 per cent of full-time. Their reduction in working hours is the equivalent of over 700 full-time employees. If the City of Gothenburg had only staff working full-time, which is common in many organisations, then the City would have 700 fewer employees. If the share of those working part-time were to double (from 7 to 14 per cent) then another 700 employees could be hired (without raising taxes). This should be considered in light of the 12,000 people unemployed in Gothenburg in 2016 (Regionfakta 2018). There is a matching problem here, however, since there is a shortage of labour in some occupations such as teaching. If many teachers were to reduce their hours, it would make these problems even worse.

Part-time work is a form of working hours reduction where the employee "pays" the cost through lower income. There are some extra employer costs for administration and for hiring new staff, but these are likely to be marginal in relation to the positive effects on public finances. More individuals employed part-time means that fewer individuals need unemployment benefits. Other potential benefits include improved job satisfaction and health, which can result in lower sick-leave costs. In addition, many of the respondents claimed that working part-time has increased their productivity in several different ways: they feel better rested when they come to work, they now organise their work more efficiently, they have more energy at the end of the day, and they feel obliged to work harder. There is also an ongoing interview study being conducted with managers, where a clear majority of them are positive to the right to work part-time, mainly because they find that what is good for the employees' health and job satisfaction is also good for their job performance.

## Effects on environmental impact

Since part-time work is a form of working hours reduction where the reduced hours directly transform into lower income, the effect on carbon footprint can be anticipated to be in line with the results described in section "Work-time reduction and quality of life." On average, part-time workers at the City of Gothenburg work 20 per cent less than full-time workers, which roughly translates into 20 per cent less GHG emissions from private consumption. However, since it can be anticipated that working part-time leads to fewer people unemployed, as described earlier, it also means that these individuals increase their consumption. We have not quantified this effect in detail, but if we assume that an individual who goes from receiving unemployment benefits to receiving a salary doubles their income and consumption level, then the initial 20 per cent reduction in carbon footprint would instead lead to a 10 per cent reduction at the community level. This illustrates that the community benefits of work-time reduction are divided between lower unemployment and lower carbon footprint.

Working part-time also enables more conscious choices for downshifting lifestyles. Downshifting can be defined as a long-term strategy to improve the quality of one's life, involving significantly less income or consumption in relation to one's potential (Larsson 2011). As shown in Table 26.4, part-timers use more time for activities that correlate clearly with well-being such as time spent with family and friends and for exercise (Killingsworth and Gilbert 2010). In the motives for choosing to work part-time, 52 per cent state that one reason is have "more own time, e.g. for exercise, hobbies, friends and recovery." Examples of what this can mean in practice can be found in the comments responses "to give scope to my artistic side," "to avoid the feeling of being in the rat-race," "to have the freedom to control my own time and life," "after cancer treatment, I no longer take for granted that I will live until retirement, I want to have time for exercise, etc."

In the literature, a downshifting lifestyle is often seen as a strategy for living more sustainably. Aspects of this can be found in the responses to the questions on motives for choosing to work part-time: "To live more sustainably, e.g. to repair clothing and other belongings, to consume less" and "To care for the environment and try to avoid increasing consumption."

With a right to work part-time, a downshifting lifestyle is facilitated rather than hindered, which can be the case if there is a strict full-time norm in the community. While a collective working hours reduction stipulates that everyone is to reduce their working hours, a part-time working hours reform instead has the effect of weakening the existing working hours structures. A right to work part-time gives better formal options to choose the number of working hours that fits one's current life situation and to prioritise between time and money more freely. The informal structures such as a full-time norm, a consumption culture, and gender ideals however will continue to limit the individual's options. The combined effect of these types of reforms will depend on the extent to which such downshifting lifestyles spread as the full-time norm weakens.

## Discussion and future research

As we have shown, work-time reduction can have positive outcomes for the individual's well-being and carbon footprint for example. However, on the contrary, politicians in many countries are introducing reforms that aim to increase working hours such as through raising the retirement age. The background to this, at least in social democratic welfare regimes (Esping-Andersen 1996), is often a need for more resources for welfare services: schools, health care, etc., and for the ageing population (The Swedish Ministry of Finance 2015). With more hours worked, incomes can rise and thus also tax revenues. To understand these issues better, more research is needed on the conflict between working hours reduction and the financing of welfare services, and how this conflict can be managed. What role might tax bases that are not work-related play, for example property taxes? Could a green tax reform involving high taxes on $CO_2$ emissions be combined with a general reduction in working hours to achieve a double environmental impact?

Another interesting area for future research is how work-time reduction can be linked to other ideas of sustainable consumption patterns, for example sharing of products. One problem with sharing and other ways of decreasing consumption in specific areas is that the money saved can lead to increased expenditure on other types of consumption. This rebound effect could be avoided if sharing activities are combined with reduced income as a result of a reduction in working hours. Seen from the opposite side, sharing could also be an enabler of a working hours reduction, since it can be a way to maintain a high material living standard despite a lower income.

To conclude, work-time reduction could provide a significant contribution to sustainability transitions such as the transformation towards long-term climate targets. It is important to realise, however, that such changes would be far from adequate and would need to be complemented with other measures. Scenario analyses for 2050 have shown that even a rather radical post-materialism scenario, where the standard working week is reduced from 40 to 30 hours, and where the sharing of services in consumption is increased substantially, could reduce GHG emissions by about a third compared to business as usual (Nässén 2015). Hence, any strategy aimed at achieving long-term climate targets would also require massive improvements in technical eco-efficiency. But in a similar way, it can also be shown that scenarios based on technical solutions alone are quite unrealistic (Alexander, this volume). The key question is probably under what conditions a reduction in working hours could be implemented at a high enough pace to play an important role. In this chapter, we have identified financing welfare services as the most important barrier, which would thus need more thought and research. Clearly work-time reduction would need to be viewed as a long-term strategy where improvements in productivity are gradually channelled into more leisure time rather than increased consumption. The right to work part-time could be an important first step in such a strategy and have the additional benefit of destabilising the norm of a 40-hour working week.

## Acknowledgements

Financial support from Formas (grant no. 254-2013-128) and the Mistra Sustainable Consumption program (2016/3) is gratefully acknowledged.

## References

Ahuvia, Aaron C. 2002. "Individualism/collectivism and cultures of happiness: A theoretical conjecture on the relationship between consumption, culture and subjective well-being at the national level." *Journal of Happiness Studies* 3 (1): 23–36.

Albertsen, Karen, Guðbjörg Linda Rafnsdóttir, Asbjörn Grimsmo, Kristin Tómasson, and Kaisa Kauppinen. 2008. "Workhours and worklife balance." *Scandinavian Journal of Work, Environment & Health* 34 (5): 14.

Alfredsson, Eva. 2002. *Green consumption, energy use and carbon dioxide emission.* Edited by cs. Umeå: GERUM Kulturgeografi.

Argyle, Michael. 1999. "Causes and correlates of happiness." In *Well-being: The foundations of hedonic psychology*, edited by Daniel Kahneman, Ed Diener and Norbert Schwarz, xii, 593 p. New York: Russell Sage Foundation.

Argyle, Michael. 2013. *The psychology of happiness.* London: Routledge.

Askenazy, Philippe. 2013. "Working time regulation in France from 1996 to 2012." *Cambridge Journal of Economics* 37 (2): 323–347.

Bannai, Akira, and Akiko Tamakoshi. 2014. "The association between long working hours and health: A systematic review of epidemiological evidence." *Scandinavian Journal of Work, Environment & Health* 40: 5–18.

Bispinck, Reinhard. 2006. "Germany: Working time and its negotiation." In M. Keune and B. Galgóczi (eds.) *Collective bargaining on working time: Recent European experiences* (pp. 111–129). Brussels: ETUI.

Bosch, Gerhard. 2000. "Working time reductions, employment consequences and lessons from Europe." In *Working time: International trends, theory and policy perspectives*, edited by Lonnie Golden and Deborah Figart. London: Routledge.

Davis, Steven J., and Ken Caldeira. 2010. "Consumption-based accounting of $CO_2$ emissions." *Proceedings of the National Academy of Sciences* 107 (12): 5687–5692.

Esping-Andersen, Gøsta. 1996. *Welfare states in transition: National adaptations in global economies.* London: Sage.

Fitzgerald, Jared B., Andrew K. Jorgenson, and Brett Clark. 2015. "Energy consumption and working hours: A longitudinal study of developed and developing nations, 1990–2008." *Environmental Sociology* 1 (3): 213–223.

Gerold, Stefanie, and Matthias Nocker. 2018. "More leisure or higher pay? A mixed-methods study on reducing working time in Austria." *Ecological Economics* 143: 27–36.

Göteborg Stad. 2015. Reviderat pm angående frågor om anställning, rörlighet och ledighet. Göteborg Stad, Stadsledningskontoret, Verksamhetsområde HR.

Hayden, Anders. 2006. "France's 35-hour week: Attack on business? Win-win reform? Or betrayal of disadvantaged workers?" *Politics & Society* 34 (4): 503–542.

Hayden, Anders, and John M. Shandra. 2009. "Hours of work and the ecological footprint of nations: An exploratory analysis." *Local Environment* 14 (6): 575–600. doi: 10.1080/13549830902904185.

Helliwell, John F. 2003. "How's life? Combining individual and national variables to explain subjective well-being." *Economic Modelling* 20 (2): 331–360.

Hertwich, Edgar G., and Glen P. Peters. 2009. "Carbon footprint of nations: A global, trade-linked analysis." *Environmental Science & Technology* 43 (16): 6414–6420. doi: 10.1021/es803496a.

Huberman, Michael, and Chris Minns. 2007. "The times they are not changin': Days and hours of work in Old and New Worlds, 1870–2000." *Explorations in Economic History* 44 (4): 538–567.

Kahneman, Daniel, and Angus Deaton. 2010. "High income improves evaluation of life but not emotional well-being." *Proceedings of the National Academy of Sciences* 107 (38): 16489–16493.

Kallis, Giorgos. 2011. "In defence of degrowth." *Ecological Economics* 70 (5): 873–880.

Kasser, Tim, and Kennon M. Sheldon. 2009. "Time affluence as a path toward personal happiness and ethical business practice: Empirical evidence from four studies." *Journal of Business Ethics* 84 (2): 243–255.

Kerkhof, Annemarie C., Sanderine Nonhebel, and Henri C. Moll. 2009. "Relating the environmental impact of consumption to household expenditures: An input-output analysis." *Ecological Economics* 68 (4): 1160–1170.

Killingsworth, Matthew A., and Daniel T. Gilbert. 2010. "A wandering mind is an unhappy mind." *Science* 330 (6606): 932.

King, Lewis C., and Jeroen C. J. M. van den Bergh. 2017. "Worktime reduction as a solution to climate change: Five scenarios compared for the UK." *Ecological Economics* 132: 124–134.

Knight, Kyle W., Eugene A. Rosa, and Juliet B. Schor. 2013. "Could working less reduce pressures on the environment? A cross-national panel analysis of OECD countries, 1970–2007." *Global Environmental Change* 23 (4): 691–700.

Larsson, Jörgen. 2011. *Encyclopedia of consumer culture.* Edited by Dale Southerton, 1991–1993. London: Sage.

Larsson, Jörgen, and Sofia Björk. 2017. "Swedish fathers choosing part-time work." *Community, Work & Family* 20 (2): 142–161. doi: 10.1080/13668803.2015.1089839.

Lenzen, Manfred, Mette Wier, Claude Cohen, Hitoshi Hayami, Shonali Pachauri, and Roberto Schaeffer. 2006. "A comparative multivariate analysis of household energy requirements in Australia, Brazil, Denmark, India and Japan." *Energy* 31 (2–3): 181–207.

Lucas, Richard E., Andrew E. Clark, Yannis Georgellis, and Ed Diener. 2004. "Unemployment alters the set point for life satisfaction." *Psychological science* 15 (1): 8–13.

Lyubomirsky, Sonja, Kennon M. Sheldon, and David Schkade. 2005. "Pursuing happiness: The architecture of sustainable change." *Review of General Psychology* 9 (2): 111.

Melo, Patricia C., Jiaqi Ge, Tony Craig, Mark J. Brewer, and Ines Thronicker. 2018. "Does work-life balance affect pro-environmental behaviour? Evidence for the UK using longitudinal microdata." *Ecological Economics* 145: 170–181.

Nässén, Jonas. 2014. "Determinants of greenhouse gas emissions from Swedish private consumption: Time-series and cross-sectional analyses." *Energy* 66: 98–106.

Nässén, Jonas. 2015. "Konsumtionens övergripande utveckling." In *Hållbara konsumtionsmönster. Analyser av maten, flyget och den totala konsumtionens klimatpåverkan idag och 2050. En forskarantologi,* edited by Jörgen Larsson. Stockholm.

Nässén, Jonas, David Andersson, Jörgen Larsson, and John Holmberg. 2015. "Explaining the variation in greenhouse gas emissions between households: Socio-economic, motivational and physical factors." *Journal of Industrial Ecology* 15 (3): 480–489.

Nässén, Jonas, and John Holmberg. 2009. "Quantifying the rebound effects of energy efficiency improvements and energy conserving behaviour in Sweden." *Energy Efficiency* 2 (3): 221–231. doi: 10.1007/s12053-009-9046-x.

Nässén, Jonas, and Jörgen Larsson. 2015. "Would shorter working time reduce greenhouse gas emissions? An analysis of time use and consumption in Swedish households" *Environment and Planning C: Government and Policy* 33: 726–745.

Pittau, M. Grazia, Roberto Zelli, and Andrew Gelman. 2010. "Economic disparities and life satisfaction in European regions." *Social Indicators Research* 96 (2): 339–361.

Pullinger, Martin. 2014. "Working time reduction policy in a sustainable economy: Criteria and options for its design." *Ecological Economics* 103: 11–19.

Regionfakta. 2018. http://www.regionfakta.com/Vastra-Gotalands-lan/Arbete/Oppet-arbetslosa/. Accessed March 1, 2018.

Roca, J., and M. Serrano. 2007. "Income growth and atmospheric pollution in Spain: An input-output approach." *Ecological Economics* 63 (1): 230–242.

Rosnick, David, and Mark Weisbrot. 2007. "Are shorter work hours good for the environment? A comparison of U.S. and european energy consumption." *International Journal of Health Services* 37 (3): 405–417.

Shao, Qing-long, and Beatriz Rodríguez-Labajos. 2016. "Does decreasing working time reduce environmental pressures? New evidence based on dynamic panel approach." *Journal of Cleaner Production* 125: 227–235.

Spiegelaere, Stan De, and Agnieszka Piasna. 2017. The why and how of working time reduction.

TCB. 2017. Output, labor, and labor productivity, 1950–2017. The Conference Board Total Economy Database.

The Swedish Ministry of Finance. 2015. Långtidsutredningen. Stockholm: Regeringskansliet. https://www.regeringen.se/contentassets/86d73b72a97345feb2a8cbc8b6700fa7/sou-2015104-langtidsutredningen-2015-huvudbetankande.

Victor, Peter A. 2012. "Growth, degrowth and climate change: A scenario analysis." *Ecological Economics* 84: 206–212.

Visser, Jelle. 2002. "The first part-time ecoomy in the world: A model to be followed?" *Journal of European Social Policy* 12 (1): 23–42.

# 27

# Decarbonisation

*Richard Lane*

## Introduction

"Decarboni[s]ation of the world's economy would bring colossal disruption of the status quo." This is how a recent editorial in *Nature* (2018) began its description of the scale of the task facing us in order to address climate change. Of course, if there's one thing more disruptive than decarbonisation, it's our as yet ongoing failure to decarbonise. 2018 saw $CO_2$ emissions from industry and fossil fuels grow 2.7% – the largest increase in seven years (Le Quere et al., 2018) – accompanied by the increasingly familiar stable of apocalyptic harbingers: historic low arctic ice levels accompanied by methane release, ocean warming and coral bleaching, collapsing global insect populations, food system breakdown, and record breaking wildfires from Queensland to the arctic circle, California to the outskirts of Athens. So this we know for certain, whether we want it to or not, decarbonisation as a key goal of global sustainability governance will be transformational in either its successes or its failures.

Decarbonisation is largely uncontested as the primary goal and institutional paradigm of current climate change governance and is derived, seemingly unproblematically, from the internationally recognised and agreed requirement to maintain global temperature increases to 2 °C and preferably 1.5 °C. In this sense, decarbonisation has simply become the de facto shorthand for global climate change policy. It is deployed within global sustainability governance as a term referring to both a goal and a broad series of processes, innovations, and regulations aimed at achieving this goal. As a goal, decarbonisation is defined by the Intergovernmental Panel on Climate Change (IPCC) in its Fifth Assessment Report (AR5) as "denot[ing] the declining average carbon intensity of primary energy over time" (IPCC, 2014). In terms of the Paris goal of holding the increase in global average temperature to well below 2 °C above pre-industrial levels and pursuing efforts to limit the temperature increase to 1.5 °C, decarbonisation is also used synonymously with the requirement to reduce global greenhouse gas (GHG) emissions to (net) zero somewhere between 2050 and 2100 (Åhman et al., 2017; European Commission, 2018; IPCC, 2014; Wesseling et al., 2017).

However, as the *Nature* editorial quoted in the first line put it: "it's clear that many policy-makers who argue that emissions must be curbed, and fast, don't seem to appreciate the scale of what's required" (2018). The programmes espoused avoid the certainty of the transformational nature of decarbonisation and its necessary restructuring of the relationships between the global economy and a sustainable global environment. From the United Nations (UN) 2030 Sustainable Development Goals (SDGs); the United Nations Framework Convention on Climate Change's (UNFCCC) Paris Climate agreement; the slew of calls for green growth and green or circular economies from the Organisation for Economic Co-operation and Development (OECD) and the EU as part of its 2050 Climate Action Plan; to the G20's renewed commitment to a raised ambition in order to address climate change at the 2018 leaders' summit in Buenos Aires, all are explicitly or implicitly focused on the minimisation of transformative disruption to the global economy (e.g. Baron, 2016).

This chapter argues that an important part of the failure to appreciate the necessary scale and speed of change to global economic systems, and their subsequent thorough restructuring, has paradoxically arisen due to the development, stabilisation, and resultant impacts of decarbonisation as it is currently understood. In section "The incoherence of decarbonisation," the current incoherence of decarbonisation is outlined, highlighting the problems of absolute decoupling of the economy from GHG emissions and the reliance on Negative Emissions Technologies (NETs) within decarbonisation scenario development and governance. Section "The exclusions of decarbonisation" shows how these incoherencies both result from, and are maintained by, a series of exclusionary processes: the exclusion of (certain) people; the exclusion of nature and particularly climate change itself; and the exclusion of systemic change. These have institutionally stabilised and propagated the current understanding of decarbonisation as a process that is neither inherently transformative nor even desirable as the primary goal of climate change governance. The chapter concludes by indicating how decarbonisation can be reconstructed as a transformatory locus of climate governance that is inclusive and coherent.

## The incoherence of decarbonisation

Coherence in global governance can be understood as the development and promotion of policies that are mutually reinforcing, reflect legitimate goals, and "is a function of how rules, policies, and arrangements across dimensions of global governance are coordinated" (Bernstein, 2017: 218). Coherence here involves both institutional and ideational components. For the former, this requires organisations to work synergistically via institutional coordination and monitoring. For the latter, the goals and purposes of institutions should reflect a common, acceptable, and legitimate framework (Bernstein, 2017: 218). Decarbonisation has developed as a goal of sustainability governance within both the institutional and ideational frameworks of a hegemonic global capitalism. Coherent decarbonisation, as it is currently understood, requires the systematic promotion of policies enabling both the mitigation of $CO_2$ emissions *and* economic growth.

In this section, I outline the broader problem of the decoupling of economic growth from material throughput and the more specific reliance upon NETS within decarbonisation scenario development and governance. Together, these render decarbonisation – currently understood – as a radically incoherent global sustainability paradigm.

### The problem of decoupling

The problem of decoupling is usefully introduced by reference to Johan Rockström et al.'s (2017) treatment of renewables, as part of their recent roadmap for rapid decarbonisation.

Extrapolating from 2005 to 2015 global trends, they argue that by keeping the historical doubling times of renewable energy share of primary energy constant at around 5.5 years, this would yield full decarbonisation in the entire energy sector by around 2050. Three immediate problems arise here. First is the issue of taking as your baseline a ten-year period which includes the largest financial crisis for nearly a century. Second is the simple assumption of the historically unprecedented constant doubling of renewable energy's share of total energy production. Third is the troubling failure to consider that a trajectory of increasing renewable share of primary energy tells us nothing about absolute energy production and therefore absolute decarbonisation trends over this period.

Looking at a longer historical period, the IPCC notes that there has been a secular trend towards decarbonisation as a reduction in intensity of primary energy production since 1850. However, the global rate of decarbonisation has averaged only around 0.3% per annum – approximately six times too low to offset the global average increase in energy use of around 2% per annum. At the same time, emissions from land use change related to traditional biomass utilisation have also far exceeded carbon releases from energy-related biomass burning (IPCC, 2014).

While there has been a significant slowing of decarbonisation trends since the energy crises of the 1970s, in terms of global emissions intensity per unit of economic output, current levels are around 33% lower than in 1970 (Janković & Bowman, 2014: 248) – driven by increasing energy efficiency. However, since the early 2000s, increasing global coal usage has further slowed reductions in carbon intensity of primary energy production (IPCC, 2014), and reductions in overall emissions intensity in the last decade have been driven largely by economic and manufacturing collapse in the wake of the 2007–2008 financial crisis (Janković & Bowman, 2014: 248).

Key here is a distinction between relative and absolute decoupling. Relative decoupling involves a reduction in material-use intensity, while absolute decoupling involves a reduction in the rate of resource use irrespective of the growth rate of economies (UNEP, 2011). There is historical evidence of modest absolute decoupling of economic growth from energy requirements and material throughputs within some countries (Knight & Schor, 2014; UNEP, 2017; von Weizsäcker et al., 2014). However, these analyses do not account for the impacts of burden shifting (the externalisation of pollution intensive activities) and emissions leakage involved in globally diffuse production chains (for a more detailed analysis of the problem of decoupling see Alexander & Rutherford, this volume). When taking into account indirect and consumption-based emissions, there is currently no evidence of absolute decoupling of economic growth and resource use on a global scale either historically or potentially (Knight & Schor, 2014; Schandl et al., 2018; Ward et al., 2016).

Simply put, "in a system based on perpetual economic growth relative decoupling does not translate into absolute decoupling" (Janković & Bowman, 2014: 248). Neither, therefore can the EU's 2015 Circular Economy Action Plan be easily squared with the reality of increasing material flows under conditions of economic growth (Haas, Krausmann, Wiedenhofer, & Heinz, 2015). As Tim Jackson has argued "nothing less than a complete decarbonisation of every single dollar" (Jackson, 2009: 81) is ultimately required if decarbonisation is to be based on the dual goals of carbon emissions mitigation and perpetual economic growth – a requirement which appears to have very little possibility of becoming a meaningful reality. Instead decoupling can be thought of as the "collaborative management of sustained unsustainability" (Blühdorn & Deflorian, this volume). Or in psychoanalytic terms as a fantasy that "serves to sustain faith in the possibility of attaining sustainable development within the context of a neoliberal capitalist economy that necessitates continual growth to confront inherent contradictions" (Fletcher & Rammelt, 2016: 450).

This fantasy is all too visible in recent trends in absolute emissions. The post-crisis return of growth to the United States meant that while 2018 saw a record number of coal plant retirements, $CO_2$ emissions actually rose by 3.4%, the biggest increase in eight years (Plumer, 2019). Figures released by the International Energy Agency (IEA) in early 2018 indicated that the 1.4% increase in GHG emissions from the global energy industry in 2017 was the result of a 2.1% increase in global energy demand, double that of 2016 and driven by economic growth in China and India (IEA, 2018). The IEA's 2018 World Energy Outlook, released in late 2018, projected these trends forward in a number of scenarios. Its main "New Policies" scenario is based on existing climate and energy policies and assumes continued improvements in energy efficiency as well as the uptake of additional targets and policies announced but not currently implemented by governments – such as Paris Agreement National Action Plans. This scenario shows global energy demand increasing by more than a quarter to 2040, a demand which is not expected to be met by increasing renewable energy share. This means oil and gas production is predicted to continue to increase in absolute terms, resulting in increased GHG emissions under this scenario (Evans, 2018; IEA, 2018).

## The reliance on NETs

The development of Shared Socioeconomic Pathways (SSPs) as part of the process of Integrated Assessment Modelling (IAM) has become a central means of understanding future decarbonisation processes within the IPCC's work, in both its AR5 and its 2018 Special Report on Global Warming of 1.5 °C (SR15). These reports feed directly into national and global policymaking efforts and therefore effectively shape and constrain the parameters of decarbonisation efforts. Kevin Anderson has powerfully argued that the vast scale of NETS assumed to be technically, socially, and economically viable in the 900 decarbonisation scenarios analysed by the IPCC in the preparation of its AR5 is simply not well-understood by policymakers (Anderson, 2015a, 2015b; Anderson & Peters, 2016: 182; see also Nature, 2018).

While NETS exist at various levels of development,[1] Carbon Capture and Storage (CCS) and Bioenergy with CCS (BECCS) are the most widely incorporated NETs within IPCC and non-IPCC models alike. The former because this enables the continued use and reliance upon fossil fuels and the latter due to its ability to combine energy generation with atmospheric $CO_2$ removal. In the case of BECCS, an imagined future global carbon pricing regime enables this technology to be considered as having the benefit of offsetting (at least in part) its own implementation costs. Rockström et al.'s (2017) roadmap for rapid decarbonisation argues that "in the absence of viable alternatives, the world must aim at rapidly scaling up $CO_2$ removal by technical means" (2017: 1269). This requires a change from a current zero giga tonnes of carbon dioxide per year ($GtCO_2$/year) removed from the atmosphere, to 0.5 by 2030, 2.5 by 2040, and $5GtCO_2$/year by 2050 (2017: 1269) – an historically unprecedented change, and given the current stage of development of CCS, one that is as much a fantasy as the concept of decoupling.

In their critical review of 17 global decarbonisation scenarios, Loftus, Cohen, Long, and Jenkins (2015) focused on what each of the modelling studies indicate about the feasibility of various decarbonisation strategies. They aimed to "examine how each study addresses the key technical, economic, and societal factors that may constrain the pace of low-carbon energy transformation." A key assumption made here was the exclusion as a priori outliers of scenarios that exclude CCS and nuclear technologies from their models, made on the basis that these require much faster low-carbon energy production capacity and energy intensity improvements than the other scenarios. Yet, the included scenarios "envision CCS technology installed on 5–40% of global power generation capacity by 2050" (Loftus et al., 2015: 105).

Aside from the frankly heroic nature of this assumption in the short-to-medium term, there are two further significant problems with BECCS. The first is that, in spite of two decades of research and pilot plant implementation, trials have struggled to prove the technical and economic viability of power generation at scale with CCS. Baik et al. (2018) found that only 30% of the projected available 2020 biomass resources in the United States can be utilised for BECCS, given the lack of suitable storage sites and transportation options of biomass and/or $CO_2$. Loftus et al. (2015) are more sanguine in their evaluation of the possibility of CCS, maintaining that the infrastructural capacity additions (drilling and completing wells, installing compression and pipeline capacity) required for the majority of scenarios are "close to the recent experience range of such activities in the global oil and gas industries" (Loftus et al., 2015: 105). However, they then highlight that this can and has been taken to actually indicate the vast scale of the task.

The second relates to the amount of land required for this technology. Loftus et al. (2015) note that estimates of the total global bioenergy potential of the planet – which crucially includes all land currently planted with food crops – fall within the range 7–11 terawatt-year (TW-yr). Where decarbonisation scenarios identify BECCS contributions to $CO_2$ reductions they fall within a range from 2.8 to 5.8 TW-yr, that is between 25% and 83% of the global bioenergy potential identified. While they only claim this would have "uncertain impacts on agriculture" (Loftus et al., 2015: 105),[2] it is rather the case that converting between one quarter to three quarters of the planet's available land into bioenergy would have very certain negative impacts, and raise "profound questions…about carbon neutrality, land availability, competition with food production, and competing demands for bioenergy from the transport, heating, and industrial sectors" (Anderson & Peters, 2016: 183).

While these issues indicate a startlingly novel understanding of the concept of viability with respect to decarbonisation mechanisms, what becomes clear here is the radical incoherence of an approach to decarbonisation that favours non-existent and largely implausible future technical solutions over current systemic change to the operation of the global economy. However, this does at least have a familiar echo of Frederic Jameson's famous quote: it is, apparently, easier to model the end of the world than is to model the end of fossil capitalism.

## The exclusions of decarbonisation

The incoherence of decarbonisation as the overarching goal and paradigm of global sustainability governance is both driven by and sustained through the removal of the contested history of its conceptual assumptions, historical implementations, and current means of control. At the Third International Science and Policy Conference on the resilience of social and ecological systems in 2014, Melissa Leach, the Director of the UK's Institute for Development Studies, highlighted a concern over issues of power with respect to the development and use of the related concept of planetary boundaries. As Leach later put it in a blog post following the debate: "The concept of planetary boundaries can be seen as a discourse that enables some things while marginalizing and excluding others" (Leach, 2014). Analogously, the apparent coherence of decarbonisation in spite of its radical incoherence is enabled and maintained through a series of exclusionary processes. These are variously oriented around: the exclusion of (certain) people; the exclusion of nature and particularly climate change itself; and the exclusion of systemic change.

## Exclusion of people

Two particular moments are observable in the exclusion of people from the goal of decarbonisation. The first is a kind of politics of depoliticisation, the subtle construction of a universal "we" that both drives the need for, and is subject to, decarbonising programmes. The second are the forms of post-political governance and control that follow from this moment of depoliticisation. Through these mechanisms, the apparently positive programme of decarbonisation (the absolute necessity to reduce global emissions) is written precisely through the silences imposed on the broad array of conflicts, oppressions, and impacts that historically lead to these emissions through the development of a globe-spanning fossil-fuel-based capitalism.

First, the politics of depoliticisation around decarbonisation operates in the same way as it has around the development of the "we" of the anthropocene (see e.g. Guenther, 2018; Hecht, 2018). That is, the construction of this universal "we" of humanity in general excludes the history of specific actions by specific individuals – a fossil-powered global capitalism (e.g. Malm, 2016; Moore, 2016) – from the role of causative agents of climate change while at the same time avoiding the overwhelming continued responsibility for climate change of global corporations and the richest 10% of the world's population (Gore, 2015; Griffin & Heede, 2017). At the same time, the politics of depoliticisation flattens out the differential and uneven impacts of climate change – which overwhelmingly affect the poorest, most insecure, and vulnerable populations both across and within nations. As Genevieve Guenther (2018) put it: "[t]o think of climate change as something that *we* are doing, instead of something we are being prevented from *undoing*, perpetuates the very ideology of the fossil-fuel economy we're trying to transform."

Second, the prior politics of depoliticisation is important for understanding the way in which the authority of decarbonisation as a global goal is dependent upon a largely post-political (Lövbrand et al., 2015; Swyngedouw, 2010) process of scenario modelling and managerial control. This involves the exclusion of problem identification, desired outcomes, and achievement mechanisms (Newell & Lane, 2017) and their replacement by "cockpit-ism" (Hajer et al., 2015). It can be observed in schemes such as the Deep Decarbonisation Pathways Project (e.g. Bataille et al., 2016) that replace a broader political engagement and questions of how, where, and who should decarbonise by a largely techno-managerial planning process legitimated by the urgency and scale of the task revealed.

These dynamics should be immediately familiar to observers of the "crisis capitalism" employed within contemporary neoliberal planning (Klein, 2007, 2014; Mirowski, 2013). Calls for further "Earth stewardship" (Steffen et al., 2015) and the strengthening of processes of earth system governance (Biermann, 2012) to enable subsequent policy initiatives and innovations to meet decarbonisation targets are typically constrained to technological choices within a rubric of intertemporal risk calculation. This similarly feeds into an increasing focus on climate change as a "threat multiplier," a term used in 2017 by UN Secretary General Antonio Guterres and one explicitly developed within and by military circles in order to facilitate military planning and management in the face of climate risks (Newell & Lane, 2018). The securitisation and technocratic management of climate change are enabled through the explicit exclusion of the world's poorest and most vulnerable people from the fora and mechanisms established to address it. In this way the current narrative on decarbonisation heightens the possibility of "environmental authoritarianism" as an ultimate outcome (Leach, 2014).

## Exclusion of nature

Decarbonisation has frequently been understood as an opportunity to reconcile capitalist accumulation with the absolute requirements of climate change mitigation (even from a critical perspective, e.g. Newell & Paterson, 2010). In a 2008 speech at the Investor Summit on Climate Risk the then UN General Secretary Ban Ki Moon reiterated "You are here today because you recognize climate change as an opportunity, as well as a threat. You understand that the shift to a low-carbon economy opens new revenue streams and creates new markets" (UN Department of Public Information 2008; cited in Janković & Bowman, 2014: 236). Here the possibilities or otherwise for decarbonisation present in e.g. United Nations Environment Programme's (UNEP) *Green Economy* proposals (2011), the Word Bank's push for *Inclusive Green Growth* (2012), the EU's *Circular Economy Action Plan* (2015), the UN's 2020 *Sustainable Development Goals initiative*, and the ten-year Framework of Programmes on Sustainable Consumption and Production Patterns adopted at the *Rio+20 Conference* in 2012 are essentially framed as an empirical question dependent upon policy alignment, effective governance, and meaningful regulation in line with scientific findings.

The problem of decoupling discussed earlier indicates that policy formation and institutional governance focused on climate change mitigation and economic growth are deeply incoherent. How then is the apparent coherence of the earlier projects maintained? Janković and Bowman (2014) argue that what has occurred is not simply a reconciliation of the two "eco's" of ecology and economy, but that climate change has, in fact, been reconstructed as "an axiomatic framework" of long-term economic strategy. That is, contemporary climate governance has been translated through a narrative of ecological modernisation (Bailey et al., 2011; Wanner, 2015) and an economic calculus of efficiency and external effects (Lane, 2012) into programmes focused on green global growth through the ontological dislocation of climate change and the exclusion of nature as the predominant concern. As the World Bank makes explicit, achieving decarbonisation means achieving *cost-effective decarbonisation* (Fay et al., 2015).

Sustainable economic activity here becomes independent of the question of whether decarbonisation initiatives *actually address climate change itself either individually or in the aggregate* (Janković & Bowman, 2014: 251), see e.g. the fate of the UN's Clean Development Mechanism and broader pre-Paris carbon markets as a prime example of decarbonisation projects driven and legitimated by their potential to generate "sustainable" economic activity in spite of their manifest failures as GHG emissions reduction mechanisms (Stephan & Lane, 2014). This exclusion of nature – specifically climate change – from the drive for decarbonisation innovations has the effect of converting green investment from a solution to climate change to an end in itself. This shift was at least rhetorically recognised by Ban Ki Moon later in his 2008 Investor Summit speech, when he stated that with respect to a shift towards a green future "…the United Nations looks to you, as leaders in the financial sector, to lead in innovative financing and technological development" (UN Department of Public Information 2008; cited in Janković & Bowman, 2014: 236).

Even if we are (in spite of the evidence of the lack of global absolute decoupling) relatively agnostic about the future possibility of decarbonisation initiatives focused on green growth to actually decarbonise the global economy,[3] the dislocation of climate change subverts the relationship between climate science and action on climate change. This raises the troubling question of how a drive towards decarbonisation that is paradoxically independent of the actual ecological impacts of climate change radically alters what it means to "solve" the climate crisis (Janković & Bowman, 2014: 252).

## Exclusion of systemic change

As discussed earlier with respect to the prevalence of NETS, modelling activities in the form of IAMs are a key process in the development of global approaches to decarbonisation. Rockström et al.'s (2017) focus on developing a "carbon law" and "roadmap for rapid decarbonisation" and the development of SSPs as part of the IPCC's AR5 in 2014 and SR15 in 2018 play crucial roles in the current formation of decarbonisation approaches, policies, and governance mechanisms. A straightforward point here is that they necessarily exclude systemic transformation given the requirement to parameterise current socio-economic realities during the process of model construction. The current system is, literally, coded in.

There is, however, a growing scholarship around questions of the socio-technical transformations (see e.g. Geels, Sovacool, Schwanen, & Sorrell, 2017) required for decarbonisation. But the reliance on economistically defined dichotomies of state versus market and private innovation versus public regulation within an overall systems-analytic framing makes it difficult for these approaches to adequately capture the political, economic, social, and cultural elements involved in sustaining the currently unsustainable global economy. For example, with respect to the widely recognised necessity of removing fossil fuel subsidies, these approaches fail to adequately capture the productive relationships between public governance and private profit (as described by Mazzucato, 2015) and how apparently cheaply exploitable nature is only made cheap through high – literal and figurative – costs (Patel & Moore, 2017).

Both current IAMs and the scholarship on socio-technical transformation focus almost exclusively on innovation. Lee Vinsel has highlighted how innovation as a buzzword and organising concept arose in prominence in the late 1960s in the United States as a more morally neutral alternative to notions of progress. At the time, progress was evidently in little supply given the apparent failures of racial integration, social development, environmental degradation, the Vietnam War, and the Kennedy and King assassinations. The focus on innovation built on the widespread acceptance in the United States of consumer technologies as proxies for societal progress, exemplified by the "Kitchen Debate" of 1959 between Nixon and Khrushchev (Russell & Vinsel, 2016).

Two analogue developments within the discipline of economics also played crucial roles in the valorisation of innovation as a pure good at this time. In the first instance, since the 1950s the development of modern growth theory added technological change to the neoclassical factors of production: capital, labour and land. In the work of economic luminaries such as Robert Solow and Kenneth Arrow, technological change was presented as the mystery factor explaining economic growth (Blaug, 1980; Vinsel, 2017). In the second instance, the postwar development of Cost-Benefit Analysis as a widely applicable economic planning technology resulted in an intensified focus on and increasingly politically motivated use of discounting rates (Andrews, 1984; Knafo et al. 2019; Kysar, 2007). The ouroboros-like relationship between the discounted costs of potential future innovations and modern growth theory's mobilisation of these innovations as necessary drivers of economic growth has helped establish a fetishised notion of innovation at the core of decarbonisation.

Through the 1980s–1990s, the rise of Silicon Valley amid the stagnation of traditional manufacturing industries resulted in a neo-Schumpeterian inflection to innovation policy, which focused on fostering economic growth via technological change. The new focus here on "disruptive innovation" was heralded by Clayton M. Christensen's now widely discredited 1997 text "The Innovator's Dilemma" (Lepore, 2014; Russell & Vinsel, 2016; Vinsel, 2015), but has become central to the way that socio-economic and socio-technical transformations are currently understood (e.g. de Coninck et al., 2018; Geels et al., 2017; Rockström et al., 2017). That is, the structural change and transformation of current global economic systems focused on growth with their resultant high carbon emissions have been replaced within decarbonisation narratives with a focus on non-disruptive "disruptive" innovation.

Decarbonisation as it is currently conceptualised, configured and manifested within particular programmes and forms of control, is largely shorn of a politically sensitive awareness of the historical relationships between societies, economies, and technologies. The exclusion and marginalisation of the voices of the world's poorest and most vulnerable, the ontological dislocation of climate change underpinning green growth, the fetishism of innovation and the subsequent exclusion of systemic change from the narratives, modelling technologies, and institutions of decarbonisation have impacts beyond the straightforward reduction of global GHG emissions. A danger here is of an increasing reliance on the Facebook mantra to "move fast and break things" in the face of an impending climate crisis. But as highlighted earlier, the thing that looks increasingly likely to be broken is a global ecosystem capable of sustaining the planet's most vulnerable inhabitants – both human and non-human.

## Conclusion: towards a coherent, inclusive decarbonisation

This chapter has explored how the development of decarbonisation as the de facto goal of global climate change governance is not as straightforward as it first appears. Decarbonisation of the global economy is an absolute necessity if we are to address the threat of climate change. Yet paradoxically, this chapter has argued that the ways in which decarbonisation is currently understood have itself obscured the necessary economic, political, and social transformation of the global economy. Currently, programmes of decarbonisation are based on a radical incoherence given the problem of decoupling economic growth from environmental impacts; and the reliance upon NETS and more specifically BECCS within scenario modelling. This incoherence is both a result of, and further sustains, a number of important exclusions from the process of defining decarbonisation and governing climate change, namely the exclusion of (certain) people, the exclusion of nature and particularly climate change itself, and the exclusion of systemic change.

However, as a core project of sustainability governance in the anthropocene, decarbonisation is arguably amenable to a similar process of conceptual and then institutional reconstruction. The arguments of Lövbrand et al. (2015) with respect to the potential for moving beyond a post-natural, post-social, and post-political account of the anthropocene can be equally applied to decarbonisation as the focus of governing climate change within this new geological epoch.

First, the post-political authority of decarbonisation as a global goal needs to be challenged not in order to replace it, but rather to invigorate it with an intersectional politics capable of fostering "a vibrant public space where manifold and divergent socio-ecological relations and nature concepts can be exposed and debated" (Lövbrand et al., 2015). To be clear, this is not an advocation of public argumentation over acceptable atmospheric $CO_2$ levels or global temperature increases. Instead, it is the means to meet these goals that have themselves been removed from critical contestation, which require debate. Questions of how, where, and who should decarbonise need to replace the central focus on cost-effective decarbonisation that is reliant upon incoherent but seemingly natural assumptions regarding the feasibility of decoupling. These ultimately favour the ecological devastation and social exploitation of developing countries over the socio-economic transformation of developed ones. Practically, this requires not just a focus on guiding ethical principles as discussed in section "The exclusions of decarbonisation" of this volume, but reconsideration of contemporary assumptions around consumer culture and values (Krogmann, this volume), the population challenge (Coole, this volume), north-south inequity (Okereke, this volume), and democracy in the anthropocene (Mert, this volume).

Second, a recognition of the ontological dislocation of climate change via its translation through an eco-modernist idiom from a primarily physical and ecological concern to a long-term economic strategy requires the fundamental reconsideration of growth as one of the twin goals of decarbonisation (see also Higgs, this volume). It is already broadly accepted that the climate change commitments put forward by countries as their Nationally Determined Contributions are insufficient to meet the internationally agreed Paris climate targets (e.g. UNEP, 2017). Or as Anderson and Bows put it "climate change commitments are incompatible with short- to medium-term economic growth" and therefore "Civil society needs scientists to do science free of the constraints of failed economics" (Anderson & Bows, 2012: 640). Indeed, there is currently a proliferation of scholarship and broader political economic concern focused on the recognition that infinite economic growth is neither possible nor desirable as a measure of human well-being (Jackson, 2009; Philipsen this volume). Calls for steady-state economies and degrowth programmes informed by ecological economics (e.g. D'Alisa, Demaria, & Kallis, 2014; Kallis & March, 2015) highlight the importance of considering issues of growth and development. Similarly, financialisation (Clapp and Stephens, this volume) and the role of debt as a colonisation of the future, as well as questions about what a sustainable economy is (Lange, this volume) should be asked as part of the redevelopment of decarbonisation as a goal of global sustainability governance.

Third, Russell and Vinsel (2016) suggest a series of ways to characterise the relationships between society and technology that avoid the fetishisation of innovation and enable a focus on the broader drivers of systemic change. This begins with appreciating that technology is not reducible to innovation and that instead there is a requirement to seriously consider both the physical and social infrastructure of already existing technologies. This implies that an historical sensitivity to the construction of already existing high-carbon socio-technical institutions would help enable the further identification of points of departure for a programme of decarbonisation. The current renaissance of Green New Deal ideas in the United States and globally (see e.g. Pollin et al., 2014) in part undertakes a reconsideration of the relationship between public and private, markets and regulation, that current accounts of innovation for decarbonisation miss. But there is the necessity to go further here and involve both the incorporation of the extensive critical analysis of the development of our current high carbon world of economic growth, consumption, waste, exploitation, and inequality, alongside consideration of working time reduction/deceleration (Larsson and Nässén, this volume) and localism, sharing, and care (Litfin, this volume).

Decarbonisation as it is currently understood is both fatally incoherent, and excludes the planet's most vulnerable inhabitants. Reconstructing decarbonisation as a transformatory locus of climate governance that is coherent and inclusive is possible, but requires a commitment to an emancipatory politics, that as the late cultural theorist Mark Fisher put it "…must always destroy the appearance of a 'natural order,' must reveal what is presented as necessary and inevitable to be a mere contingency, just as it must make what was previously deemed to be impossible seem attainable" (Fisher, 2009: 17).

## Notes

1 With technologies such as direct air capture, biochar, and soil sequestration, enhanced weathering of soils, and ocean fertilisation at various stages of trial development and experimentation.
2 They also note that scenarios typically do not consider indirect carbon emissions from bioenergy associated with land use change.
3 Indeed, Janković and Bowman (2014) highlight the influx of private sector investment following this ontological dislocation.

## References

Åhman, M., Nilsson, L. J., & Johansson, B. (2017). Global climate policy and deep decarbonization of energy-intensive industries. *Climate Policy, 17*(5), 634–649.

Anderson, Kevin. (2015a). Duality in climate science. *Nature Geoscience, 8*(12), 898.

Anderson, Kevin. (2015b). Talks in the city of light generate more heat. *Nature, 528*(7583), 437–437.

Anderson, Kevin, & Bows, A. (2012). A new paradigm for climate change. *Nature Climate Change, 2*(9), 639–640.

Anderson, K., & Peters, G. (2016). The trouble with negative emissions. *Science, 354*(6309), 182–183.

Andrews, R. N. L. (1984). Economics and environmental decisions, past and present. In V. K. Smith (Ed.), *Environmental policy under Reagan's executive order: The role of benefit-cost analysis*. Chapel Hill and London: University of North Carolina Press.

Baik, E., Sanchez, D. L., Turner, P. A., Mach, K. J., Field, C. B., & Benson, S. M. (2018). Geospatial analysis of near-term potential for carbon-negative bioenergy in the United States. *Proceedings of the National Academy of Sciences*, 201720338.

Bailey, I., Gouldson, A., & Newell, P. (2011). Ecological modernisation and the governance of carbon: a critical analysis. *Antipode, 43*(3), 682–703.

Baron, R. (2016) Energy transition after the Paris agreement: policy and corporate challenges. Background paper for the 34th Round Table on Sustainable Development 28–29 September 2016, WindEurope Summit, Hamburg.

Bataille, C., Waisman, H., Colombier, M., Segafredo, L., Williams, J., & Jotzo, F. (2016). The need for national deep decarbonization pathways for effective climate policy. *Climate Policy, 16*(supp 1), S7–S26.

Bernstein, S. (2017). The United Nations and the Governance of Sustainable Development Goals. In N. Kanie & F. Biermann (Eds.), *Governing through goals: Sustainable Development Goals as governance innovation*, 213–240. Cambridge, MA: MIT Press.

Biermann, F. (2012). Planetary boundaries and earth system governance: Exploring the links. *Ecological Economics, 81*, 4–9.

Blaug, M. (1980). *The methodology of economics*. Cambridge: Cambridge University Press.

D'Alisa, G., Demaria, F., & Kallis, G. (2014). *Degrowth: A vocabulary for a new era*. London: Routledge.

de Coninck, H., Revi, A., Babiker, M., Bertoldi, P., Buckeridge, M., Cartwright, A. … Ley, D. (2018). Strengthening and implementing the global response. In V. Masson Delmotte et al., *Global warming of 1.5 °C. An IPCC special report on the impacts of global warming of 1.5 °C above pre-industrial levels and related global greenhouse gas emission pathways, in the context of strengthening the global response to the threat of climate change, sustainable development, and efforts to eradicate poverty*. Geneva: Intergovernmental Panel on Climate Change.

European Commission. (2018). A clean planet for all a European strategic long-term vision for a prosperous, modern, competitive and climate neutral economy. *COM (2018) 773 Communication from the Commission to the European Parliament, the European Council, the Council, the European Economic and Social Committee, the Committee of the Regions and the European Investment Bank*. Brussels, November 28.

Evans, S. (2018). Global coal use may have peaked in 2014, says latest IEA World Energy Outlook. *CarbonBrief*. Retrieved December 12, 2018, from https://www.carbonbrief.org/global-coal-use-may-have-peaked-iea-world-enery-outlook

Fay, M., Hallegatte, S., Vogt-Schilb, A., Rozenberg, J., Narloch, U., & Kerr, T. M. (2015). *Decarbonizing development: Three steps to a zero-carbon future*. Washington, DC: World Bank Group.

Fisher, M. (2009). *Capitalist realism: Is there no alternative?*. London: John Hunt Publishing.

Geels, F. W., Sovacool, B. K., Schwanen, T., & Sorrell, S. (2017). Sociotechnical transitions for deep decarbonization. *Science, 357*(6357), 1242–1244.

Gore, T., (2015). Extreme carbon inequality: Why the Paris climate deal must put the poorest, lowest emitting and most vulnerable people first. Oxfam Media Briefing. Retrieved December 18, 2018, from https://oxfamilibrary.openrepository.com/bitstream/handle/10546/582545/mb-extreme-carbon-inequality-021215-en.pdf?sequence=9

Griffin, P., & Heede, R. (2017). *The carbon majors database: CDP carbon majors report 2017.* Inglaterra: CDP Worldwide.

Guenther. (2018). Who is the *We* in 'We are causing climate change'?. *Slate.* Retrieved December 18, 2018, from https://slate.com/technology/2018/10/who-is-we-causing-climate-change.html

Haas, W., Krausmann, F., Wiedenhofer, D., & Heinz, M. (2015). How circular is the global economy?: An assessment of material flows, waste production, and recycling in the European Union and the world in 2005. *Journal of Industrial Ecology, 19*(5), 765–777.

Hecht, G. (2018). The African Anthropocene. *Aeon.* Retrieved March 22, 2018, from https://aeon.co/essays/if-we-talk-about-hurting-our-planet-who-exactly-is-the-we

IEA. (2018). *World Energy Outlook 2018.* IEA: Paris.

IPCC. (2014). *Climate change 2014: Impacts, adaptation and vulnerability.* Cambridge: Cambridge University Press.

Jackson, T. (2009). *Prosperity without growth: Economics for a finite planet.* London and New York: Earthscan.

Janković, V., & Bowman, A. (2014). After the green gold rush: The construction of climate change as a market transition. *Economy and Society, 43*(2), 233–259.

Kallis, G., & March, H. (2015). Imaginaries of hope: The Utopianism of degrowth. *Annals of the Association of American Geographers, 105*(2), 360–368.

Klein, N. (2007). *The shock doctrine: The rise of disaster capitalism.* New York: Alfred A. Knopf Canada.

Klein, N. (2014). *This changes everything: Capitalism vs. the climate.* New York: Simon and Schuster.

Knafo, S., Dutta, S. J., Lane, R., & Wyn-Jones, S. (2019). The managerial lineages of neoliberalism. *New Political Economy, 24*(2), 235–251.

Knight, K. & Schor, J. (2014). Economic growth and climate change: A cross-national analysis of territorial and consumption-based carbon emissions in high-income countries. *Sustainability, 6,* 3722–3731.

Kysar, D. A. (2007). Discounting … on stilts. *The University of Chicago Law Review, 74*(1), 119–138.

Lane, R. (2012). The promiscuous history of market efficiency: The development of early emissions trading systems. *Environmental Politics, 21*(4), 583–603.

Leach, M. (2014). Resilience 2014: Limits revisited? Planetary boundaries, justice and power. Retrieved March 18, 2018, from https://steps-centre.org/blog/resilience2014-leach/.

Lepore, J. (2014). The disruption machine: What the gospel of innovation gets wrong. *The New Yorker Magazine.* Retrieved March 20, 2018, from https://www.newyorker.com/magazine/2014/06/23/the-disruption-machine

Le Quere, C., Andrew, R. M., Friedlingstein, P., Sitch, S., Hauck, J., Pongratz, J., … Zheng, B. (2018). Global carbon budget 2018. *Earth System Science Data, 10,* 1–54.

Loftus, P. J., Cohen, A. M., Long, J. C. S., & Jenkins, J. D. (2015). A critical review of global decarbonization scenarios: What do they tell us about feasibility? *Wiley Interdisciplinary Reviews: Climate Change, 6*(1), 93–112.

Lövbrand, E., Beck, S., Chilvers, J., Forsyth, T., Hedrén, J., Hulme, M., … Vasileiadou, E. (2015). Who speaks for the future of Earth? How critical social science can extend the conversation on the Anthropocene. *Global Environmental Change, 32,* 211–218.

Malm, A. (2016). *Fossil capital: The rise of steam-power and the roots of global warming.* London and Brooklyn, NY: Verso.

Mazzucato, M. (2015). *The entrepreneurial state: Debunking public vs. private sector myths* (Vol. 1). New York: Anthem Press.

Mirowski, P. (2013). *Never let a serious crisis go to waste: How neoliberalism survived the financial meltdown.* London: Verso.

Moore, J. W. (2016). *Anthropocene or Capitalocene?: Nature, history, and the crisis of capitalism.* Oakland, CA: PM Press.

Nature. (2018). Why current negative-emissions strategies remain 'magical thinking'. *Nature, 554,* 404.

Newell, P. & Lane, R. (2017). IPE and the environment in the age of the Anthropocene. In *Traditions and trends in global environmental politics* (pp. 148–165). London: Routledge.

Newell, P., & Lane, R. (2018). A climate for change? The impacts of climate change on energy politics. *Cambridge Review of International Affairs,* 1–18.

Newell, P., & Paterson, M. (2010). Climate capitalism: Global warming and the transformation of the global economy. Cambridge and New York: Cambridge University Press.

Patel, R., & Moore, J. W. (2017). A history of the world in seven cheap things: A guide to capitalism, nature, and the future of the planet. Oakland: University of California Press.

Plumer, B. (2019). U.S. carbon emissions surged in 2018 even as coal plants closed. New York Times. Retrieved January 9, 2018, from https://www.nytimes.com/2019/01/08/climate/green house-gas-emissions-increase.html

Pollin, R., Garrett-Peltier, H., Heintz, J., & Hendricks, B. (2014). Green growth a US program for controlling climate change and expanding job opportunities. Political Economy Research Institute, University of Massachusetts.

Rockström, J., Gaffney, O., Rogelj, J., Meinshausen, M., Nakicenovic, N., & Schellnhuber, H. J. (2017). A roadmap for rapid decarbonization. Science, 355(6331), 1269–1271.

Russell, A., & Vinsel, L. (2016). Hail the maintainers. Aeon. Retrieved March 25, 2018, from https://aeon.co/essays/innovation-is-overvalued-maintenance-often-matters-more

Schandl, H., Fischer-Kowalski, M., West, J., Giljum, S., Dittrich, M., Eisenmenger, N., ... & Krausmann, F. (2018). Global material flows and resource productivity: forty years of evidence. Journal of Industrial Ecology, 22(4), 827–838.

Steffen, Will, Richardson, K., Rockström, J., Cornell, S. E., Fetzer, I., Bennett, E. M., ... de Wit, C. A. (2015). Planetary boundaries: Guiding human development on a changing planet. Science, 347(6223), 1259855.

Stephan, B., & Lane, R. (Eds.). (2014). The politics of carbon markets. London: Routledge.

Swyngedouw, E. (2010) 'Apocalypse forever? Post-political populism and the spectre of climate change. Theory, Culture and Society, 27(2–3), 213–232.

UNEP. (2011). Towards a green economy: Pathways to sustainable development and poverty eradication. Nairobi: UNEP.

UNEP. (2017). The emissions gap report 2017: A UN environment synthesis report. Nairobi: UNEP.

Vinsel, L. (2015). Snake oil for the innovation age: Christensen, Forbes, and the problem with disruption. Retrieved March 25, 2018, from http://leevinsel.com/blog/2015/9/16/snake-oil-for-the-innovation-age-christensen-forbes-and-the-problem-of-disruption

Vinsel, L. (2017). Regulatory enforcement as sociotechnical systems maintenance. In B. Godin & D. Vinck (Eds.), Critical studies of innovation (pp. 257–275). Cheltenham: Edward Elgar Publishing.

Wanner, T. (2015). The new 'passive revolution' of the green economy and growth discourse: Maintaining the 'sustainable development' of neoliberal capitalism. New Political Economy, 20(1), 21–41.

Ward, J. D., Sutton, P. C., Werner, A. D., Costanza, R., Mohr, S. H., & Simmons, C. T. (2016). Is decoupling GDP growth from environmental impact possible? PLoS One, 11(10), e0164733.

Wesseling, J. H., Lechtenböhmer, S., Åhman, M., Nilsson, L. J., Worrell, E., & Coenen, L. (2017). The transition of energy intensive processing industries towards deep decarbonization: Characteristics and implications for future research. Renewable and Sustainable Energy Reviews, 79(Supplement C), 1303–1313.

# 28

# Localism, sharing, and care

*Karen Litfin*

An ideologically diverse array of voices from around the world is calling for the localisation of economics, politics, and culture. Common to most of these discourses of localism is a strong critique of globalisation, with its associated social alienation, political domination, cultural homogeneity, and ecological destruction. In contrast to the placeless and faceless global, the local holds out the promise of real relationships with real people and places.

Localism is grounded in economic, ecological, political, and social claims. The economic and ecological rationale is simple: all things being equal, an economy with lower energy and resource requirements will generate fewer market externalities. From a political perspective, localisers argue that decades of "green diplomacy" in a neoliberal world have accomplished too little, too late. The Anthropocene is upon us and we must prepare for "an era of upheaval" (Lerch 2017). Communities should therefore redesign themselves for resilience in a postcarbon world. From a social perspective, localisers believe that a world where we make decisions together and know who grows our food, who produces our goods, and where our waste goes will be a more just and convivial world – one that is also better prepared to face crises in an era of climate consequences.

Eco-localism, however, is only one amongst a complex array of localism discourses. In the words of Simin Davoudi and Ali Madanipour (2015: 1), "The motivations for localism range from communitarian intents to liberal and libertarian agendas and are riddled with tensions between progressive and regressive potentials." Globalisation and its discontents have spawned the resurgence of the local across the political spectrum. Architects of local sustainability governance must therefore place themselves in relation to an array of localist agendas. How, for instance, should proponents of eco-localism respond to the separatist impulses of anti-immigration localisers or other populist appeals to community identity?

At the same time, a localist sustainability agenda finds itself in tension with global institutional and planetary ecological forces. First, it risks ceding the vast territory of the global to those forces that currently occupy that ground – most obviously multinational corporations and the political institutions serving their agenda. Second, the lifestyles of a growing portion of humanity have a global reach. Cities in particular are cited as a primary arena for relocalisation but (at least in their current form) they function as highly consumptive nodes in global commodity chains. The jury is out as to whether cities can shift from their current status as ecological parasites to serving primarily as laboratories and brains for sustainability innovation (Liu et al. 2016). Third, humanity is now operating as a geophysical force on a planetary scale. While an eco-localist approach to sustainability governance may provide a salutary counterbalance to globalisation by advancing local resilience in precarious times, it sidesteps the pressing questions of planetary governance in the Anthropocene. How, for instance, will disparate localities coordinate and collaborate amongst themselves in response to challenges like climate refugees, species migration, geoengineering, Internet governance, etc.? While the requisite political and economic institutions are nowhere in sight, this is the challenge before us.

This chapter argues that a well-considered localism can make an important contribution to global sustainability governance but should not be understood as a replacement for it. The first section describes the theoretical grounding of localism as an approach to sustainability governance, tracing the work of key thinkers since the 1970s. Recent trends reinforce the case for eco-localism: urbanisation, disillusionment with national and international sustainability initiatives, the rise of the sharing economy, new possibilities for decentralised energy production, and the realities of climate change.

The second section investigates the promise of localism in practice and inquires into the underlying assumptions and motivations that characterise eco-localist initiatives. What, if anything, unifies farmers' markets, community-supported agriculture (CSA), local currency systems, car-sharing schemes, ecovillages, community energy systems, municipal climate plans, and local business alliances into a coherent approach to sustainability governance? This is not merely an academic inquiry; it takes on some practical urgency because eco-localism is not the only game in town. Within the past decade, populists, libertarians, conservatives, and liberals have also seized the discursive terrain of the local. To the extent that eco-localism succumbs to anti-politics, it risks ceding the local to those who deploy sustainability language in support of exclusionary, anti-democratic, and potentially environmentally destructive agendas. Localism's appeal across the political spectrum points to widespread dissatisfaction with the forces of globalisation. Despite their ideological differences, proponents tout localism's potential to foster citizen participation, place making, social cohesion, and problem solving. If eco-localism is to actualise its potential as more than a reactive retort to toxic globalisation and make a meaningful contribution to sustainability governance, it must distinguish itself from and/or find common cause with other localisms. Perhaps the greatest power of localism is its ability to inspire agency, collective action, and innovation in the face of potentially overwhelming complexity.

The third section presents a higher-order synthesis. While localism is a healthy adaptive response to a rapacious and dysfunctional globalism, we cannot so easily disentangle ourselves from the global so long as we have the Internet, global commodity supply chains, and planetary ecological problems. If localist approaches to sustainability governance are to realise their full potential, they should therefore move beyond – not just against – globalism as presently constituted. Organic globalism (Litfin 2014) envisions a world of locally based, globally networked citizens' initiatives. In contrast to the hegemonic practices associated with corporate-driven globalisation, organic globalism understands the world as a nested hierarchy of living systems, from the cell to the Earth system, seeking to harmonise human systems with living systems at every level. Such a higher-order synthesis is evident in a growing number of global activist networks rooted in strong localist agendas. Implicit in their work is the very practical question: what should be localised and what should be globalised? This translates into eminently political questions about the limits and range of our caring and sharing.

## Localism in theory

The ecological rationale for localism is straightforward: local self-sufficiency places fewer demands on ecosystems than global networks of extraction, production, transportation, and consumption. Localisers also point to significant psychological, social, cultural, political, and moral benefits. No doubt, the social alienation engendered by up-scaling trends like suburbanisation and mega-urbanisation has catalysed a localist response, often with a communitarian sensibility. Localisers often speak of the "human-scale," implying that large-scale enterprises – multinational corporations, global supply chains, nuclear reactors, and the like – are fundamentally dehumanising. This was E.F. Schumacher's point when he declared over 40 years ago that "small is beautiful" (1973).

In the intervening years, the socio-ecological consequences of "bigness" have become more consequential. The second half of the 20th century saw a phenomenal rise in the speed, volume, and geographic scope of commerce, spurred on by new technologies and international trade, finance, and development institutions. Most governments saw their key function as facilitating the movement of goods and capital. Indeed, the very notion of human progress came to be associated with bigger, faster, farther, and more – all propelled by fossil fuels, with petroleum at the helm (see also Lane, this volume). This extravaganza of energy consumption generated unprecedented growth in both human numbers and material wealth. The attendant "negative externalities" – pollution, capital flight, and social displacement – however, left in their wake devastated communities and ecosystems on every continent. No wonder, then, that the trickle of complaints in the 1980s became, by the millennium, a diverse movement for global sustainability and justice. As many as a million initiatives, from organic farms to labour unions to indigenous people's organisations, constitute this far-flung movement of movements (Hawken 2007). Woven through the global tapestry of these movements is a strong thread of localism.

In their anthology on localisation, Raymond DeYoung and Thomas Princen suggest that the primary concern is "how to adapt institutions and behaviors to live within the limits of natural systems" (2012a: xvii). They predict that industrial societies will experience a shift from the *centrifugal* forces of globalisation to the *centripetal* forces of localisation. Whereas the former rely upon concentrated economic and political power, cheap and plentiful resources, intensive commercialisation, displaced wastes, and abstract modes of communication, the latter are associated with diffuse leadership, sustainable production and consumption, personal know-how, and community self-reliance.

Localisation is at or near the top of the agenda for many who warn of the potential for social and economic collapse as a consequence of ecological catastrophe (Tainter 1990; Homer-Dixon 2006). Among localist concerns are climate change, the precipitous loss of biodiversity, and freshwater depletion. None, however, received more attention during the first decade of the new century than "peak oil." For many localisers, small was not only beautiful but it became inevitable with the zenith of global petroleum production sometime around 2007. A host of books like *The Party's Over, The Last Hours of Ancient Sunlight*, and *Out of Gas* drive the point home. In laying out why "small is inevitable," Rob Hopkins, a prominent localiser and founder of the Transition Towns movement, focused on peak oil (2008, 68–78). As the master resource, energy propels every aspect of the economy: building, manufacturing, heating and cooling, and so on. And because the entire global economy – especially mining, transportation, and agriculture – is tied to petroleum, peak oil meant *peak everything* (Heinberg 2007).

With oil as the lifeblood of the global economy, "peak oil" should make localisers of us all – unless petroleum can be replaced by other cheap hydrocarbons or the melting of Arctic sea ice renders the "peak oil" thesis a moot point, both of which seem to be the case (Schiermeier 2012).

363

Given the apparent abundance of unconventional hydrocarbons (shale oil and gas, oil sands, and methane hydrates), the "peak oil" case for localism requires reconsideration: at least for the near term, energy descent is not inevitable. With the falling price of renewables, however, "peak oil demand" appears to be a more compelling reality (Brandt et al. 2013), opening up new possibilities for local control of energy resources. We shall elaborate upon this point in the next section.[1]

Aside from energy, there are other powerful arguments for localisation. Again, the trickle of voices from decades past seems to be swelling into a flood. With growing concerns about climate and peak oil, for instance, Kirkpatrick Sale's bioregional writings from the 1980s enjoyed a revival in the new century (Thayer 2003). While Sale's bioregionalism, emphasising both the psycho-social and ecological value of place-based identity, resonates well with contemporary localist movements, it lacks the thoroughgoing critique of global capitalism articulated during the same period by Murray Bookchin. In contrast to deep ecologists who tended to downplay social injustice, Bookchin's social ecology roots environmental problems in social domination and hierarchy. Likewise, Bookchin's notions of communalism and sustainable cities presage important elements of Transition Towns and other eco-localist movements. Bookchin's seminal works, *Post-Scarcity Anarchism* (1971) and *The Ecology of Freedom* (1982), were reprinted in 2004 and 2005, respectively, as valuable resources for both critiquing global capitalism and articulating a theoretical basis for the new localism.

Most importantly, localism emerges in response to the shortcomings of international and national institutions. It is said that nature abhors a vacuum and this is no less true in the realm of politics and social action. We might have hoped that the World Trade Organization (WTO) would make good on its mission of promoting sustainable development or that decades of negotiations would have stabilised our home planet's climate and forestalled the sixth mass extinction event, but things are as they are. One veteran of United Nations (UN) conference diplomacy called the international declaration from the 2012 Rio+20 Earth Summit "the longest suicide note in history" (quoted in McDonald 2012). The policy vacuum left by governments and international organisations is giving rise to a host of local initiatives. National governments may have failed to reduce greenhouse gas emissions, for instance, but thousands of cities and businesses are taking up the effort. Ironically, subnational initiatives are proliferating in the world's most laggard country on international climate governance: the United States (Hoffman 2011; Wilbanks and Fernandez 2014). In the face of a disenchantment with the top-down politics of globalisation, the ecological case for localisation is compelling.

Beyond the ecological case for localism, many localisers argue on anthropological grounds that people are more likely to flourish in place-based communities. Others promote local businesses and community ownership on the grounds of economic efficiency and social accountability (Shuman 2000). Still others see localisation as a healthy response to the neocolonial model of development associated with globalisation (Goldsmith and Mander 1997). In their quest to green and humanise contemporary cities, urban planners are lending their voices to the growing choir (Simin and Ali Madanipour 2015). In addition to ecological sustainability, localisation is said to simultaneously foster social resilience, economic wellbeing, democratic participation, community values, and psychological health. With all of these advantages, one would be hard pressed to register serious objections to localism. As they say, though, the devil's in the details.

## The promise and limits of eco-localism

Just as globalisation was shifting into high gear in the 1990s, the slogan, "Think globally, act locally," became popular. If we truly care about global problems, the slogan suggests that we should set our own house in order. The implication is that local actions like investing in community banks and buying local organic food will get us out of the planetary mess we are in. Even more: a global perspective *compels* us to act locally: where else can we act? There are at least four powerful moments of truth to this claim. First, as a response to the destructive legacy of globalisation, the slogan communicates a healthy wariness of large-scale action in challenging speed and convenience, the core values of consumerist culture (see also Fuchs, this volume). Second, the consequences of globalisation – whether they be floods, droughts, refugee migrations, or financial meltdowns, are always experienced locally. Building local resilience is therefore a high priority. Third, the levers of power are more visible and accessible at the local level than at the global level. Banding together with our neighbours is more feasible than influencing the WTO. Fourth, if we profess a great concern for the human and ecological wreckage wrought by globalisation yet persist in externalising the negative consequences of our consumptive lifestyles, then we are essentially hypocrites. On this view, internalising costs requires local economies.

An impressive range of local sustainability initiatives has emerged to put the slogan into practice: complementary currency systems, farmers' markets, permaculture, CSA farms, car-sharing schemes, ecovillages and intentional communities, Transition Towns, school and community gardens, maker movements, community corporations and banks, and local business alliances. While these initiatives may not give voice to a coherent political agenda, they do share a common vocabulary and, arguably, a common cultural orientation. The watchwords for eco-localism include community, slower pace, resource efficiency, accountability, pragmatism, solidarity, responsibility, participatory democracy, preparedness, self-reliance, and integrity. Returning to the theme of "small is beautiful," the common ground of eco-localism seems to be as much aesthetic and emotional as political and strategic. For many, an actual village has a more positive valence than the global village.

Food is particularly visible in the new localism, with local food serving as a marker for cultural identity, sustainability, economic justice, and community solidarity. Yet reality falls short of the rhetoric: even on its own terms, local is not necessarily healthier, more just, or more ecologically sustainable (DeLind 2011). Valorising the local does not help us to weigh tradeoffs. All things being equal, local goods will require fewer resources but all things are never equal. Food-miles campaigns, for instance, can increase global inequality without necessarily improving environmental outcomes (Ballingall and Winchester 2010). Should not localisers in affluent countries consider their impact on poor countries – particularly if their own national governments helped to bring those countries into the global food economy?

Energy offers another arena for local ownership and community-scale projects. While "peak oil" has not made localism inevitable, "peak fossil fuel demand" is a different animal. Declining prices for renewable energy, especially solar and wind, are opening up new possibilities for "energy democracy" (Fairchild and Weinraub 2017). These projects have been a central element in the transition to renewables across Europe, particularly in Germany and Denmark. Community energy projects capture the economic benefits of renewables while boosting the bottom-up movement to address climate change. In the United States, community energy faces an uneven patchwork of federal and state regulations; nonetheless, a range of policy instruments is making "energy democracy" an increasingly viable option (Burke and Stephens 2017). Ironically, though, "community energy" infrastructure is likely to bear a "Made in China" label.

Complicating the matter is the fact that localism comes in many flavours, some having little or nothing to do with sustainability governance. In 2010, for instance, the Localism Act was passed in Britain by a coalition of Conservatives and Liberal Democrats in order to shift from "big government" to "big society." Beyond the paradox of a nationally mandated localism, the act devolved responsibilities to local governments to a far greater extent than power or resources. This form of "managerial localism" stands in contrast to "community localism," which empowers citizens in policymaking and implementation (Evans, Marsh, and Stoker 2013). In the United States, conservatives and libertarians have also jumped on the localism bandwagon. The American Enterprise Institute's *Localism in America* (2018) makes the case for "why we should tackle our big challenges at the local level." The report says very little about sustainability as one of those challenges but is very clear about the need to devolve responsibilities for education, welfare, and health care to the local level. While eco-localists may or may not agree with these positions, they should consider their implications and understand that localism's discursive territory is incontrovertibly political.

Despite localism's appeal across the political spectrum, it faces strong ethical and practical challenges. First, today's most powerful institution is arguably the multinational corporation, with nation-states and international institutions like the WTO, the International Monetary Fund (IMF, and the World Bank operating at its behest. Simply localising leaves global action – and hence the primary levers of economic and political power – in the hands of these players. As Nicholas Low and Brendan Gleeson argue in their proposal for a nested hierarchy of cosmopolitan democratic governance, so long as global capitalism persists, there must be a countervailing power on a global scale. In their words,

> The slogan 'think globally, act locally' is no longer appropriate. Local action within an unchanged global order of production and governance rapidly reaches its limits. It is necessary today not only to think about the global consequences of local action, but to act to change the global context of local action: 'Think and act, globally and locally'.
>
> *(1998: 189)*

Second, the localist impulse often discloses an underlying nostalgia for purity. While localism's emphasis on community values offers a healthy corrective to the values of speed, efficiency, and convenience associated with globalisation, local producers are not necessarily any more deserving or trustworthy than peasants or factory workers overseas. Indeed, the anonymity and lack of accountability associated with globalisation have been far more damaging in the Third World than the First. Affluence is highly concentrated but its shadow ecologies are spread across the globe, which places the image of smart-phone addicts waxing eloquent about local food in an unsettling light. "Going local" can serve as a kind of purification ritual, one that denies the human and material consequences of one's own lifestyle. If our smartphones contain coltan from the Congo, the batteries on our electric bicycles contain lithium from Bolivia, and our locally produced solar panels are made with Chinese and German components, how local can we really be?

Third, like it or not, billions of people are now highly dependent upon the global economy. A mass conversion to local diets, for instance, would have far-flung negative consequences. Consider a recent book whose title speaks for itself: *The Locavore's Dilemma: In Praise of the 10,000-Mile Diet* (Desrochers and Shimizu 2012). The authors view the local food movement as an elite-driven fad and a potentially dangerous distraction from serious global food issues. While they sidestep thorny issues regarding the environmental impact of industrial agriculture, they make a valid point: efficiencies of production have created a global food system that feeds more people than any other system in the past. This system, no doubt, is deeply flawed but we should not dismiss it out of hand.

Fourth, given that the affluence of the global North was amassed through access to foreign natural and human resources, a fetishism of the local just as planetary systems are approaching a tipping point is an awkward moral and pragmatic strategy. From an ethical perspective, if we retreat to our fortresses after wrecking the climate, we hardly have a leg to stand on. Should our circles of caring and sharing include only those who happen to live near us when our actions ramify across the planet? From a practical perspective, humanity seems to be approaching "peak everything" just as the global South is beginning to "catch up." The 80% of humanity living in developing countries are unlikely to change their trajectories in the absence of a compelling exemplar – nor without assistance from the wealthy countries. Global justice, therefore, becomes a matter of "geoecological realism" (Athanasiou and Baer 2002: 74). In this context, localisation is a viable strategy only if it is pursued under the umbrella of global solidarity. Such a strategy requires not only *thinking* globally but also *acting* globally on an institutional level. A localist retreat at the dawn of the Anthropocene is a moral and practical chimera. We might ignore the looming issues of climate refugees, geo-engineering, and species triage but they won't go away.

We have entered a new era. Humanity is operating as a geophysical force; yet most of us are utterly unaware of our perilous entry into the Anthropocene. For the few who have registered this fact, there is a mighty temptation to see human survival itself as dependent upon relocalisation. The threat of human extinction is like a dark cloud hanging over the discourse of localism, one that is rarely acknowledged but one that can also be fairly easily dispelled. A weedy species can inhabit and spread across a wide range of ecosystems, and humans are arguably the weediest species on the planet. While anything is possible, human extinction is probably not in the cards for the foreseeable future. Precluding a nuclear winter or an asteroid impact, in the event of global catastrophe, we can expect human cultures to revert to their *modus operandi*: the local.

The local, then, is a given. For the time being, globalisation is also a given – albeit a lopsided and unsustainable one. The question at the dawn of the Anthropocene is therefore whether we can devise a viable way of inhabiting the planetary, which requires extending our circles of caring and sharing not only to distant peoples but the Earth system itself. While we are very far from the requisite political and economic institutions, this is the challenge. It is at once a social, economic, political, ecological, and deeply personal challenge.

## Towards a local/global synthesis

Localism is no doubt a reaction to globalism, but its growing popularity does not necessarily represent a mere swing of the pendulum. "Think globally, act locally" always evoked more than a simplistic dichotomy, implying rather that localism could serve as a bottom-up strategy for implementing a global agenda. Otherwise, why would it matter that we think globally? When I consider my interviews with hundreds of ecovillagers from around the world, for instance, many of whom framed their lifestyle choices as responses to planetary exigencies and cosmopolitan values, I find myself questioning the local/global dichotomy. And the fact that ecovillagers have organised themselves into the Global Ecovillage Network, which has UN consultative status, points to an organic, grassroots globalism that is quite distinct from conventional understandings of globalism (Litfin 2014).

In his study of local watershed management experiments and the evolution of sustainable development, Craig Kauffman comes to the same conclusion. As he argues, "the grassroots level should not be viewed merely as an object of global governance, but rather a terrain (hidden in most analysis) where global governance is constructed" (2017, 6). While Integrated

Watershed Management was first promulgated as a global discourse by transnational governance networks of international organisations, non-governmental organisations (NGOs), state agencies, and scientific experts, local coalitions make it their own by infusing it with local norms, institutions, and practices. In response to these local experiences, international actors change their discourses and strategies, with the surprising result that farmers and indigenous people become "grassroots governors." Grassroots global democracy, as Kauffman calls it, is a vibrant expression of organic globalism. Rather than being either top-down or bottom-up, global sustainability governance is an iterative evolutionary process, with the kinds of feedback loops we might expect to find in a living system.

Likewise, there are strong elements of the local food movement both in the Global North and the Global South that are scaling themselves up. A GRAIN report notes this trend among local food groups, citing Growing Power, Slow Food, and La Via Campesina (GRAIN 2014). For the most part, while local food campaigns do not begin with global ambitions, they often arrive there by consistently taking the next logical step. For instance, La Via Campesina was founded in 1993 by local farmers' organisations across Europe, Latin America, Asia, North America, Central America, and Africa in order to protest the globalisation of industrialised agriculture under the emerging WTO. This global collective is now a trusted advisor to the Food and Agricultural Organization of the United Nations (FAO) and United Nations Human Rights Council (UNHRC).

"Taking the next logical step" seems to be precisely what propelled NIMBY activists in the 1990s to shift from "Not in my backyard!" to declaring NOPE, "Not on planet Earth!" (Robinson 1999). Given global socioeconomic inequities and planetary exigencies, eco-localism will be necessary but not sufficient to the task of forging a path towards global sustainability governance.

Organic globalism understands the world as a nested hierarchy of living systems, from the cell to the Earth system, and seeks to harmonise human systems with living systems at every level. Harmonious integration is more straightforward in local economies. There may be greed and deception in a village, but it is more visible and the community has more power in the equation. If we are to persist as a global species, then, we must devise economies of care and connection that transcend the local and we must do some serious number crunching. What, for instance, do we acquire locally and what from afar? If I live in the western United States, for instance, I may need to consider that grass-fed beef from New Zealand might be more ecologically benign than corn-fed beef from California. And then, we face an even more radical question: what do we forego? Beef, perhaps.

Besides rigorous ecological footprint analysis, economies of care and connection will also require relational modes of production and consumption that supplant the current norms of exploitive distancing. These relational networks are growing, with fair trade being the most obvious, but these account for only a tiny fraction of world markets (Stiglitz and Charlton 2007) and are not without their own inequities (Sylla and Leye 2014). For localists who see a role for international trade, governance and production decisions would be guided by the subsidiarity principle (DeYoung and Princen 2012b: 333). Localisers and organic globalists could find common cause in mapping out how the subsidiarity principle would be implemented in practice. A key element of this mapping project would be determining the energetic requirements for a global civilisation, a possibility that sociologist Stephen Quilley (2011) labels "low-energy cosmopolitanism."

First, however, they would have to grapple with the myopic approach of prevailing global institutions. The WTO, for instance, has been a lightning rod for localist sentiments, with many arguing for its elimination. Yet, as the most powerful global political institution, a democratically restructured WTO would be the most likely candidate for an institutionally grounded organic globalism. Here, localism and organic globalism find themselves on common ground, recognising that the nation-state is neither large enough to inspire a planetary identity nor small enough to nurture the place-based identities essential to participatory governance. The nation-state would not necessarily disappear; rather, it would be incorporated into broader cross-cutting networks of supranational, regional, and local forms of governance.

Indeed, we can already see evidence of these cross-cutting networks. Consider the International Consortium of Local Environmental Initiatives (ICLEI), a bottom-up network that emerged in the wake of failed international climate negotiations. The consortium serves as a forum for cities to not only respond to the international policy vacuum left but also to share their best practices on a host of other environmental concerns. Much of the environmental movement itself is organised on a network model, spanning geographic and political scales from the local to the global. Global action networks are in place for a range of issues, including rainforests, climate, pesticides, and hazardous waste. The global action network model is cropping up for other issues as well. Many of these local-to-global networks have a presence both at international civil society gatherings like the World Social Forum and at intergovernmental gatherings like Rio+20. These bottom-up networks, aptly dubbed by Joshua Karliner (1997) as "grassroots globalisation," reflect the kind of higher-order synergism that can help eco-localism to realise its full potential. Like organic globalism, cosmopolitan localism is based in "the valuing of diversity as a universal right" (McMichael 2017) and envisions a world of locally based and globally networked citizens' initiatives.

Ultimately, organic globalism is founded on an emerging form of identity: a sense of global citizenship that simultaneously transcends and includes our bounded self, our local, and our national identities. While a resurgence of the local is a healthy response to the destructive legacy of mechanical globalism, the planetary phase of civilisation calls for a larger sense of global identity and responsibility. In stretching our loyalties, we are simultaneously enlarged. As Robert Nozick says, "The size of a soul, the magnitude of a person, is measured in part by the extent of what that person can appreciate and love" (1989: 258). Globalisation has given us the material infrastructure for planetary connectivity. The question now, as we cross the threshold into the Anthropocene, is whether we can develop the inner sense of connectivity to live as one species on our one Earth. For all its maladies, globalisation is an unprecedented human achievement – one that may be far from complete. With greater integrative synergies, the resurgence of the local can shift from being a reflexive recoil from the global to its higher calling as a cornerstone of global sustainability governance. In some important ways, it already is; my aim here has been to amplify these efforts.

## Note

1 Because extraction costs increase as accessibility declines, "peak oil" may be delayed but most likely will not be forestalled altogether. In the meantime, preventing runaway climate change must therefore be a matter of collective choice – a choice that will require some degree of global governance.

## References

Athanasiou, T. and Baer, P. 2002, *Dead Heat: Global Justice and Global Warming: Seven* Stories Press, New York.

Brandt, A. et al. 2013, "Peak Oil Demand: The Role of Fuel Efficiency and Alternative Fuels in a Global Oil Production Decline." *Environmental Science and Technology* 47, 14: 8031–8041.

Burke, M. and Stephens, J. 2017, "Energy Democracy: Goals and Policy Instruments for Sociotechnical Transitions." *Energy Research & Social Science* 33: 35–48.

Davoudi, S. and Madanipour, A. 2015, *Reconsidering Localism*. Routledge, New York.

DeLind, L. 2011, "Are Local Food and the Local Food Movement Taking Us Where We Want to Go? Or Are We Hitching Our Wagons to the Wrong Stars?" *Agriculture and Human Values* 28, 2: 273–83.

Desrochers, P. and Shimizu, H. 2012, *The Locavore's Dilemma: In Praise of the 10,000-Mile Diet*. PublicAffairs, New York.

DeYoung, R. and Princen, T. (eds.) 2012a, *The Localisation Reader: Adapting to the Coming Downshift*. The MIT Press, Cambridge, MA.

———. 2012b, "Downshift/Upshift: Our Choice" in De Young and Princen, *The Localisation Reader*.

Evans, M., Marsh, D. and Stoker, M. 2013, "Understanding Localism." *Policy Studies* 34, 4: 401–07.

Fairchild, D. G. and Weinrub, A. 2017, *Energy Democracy: Advancing Equity in Clean Energy Solutions*. Island Press, Washington, DC.

Goldsmith, E. and Mander, J. 1997, *The Case Against the Global Economy: And for a Turn Toward the Local*. Sierra Club, San Francisco.

GRAIN 2014, www.grain.org/article/entries/4929-hungry-for-land-small-farmers-feed-the-world-with-less-than-a-quarter-of-all-farmland.

Hawken, P. 2007, *Blessed Unrest: How the Largest Movement in the World Came Into Being and Why No One Saw It Coming*. Viking, New York.

Heinberg, R. 2007, *Peak Everything: Waking Up to the Century of Declines*. New Society, Gabriola Island, BC.

Hoffman, M. 2011, *Climate Governance at the Crossroads: Experimenting with a Global Response after Kyoto*. Oxford University Press, New York.

Homer-Dixon, T. 2006, *The Upside of Down: Catastrophe, Creativity and the Renewal of Civilization*. Island Press, Washington, DC.

Hopkins, R. 2008, *The Transition Handbook: From Oil Dependency to Local Resilience*. Green Books, Totnes, UK.

———. 2012, "The Arc of Scenarios" in DeYoung and Princen, pp. 59–68.

International Energy Agency 2010, *International Energy Outlook 2010*. www.worldenergyoutlook.org/.

Karliner, J. 1997, *The Corporate Planet: Ecology and Politics in the Age of Globalisation*. Sierra Club Books, San Francisco.

Kauffman, C. 2017, *Grassroots Global Governance: Local Watershed Management Experiments and the Evolution of Sustainable Development*. Oxford University Press, New York.

Lerch, D. 2017, *The Community Resilience Reader: Essential Resources for an Era of Upheaval*. Island Press, Washington, DC.

Litfin, K. 2014, *Ecovillages: Lessons for Sustainable Community*. Polity Press, Cambridge.

Liu, G., Amakpah, S. W., Yang, Z., Chen, B., Hao, Y. and Ulgiati, S. 2016, "The Evolution of Cities: 'Brains' or 'Parasites' of Sustainable Production and Consumption Processes in China." *Energy Procedia* 88: 218–23.

Low, N. and Gleeson, B. 1998, *Justice, Society and Nature: An Exploration of Political Ecology*. Routledge, London.

McDonald, M. 2012, "U.N. Report from Rio on Environment 'a Suicide Note,'" *The New York Times*, http://rendezvous.blogs.nytimes.com/2012/06/24/u-n-report-from-rio-on-environment-a-suicide-note/.

McMichael, P. 2017, *Development and Social Change: A Global Perspective*. Sixth ed. SAGE, Los Angeles.

Nozick, R. 1989, *The Examined Life*. Simon and Schuster, New York.

Quilley, S. 2011, "Entropy, the Anthroposphere and the Ecology of Civilization: An Essay on the Problem of 'Liberalism in One Village' in the Long View." *Sociological Review* 59: 65–90.

Robinson, A. 1999, "From NIMBY to NOPE: Building Eco-bridges." *Contemporary Politics* 5, 4: 339–64.

Sale, K. 1985, *Dwellers in the Land: The Bioregional Vision*. Random House, New York.

Schiermeier, Q. 2012, "The Great Arctic Oil Race Begins." *Nature* 482, 7383: 13–14.

Schumacher, E. F. 1973, *Small Is Beautiful: A Study of Economics as If People Mattered*. Harper and Row, New York.

Shuman, M. 2000, *Going Local: Creating Self-reliant Communities in a Global Age*. Routledge, New York.

———. 2012, "Locally Owned Business" in DeYoung and Princen, pp. 85–108.

Stephens, J. C., Wilson, E. J. and Peterson, T. R. 2014, *Smart Grid (R)evolution: Electric Power Struggles.* Cambridge University Press, Cambridge.

Stiglitz, J. and Charlton, A. 2007, *Fair Trade for All: How Trade Can Promote Development.* Oxford University Press, New York.

Sylla, N. S. and Leye, D. C. 2014, *The Fair Trade Scandal: Marketing Poverty to Benefit the Rich.* Ohio University Press, Athens, OH.

Tainter, J. A. 1990, *The Collapse of Complex Societies.* Cambridge University Press, Cambridge, UK.

Thayer, R. 2003, *LifePlace: Bioregional Thought and Practice.* University of California Press, Berkeley.

Wilbanks, T. and Fernandez, S. 2014, *Climate Change and Infrastructure, Urban Systems, and Vulnerabilities.* Island Press, Washington, DC.

# Conclusion

## Global sustainability governance – really?

*Doris Fuchs, Anders Hayden, and Agni Kalfagianni*

Four fundamental questions exist, when it comes to global sustainability governance. The first one relates to the politics of global sustainability governance: why are we not seeing real action in the face of the sustainability challenges humankind faces? To a bystander, it must seem like we are driving full speed into the wall, all the time having our gazes strictly fixed on entertaining stuff happening on the sidewalk. Do not look ahead! appears to be the motto of our time. Against the background of this question, the second one develops: where can we identify potential sources of hope for change, if any exist at all? Underlying these questions are two additional ones that students (in the broadest sense) of global sustainability governance need to ask of any representations of sustainability challenges and solutions: what analytical lens does the argument take, i.e. what does it focus on, but just as importantly, what does it ignore? The answer to this question provides a first basis for evaluating the plausibility of the representation of the problem and solution. The final question concerns the normative frame explicitly or implicitly applied in the respective representation of sustainability challenges and solutions. Taking this frame into account allows us to evaluate the suggested solution in terms of its fit with broader norms and values, so as to avoid potential trade-offs between sustainability and other societal goals. This Handbook and its contributions speak to these four questions. In fact, while the chapters are placed in four sections corresponding to the earlier questions, all chapters speak to more than one of them.

Various chapters in the Handbook highlight structural barriers to transformative change, pinpointing both opposing material interests as well as underlying ideational paradigms that promote unsustainability. The role of the latter is noteworthy in terms of the hegemony of ideas of growth (Higgs), mining (Princen), and consumption (Krogman), for instance. But ideational paradigms also include the definitions we use and stories we tell about decarbonisation (Lane), technological innovation (Alexander and Rutherford), local participation (Litfin), governance (Blühdorn and Deflorian), and democracy (Mert), or more broadly our "magical thinking" (Maniates) as well as the meaning we attach to everyday practices (Meyer). Material structures create barriers to transformation in terms of asymmetries in influence on public opinion and policy debates (Brulle and Aronczyk), by expanding and simultaneously disguising the reach and impact of financial logics (Clapp and Stephens), and by advantaging the interests of present generations over future ones (Lawrence) and humankind over nature (Gumbert). Meaning and material structures also come together in shaping our everyday practices (Meyer).

A number of chapters specifically try to look for promising strategies and approaches to sustainability governance, in the midst of the existing material and ideational barriers. They suggest there is a need and opportunities to reorganise patterns of work time (Larsson et al.), consumption (Fuchs), participatory focus and space (Litfin), as well as the economy (Lange, Princen), governing institutions (Mert), and their goals and steering measures (Higgs, Philipsen). They propose that moving forward may become possible by stepping away from our dualistic and anthropocentric worldviews (Inoue et al., Gumbert), as well as by developing a new consensus on a vision of the future to strive for (Di Giulio and Defila, Vanhulst and Beling) and the normative foundations of such a future (Hayden, Kalfagianni et al.). In this vein, we may also want to provide more room for spiritual foundations (Glaab), and need to provide more voice to underrepresented groups (Inoue et al., Lawrence, Okereke).

Other chapters present a range of analytical lenses against which to evaluate the stories told about global sustainability challenges and potential solutions in research and practice. With their help, the importance of paying attention to the role of structural forces as constraints and potential enablers of sustainability transformations becomes clear. Thus, authors highlight the important role of both agent-centred and structural power (Bexell), of critical inquiries into the underlying and potentially counter-intuitive intent and character of sustainability governance (Blühdorn and Deflorian), or of evaluations of the transformative potential of governance solutions related to everyday practices and their foundations in terms of meaning and material resources (Meyer). Other chapters ask us to critically reflect on whom and what we actually consider when we talk about the goal of sustainability and associated norms and strategies (Inoue et al., Gumbert, Lawrence, Okereke). In a similar vein, we can approach narratives about sustainability problems and solutions by inquiring into their underlying assumptions about the potential and need for continued economic growth, the limits to growth, and the governability of such limits (Higgs, Lange) or optimism regarding technological innovation and decoupling (Alexander and Rutherford, Lane).

Finally, a number of chapters concentrate on normative lenses and ethical questions one can apply to sustainability challenges and especially to suggested solutions. They ask us to reflect on the fundamental goals of sustainability and sustainability governance (Di Giulio and Defila, Vanhulst and Beling), and suggest normative yardsticks against which we can evaluate developments in sustainability (governance) (Kalfagianni et al., Pellizzoni, Lawrence). They also develop normative requirements for a sustainability transformation (Hayden) and discuss the potential to draw on specific normative resources in its pursuit (Glaab). Finally, the identified norms allow us in research and practice to evaluate sustainability-related developments, from financialisation processes (Clapp and Stephens) and indicators of well-being (Philipsen) to questions of consumer rights and responsibilities (Fuchs), economic organisation (Lange, Larsson et al.), or even population growth (Coole) in terms of their potential to provide (or endanger) the foundations of a world we want to live in.

As pointed out earlier, barriers to and hopes for global sustainability governance, the analytical lenses applied, and ethical questions asked are interlinked. They come together, for instance, in our understanding of consumption, the way we have set up our economic and political systems as well as societies in terms of fostering overconsumption by the global consumer class, the questions we fail to ask about the meaning of and drivers behind this overconsumption, alternative visions to strive for, and potential strategies for reorganisation. Their interaction is also reflected in the questions we, as researchers or practitioners, ask or fail to ask, the normative characteristics of the transformation we aspire to, and the transformative potential we see or fail to see in governance and, more specifically, our participatory practices and democratic institutions. Altogether, then, the contributions to this Handbook tell the story of barriers to and hopes for global sustainability governance, and provide tools to analyse and evaluate it. They share scepticism regarding easy fixes and point to the complex and fundamental political, economic, and societal structures that have brought us to this point and continue to hinder a turn towards transformation. They also show a way forward, however, drawing our attention to the question of what really matters both in terms of a vision to strive for and structural barriers that we need to actively deconstruct and remove.

## The future of global sustainability research

In consequence, this Handbook is as much a summary of the state of the art of research on sustainability governance as it is a springboard for further research. Sustainability challenges are constantly evolving and answers to these challenges are continuously changing – even if woefully inadequate overall to date. Research into and knowledge of the various requirements, potentials, foundations, conditions, and implications of global sustainability governance are simultaneously constantly developing as well. The individual chapters therefore highlight specific necessities and promising targets for future research. As editors, we would therefore like to raise a couple of overarching priorities for research on global sustainability governance that go beyond the foci of the individual chapters: the sustainability of sustainability research and communication, and the role of money in (sustainability) governance.

When studying developments in sustainability research (and its uptake in the political realm) over time, we cannot avoid asking ourselves: why do we see the same suggestions for solutions being repeated in fashion cycles, even though they have been debunked on theoretical and empirical grounds? Or framed differently, what are the challenges for learning at least on the part of researchers? Examples of such reoccurring fashionable stories are the all too optimistic hopes placed in grassroots initiatives, public participation, or changes in consumer values. While the idea that we can change the system from the bottom-up, that we can individually learn and thereby collectively exercise control and transform our world in the direction of sustainability is certainly extremely attractive (which is already one part of the answer to why these perspectives keep reappearing), the empirical reality simply does not provide sufficient evidence of such successes. Grassroots initiatives tend to face serious challenges in terms of the scale and reach of their impact; participatory governance often is not inclusive, transparent, and a secure basis for a transformation of interests from individual to collective objectives; and consumers may purchase greener products but only under certain conditions, and they, most importantly, are almost always invited to shop more rather than less (e.g. chapters by Fuchs, Krogman, Litfin, and Maniates). This is not to deny meaning and (some) impact to such efforts, but to caution strongly against being too naïve in attributing transformative potential. The same applies to placing hopes in the "green state" or top-down approaches. Given what we know about opposing interests, both in their concentrated and individually powerful form (e.g. corporations in general and institutional investors in particular) as well as in their decentralised and collectively powerful form (e.g. consumers and employees embedded in incentive structures that prioritise if not consider exclusively economic factors), and given the failure of national governments and (often toothless) international organisations and agreements to achieve the necessary decoupling of human well-being and resource consumption (e.g. chapters by Bexell, Blühdorn and Deflorian, Brulle and Aronczyk, Mert), why should one expect transformative change from this source? And yet, we find well-meaning research all too easily touting both stories again and again, without critically reflecting on their limits.

To be clear, we are not arguing that we should toss the hopes of bottom-up or top-down change out of the window entirely. If change is not coming from the top and not from the bottom, where would it come from? What we want to strongly caution against is continuously reinventing these stories, as if they had not been told (and – in their most simplistic forms – debunked) before. We will need both bottom-up *and* top-down impulses for change; however, to identify realistic conditions for success on both accounts, we have to learn from what research on global environmental and sustainability governance has already found over the course of the last four to five decades. And this leads us to the sustainability challenges that sustainability research faces today.

## The unsustainability of sustainability research?

Research currently is characterised by numerous unsustainable practices on different levels. Similar to other societal realms, governance by numbers has become a dominating steering strategy by university administrations and governments. Relevant "numbers" include grant acquisitions by researchers, impact scores achieved with publications, and international "visibility" demonstrated with publications, conference participation, and media presence. The increasing dependence of universities and scholarly careers on grant acquisitions promotes a short-term focus on research topics for which funding happens to be available and on research questions and hypotheses attractive to funders. To acquire governmental funding, for instance, research emphasising win-win solutions, participation as a basis for sustainability transformations, or technological transformation potentials (bioengineering, digitalisation, etc.) tends to be more helpful than focusing on overconsumption or power asymmetries in politics. Increasingly co-funding with "societal actors," largely translated as "big corporations," is a requirement for governmental research funding. This has serious implications for the type of research pursued, which typically involves problem-solving in a way demonstrably benefiting businesses (in the hope that this will also benefit society) as opposed to critical, reflective, and normative research. For corporate funding, research analysing the potential contribution of corporate responsibility programs to sustainability governance, for instance, is much more likely to receive funding than research critically inquiring into the intent and impact of such programs.

In terms of publications, similar problems with respect to the fashionability and mainstream character of research topics exist when it comes to the need to get into journals with high impact factors. This dynamic is aided by a demand for quantity in publications, which exists for both junior and senior researchers in many academic systems today, creating a constant quest for new output. Obviously, a high frequency of publications has to reduce the amount of research going into each individual publication and therefore its innovative potential, as researchers are only human, too.[1] The original motivation behind the publication of research, i.e. the communication of new and relevant research findings to peers, has lost weight relative to the need to simply publish. This need to publish and publish felt by researchers is accompanied by a need to be present at international conferences. Here, too, the initial motivation of being able to engage in direct exchange on topics of joint interest with other researchers appears to have been pushed back relative to the need to increase one's visibility and therefore invitations to joint grant applications and publications. Given the number of paper proposals submitted to many international conferences and the recognition of the importance of "active participation" by researchers, many associations have started to organise so many panels parallel to each other that the audience for the individual panel sometimes barely matches the size of the panel. The resulting quantity of publications and presentations, in turn, means that every researcher wanting to be heard amidst the noise needs to be even more omnipresent. The result of these trends then is that we publish more than we read, we speak more than we listen, and we have too little time to really take into account what decades of research on a given topic may already be able to tell us, while simultaneously using up energy for computer time and storage, paper (though perhaps increasingly less so), and creating large ecological footprints via air travel.

These trends apply to most fields of research, including sustainability research. And perhaps it is because of the perversity of the effects described earlier for researchers trying to foster a sustainability transformation that they become most noticeable here. Accordingly, relevant scientific associations or subgroups within these associations have debated questions of moving to paperless publications or virtual conferences for some time. Many sustainability scholars are struggling with related questions and issues of personal responsibility versus systemic constraints, and the associated costs of individual counter-agency. Slow science movements are emerging. Among the contributors to this volume, one can identify a range of personal choices regarding the situation ranging from attempts to avoid air travel to the refusal to participate in the grant application game. Senior (tenured) researchers can make those decisions more easily than junior scholars and untenured faculty, of course.

Still, sustainability research needs to pay particular attention to the question of what sustainable research is. We certainly should not waste resources of reinventing (all too imperfect) wheels again and again. We need to take sufficient time to make sure that we listen to and learn from each other. We need to focus on questions and try to develop answers that really matter, rather than those for which funding can be obtained. And we need to call out systemic constraints destroying the foundations for sound and therefore societally valuable research.

## Communicating sustainability issues

Another issue to which we need to pay more attention as sustainability scholars is what happens with sustainability-relevant knowledge in public discourse and politics. An underlying problem here may be that many of us too easily assume that everybody knows the seriousness of the sustainability challenges humankind faces (and therefore wonders why nobody takes action – see above). Surely, a considerable share of the population in high income countries is somewhat aware of these challenges and chooses to use humanity's incredible ability to close our eyes to unattractive news and to turn our thoughts away from the consequences. Another group, however, simply is not convinced that the challenges are all that real or that easy fixes are not around the corner. And why should they be? Wouldn't they hear a lot more about these challenges – relative to other news – if they were so serious? Indeed, in the overall noise, sustainability-related messages are simply drowned out. Media logics and power relations behind the distribution of media time, as well as the human inclination to focus on easier or more entertaining information (especially when faced with challenges that seem to be beyond one's control) mean that we receive hundreds if not thousands of messages about other "stuff" for every sustainability-related message that we receive. Politicians also send contradictory messages to citizens, given how rarely they talk about sustainability problems – even if they point out their seriousness when they do – compared to other issues, and how frequently they promote opposing interests. Angela Merkel may enjoy appearing as the "Klimakanzlerin" (climate chancellor) in preparation for the Climate Summit in Paris, but German governments under Merkel also have opposed stricter EU emission standards for cars, rolled back support for renewable energy sources, been complicit in the car industry's efforts to cheat on emissions controls, and delayed exiting coal, to name just a few countervailing efforts.[2] The resulting message to the public, then, is that climate governance is nice to have, but not as important as "…." And then, there are highly visible individuals who explicitly oppose any ideas of a seriousness of sustainability challenges and call climate change a hoax, of course (Brulle and Aronczyk). Efforts to discredit science are all too evident today. A recent manifestation comes from the Dutch Senatorial elections of 20 March 2019, when Thierry Baudet, leader of the populist Forum for Democracy (FvD), attacked universities in his victory speech[3] and later provided an online mechanism inviting students to report "political indoctrination" by lecturers and professors.[4] In sum, questions of how to deal with information overflow in general and fake news in particular cannot be ignored by sustainability scholars.

The rising relevance of social media and the decreasing relevance of facts for political and media communication are relatively new phenomena, relevant not just for sustainability research. Again, however, they are particularly relevant for sustainability research given the need for real and fast action. What does this mean for sustainability researchers? Many of us are already trying to insert our voices into the debate via various means. The sustainability scholar in the ivory tower is probably a rare phenomenon. Moreover, critical sustainability scholars do not shy away from communicating tough messages. Beyond such individual efforts, however, the situation described earlier shows that sustainability scholars cannot leave questions of the role of communication and information in politics to others. We need a better understanding of how to enable sustainability messages to survive in the media jungle, to be heard and to be trusted.

## Money and governance

This brings us to a related problem: the role of money in (democratic) governance. The potential for serious sustainability governance cannot be thought of independently from the distribution of power in the political system. We need to open creative spaces for the development and adoption of policy options by vastly reducing the influence of powerful market actors on politics. Addressing power imbalances in democracies implies reducing the role of external money in politics in as many forms as possible, which is much easier said than done, unfortunately. Lobbying could be strongly curtailed and the remaining parts balanced among different types of interests as well as made as transparent as possible. Private sponsoring and campaign finance could be abolished. But money also exercises influence on politics via the media and the shaping of public discourse. In consequence, the dependence of media on advertising and the political connotations if not explicit content of advertising and other communications in the media also need to be considered. A possible strategy in this respect would be a massive reduction in advertising via the creation of advertising-free spaces (important in schools as well), as well as the taxation of advertising (rather than its subsidisation via tax reductions). Given the contribution of advertising to (over)consumption, its reduction would even garner a double-dividend.

None of these ideas is new and not just sustainability scholars have proposed them. There is a considerable amount of research about the increasing dysfunction of our democracies. Currently, such research tends to focus more on the rise of populism and polarisation, while questions regarding the influence of corporate power and money in general are pushed to the background. Clearly, populism is a serious phenomenon deserving attention, for reasons including its implications for sustainability governance. But as sustainability researchers we should also not lose sight of the influence of imbalances in material power (and the discursive power it can buy) on politics. Admittedly, this is a field of research where laurels are not easy to gain. "Proving" the influence of a particular actor or group of actors on specific policies is difficult, to say the least, given problems of data accessibility and methodological standards regarding the demonstration of causality that have taken hold in much of the social sciences. Still, it is a crucial field for sustainability researchers if the ground for a transformative potential in democracies is to be laid. Demonstrating undue influence and revealing cases in which the interests of a rich or powerful few trumped the interests of many can serve as a basis for contestation by the public of political misrule and a reconsideration of policies and regulations. Such research would promote the (re)politicisation of sustainability governance.

## Concluding thoughts

The good news is that many open questions remain for global sustainability governance in research and practice. Well, actually, that may be indicative of rather bad news. Certainly, the sustainability challenges we face are not small. As scholars, we struggle with their size, reach, and complexity as well as with the multifaceted forces hindering their effective targeting. The research and teaching by the contributors to this Handbook, however, show that we have not given up on hope and that we continue to invest in the future of humankind. In these efforts, we are convinced that critical perspectives on global sustainability governance provide a particularly powerful vantage point, allowing us to question the material and discursive status quo, reflect on the potential for alternative realities, and develop promising ideas for transformative strategies. We hope that you will join us on this journey!

## Notes

1 In fact, one may be critical of handbooks like this one, for which contributors tend to be asked to contribute chapters on a core aspect of their research, on which they will naturally already have published before. As editors, we carefully weighed the idea of publishing this Handbook, therefore. However, we felt that a book combining drawing together the included critical perspectives on sustainability governance, showing their breadth and interaction, and thereby underlining the relevance and combined value of these perspectives was direly missing.
2 www.greenpeace.de/sites/www.greenpeace.de/files/publications/20171026-greenpeace-bilanz-klima-merkel.pdf.
3 www.nrc.nl/nieuws/2019/03/21/de-uil-van-minerva-spreidt-zijn-vleugels-bij-t-vallen-van-de-avond-a3954103 (in Dutch).
4 www.renaissanceinstituut.nl/actueel/de-nieuwe-schoolstrijd-meldpunt-indoctrinatie-op-scholen-en-universiteiten (in Dutch).

# Index

Note: **Bold** page numbers refer to tables; *italic* page numbers refer to figures; page numbers followed by "n" denote endnotes and page number followed by "b" denote boxes.

Printed in the United States
by Baker & Taylor Publisher Services